Shape Memory Alloys

Dimitris C. Lagoudas
Editor

Shape Memory Alloys

Modeling and Engineering Applications

Springer

Editor
Dimitris C. Lagoudas
Department of Aerospace Engineering
Texas A&M University
College Station
TX, USA

ISBN 978-0-387-47684-1 ISBN 978-0-387-47685-8 (eBook)
DOI 10.1007/978-0-387-47685-8

Library of Congress Control Number: 2007942944

Printed on acid-free paper

9 8 7 6 5 4 3 2 1

springer.com

To all of our loved ones.

Preface

It all started with a trip to Red River...

Coauthors, families, and colleagues enjoy a working vacation in the Sangre de Cristo Mountains of New Mexico, March 2006.

As technical conversations on modeling, characterization and applications of shape memory alloys (SMAs) were blending with the view of the white snowy peaks surrounding Red River, New Mexico, it became clear to our research group that a consistent and comprehensive text on SMAs would be very helpful to future students interested in performing research in this field. Many communication barriers could be eliminated and access to the substantial body of research discussed in the literature would be increased. In this way, a working vacation became the motivating factor behind a challenging research project.

This book has been written with contributions from three of my current Ph.D. students, Luciano Machado, Parikshith Kumar and Darren Hartl, and three former Ph.D. students, Pavlin Entchev, Peter Popov and Björn Kiefer. These latter three coauthors were still members of the Shape Memory Alloy Research Team (SMART), or in close proximity, when we started the project of writing this book more than a year and a half ago. The work of a seventh former Ph.D. student, Siddiq Qidwai, is also included in this book. The task of putting forth a sequence of topics on shape memory alloys (SMAs) that forms a coherent learning path seemed natural, given the diversity of topics covered by their Ph.D. work.

In the first chapter, Parikshith describes the basic properties and applications of SMAs, followed by the second chapter on thermomechanical characterization and material parameter identification presented by Darren. The thermomechanical constitutive modeling and closed form solutions are covered in the third chapter by Luciano, while the numerical implementation and finite element analysis examples are presented in the fourth chapter by Siddiq and Darren. The incorporation of transformation induced plasticity is discussed in chapter five by Pavlin and an extended model for SMAs accounting both for phase transformation and reorientation is described in the sixth chapter by Peter. Finally, Björn introduces modeling of magnetic SMAs in the seventh chapter.

Even though the seven chapters cover a wide variety of topics and discuss different aspects of modeling of SMAs, there are many specialized considerations that have been left out due to space limitations. The reader will hopefully gain enough background to be able to seek additional sources of information and appreciate the complexity of the constitutive response of SMAs and the importance of modeling in the design and analysis of engineering systems.

The work of many other graduate and undergraduate student members of my research group has been valuable in writing this book. In particular, the help provided by Amnaya Awasthi, Krishnendu Haldar, Jesse Mooney, Justin Schick and Francis Phillips is greatly appreciated. The many various tasks performed by these individuals were coordinated in a large part by Darren Hartl, who also helped with the overall compilation of the manuscript. The proofreading services of Gary Seidel, Olivier Bertacchini, Brent Volk, Matthew Kuester, Jack Vincent, Pam McConal, Alex McCord and Natasha, Georgia and Magda Lagoudas are also appreciated. Finally, the inspiration provided by my colleagues at Texas A&M University and elsewhere, and the financial support provided by the Department of Defense, NASA, NSF and industry partners over the years to sustain the SMART research team is gratefully acknowledged.

Texas A&M University
Dimitris C. Lagoudas
Editor

College Station, Texas
December 2007

Contents

Appendices

List of authors

Darren J. Hartl
Department of Aerospace Engineering
Texas A&M University

Parikshith K. Kumar
Department of Aerospace Engineering
Texas A&M University

Luciano G. Machado, Ph.D.,
Department of Aerospace Engineering
Texas A&M University

Dimitris C. Lagoudas, Ph.D.,
John and Bea Slattery Chair
Department of Aerospace Engineering
Texas A&M University

Björn Kiefer, Ph.D.,
Institute of Applied Mechanics
University of Stuttgart

Peter Popov, Ph.D.,
Institute of Scientific Computation
Texas A&M University

Pavlin B. Entchev, Ph.D.,
ExxonMobil Upstream Research Company

M. A. Siddiq Qidwai, Ph.D.,
Naval Research Laboratory Operations
Science Applications International Corporation (SAIC)

List of Symbols

α	Magnetic domain volume fraction (MSMAs)
α^A	Thermal expansion coefficient of austenite
α^M	Thermal expansion coefficient of martensite
$\boldsymbol{\Theta}$	Tangent thermal moduli tensor
$\boldsymbol{\alpha}$	Thermal expansion coefficient tensor
$\boldsymbol{\alpha}^A$	Thermal expansion coefficient tensor of austenite
$\boldsymbol{\alpha}^M$	Thermal expansion coefficient tensor of martensite
$\boldsymbol{\beta}$	Back stress tensor
$\boldsymbol{\varepsilon}$	Infinitesimal strain tensor
$\boldsymbol{\varepsilon}^p$	Plastic strain tensor
$\boldsymbol{\varepsilon}^{in}$	Inelastic strain tensor
$\boldsymbol{\varepsilon}^r$	Reorientation strain tensor
$\boldsymbol{\varepsilon}^{t-r}$	Transformation strain at the reversal of phase transformation
$\boldsymbol{\varepsilon}^{th}$	Thermoelastic strain tensor
$\boldsymbol{\varepsilon}^t$	Transformation strain tensor
$\boldsymbol{\sigma}$	Cauchy stress tensor
$\boldsymbol{\sigma}^{\text{eff}}$	Effective stress tensor
$\boldsymbol{\sigma}'$	Deviatoric stress tensor
$\boldsymbol{\sigma}^{\text{eff}'}$	Deviatoric effective stress tensor
$\bar{\sigma}^{\text{eff}}$	Mises equivalent effective stress
$\bar{\sigma}'$	Effective (von Mises equivalent) stress
$\boldsymbol{\zeta}$	Set of internal state variables
$\boldsymbol{\mathcal{S}}$	Compliance tensor
$\boldsymbol{\mathcal{S}}^A$	Compliance tensor of austenite
$\boldsymbol{\mathcal{S}}^M$	Compliance tensor of martensite
$\boldsymbol{\mathcal{L}}$	Tangent stiffness tensor
η	Drag stress
$\boldsymbol{\Lambda}$	Transformation tensor
$\boldsymbol{\Lambda}^p$	Transformation-induced plasticity tensor
$\boldsymbol{\Lambda}^r$	Reorientation tensor
$\mathbf{b}$	Body force vector

$\mathbf{E}$	Green-Lagrange strain tensor
$\mathbf{F}$	Deformation gradient tensor
$\mathbf{H}$	Magnetic field strength vector
$\mathbf{M}$	Magnetization vector
$\mathbf{M}^{(i)}$	Local magnetization vectors
$\mathbf{q}$	Heat flux vector
$\mathbf{r}$	Position vector of a material point
$\mathbf{t}$	Traction vector
$\mathbf{u}$	Displacement vector
$\mathbf{v}$	Velocity vector
$\mathbf{X}$	Reference position of a material point
$\mathbf{x}$	Current position of a material point
Φ	Transformation function
μ_i^c	Model parameters for cosine model
μ_i^e	Model parameters for exponential model
μ_0	Permeability of free space
μ_i	Model parameters for polynomial model
ν^A	Poisson's ratio of austenite
ν^M	Poisson's ratio of martensite
Φ^ξ	Reorientation function
π	Thermodynamic force conjugated to ξ
π^α	Thermodynamic driving force for domain wall motion (MSMAs)
π^ξ	Thermodynamic driving force for variant reorientation (MSMAs)
π^{θ_i}	Thermodynamic driving force for rotation of local magnetization vectors (MSMAs)
ψ	Specific Helmholtz free energy
ρ	Mass density
σ^{Af}	Completion stress for reverse transformation into austenite
σ^{As}	Initiation stress for reverse transformation into austenite
σ^b	Blocking stress
σ^{Mf}	Completion stress for forward transformation into martensite
σ^{Ms}	Initiation stress for forward transformation into martensite
σ	Uniaxial stress
σ_f	Stress for completion of martensitic reorientation (detwinning)
σ_s	Stress for initiation of martensitic reorientation (detwinning)
σ_Y^A	Yield stress of austenite
σ_Y^M	Yield stress of martensite
θ_i	Local magnetization vector rotation angles
ε	Uniaxial total strain
ε^e	Uniaxial elastic strain
ε^t	Uniaxial transformation strain
ε^{max}	Maximum attainable reorientation strain (MSMAs)
ε^{th}	Uniaxial thermal strain
ξ	Total martensitic volume fraction
ξ	Volume fraction of second martensitic variant (MSMAs)

ξ^d	Detwinned martensitic volume fraction
A_s	Austenitic start temperature at zero stress
A_s^σ	Austenitic start temperature under stress
A_f	Austenitic finish temperature at zero stress
A_f^σ	Austenitic finish temperature under stress
M_s	Martensitic start temperature at zero stress
M_s^σ	Martensitic start temperature under stress
M_f	Martensitic finish temperature at zero stress
M_f^σ	Martensitic finish temperature under stress
C^A	Stress influence coefficient of austenite
C^M	Stress influence coefficient of martensite
a_c^A	Model parameter for cosine model
a_e^A	Model parameter for exponential model
a_c^M	Model parameter for cosine model
a_e^M	Model parameter for exponential model
b^A	Model parameter for polynomial model
b^M	Model parameter for polynomial model
c	Specific heat capacity
c^A	Specific heat capacity of austenitic phase
c^M	Specific heat capacity of martensitic phase
D_i^b	Parameters for the evolution of back stress
D_i^d	Parameters for the evolution of drag stress
E	Elastic (Young's) modulus
E^A	Elastic modulus of austenite
E^M	Elastic modulus of martensite
f	Hardening function
f^α	Hardening function for evolution of magnetic domains (MSMAs)
f^ξ	Hardening function for evolution of martensitic variants (MSMAs)
G	Specific Gibbs free energy
$G^{\alpha-mix}$	Gibbs free energy due to mixing of magnetic domains (MSMAs)
$G^{\xi-mix}$	Gibbs free energy due to mixing of martensitic variants (MSMAs)
G^{an}	Magnetocrystalline anisotropy energy (MSMAs)
h	Specific enthalpy
H^d	Maximum detwinning strain
H^{cur}	Maximum transformation strain at current stress
H^{max}	Maximum attainable transformation strain
H_y^{crit}	Magnetic field for full alignment of magnetization vectors
$H_y^{f(1,2)}$	Magnetic field for termination of forward variant reorientation
$H_y^{f(2,1)}$	Magnetic field for termination of reverse variant reorientation
$H_y^{s(1,2)}$	Magnetic field for activation of forward variant reorientation
$H_y^{s(2,1)}$	Magnetic field for activation of reverse variant reorientation
K_n	n-th magnetocrystalline anisotropy energy coefficient
M^{sat}	Saturation magnetization
m_1	Parameter for the evolution of drag stress

R	Residual for closest-point projection algorithm
r	Heat source
s	Specific entropy
s_0	Specific entropy at reference state
s_0^A	Specific entropy of austenite at reference state
s_0^M	Specific entropy of martensite at reference state
T_0	Reference temperature
u	Specific internal energy
u_0	Specific internal energy at reference state
u_0^A	Specific internal energy of austenite at reference state
u_0^M	Specific internal energy of martensite at reference state
Y	Critical value for thermodynamic force to cause transformation

1

Introduction to Shape Memory Alloys

P. K. Kumar and D. C. Lagoudas

Shape Memory Alloys (SMAs) have been on the forefront of research for the last several decades. They have been used for a wide variety of applications in various fields. This chapter introduces the unique behavior that is observed in SMAs. Their characteristic properties and associated microstructural behavior will be discussed in detail. The different types of SMAs and some common applications will also be reviewed.

1.1 Introduction: Overview of Active Materials

For centuries, metals have played an important role as structural materials. Techniques of alloying, smelting, and forging have been evolving since the bronze and iron ages. With advancements in science and technology, and a deeper understanding of the effects of microstructure and processing techniques on the material behavior, the field of material science has radically improved through the past decades. The capability to engineer different material properties (mechanical, thermal, electrical, etc.) for a variety of applications has enabled the development of new alloys and composites. The demand for lighter, stronger materials with tailored properties that address both stringent structural requirements and provide additional engineering functionality (e.g., sensing, actuation, electromagnetic shielding) has spawned a new branch of materials called *multifunctional materials*. A specialized subgroup of multifunctional materials exhibiting sensing and actuation capabilities is known as *active materials*.

In sensing, a mechanical signal is converted into a non-mechanical output (e.g., voltage), while an actuator converts a non-mechanical input (e.g., electrical power) into a mechanical output. Active materials in general exhibit a mechanical response when subjected to a non-mechanical field (thermal, electrical, magnetic, optical, etc.). The mechanical response of these materials is typically one or more orders of magnitude greater than the response resulting from conventional material behavior such as thermal expansion. Some examples of active materials include piezoelectrics and electrostrictives (coupling of mechanical with electric fields), piezomagnetics and magnetostrictives (coupling of mechanical with magnetic fields), and shape memory materials

D.C. Lagoudas (ed.), *Shape Memory Alloys*, DOI: 10.1007/978-0-387-47685-8_1,

(coupling of thermal with mechanical fields). Active materials can be further subdivided into materials that exhibit direct or indirect coupling. Piezoceramics, piezoelectric polymers, magnetostrictive ceramics, shape memory alloys and magnetic shape memory alloys are examples of active materials that exhibit a direct coupling. This implies that either the mechanical or the non-mechanical field can serve as an input while the other as the output. In contrast, for active materials such as electro-rheological fluids (ERF) and magneto-rheological fluids (MRF), a change in the electric field or the magnetic field can indirectly couple with the mechanical behavior through a change in the viscosity of the fluid. This indirect, or one-way, coupling usually lacks the reciprocity of the two-way coupling exhibited by active materials that directly couple two fields.

The suitability of an active material with direct coupling for actuation applications depends on many factors. Two key design drivers are the actuation energy density (available work output per unit volume) and the actuation frequency of the material. An ideal active material would have both a high actuation energy density and a high actuation frequency. Figures 1.1 and 1.2 compare the actuation energy densities and the actuation frequencies, respectively, of some common active materials. The actuation energy density is denoted in Fig. 1.1 by the dotted lines and is defined as the product of the actuation strain (related to the stroke of an actuator) with the actuation stress, assuming here that the active material is operating under constant stress. The specific actuation energy density (work output per unit mass)

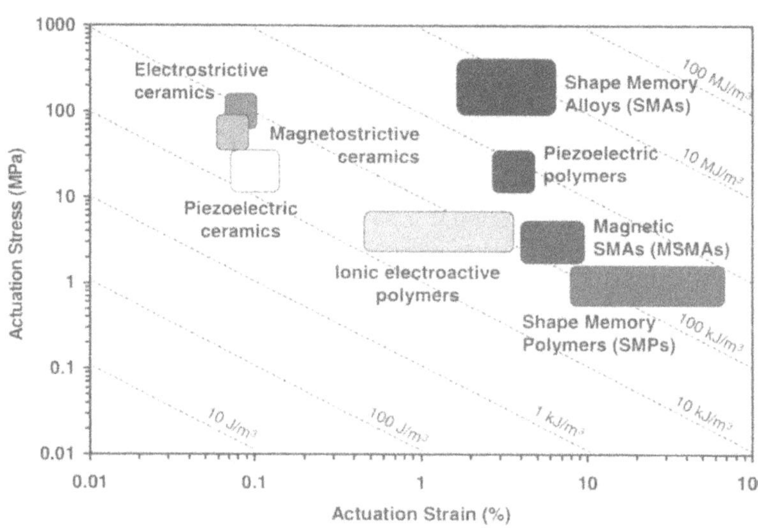

Fig. 1.1. Actuation energy density diagram indicating typical ranges of actuation stress, actuation strain, and the actuation energy densities of different active materials that exhibit direct coupling.

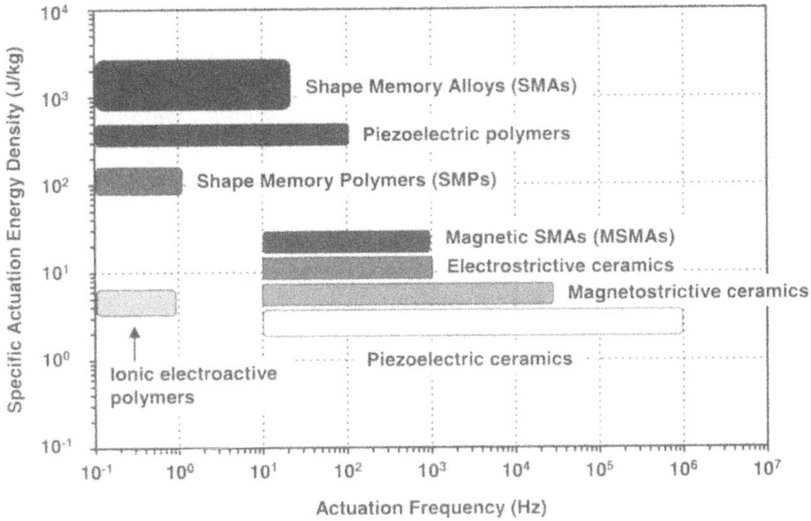

Fig. 1.2. Actuation frequency diagram comparing the actuation frequency ranges of different active materials that exhibit direct coupling.

for a specific active material can be calculated from Fig. 1.1 by dividing the actuation energy density by the mass density.

Shape Memory Alloys (SMAs) are a unique class of shape memory materials with the ability to recover their shape when the temperature is increased. An increase in temperature can result in shape recovery even under high applied loads therefore resulting in high actuation energy densities as shown in Fig. 1.1. In addition, under specific conditions, SMAs can absorb and dissipate mechanical energy by undergoing a reversible hysteretic shape change when subjected to applied mechanical cyclic loading. These unique characteristics of SMAs have made them popular for sensing and actuation, impact absorption and vibration damping applications. SMAs do, however, exhibit low frequency response, as shown in Fig. 1.2. Higher actuation frequencies are achievable for a class of SMAs called magnetic shape memory alloys, which have recently been investigated.

The application of SMAs spans a wide variety of industrial sectors such as aerospace, automotive, biomedical, and oil exploration. Over the past few decades, several key works have explored the microstructural mechanisms, engineering effects, and applications of shape memory alloys, including the experimental work of Jackson and coworkers [1], the application considerations of Duerig and others [2], and the comprehensive summaries of Perkins, Funakubo, and Otsuka and Wayman [3–5]. In the context of the current textbook, this chapter will provide insights into the history of SMAs, their properties, their microstructural behavior, and their varied industrial applications.

1.2 Shape Memory Alloys - A Brief History

The discovery of martensite in steels in the 1890s by Adolf Martens was a major step toward the eventual discovery of shape memory alloys. The martensitic transformation was perhaps the most widely studied metallurgical phenomenon during the early 1900s. The martensitic transformation, as observed in the Fe-C system, was established as an irreversible process. The concept of thermoelastic martensitic transformation, which explained the reversible transformation of martensite, was introduced in 1949 by Kurdjumov and Khandros [6], based on experimental observations of the thermally reversible martensitic structure in CuZn and CuAl alloys. By 1953, the occurance of thermoelastic martensitic transformation was demonstrated in other alloys such as InTl and CuZn.

The reversible martensitic transformation and the alloys that exhibited them remained unutilized until 1963. The breakthrough for engineering applications occurred with the discovery of NiTi by Buehler and coworkers while investigating materials useful for heat shielding [7]. It was noticed that in addition to its good mechanical properties, comparable to some common engineering metals, the material also possessed a shape recovery capability. Following this observation, the term "NiTiNOL" was coined for this NiTi material in honor of its discovery at the Naval Ordnance Laboratory (NOL). The term Shape Memory Effect (SME) was given to the associated shape recovery behavior. The discovery of Nitinol spearheaded active research interest into SMAs. The effects of heat treatment, composition and microstructure were widely investigated and began to be understood during this period [1].

In 1965, studies [8] showed that the addition of a third alloying element such as Co or Fe to the existing NiTi system caused a dramatic decrease in the SMA transformation temperatures. The new alloys inspired the first commercial SMA application, known as Cryofit, where SMA material was used for pipe couplings in F-14 fighter aircraft [9, 10]. The transformation temperatures for Cryofit were so low that, to prevent actuation from occurring before the assembly, the pipe couplings were transported in liquid nitrogen. Continued research to address this issue led to the development of the NiTiNb system in 1989, which was easier to handle due to its larger temperature hysteresis, and found widespread applications in battle damage repairs and in repairs for nuclear reactors [11]. *High Temperature SMAs* (HTSMAs), such as TiPd, TiPt and TiAu (with transformation temperatures greater than 100 °C), were also developed as early as 1970 [12]. Meanwhile, Melton and Mercier [13], while studying the fatigue properties of NiTi in 1978, showed that alloying the material with Cu did not change the transformation temperatures considerably, but narrowed the stress hysteresis. Later in 1999, Miyazaki showed improved fatigue life for NiTiCu alloys [14]. The improved fatigue life and the low cost associated with this material system made it suitable for a wide variety of engineering applications.

Since the initial discovery of Nitinol in 1963, many commercial applications have been developed. During the 1970s, several uses of NiTi in biomedical applications appeared, but it wasn't until the 1990s that NiTi stents made their commercial breakthrough. By this time, SMAs had found additional applications in air conditioning vents, electronic cable connectors, valves and a variety of other products. In addition, over the last decade the demand for actuation under high temperature operating conditions, driven by the aerospace and oil industries, has revived a great deal of interest in the development of HTSMAs. Finally, alloys that exhibit shape change characteristics similar to SMAs but under the influence of a magnetic field have recently been under investigation [15, 16]. The high actuation frequencies and the large strains generated in *Magnetic SMAs* (MSMAs) have made these materials a strong candidate for high frequency actuation devices.

1.3 Phenomenology of Phase Transformation in Shape Memory Alloys

A metallurgical phase diagram for a metallic alloy is a schematic representation of the equilibrium conditions between distinct phases. Phase diagrams consist of equilibrium lines or phase boundaries that separate different phases from each other. For an alloy consisting of at least two elements, the concentration becomes an important variable and is generally represented along the abscissa axis. The other variable commonly used is the temperature, represented along the ordinate axis. A phase diagram can have different control variables (stress, temperature, concentration, electric field etc.) within the bounds of which the equilibrium phases can be represented. For an alloy at a fixed composition (i.e. any vertical line parallel to the ordinate axis), the formation and disassociation of phases with the change in temperature is shown. Similarly, within the typical operating temperature range, SMAs have two phases, each with a different crystal structure and therefore different properties. One is the high temperature phase called *austenite* (A) and the other is the low temperature phase called *martensite* (M). Austenite (generally cubic) has a different crystal structure from martensite (tetragonal, orthorhombic or monoclinic). The transformation from one structure to the other does not occur by diffusion of atoms, but rather by shear lattice distortion. Such a transformation is known as martensitic transformation. Each martensitic crystal formed can have a different orientation direction, called a *variant*. The assembly of martensitic variants can exist in two forms: *twinned* martensite (M^t), which is formed by a combination of "self-accommodated" martensitic variants, and *detwinned* or reoriented martensite in which a specific variant is dominant (M^d). The reversible phase transformation from austenite (parent phase) to martensite (product phase) and vice versa forms the basis for the unique behavior of SMAs.

Upon cooling in the absence of an applied load, the crystal structure changes from austenite to martensite. The phase transition from austenite to martensite is termed the *forward transformation*. The transformation results in the formation of several martensitic variants, up to 24 for NiTi. The arrangement of variants occurs such that the average macroscopic shape change is negligible, resulting in *twinned martensite*. When the material is heated from the martensitic phase, the crystal structure transforms back to austenite, and this transition is called *reverse transformation*, during which there is no associated shape change.

A schematic of the crystal structures of twinned martensite and austenite for an SMA and the transformation between them is shown in Fig. 1.3. There are four characteristic temperatures associated with the phase transformation. During the forward transformation, austenite, under zero load, begins to transform to twinned martensite at the *martensitic start temperature* (M_s) and completes transformation to martensite at the *martensitic finish temperature* (M_f). At this stage, the transformation is complete and the material is fully in the twinned martensitic phase. Similarly, during heating, the reverse transformation initiates at the *austenitic start temperature* (A_s) and the transformation is completed at the *austenitic finish temperature* (A_f).

If a mechanical load is applied to the material in the twinned martensitic phase (at low temperature), it is possible to detwin the martensite by reorienting a certain number of variants (see Fig. 1.4). The detwinning process results in a macroscopic shape change, where the deformed configuration is retained when the load is released. A subsequent heating of the SMA to

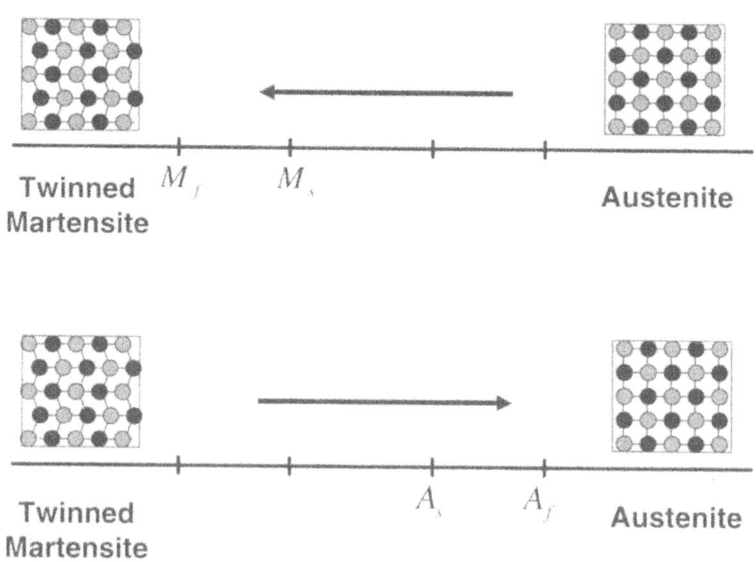

Fig. 1.3. Temperature-induced phase transformation of an SMA without mechanical loading.

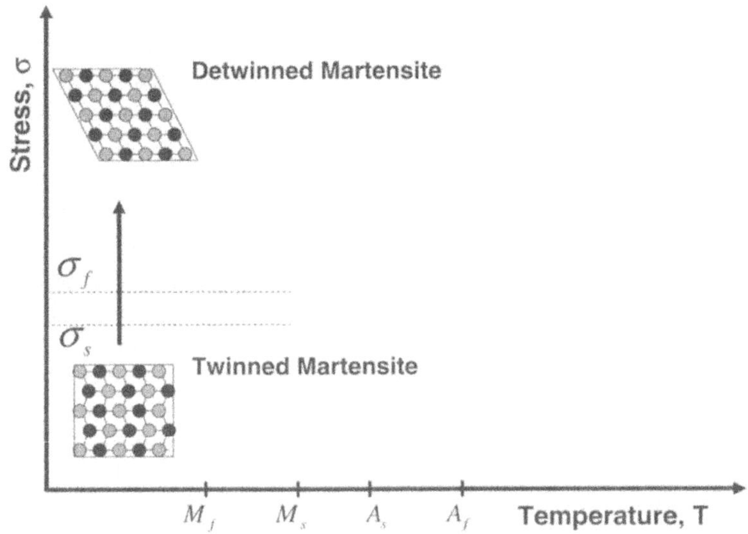

Fig. 1.4. Schematic of the shape memory effect of an SMA showing the detwinning of the material with an applied stress.

a temperature above A_f will result in a reverse phase transformation (from detwinned martensite to austenite) and will lead to complete shape recovery (see Fig. 1.5). Cooling back to a temperature below M_f (forward transformation) leads to the formation of twinned martensite again with no associated

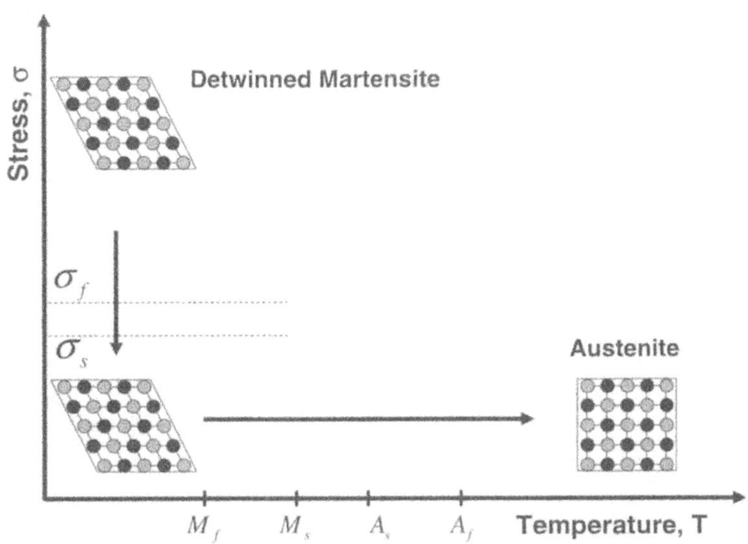

Fig. 1.5. Schematic of the shape memory effect of an SMA showing the unloading and subsequent heating to austenite under no load condition.

shape change observed. The process described above is referred to as the *Shape Memory Effect* (SME). The load applied must be sufficiently large to start the detwinning process. The minimum stress required for detwinning initiation is termed the *detwinning start stress* (σ_s). Sufficiently high load levels will result in complete detwinning of martensite where the corresponding stress level is called the *detwinning finish stress* (σ_f).

When the material is cooled with a mechanical load greater than σ_s applied in the austenitic phase, the phase transformation will result in the direct formation of detwinned martensite, producing a shape change. Reheating the material will result in shape recovery while the load is still applied. A schematic of the above-described loading path is shown in Fig. 1.6. Recognizing that the forward and reverse transformations occur over a range of temperatures (M_s to M_f, A_s to A_f) for a given SMA composition, we can construct transformation regions in the stress-temperature space. The transformation temperatures strongly depend on the magnitude of the applied load, with higher values of applied load leading to higher transformation temperatures. As a consequence, the transformation regions representing the A→M^d and M^d→A transformations have a positive slope in stress-temperature space. Irrespective of the nature of applied load (tension or compression), the transformation temperatures increase with an increase in the magnitude of the load. Under an applied uniaxial tensile load with a corresponding stress, σ, the new transformation temperatures are represented as M_f^σ, M_s^σ, A_s^σ and A_f^σ for martensitic finish, martensitic start, austenitic start and the austenitic finish temperatures, respectively. It should be noted that σ refers to the magnitude

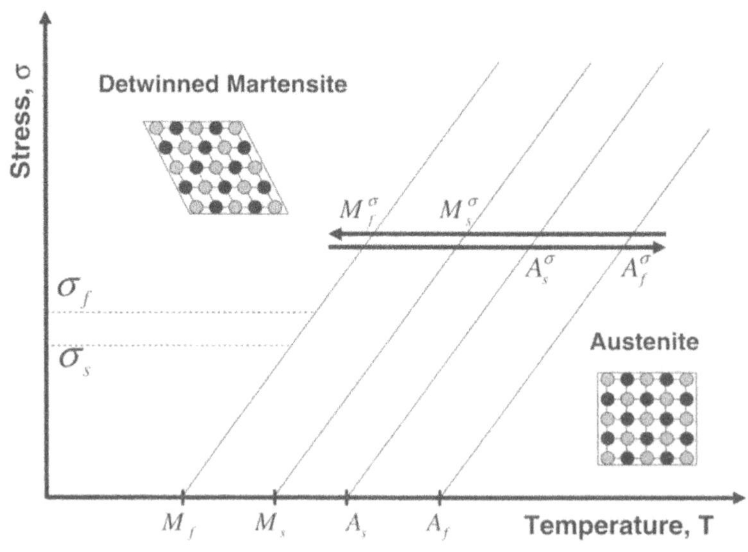

Fig. 1.6. Temperature-induced phase transformation in the presence of applied load.

of a uniaxial stress state or an appropriate scalar measure for a multiaxial stress state, as will be explained in Chapter 3.

In addition to thermally induced phase transformation, transformation can also be induced by applying a sufficiently high mechanical load to the material in the austenitic phase. The result of this load is fully detwinned martensite created from austenite. If the temperature of the material is above A_f, a complete shape recovery is observed upon unloading to austenite. This material behavior is called the *pseudoelastic effect.* A loading path demonstrating the pseudoelastic effect is shown schematically in Fig. 1.7, while the associated macroscopic shape change due to the applied load is captured in the resulting stress-strain diagram, as shown schematically in Fig. 1.8. The stress levels at which the martensite transformation initiates and completes are denoted by σ^{Ms} and σ^{Mf}, respectively. Similarly, as the SMA is unloaded, the stress levels at which the material initiates and completes its reverse transformation to austenite are denoted by σ^{As} and σ^{Af}, respectively. If the material in the austenitic phase is tested above the M_s temperature, but below the A_f temperature, only partial shape recovery is observed.

Figure 1.9 shows a schematic representation of the different phases of the SMA, which include the austenitic phase and both the twinned (self-accommodated) and detwinned martensite, along with the transition zones, in a stress-temperature diagram. Such a diagram, illustrating the different phases in a stress-temperature space for a given SMA with fixed composition, is called the *phase diagram.* Note that the phase diagram of Fig. 1.9 is a special case of the metallurgical phase diagram introduced in the beginning of Section 1.3, which involves composition as another variable. Construction of the

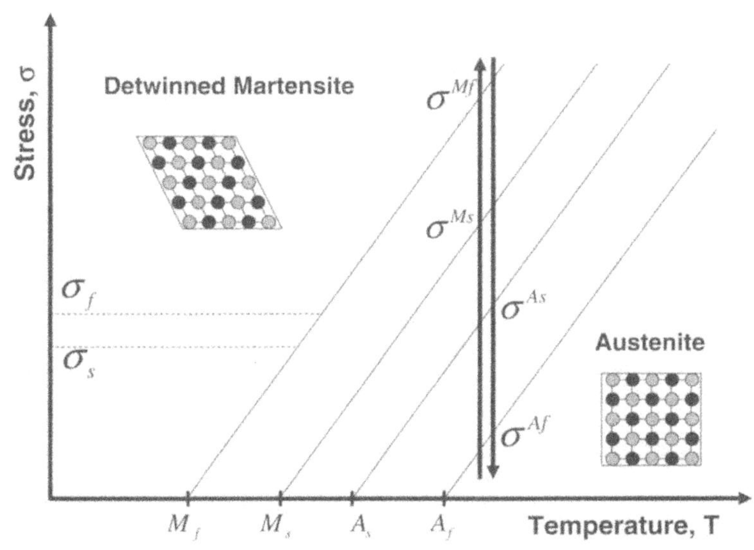

Fig. 1.7. A pseudoelastic loading path.

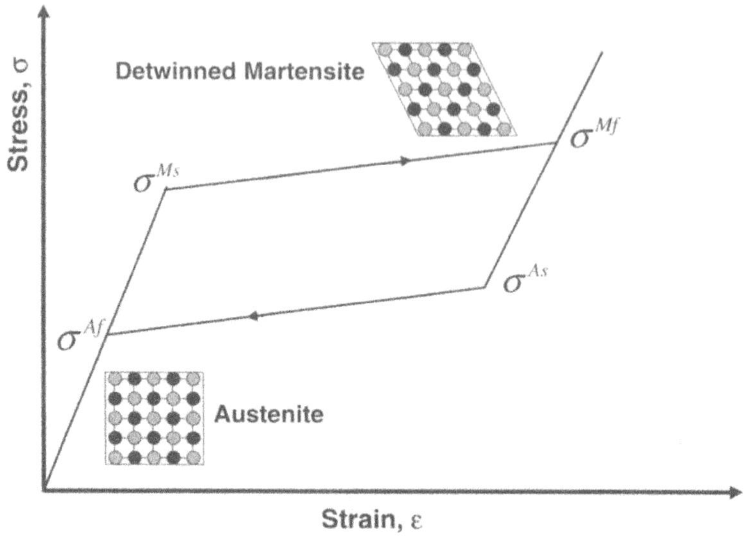

Fig. 1.8. Schematic of a pseudoelastic stress-strain diagram.

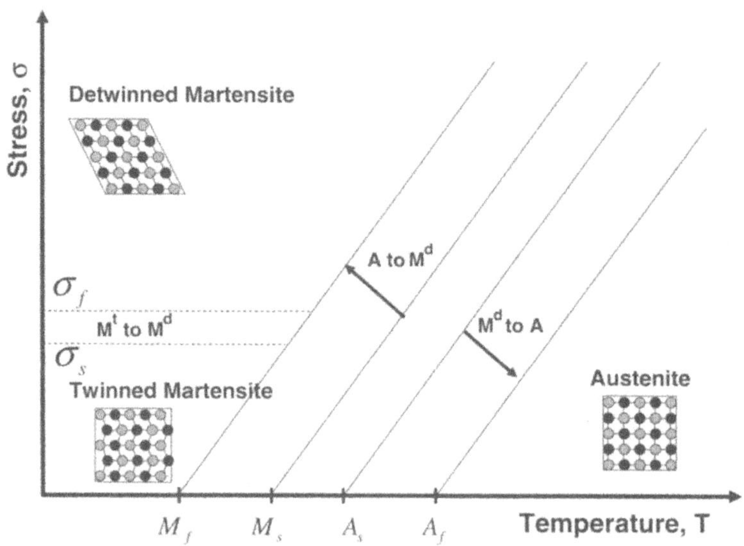

Fig. 1.9. Schematic of a stress-temperature phase diagram for an SMA.

phase diagram involves the interpretation of the SMA material response subjected to various thermomechanical loading paths resulting in shape memory thermal actuation under load and pseudoelastic behavior. In the following sections, the two important characteristics of SMAs, namely the shape memory effect and pseudoelasticity, will be discussed in more detail.

1.4 Shape Memory Effect

An SMA exhibits the *shape memory effect* (SME) when it is deformed while in the twinned martensitic phase and then unloaded while at a temperature below A_s. When it is subsequently heated above A_f, the SMA will regain its original shape by transforming back into the parent austenitic phase. The nature of the SME can be better understood by following the thermomechanical loading path in a combined stress-strain-temperature space as shown in Fig. 1.10. Figure 1.10 represents experimental data for a typical NiTi specimen tested under uniaxial loading. The stress σ is the uniaxial stress on the specimen due to an applied load. The corresponding strain ε is the change in the length of the specimen along the direction of applied load, normalized by the original length.

Starting from the parent phase (point A in Fig. 1.10), the stress-free cooling of austenite below the forward transformation temperatures (M_s and M_f) results in the formation of twinned martensite (point B). When the twinned martensite is subjected to an applied stress that exceeds the start stress level (σ_s), the reorientation process is initiated, resulting in the growth of certain favorably oriented martensitic variants that grow at the expense of other less favorable variants. The stress level for reorientation of the variants is far lower than the permanent plastic yield stress of martensite. The detwinning process is completed at a stress level, σ_f, that is characterized by the end of

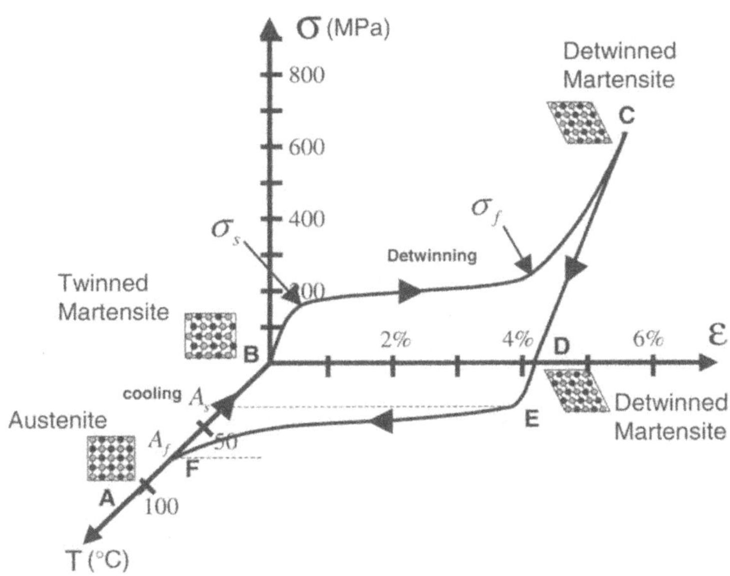

Fig. 1.10. Stress-strain-temperature data exhibiting the shape memory effect for a typical NiTi SMA.

the plateau in the σ-ε diagram in Fig. 1.10. The material is then elastically unloaded from C to D and the detwinned martensitic state is retained. Upon heating in the absence of stress, the reverse transformation initiates as the temperature reaches A_s, (at E) and is completed at temperature A_f (point F), above which only the parent austenitic phase exists. In the absence of permanent plastic strain generated during detwinning, the original shape of the SMA is regained (indicated by A). The strain recovered due to the phase transformation from detwinned martensite to austenite is termed as the transformation strain (ε^t). Subsequent cooling to martensite will again result in the formation of self-accommodated twinned martensitic variants with no associated shape change, and the whole cycle of the SME can be repeated. The above described phenomenon is called *one-way shape memory effect*, or simply SME, because the shape recovery is achieved only during heating after the material has been detwinned by an applied mechanical load.

Example 1.1. In order to review the SME, let us consider an SMA wire whose behavioral characteristics are represented by the stress-strain-temperature diagram shown in Fig. 1.10. The SME of the wire is used for a one time actuation application. The wire, held at a temperature below the M_f (twinned martensitic state), is stretched along the axial direction. Under the applied loading, the material exhibits an elastic behavior and continues to elastically deform as the stress is increased. When the stress due to the applied loading reaches approximately 150MPa, the SMA wire begins to elongate significantly with a small increment in the stress level. This point marks the beginning of the martensitic detwinning in the wire. The detwinning continues until the total strain reaches approximately 4% and the entire wire has detwinned. At this point, the wire begins to stiffen again as the stress is increased. The end of the detwinning process is marked by the change in the slope during loading. The detwinned wire elastically unloads as the stress is released and the strain induced due to the detwinning is not recovered. This detwinned wire is then attached on the structure for the actuation application. During the actuation process, the wire is heated using a thermal source such as resistive heating. As the temperature of the wire increases, the wire initially undergoes thermal expansion. However, as the temperature reaches the austenitic start temperature of approximately 30 °C, the detwinned martensite in the material begins transformation into austenite. This results in the contraction (i.e., actuation) of the SMA. As the temperature reaches a value above 70 °C, the transformation (actuation) is complete and the wire is in the austenitic state. The exact austenitic finish temperature will depend on the stress of the SMA wire during the reverse transformation. Subsequent cooling returns the wire to the twinned state in the absence of a recovery stress applied by the structure on which the SMA is attached. The wire would then have to be detwinned for the next actuation cycle unless the structure can provide sufficient stress for detwinning upon cooling.

1.5 Pseudoelasticity

The *pseudoelastic* behavior of SMAs is associated with stress-induced transformation, which leads to strain generation during loading and subsequent strain recovery upon unloading at temperatures above A_f. A pseudoelastic thermomechanical loading path generally starts at a sufficiently high temperature where stable austenite exists, then develops under an applied load to a state at which detwinned martensite is stable, and finally returns to the austenitic phase when returned to zero stress state. An example of this path ($a \to b \to c \to d \to e \to a$) is shown in Fig. 1.11 as path 1. Most commonly, a pseudoelastic test is performed at a nominally constant temperature above A_f. The loading path for such a test is shown as path 2 in Fig. 1.11.

To illustrate the pseudoelastic behavior in greater detail, let us consider the thermomechanical loading path ($\mathsf{A} \to \mathsf{B} \to \mathsf{C} \to \mathsf{D} \to \mathsf{E} \to \mathsf{F} \to \mathsf{A}$) in Fig. 1.11, which starts at zero stress at a temperature above A_f. The corresponding σ-ε experimental data for the loading path is shown in Fig. 1.12. When a mechanical load is applied, the parent phase (austenite) undergoes elastic loading ($\mathsf{A} \to \mathsf{B}$). At a specific load level, the loading path intersects the surface for initiation of martensitic transformation on the phase diagram. This marks the stress level (σ^{Ms}) for the onset of transformation into martensite. Note that the stress-induced transformation from austenite to martensite is accompanied by the generation of large inelastic strains as shown in the stress-strain diagram of Fig. 1.12. The transformation proceeds ($\mathsf{B} \to \mathsf{C}$), to the stress level ($\sigma^{Mf}$) where the loading path intersects the M_f transformation surface, indicating the end of the transformation.

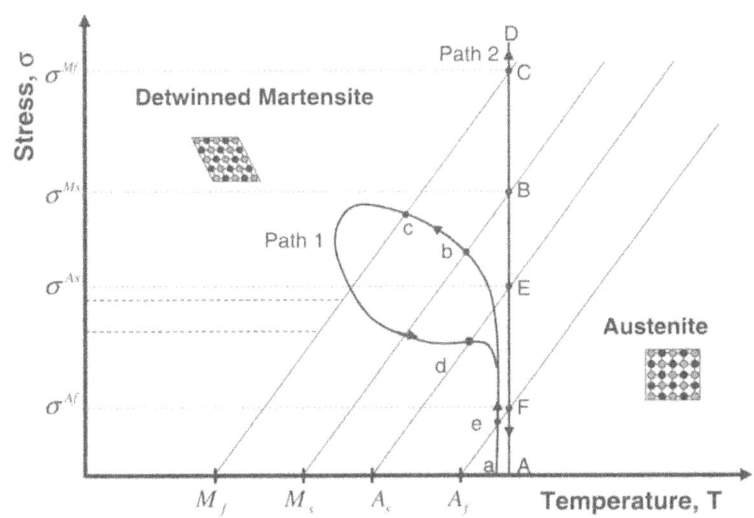

Fig. 1.11. Phase diagram and two possible pseudoelastic loading paths.

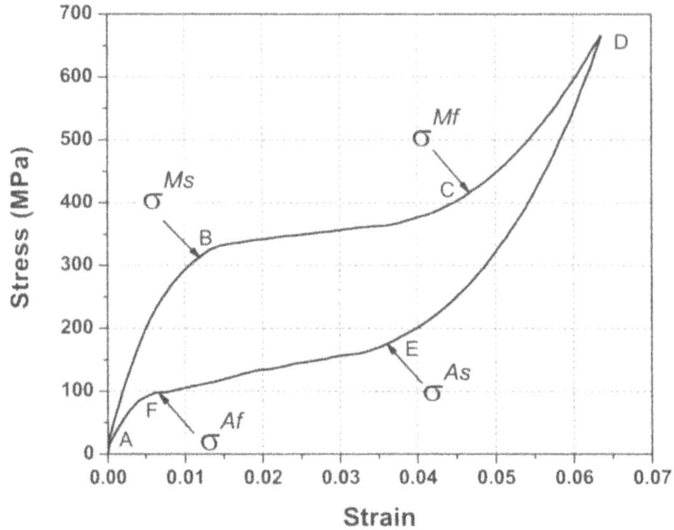

Fig. 1.12. A typical SMA pseudoelastic loading cycle.

The completion of martensitic transformation is indicated by a distinct change in slope on the σ-ε curve, which is associated with the elastic loading of the martensitic phase. A subsequent increase in the stress causes no further transformation and only the elastic deformation of detwinned martensite occurs (C $\rightarrow$ D). When the stress is released gradually by unloading, the martensite elastically unloads along the path (D $\rightarrow$ E). At point E, the unloading path intersects the austenitic start surface (at σ^{As}), which causes the martensite to revert to austenite. The process is accompanied by the recovery of the strain due to phase transformation at the end of unloading. The end of the transformation back into austenite is denoted by the point at which the σ-ε unloading curve rejoins the elastic region of austenite (point F corresponding to stress σ^{Af}). The material then elastically unloads to A. The forward and reverse phase transformation during a complete pseudoelastic cycle results in a hysteresis, which, in the σ-ε space, represents the energy dissipated in the transformation cycle. The transformation stress levels and the size of the hysteresis vary depending on the SMA material and testing conditions.

The detwinned martensite that forms from austenite as a result of the applied stress during Path 1 or 2 in Fig. 1.11 is one form of *stress-induced martensite* (SIM). SIM, in general, is martensite that forms from austenite in the presence of stress. There are many thermomechanical loading paths that can result in the formation of SIM.

Generally, the term *pseudoelasticity* describes both *superelastic* behavior and so-called *rubber-like* behavior [5]. The reversible phase transformation (described in the previous paragraph) caused by a thermomechanical loading path is strictly called the superelastic behavior. The rubber-like effect is an exclusive behavior of the martensite phase only and occurs due to the reversible reorientation of martensite. In some cases, aging the martensitic phase can enable the reversal of the martensitic detwinning process upon unloading at temperatures below M_f. The resulting σ-ε curve is similar to the superelastic curve, and this phenomenon is called the *rubber-like* effect to emphasize the similarities with the nonlinear elastic behavior of rubber. In SMAs exhibiting the rubber-like effect, the stress required to detwin martensite is very small compared to σ^{Ms}. We will not consider the rubber-like effect any further, and the term pseudoelasticity will refer to the superelastic behavior of SMAs only.

1.6 Cyclic Behavior of SMAs

We have studied the one-way SME behavior in SMAs. Sometimes an SMA can exhibit repeatable shape changes under no applied mechanical load when subjected to a cyclic thermal load. This behavior is termed *two-way shape memory effect* (TWSME). The TWSME can be observed in a SMA material which has undergone repeated thermomechanical cycling along a specific loading path (*training*). Repetition along a loading path for a large number of cycles can induce changes in the microstructure, which causes macroscopically observable permanent changes in the material behavior.

Training an SMA refers to a process of repeatedly loading the material following a cyclic thermomechanical loading path until the hysteretic response of the material stabilizes and the inelastic strain saturates. Let us consider the case of cyclic thermal loading of an SMA specimen under a constant applied stress (Fig. 1.13). During the first thermal cycle, only a partial recovery of the strain generated during cooling is observed upon heating with some permanent (irrecoverable or plastic) strain generated during the cycle. A small, permanent strain remains after each thermal cycle is completed. The additional permanent strain associated with each consecutive cycle begins to gradually decrease until it practically ceases to further accumulate (Fig. 1.13). A similar behavior can be noticed in the case of mechanically cycling an SMA repeatedly in its pseudoelastic regime, until saturation takes place (Fig. 1.14). The TWSME behavior can also be achieved by adopting different training sequences [17, 18]. A more recent technique that leads to TWSME deals with aging the material under stress in the martensitic state [19].

TWSME is a result of defects introduced during training. These permanent defects create a residual internal stress state, thereby facilitating the formation of preferred martensitic variants when the SMA is cooled in the

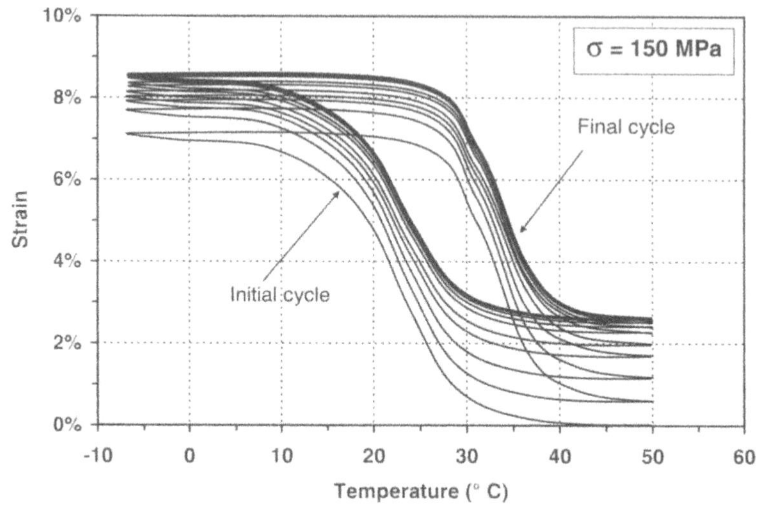

Fig. 1.13. Thermal cyclic loading (50 cycles) of a NiTi shape memory alloy wire under constant load of 150 MPa [18].

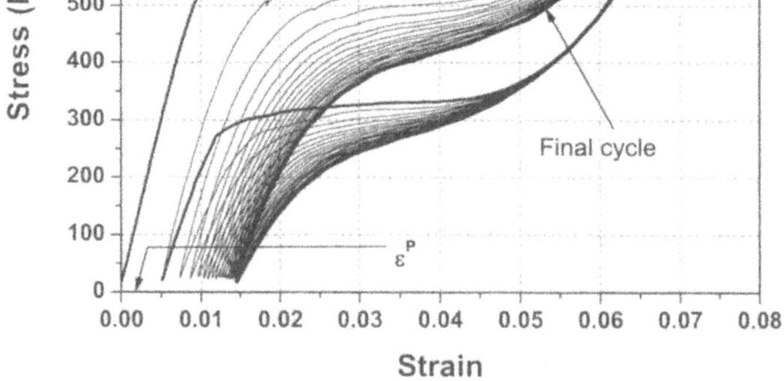

Fig. 1.14. Pseudoelastic response of an as-received NiTi wire with $A_f = 65\,°C$, tested at a temperature of 70 °C. Also shown is the stabilized pseudoelastic hysteresis loop after 20 cycles.

absence of external loads. If the internal stress state is modified for any reason (e.g., aging at high temperature or mechanical overload), the TWSME will be perturbed [20].

1.7 Transformation Induced Fatigue in SMAs

Since SMAs are widely preferred for actuation and sensing applications that require multiple cycles, it is now important to extend what we learned about cyclic effects to the topic of transformation induced fatigue behavior. The fatigue behavior of SMAs depends on the material processing (fabrication process, heat treatment, etc.), the type of loading conditions (applied stress, strain, temperature variations, environment, etc.), and transformation induced microstructural modifications (e.g., defects on grain boundaries due to strain incompatibilities). In most SMA applications, a large number of transformation cycles are induced by repeating a loading path that exhibits either pseudoelasticity or thermally induced phase transformation under applied load. As discussed in the previous section, repeated loading along a thermomechanical path causes gradual microstructural changes. These changes cause the degradation of the SMA behavior leading to low cycle fatigue as opposed to high cycle fatigue most commonly observed in loading paths operating in a purely elastic regime of a material. This section will provide a brief review of mechanically and thermally induced transformation fatigue behavior of SMAs.

Mechanically induced fatigue behavior of SMAs is typically examined by performing rotating bending tests or by mechanically cycling the material on the load frame between two stress or strain levels along a given loading path [18, 21]. Such mechanical cycling can be performed to induce a complete transformation (i.e., cycled between complete austenite and martensite phase) or partial transformation (cycling between states where one or both end limits are not purely martensite or austenite but a mixture). If the deformation or stress level applied to the SMA specimen remains within the elastic regime, this can lead to fatigue life as high as $\sim 10^7$ cycles. However, in some cases, the material can be taken through detwinning or stress induced martensitic transformation by applying sufficiently high load levels. In such cases, the material fails considerably earlier in what is termed "transformation-induced low cycle fatigue," with a fatigue life of the order of thousands of cycles [22, 23].

Similar to mechanically induced transformation fatigue, thermally-induced transformation fatigue behavior of SMAs is extremely important to study for actuation applications. Fatigue life for SMAs undergoing thermally-induced transformation cycles under applied load is dictated by the amount of transformation strain allowed to occur (partial or complete transformation) as well as the stress level under which the material is cycled. The amount of cyclic transformation strain allowed in the material can significantly affect the number of cycles to failure [24]. Under conditions of partial transformation, SMAs can exhibit much higher fatigue lives. A partial transformation limits the generation of martensite and the associated transformation strain, which causes a significant improvement in the fatigue life of the alloy [25–27].

Typical fatigue test data for complete and partial transformation in a NiTiCu material under a fixed stress level of 200 MPa is shown in Fig. 1.15. In the figures, the line A represents the strain in the SMA wire specimen at the end of each cycle after cooling to a certain level. Similarly, line B represents the strain in the SMA after heating to a certain level. In the complete transformation case, Fig. 1.15a, these levels pertain to complete martensitic transformation. In Fig. 1.15b, lines A and B represent strains generated during partial martensitic transformation. Taking the difference between these strain values at a given cycle and accounting for the effects of elastic moduli, one can obtain the transformation strain for the complete and partial transformations as shown in Fig. 1.15a and Fig. 1.15b respectively. The fatigue life is improved by approximately a factor of seven for a partial transformation and the amount of stable maximum transformation strain is reduced by a factor of three.

Other microstructural characteristics such as precipitate size and crystallographic orientation can also have a significant impact on the fatigue life of SMAs [28]. Heat treating the material under optimal conditions can also improve the fatigue behavior of SMAs. However, high annealing temperatures or chemically active environments can result in oxidation and corrosion, leading to crack nucleation and growth, thus reducing the fatigue life of SMAs [23, 27, 29, 30].

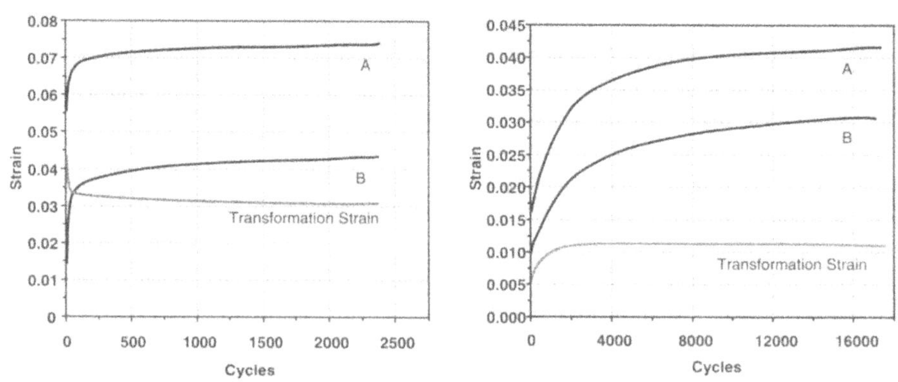

(a) Complete transformation (b) Partial transformation

Fig. 1.15. Thermally-induced transformation fatigue results for 200 MPa applied stress level. Curves indicate strain levels in martensite (A) and complete or partial austenite (B), in addition to recoverable transformation strain.

1.8 Crystallography of Martensitic Transformation

The transformation from austenite to martensite is a diffusionless transformation that occurs by shear distortion of the lattice structure (movement of atoms from their original position). This martensitic transformation possesses well-defined characteristics that distinguish it from other transformations. Within a single crystal (i.e., in a single grain in a polycrystalline material), the shear distortion occurs along a specific plane called the *habit plane*, which forms the interface between the martensitic and the austenitic phases. Since the habit plane does not rotate or deform during the course of the transformation, the plane is also referred to as the *lattice invariant plane*. Figure 1.16 schematically shows an austenite/martensite interface with its associated lattice invariant plane that separates austenite from a twinned martensite region. The transformation to martensite can occur along the lattice invariant plane by two different mechanisms, called *lattice invariant shear* mechanisms. The first one is through slip (i.e., atoms moving by one or more atomic space) and the second occurs by twinning (i.e., atoms moving through a fraction of an atomic space). Both of these mechanisms can aid formation of martensite with little or no volumetric change in the material. The strain obtained by such a cooperative movement of atoms is referred to as a *lattice invariant strain*.

In SMAs, twinning is the common mechanism of lattice invariant shear. The detwinning process results in a relative displacement of atoms that can eventually cause a macroscopic shape change while retaining their original

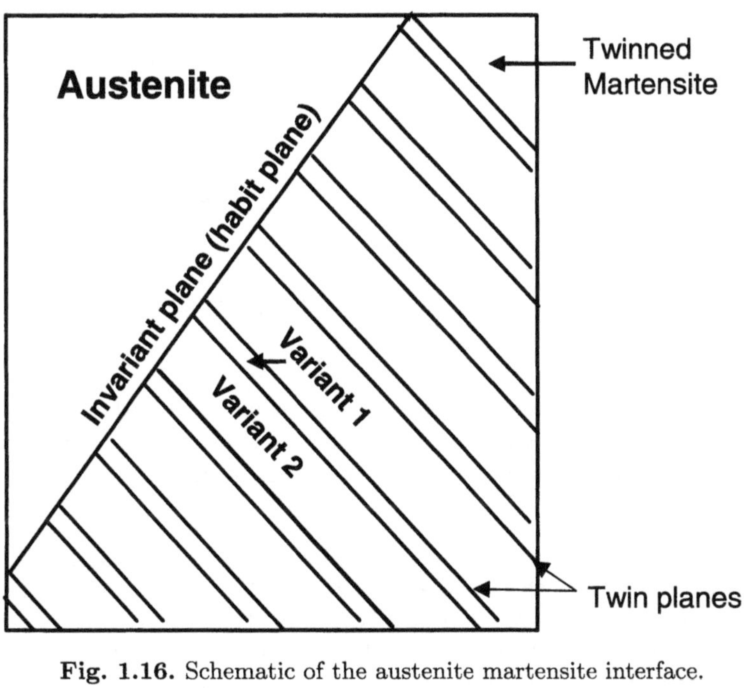

Fig. 1.16. Schematic of the austenite martensite interface.

atomic bonds, which also allows reversibility to the original crystallographic structure when heated to austenite.

Before proceeding with the detailed description of the transformation in SMAs, it is essential to have a preliminary understanding of the crystal structure of the associated phases. It is clear that there are two primary phases, austenite and martensite. The parent austenitic phase typically has a cubic (B2)[1] structure. The martensite that forms from austenite can have different crystallographic structures depending on the composition or the alloying element added. In equiatomic NiTi alloys for instance, the martensite that forms has a monoclinic (B19′) structure. Addition of an alloying element such as Cu or Pd can change the martensitic structure from a monoclinic to orthorhombic (B19), or create an intermediate R-phase (rhombohedral). In SMAs, during the transformation from austenite to martensite, every martensitic unit cell that forms can have different crystallographic orientations with respect to the cubic parent phase, and each unit cell having a different orientation is called a *variant*. Several such variants can form when the parent phase transforms to martensite. The number of variants that can form is dependent on the crystal structure of the martensite and its lattice correspondence with the parent phase unit cell.

An example of this can been seen in the NiTi system. Just as we have slip planes and slip systems in different crystal structures, there are "twin" planes along which twinning can occur. In the cubic lattice of the parent austenitic phase, the twinning shear can occur along the $\{011\}$ planes to obtain the crystallographic equivalent martensite twins. Since there are six such $\{011\}$ planes and two directions for the twinning to occur along each plane, there exist 12 equivalent martensitic twins. Each twin is composed of two martensitic variants, which, due to the shear distortion during transformation, assume a mirror symmetry. As a consequence of this, there are 24 total martensitic variants in most NiTi systems.

Recall that when austenite, under zero stress, transforms to twinned martensite, there is no associated shape change. The martensite that forms along the habit plane, under zero stress, occurs by twinning and the twins that form arrange themselves in a unique patterns to minimize the overall strain energy due to transformation. This behavior is familiarly known as the "self-accommodation" of twinned martensite.

An example of such a microstructure, as observed in NiTi is shown in Fig. 1.17[2]. In this micrograph three sets of twins A, B and C (six variants in total) together form a unique triangular morphology along the junction of three twin planes (the twin planes are shown by highlighted white lines).

[1] B2, B19′ and B19 are symbols that are partly systematic designations of the cubic, monoclinic and the orthorhombic crystal lattice structures respectively. This designation is known as the *Strukturbericht* designation.

[2] Reprinted from *Acta Materialia*, Vol. 45, Issue 12, Madangopal, K., pp. 5347–5365, Copyright 1997, with permission from Elsevier.

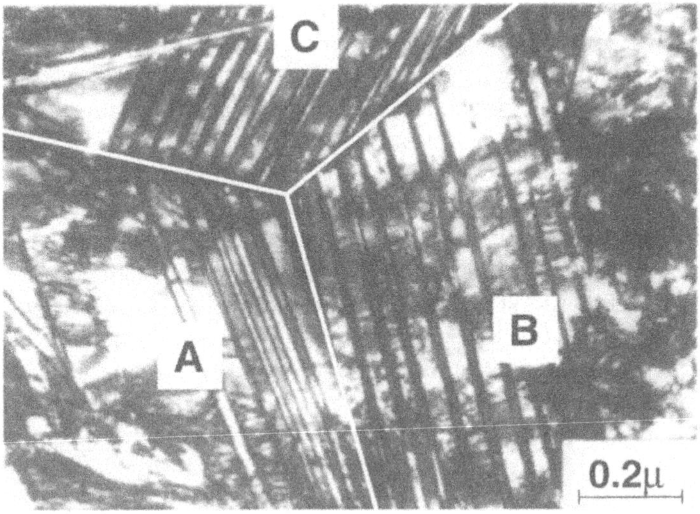

Fig. 1.17. Micrograph of self-accommodation monoclinic (B19′) structure in NiTi [31].

Similar to the self-accommodation as observed for the monoclinic structure, other martensitic crystal structures (i.e. orthorhombic, tetragonal, R-phase) can undergo self-accommodation. When the self-accommodated martensitic structure (monoclinic, orthorhombic or tetragonal) is subjected to an applied load, a resolved shear stress acts on the twin plane. When the resolved shear stress reaches a critical value, the most preferred variant (based on orientation to the applied stress) will evolve at the expense of other variants. This process of the evolution of the favorable variant and the associated generation of inelastic strain is known as the "detwinning" process. In the case of pseudoelasticity, a resolved shear stress reaches a critical value along the habit plane that leads to the formation of stress induced martensite.

The transformation from austenite to martensite and vice versa is associated with the release and absorption of latent heat. The heat of transformation and the associated transformation temperatures are most commonly determined using the *Differential Scanning Calorimeter* (DSC). The DSC is a popular thermal analysis technique that can be used to measure the phase transformation temperatures, the latent heat due to transformation, and the specific heat capacity of different phases in a material. This technique is also widely used to study the transformation temperatures of SMAs and has the advantage of requiring only a small quantity of material. The principle behind the operation of the DSC is the measurement of the rate at which heat energy is supplied to a specimen to maintain a constant heating or cooling rate. A DSC operating on such a principle is called a power compensated DSC. The device is termed differential because it has the ability to monitor the response of two specimens and to subtract the results. This is most useful when a material specimen placed in a holding pan is loaded opposite an empty pan. The result of the differential scanning calorimetry is then the net response of the material sample only. The sample material can be encapsulated in an inert atmosphere to prevent oxidation.

Figure 1.18 shows a typical DSC curve for SMAs. The power (mW) required to maintain a constant heating or cooling rate for the SMA specimen is represented on the ordinate axis and the temperature of the chamber is shown on the abscissa axis. When the specimen is heated from the twinned martensitic state, the transformation to austenite initiates at A_s. The endothermic reaction during the reverse transformation requires that additional heat power be supplied to the specimen to maintain the prescribed constant heating rate. This change in the power supplied as the temperature increases is recorded as a transformation "peak" during heating. A similar peak is also recorded during the cooling process during which the exothermic transformation from austenite to martensite takes place. The transformation temperatures from the acquired data are generally measured by drawing tangents to the start and end regions of the transformation peak and the baseline of the heating and cooling curves. The specific heat capacity can be computed by normalizing the power by the heating rate and the weight of the specimen. The associated latent heat for phase transformation can be calculated by integrating the specific heat over the range of the transformation temperatures. Most DSC software packages include the provision to directly compute the latent heat due to transformation.

The transformation temperatures can also be significantly affected by stored mechanical energy (i.e. precipitates, dislocations introduced due to processing/cutting, detwinning), therefore making the initial state of the material very important. This stored mechanical energy can cause a shift in and/or widen the transformation temperatures, or even cause an intermediate phase

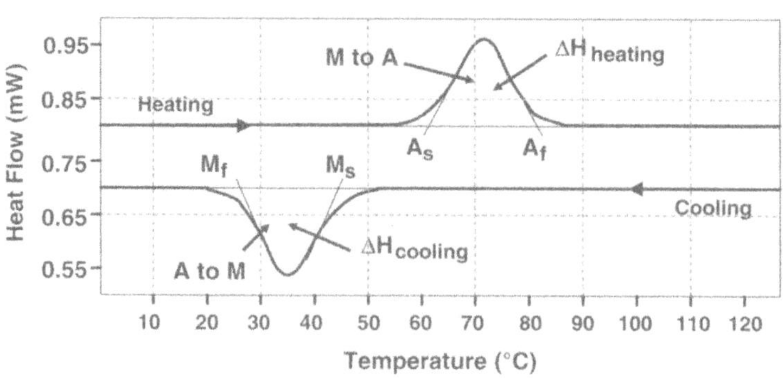

Fig. 1.18. A schematic of a DSC curve for an SMA showing the transformation temperatures and the associated latent heat of transformation during heating and cooling.

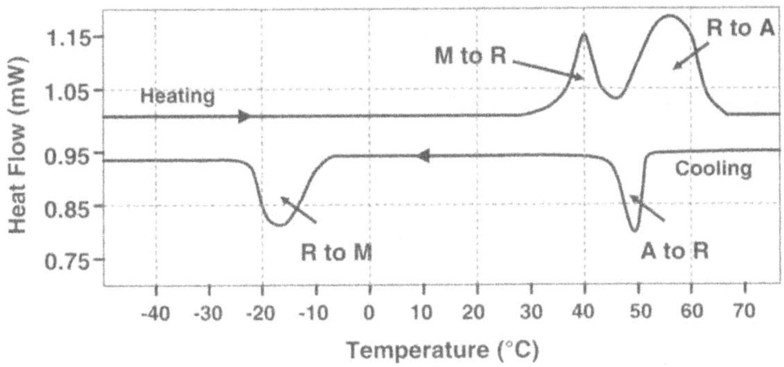

Fig. 1.19. A DSC curve for a Ni-rich NiTi SMA showing the two-stage transformation from austenite to martensite.

transformation. A DSC curve of a Ni-rich SMA showing an intermediate phase, called the R-phase, in the cooling curve is shown in Fig. 1.19. Further details about the R-phase will be discussed in Section 1.9.

1.9 Effect of Alloying on the Transformation Behavior of SMAs

Starting from the initial investigation of AuCd and AgCd alloys in the 1930s, to the discovery of Nitinol in 1963, and to the newest compositions being researched today, a wide variety of SMAs have been investigated over the last seven decades. New compositions have been fabricated by adding different alloying elements to existing alloys, providing a catalog of SMAs with a variety of properties to choose from. Such a wide selection gives designers great flexibility in tailoring SMA properties to match the constraints of a given commercial application. Shape memory alloys can be classified based on a wide variety of categories: primary alloying elements, mode of actuation (magnetic, thermal), operating temperature, or desired behavior. Some of the most commonly used SMAs, their properties, and the effect of alloying on them are discussed in this section.

1.9.1 NiTi-Based Alloys

Of the known SMA compositions, the NiTi alloy system has been studied most extensively and is used in the greatest number of commercial applications. This alloy exhibits strong SME, TWSME, and pseudoelastic behavior under the right conditions, which makes this material ideal for a variety of applications. It also exhibits resistance to corrosion and is biocompatible, making it suitable for use in biomedical applications. Compared to the less widely used alloys, the crystallography and thermomechanical response of NiTi are well understood, as are the the effects of heat treatment and the variation of transformation temperatures with changes in composition. In this section we briefly discuss the NiTi system, including the NiTi based ternary alloys.

NiTi - In the early 1960s, Buehler and his coworkers discovered the shape memory effect in an equiatomic NiTi alloy [7, 32]. The equiatomic composition (i.e. 50 at% of Ni and Ti) exhibits the maximum A_f temperature (120 °C) of all NiTi compositions studied. Decreasing the Ni atomic percentage (at.%) from the equiatomic composition does not change the transformation temperatures. If the composition of nickel is increased above 50 at.%, the transformation temperature begins to decrease, with A_f becoming as low as −40 °C for 51 at.% nickel. This variation in composition can change the ambient room temperature (23 °C) characteristics from SME to pseudoelasticity.

In Ni-rich alloys, aging at the temperature of 400 °C results in the formation of lenticular Ti_3Ni_4 precipitates. The stress fields due to the precipitates formed can result in the formation of an intermediate phase known as the *R-phase* between the austenite and martensite phases. The name "R-phase" is associated with the rhombohedral structure of the crystal. This phase generally vanishes with heat treatments at high temperatures and thus its existence is associated with certain specific conditions [33]. These conditions include substituting small amounts of Ni with Fe or Al, stress fields due to precipitates in Ni-rich NiTi, and stress fields due to dislocations in cold worked/heat treated alloys. NiTi alloys exhibit fully recoverable transformation strains of up to 8% and can be commercially obtained in various forms (e.g., wires, strips, rods, tubes and plates).

Recent studies have also investigated 55 at.% NiTi composition [34]. This composition exhibits transformation temperatures in the range of −10 °C to 60 °C. The alloy is a chemically multi-phased alloy, which is one of the primary reasons why it exhibits low transformation strains. However, 55 at.% NiTi alloy has been proven to show superior corrosion resistance as compared to stainless steels in harsh environments such as a salt water bath or salt fog [35]. The alloy also shows excellent thermomechanical stability, easier control of transformation temperatures through heat treatment, and can be hot formed into various complex shapes as these alloys do not require cold working [36, 37].

NiTiCu - The addition of Cu to NiTi preferentially replaces Ni to form NiTiCu alloys. The unique property of these alloys is that addition of Cu reduces the hysteresis of the SMA response. However, this also results in a decrease in the transformation strain. In NiTi 10 at.% Cu, the transformation hysteresis is much smaller than for the binary alloy at the expense of the total transformation strain which is reduced to approximately 4.0% [38]. The addition of Cu also reduces the pseudoelastic hysteresis. The width of the pseudoelastic hysteresis is less than 100 MPa for NiTi 10 at.% Cu when compared to a width greater than 200 MPa for the binary alloy. The addition of Cu to the binary alloy also greatly reduces the sensitivity of the martensitic start temperature to composition [39]. This change in the material behavior is associated with the change in the phase transformation. Studies on the phase transformation behavior of the NiTiCu system have shown that for a composition of 7.5 at.% ≤ Cu ≤ 15.0 at.%, the material undergoes a clear three stage cubic to orthorhombic to monoclinic transformation [40]. The low stress level required to transform the cubic phase to the orthorhombic phase results in the lowering of detwinning stress level.

The small hysteresis associated with the transformation makes TiNiCu an ideal choice for actuators. Among the different compositions of TiNiCu, 5.0 at.% ≤ Cu ≤ 10.0 at.% is most preferred. Addition of Cu greater than 10 at.% embrittles the material. Recent studies [41] on 25 at.% Cu fabricated by the melt spun technique have shown to have small hysteresis and stable pseudoelastic total strains of up to 7.0 %. DSC and Transmission Electron Microscopy (TEM) studies of this composition have also shown the $B19 \rightarrow B19'$ transformation occurring in the material [42].

NiTiNb - Unlike actuation devices that require a small hysteresis, a device used for the purpose of coupling must meet a different set of requirements in terms of SMA properties. One such requirement would be for the material to show minimal response to wide temperature changes. This can be achieved by widening the hysteresis of binary NiTi. The alloying element that facilitates this characteristic is Niobium (Nb). The effect of adding Nb was first studied in 1986 [43, 44]. It was noted that one consequence of adding Nb to NiTi was the widening of the thermal hysteresis. The wide hysteresis has important practical use in the field of SMAs, facilitating the engineering of material properties for which room temperature lies within the regions of the hysteresis. This allows the material to be deformed at low temperatures and yet be safely transported at ambient temperatures.

In the early work on $Ni_{47}Ti_{44}Nb_9$ [45], it has been shown that the alloy primarily consists of the NiTi phase with dispersed insoluble elliptical or globular precipitates of nearly pure Nb. These precipitates are extremely soft with a deformation stress equivalent to the detwinning stress of martensite. The large thermal hysteresis of the material is associated with the partitioning of the strain into a recoverable part (due to the NiTi phase) and an irrecoverable part (due to the Nb precipitates). This also explains why the material does not exhibit complete recovery during the deformation, and why a deformation corresponding to approximately 4 % strain can induce permanent strain in the material. Techniques such as pre-deformation can further increase the hysteresis width in these alloys by increasing the A_s temperature [46, 47]. Newer compositions with lower Nb concentration (3 at.%) have shown promising SME behavior [48].

NiTiX (X = Pd, Pt, Hf or Zr) alloys - Development of SMAs for commercial applications over the last four decades has primarily concentrated on applications with operating temperatures ≤ 100 °C. However applications involving high operating temperatures, such as in the core region of an aircraft engine or down hole applications in the oil industry require SMAs with high transformation temperatures and stable material properties. This demand for SMAs with high transformation temperatures has led to the development of a new class of SMAs known as *High Temperature Shape Memory Alloys* (HTSMAs). HTSMAs are a unique class of SMAs that have transformation temperatures greater than 100 °C and are capable of actuating under high temperature conditions. These alloys are produced by adding ternary elements such as palladium, platinum, hafnium, gold, and zirconium to NiTi, for which the transformation temperatures can be shifted anywhere in the range of 100–800 °C [4, 49, 50].

Similar to conventional SMAs, HTSMAs have austenitic and martensitic phases that can transform from one state to the other. The earliest studies on HTSMAs were performed in 1969 [12] on alloys such as Au-Ti, Pd-Ti and Pt-Ti to investigate whether these materials undergo martensitic transformation similar to (Au, Ag)-Cd alloy systems. In these studies, it was also revealed that these materials exhibit very high transformation temperatures. In 1981, detailed studies were performed on the TiNiPd alloy system, and it was determined that the temperatures for phase transformation were composition dependent [51]. Since then, several studies have been performed on TiNiPd, TiNiPt and TiNiAu alloy systems [49, 52]. Recent studies on various compositions of TiNiPd and TiNiPt alloys have investigated the work characteristic and workability of these materials [53, 54]. However, due to the extremely high cost associated with palladium and platinum, such alloys have limited commercial viability. Other alloy systems, such as TiNiHf [50] and TiNiZr [55], are also being widely investigated. Although these materials do not possess transformation temperatures as high as NiTiPd and NiTiPt, they are more cost effective. The primary limitations of HTSMAs are that the transformation strains associated with the material are approximately 3%, and that these materials have a low critical stress for slip. A sampling of NiTi-based ternary alloys and their associated transformation temperatures is shown in Table 1.1. Note that these values are highly dependent on material history, including heat treatment, and are given to indicate general trends.

1.9.2 Copper-Based Alloys

Although NiTi SMAs offer excellent pseudoelastic and SME properties and are biocompatible, they are relatively expensive compared to Cu-based SMAs. Good electrical and thermal conductivity along with their formability makes Cu-based SMAs an attractive alternative to NiTi. Copper-based alloys generally exhibit less hysteresis than NiTi, with the transformation temperatures in Cu-based alloys highly dependent on the composition. A precise change from 10^{-3} to 10^{-4} at.% is sometimes necessary to achieve reproducible transformation temperatures within a 5 °C range. The main Cu-based alloys are found in the Cu-Zn and Cu-Al systems. In this section some of the most commonly used Cu based SMAs will be discussed.

CuZnAl - The CuZn binary alloys are very ductile and have resistance to intergranular fracture as compared to other Cu-based alloys [5]. These alloys

Table 1.1. Representative transformation temperatures for SMAs with different compositions and heat treatments.

NiTi Based SMAs	M_f	M_s	A_s	A_f	**Reference**
$Ti_{50}Ni_{50}$	15	55	80	89	[49]
$Ti_{49.5}Ni_{50.5}$	−78	−19	9	53	[56]
$Ti_{49}Ni_{51}$	−153	−114	−89	−40	[4]
$Ti_{49}Ni_{51}Cu_{10}$	8	30	35	50	[56]
$Ti_{50}Ni_{40}Cu_{10}$	21	41	53	67	[56]
$Ti_{44}Ni_{47}Nb_{9}$	−175	−90	−85	−35	[45]
$Ti_{42.2}Ni_{49.8}Hf_{8}$	50	69	111	142	[57]
$Ti_{40.7}Ni_{49.8}Hf_{9.5}$	61	90	118	159	[57]
$Ti_{40.2}Ni_{49.8}Hf_{10}$	103	128	182	198	[57]
$Ti_{35.2}Ni_{49.8}Hf_{15}$	95	136	140	210	[57]
$Ti_{30.2}Ni_{49.8}Hf_{20}$	127	174	200	276	[57]
$Ti_{48}Ni_{47}Zr_{5}$	20	65	75	138	[58]
$Ti_{43}Ni_{47}Zr_{10}$	45	100	113	165	[58]
$Ti_{38}Ni_{47}Zr_{15}$	100	175	175	230	[58]
$Ti_{33}Ni_{47}Zr_{20}$	205	275	265	330	[58]
$Ti_{50}Pd_{50}$	550	563	580	591	[49]
$Ti_{50}Ni_{20}Pd_{30}$	208	241	230	241	[49]
$Ti_{50}Ni_{10}Pd_{40}$	387	403	419	427	[49]
$Ti_{50}Ni_{5}Pd_{45}$	467	486	503	509	[49]
$Ti_{50}Ni_{45}Pt_{5}$	10	29	36	49	[49]
$Ti_{50}Ni_{40}Pt_{10}$	−8	18	−27	36	[49]
$Ti_{50}Ni_{30}Pt_{20}$	241	300	263	300	[49]
$Ti_{50}Ni_{20}Pt_{30}$	537	619	626	702	[49]

transform to the martensitic state at a temperature below room temperature. Addition of aluminum to the binary alloy can considerably increase the transformation temperatures. Varying the composition of aluminum between 5 wt.% and 10 wt.% can shift the M_s temperature from −180 °C to 100 °C. However, the parent phase exhibits a strong tendency to decompose into its equilibrium phases when overheated or aged. Due to this, the operating temperatures are typically restricted to approximately 100 °C. The transformation temperatures of the alloy are extremely sensitive to composition, and zinc can be lost during the melt process. Due to these factors, the fabrication process of the alloy needs to be precisely controlled. CuZnAl alloys are also very sensitive to heat treatments such that the quenching rate can lead to phase dissociation or change in transformation temperatures. Their mechanical behavior is limited to stress levels of approximately 200 MPa due to the low critical stress for slip. Within the operational range of stress, the alloy exhibits perfect SME and pseudoelasticity, but the transformation strain is limited to about 3-4% [5]. Since CuZnAl is very ductile as compared to other Cu-based alloys, they are mostly chosen for use in applications.

CuAlNi - CuAlNi is less sensitive to stabilization and aging phenomena. Similar to CuZnAl, the transformation temperatures of CuAlNi can be varied by changing the aluminum or nickel content. Changing the aluminum composition between 14 at.% and 14.5 at.% can change the M_s temperature from −140 °C to 100 °C. The relative change in transformation temperatures is not significant and the hysteresis remains fairly constant. Since this alloy is harder to produce, manganese is often added to improve its ductility and titanium is added to refine its grains. However, the primary limitation of the CuAlNi system is the poor ductility due to intergranular cracking [4]. This phenomenon also affects the mechanical behavior such that the material typically fractures at a stress level of about 280 MPa. Transformation strain in these materials is limited to 3%. The material also exhibits very poor cyclic behavior [4].

Developed later (1982), the CuAlBe alloy has been studied during the last few years. Recently, several other Cu-based SMAs are in development, such as CuAlMn which has good ductility and CuAlNb which is suitable for high temperature applications.

1.9.3 Iron-Based Alloys

FeNiCoTi and FeMnSi are the main ferrous SMAs. $FeNi_{31}Co_{10}Ti_3$ after specific thermomechanical treatment, exhibits SME. The alloy exhibits a thermal hysteresis of approximately 150 °C.

Another ferrous alloy with good commercial prospects is FeMnSi. Si is primarily added to improve the shape memory effect and raise the critical stress for slip in austenite. When subject to training under a specific thermomechanical loading path, these SMAs exhibit complete SME. The transformation strains in these alloys are in the range of 2.5-4.5% [5].

1.9.4 Additional SMAs

CoNiAl - CoNiAl alloys are created by addition of Co to NiAl or Ni to the CoAl binary alloy systems, both of which exhibit very good corrosion and oxidation resistance at high temperatures. The CoNiAl alloy undergoes a transformation from the parent cubic structure phase to a tetragonal martensitic phase. Although the martensitic transformation in this alloy system was discovered as early as 1971 [59], the alloy system was not extensively investigated due to its brittle nature. However, recent efforts of forming and controlling the quantity of intermetallic phases by heat treatments has helped improve the ductility of the material [60, 61]. A typical composition of $CoNi_{33}Al_{29}$ has transformation temperatures of M_f approximately −57 °C and A_f approximately −26 °C. The material also exhibits a transformation strain of approximately 4.0% due to thermal cycling. In addition, the pseudoelastic behavior of the alloy has shown stability over a temperature range of 150 °C above A_f [62]. While the conventional SME and pseudoelastic behavior in this ternary alloy has been investigated, the CoNiAl composition has also been widely studied for its magnetic properties and the possibility of a magnetic-field-induced re-orientation. In such a transformation, the martensite reorients when subject to an external magnetic field. As a result, the material generates a magnetic-field-induced re-orientation strain. Further, these materials can operate at much higher actuation frequencies ($\simeq$ 1 kHz) as compared to thermally activated SMAs. Studies on $CoNi_{33}Al_{29}$ have shown the material to exhibit strains of 0.06% due to magnetic field reorientation [63].

NiMnGa - The most widely investigated *Magnetic Shape Memory Alloys* (MSMAs) are NiMnGa alloys. The first conclusive report of martensitic transformations in Ni_2MnGa alloys was given in 1984 [64]. However, it was only in 1996 that the first suggestion on the possibility of a magnetic field-controlled shape memory effect in these materials was made [15]. A field induced reorientation strain of nearly 0.2% was observed in the stress-free experiments on martensitic Ni_2MnGa single crystals. Further work on off-stoichiometric intermetallic compounds near the composition Ni_2MnGa, in combination with thermomechanical treatments and the utilization of a better understanding of the crystallographic structure of these alloys, have yielded larger field-induced strains of 6% [65], and up to 10% [66, 67], in single crystals. The main limitation of Magnetic Shape Memory Alloys (MSMAs) is the relatively low blocking stress (the stress at which the magnetic reorientation strain is completely suppressed). The typical blocking stress level for MSMAs is 6-10 MPa. Ni_2MnGa also undergoes the typical pseudoelastic transformation with compressive strains of up to 4.0%. Recent studies have shown that a unique combination of the pseudoelastic transformation behavior combined with a magnetic field can produce a field induced phase transformation of approximately 0.5%. This field induced transformation has also increased the blocking stress of the material to 20 MPa [68].

Other magnetic shape memory alloys have been studied including Fe-Pd [69–72], Fe-Ni-Co-Ti, Fe-Pt, Co-Ni-Ga, Ni-Mn-Al [73–75], and Co-Ni-Al [63]. These alloys exhibit lower field-induced strains, but can have other advantages. The largest field-induced strains that have been observed in Fe-Pd, for example, reach 3.1% [76]. Fe-Pd, however, is more ductile than Ni-Mn-Ga [69]. A much more detailed explanation of the magnetic field induced reorientation and transformation behavior will be discussed in Chapter 7.

1.10 SMAs as Active Materials — Applications

Active materials are quickly gaining the attention of engineers and scientists worldwide as more emphasis is placed on both reliability and multifunctionality. For the past several decades, engineers and other designers in many fields have been developing ways to convert thermal energy into mechanical work through the use of SMAs and apply these solutions to real-world applications. One of the most well-known examples was the hydraulic tube coupling used on the F-14 in 1971 [77]. Since then engineers in various industries have continued to use the unique properties of SMAs in solving engineering problems. SMAs have attracted a great deal of interest in various fields of application ranging from aerospace [78] and naval [79] to surgical instruments [80], medical implants and fixtures [81]. This section reviews some of the applications of SMAs in these different fields.

1.10.1 Aerospace Applications

SMA technology implementation in the aerospace industry has spanned the areas of fixed-wing aircraft, rotorcraft, spacecraft and work in all these areas is still progressing. Some of the more recent applications of SMAs and research on their potential uses in these areas are described in the following sections.

Fixed-Wing Aircraft Applications - Perhaps two of the most well-known fixed-wing programs are the Smart Wing program and the Smart Aircraft and Marine Propulsion System demonstration (SAMPSON) [82]. The Smart Wing program represented a collaboration between DARPA, AFRL, and Northrop Grumman, and its purpose was to implement active materials, such as SMAs, to optimize the performance of lifting bodies [83]. In this program, SMAs were used as wire tendons to actuate hingeless ailerons and were also formed into torque tubes that initiated spanwise wing twist of a scaled-down F-18 aircraft wing. Although satisfactory actuation was provided by the SMAs at 1:6 scale, it was found that the SMA torque tube was not strong enough to actuate a full-scale wing. The as-tested torque tube installation is shown in Fig. 1.20.

There have been a number of other efforts to integrate SMA elements into aerostructures. One such study led to the development of a variable geometry airfoil. Through SMA actuation, this airfoil effectively changed its configuration from symmetric to cambered [84]. Many other studies on the utilization of SMAs in aerostructures have focused on actuating smaller elements. SMAs can be used in smaller elements because their behavior is exhibited across a large range of sizes. One includes looking into the possibility of pairing SMAs and Micro-Electromechanical Systems (MEMS) to decrease the turbulent drag of an aerodynamic surface [85]. When activated appropriately, the MEMS skin would create a traveling wave to energize the boundary layer and thereby decrease turbulent drag. Some research has also been performed in the area of dynamic property optimization of aircraft structural panels using SMA elements, which provide a changing elastic stiffness via the martensitic transformation. It was found that the thermally-induced post-buckling deflection of a structure could be decreased by pre-straining the SMA or increasing the volume fraction of the SMA fibers [86]. The concept of a tunable SMA "Smart Spar" represented another attempt to alter the dynamic properties [87].

The usefulness of active materials in tailoring propulsion systems was demonstrated through the SAMPSON program [88]. One of the uses of SMAs in this program was to change the geometry of an F-15 engine inlet. This

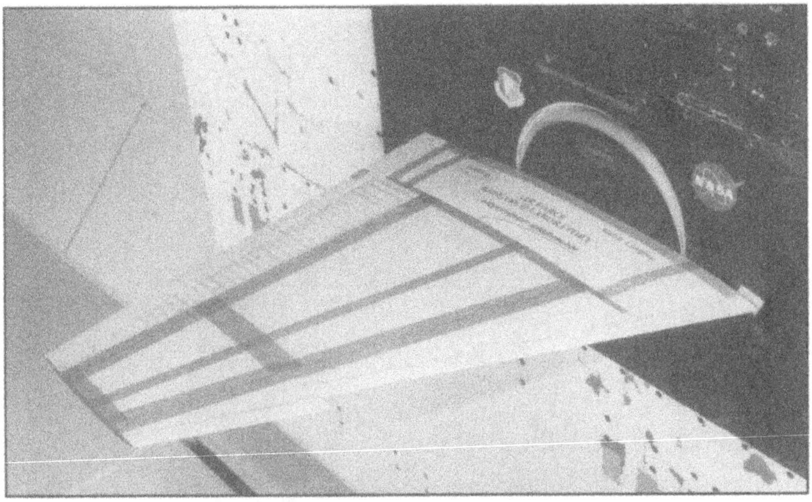

(a)

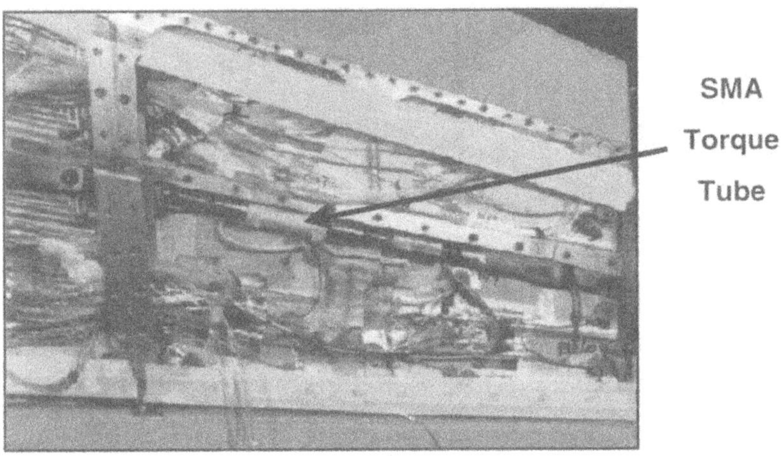

(b)

Fig. 1.20. (a) Total view of the SMART Wing model. **(b)** Cut away view of the SMA torque tubes as installed in the model during Phase 1 of the SMART Wing project [82].

experiment was done on a full scale inlet and the experimental setup can be seen in Fig. 1.21[3] [89]. A total force of approximately 26,700N was achieved through the use of SMA bundles containing 34 wires/rods. This generated

[3] Reprinted from *Journal of the Minerals, Metals, and Materials Society*, Vol. 55, No. 12, Wax, S.G., Fischer, G.M., and Sands, R.R., pp. 17–23, Copyright 2003, with kind permission from Springer Science and Business Media.

Fig. 1.21. The SAMPSON F-15 inlet cowl as installed in the NASA Langley Transonic Wind Tunnel [82, 89].

force rotated the inlet cowl through 9°. Another concept tested by the SAMPSON program was the concept of changing the shape of the inlet lip through a more complex system of SMA actuation.

Engine noise levels during take off and landing have become more highly regulated worldwide. To reduce this noise, some designers are installing chevrons onto engines to mix the flow of exhaust gases and reduce engine noise. Research is being performed into methods by which SMA beam components can be embedded inside chevrons. The SMA beams bend the chevrons into the flow during low-altitude flight or low speed flight, thereby increasing mixing and reducing noise. During high-altitude, high speed flight, these

Fig. 1.22. Boeing variable geometry chevron, flight testing [92].

SMA beam components will cool into martensite, thereby straightening the chevrons and increasing engine performance [90]. The current Boeing design for these variable geometry chevrons can be seen in Fig. 1.22.

A different solution to the active chevron problem has been proposed by NASA. In this design, SMA strips are installed on each side of the chevron centroid during the fabrication process [91]. Upon heating, the SMA strips contract alternately, leading to asymmetric stresses within the chevrons and therefore create a bending moment.

Rotorcraft - The role of SMAs in rotorcraft applications has been focused on the main rotor [93]. One active research area is SMA blade twist actuation [94]. SMAs are ideally suited for such applications because of their high actuation energy density and forces required in the small available volume within a rotor blade. One study proposed the use of SMA torque tubes to vary the twist of rotor blades, as found on tiltrotor aircraft [95]. These SMA torque tubes, when actuated, could facilitate the formation of different blade configurations and thereby optimize performance of such aircraft in both the hover and forward flight regimes. Recent work has also been performed on developing SMA-actuated tabs to improve tracking [96]. A trailing edge tab actuated by SMA wires was built into an airfoil section to accomplish this improvement. Alternately, the Smart Material Actuated Rotor Technology (SMART) Rotor project team [97] proposed to link an SMA torque tube to the tracking tab. Other rotorcraft applications include using SMA wire components for collective control [98] and to provide rotor blade tip anhedral [99], which minimized the blade vortex interaction noise by moving the blade tip vortex away from the rotor plane.

Spacecraft Applications - SMAs have been used in space applications to address problems related to actuation and release in zero atmosphere environment as well as vibration damping during spacecraft launch. Most of the applications and systems are typically designed by careful experimentation. One such application that uses SMAs is for the low-shock release mechanism in satellites [100]. Until 1984, it was estimated that nearly 14% of space missions experienced failure due to shock, and, in some cases, caused the mission to be aborted [101]. The shocks were caused due to pyrotechnic release mechanisms. The slow actuation due to gradual heating in SMAs makes them suitable for low shock release mechanisms in space applications. The scalability of SMA actuator designs also facilitates fabrication of smaller release devices for smaller satellites [102] in need of compact release mechanisms. Some of the devices developed for this purpose include the Qwknut [103] and the Micro Sep-Nut [102]. In both of these devices, the shape memory effect is used. Another miniature release device for space applications utilized a rotary actuation. The device, with a maximum dimension of 5mm, could provide a rotary actuation through an angle of 90°.

SMAs are also used in actuation of various components such as solar panels. An early design of this used SMA torsional elements to actuate solar collectors [93]. The Lightweight Flexible Solar Array (LFSA) [104] used thin SMA strips as hinges, which deploy the folded solar panels upon heating in approximately 30 seconds. The proof of concept design is shown in Fig. 1.23.

A different SMA space actuation application was utilized on the Mars Pathfinder mission in 1997. An SMA actuator was used to rotate a dust cover from a specific region of a solar cell so that the power output of this protected,

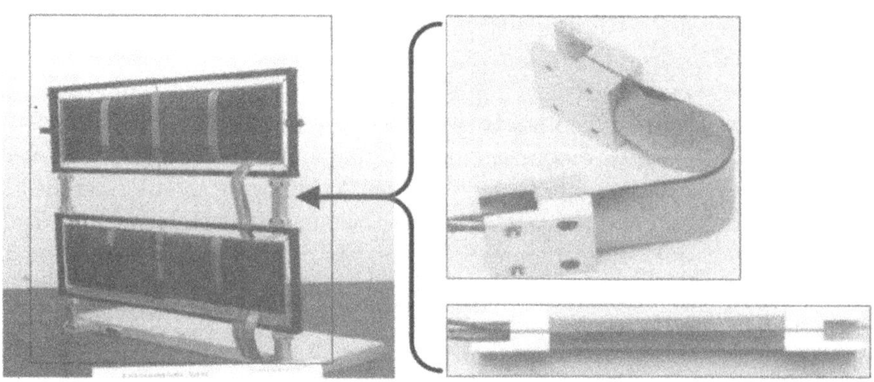

Fig. 1.23. The LFSA and the SMA hinges shown in the folded and the deployed configurations [104].

and therefore clean area of the panel could be compared to the power output of non-protected regions. This determined the negative effects of dust settlement on the solar panels. In addition to actuation, another attractive application for SMAs is vibration isolators and dampeners [101]. The hysteresis in the pseudoelastic behavior is representative of the mechanical energy that an SMA can dissipate during a cycle. Further, the change in the stiffness from the initial elastic region to that in the transformation region makes it an effective tool to isolate vibrations.

1.10.2 Medical Applications

The shape memory and pseudoelastic characteristics coupled with the biocompatibility of NiTi make them an attractive candidate for medical applications. The combination of these unique characteristics has led to the development of various applications such as stents, filters, orthodontic wires as well as devices for minimally invasive surgery (MIS).

An important requirement for an SMA, or any other material to be used in the human body, is that it be biocompatible. Biocompatibility is a property of the material to remain nontoxic through its functional period inside the human body. A biocompatible material can not produce any allergic reaction or inflammatory response in the host. The other requirement for the material is its biofunctionality, which is the ability to function desirably for its expected service life in the human body environment. These two requirements are crucial for the application of SMAs in the medical industry.

Several investigations have been performed to study the biocompatibility and biofunctionality of NiTi alloys [105, 106]. Analysis has focused on each individual element that constitutes the alloy, namely, nickel and titanium. Nickel intake occurs in a regular lifestyle [107], and most often its impact at trace levels is minimal. However, excessive intake of nickel can be poisonous to the human body. Unlike nickel, titanium and its compounds are intrinsically biocompatible and are commonly used in orthopedic and orthodontic implants [108]. The oxidation of titanium results in a coating of TiO_2, which provides a corrosion-resistant layer, making such a device stable within the human body. Corrosion studies performed on NiTi alloys have shown better stability than most alloys used in medicine and dentistry [109]. It has also been shown that surface coating NiTi with TiN or TiCN is effective in further improving the corrosion resistance of this material.

Orthodontic Applications - The properties of SMAs have been successfully implemented in a variety of dental applications. Nitinol orthodontic archwires have been used since the 1970s [110], and are more effective than other alternative materials. In a linear elastic material like stainless steel, there is a large increment in stress, for a small increment in strain which results in a large amount of force on the tooth for a small amount of corrective motion. The advantage of pseudoelastic arch wires is the ability to operate in the pseudoelastic plateau, during which the material has a near-zero stress change over a large strain increment. As a result, they provide a nearly constant, moderate force to actively move the teeth over a longer period of time compared with stainless steel. Further, the material composition and processing can be engineered to produce different levels of optimal force. An example of Nitinol orthodontic braces is shown in Fig. 1.24a.

Another key dental application for SMAs involves the use of Nitinol drills used in root canal surgery, which involves careful drilling within the tooth (Fig. 1.24b). The advantage of these Nitinol drills is that they can bend to rather large angles, which induce large strains, yet still withstand the high cyclic rotations [111].

Cardiovascular Applications - An early cardiovascular SMA device was the Simon Filter. The device acts as a filter that traps clots traveling in the blood stream. The trapped clots are then eventually dissolved. From the original expanded state, the device is deformed and constrained within a catheter. At the time of implantation, the filter is deployed in the blood vessel where the ambient temperature exceeds A_s of the filter material. The release from the constraint and the active properties of the SMA filter cause the filter to expand and it assumes its original shape as shown in Fig. 1.25a [112].

Another application of SMAs in the cardiovascular field is the atrial septal occlusion device. This device is used to seal an atrial hole that is located on the surface dividing the upper heart chambers. The traditional surgical technique used to address this problem is highly invasive and dangerous. The atrial septal occlusion device provides a suitable alternative to such a surgery. This device also exploits the shape memory characteristics exhibited by SMAs, and consists of two umbrella shape halves that can be screwed together in the center. Initially, these halves are folded, constrained and introduced into the heart. The two halves are deployed on either side of the hole and are connected using a screw. The resulting "sandwich" configuration forms a patch that seals the hole.

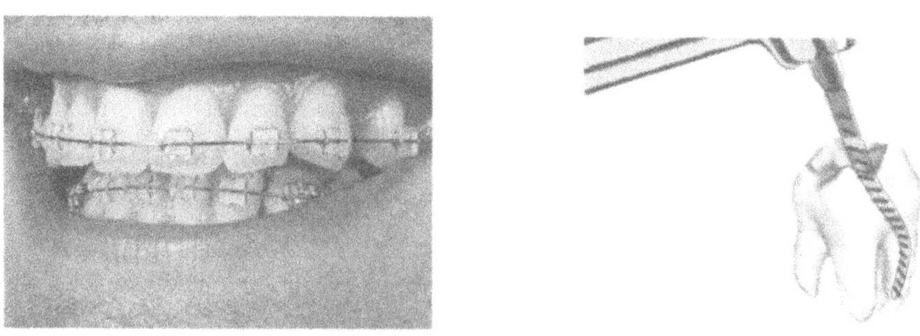

(a) (b)

Fig. 1.24. Orthodontic application of SMAs: (a) Nitinol braces used for alignment purposes in dental applications. (b) A schematic showing a NiTi drill used for root canal surgery.

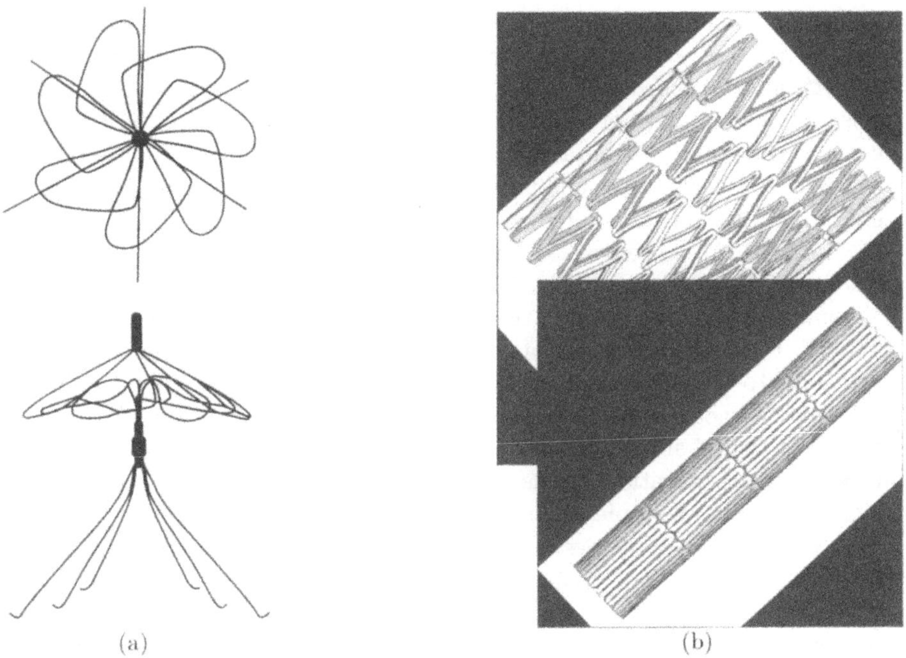

Fig. 1.25. Cardiovascular devices that utilize the engineering properties of SMAs: (a) Top view (above) and side view (below) of the Simon filter in the deployed configuration [112]. (b) A self-expanding Nitinol stent shown in the deployed configuration (above) and constrained state (below).

A more common cardiovascular application is the "self-expanding" NiTi stent. Like other conventional stents, this device is used to support the inner circumference of tubular passages in the body such as blood vessels. Traditionally, stents are made using stainless steel. These stents are expanded from the size of the introduced catheter to the size of the artery walls by an inflatable balloon. As the balloon is deflated, the steel stent undergoes elastic unloading, often resulting in a loose fit. Furthermore, to reach a particular nominal diameter, it is necessary to over-expand the stent to account for this unloading. This process of over-inflation can damage the vessels and can cause a condition where the blood vessel collapses after the procedure due to weakening of the walls. The self-expanding NiTi stents provide an attractive alternative to the traditional method. The device is generally laser cut from sheets or tubing and is then shape set to the appropriate diameter. After being constrained, the NiTi stent is introduced into the body where the temperature exceeds A_s of the stent material. It is then released in the artery where it expands to its original larger diameter and gently pushes outward on the walls. Furthermore, the device can adapt to any oblong passage as compared to the balloon inflated steel stents that are biased towards a circular shape. Fig. 1.25b shows an illustration of a Nitinol stent in the constrained and deployed configuration. Engineering analysis of such a device will be considered in Chapter 4.

Orthopedic Applications - The devices developed for orthopedic applications are used to support injured, weakened or fractured bones. One such device is the spinal vertebra spacer (Fig. 1.26a), used to provide local reinforcement to the vertebrae and prevent motion during the healing process. The device applies a constant force on the joint while providing flexibility [107].

Porous SMAs represent a different kind of material form and can be used as artificial bone implants (Fig. 1.26b) [113]. The porous nature of the material enables the existing bone tissue to migrate inward, increasing bonding strength. Furthermore, the implant properties (stiffness and porosity) can be engineered to match those of the bone. In a separate application, SMAs fasten to broken or fractured bones to facilitate healing. These devices include orthopedic staples and shape memory plates. The staple, for example, is installed in an open configuration at the fractured joint. An external heating source is used to heat the staple causing it to return to its original form via SME, which consequently provides a compressive force at the interface of the separated bones [107]. The shape memory plate is a device used when a cast cannot be applied over the fracture surface (i.e. facial areas, jaw, nose). The plate is also "shape set", deformed, installed and then actuated via (external) heating, providing a force which holds the fractured joints together [107].

Surgical Instrument Applications - Advances in medicine continue to enhance the use of minimally invasive surgery (MIS). Some of the enabling technologies advancing MIS includes instruments that can be inserted through these small openings followed by expansion to a desired size for the particular function. The pseudoelastic and shape memory effect properties of SMAs allow for more creative design options compared to conventional materials. One such device is the SMA basket used to remove stones in the bile duct. Other instruments using the shape memory effect behavior include surgical tools with grippers, scissors and tongs used in laparoscopy procedures. Pseudoelastic guide wires are widely used in surgery due to their kink resistance and superior flexibility [112].

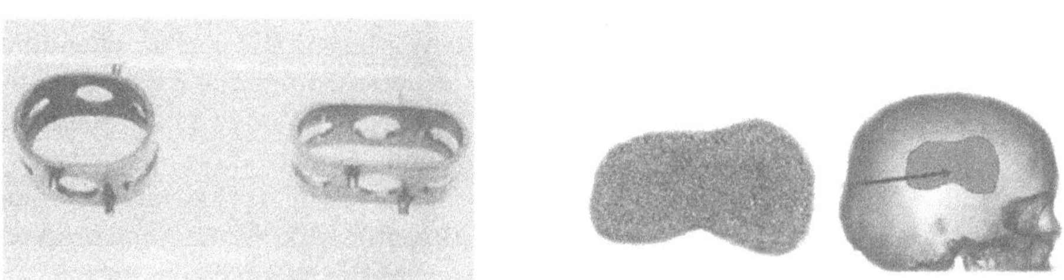

(a) (b)

Fig. 1.26. Orthopedic applications of SMAs. **(a)** Spinal vertebrae spacers showing the device in the martensitic and the deployed austenitic state [112]. **(b)** A schematic showing the prospective use of porous SMAs as artificial bone implants.

1.10.3 Transportation Applications

Shape memory alloys have been used in automobiles for applications ranging from impact absorption to sensing and actuation. The pseudoelastic behavior hysteresis provides an effective system to dissipate vibrations and impact. This property has been used for impact absorption on armor vehicles in military [114] and commercial applications [115]. One design for an impact absorption application required the deployment of a protective panel within 5-7 ms. The limited response time of commercially available actuation devices (10 ms) is overcome by the use of an SMA element. The device can be released in 3 ms and then reset for another actuation. The SME has also been implemented for actuating blinds that cover the fog lamp to prevent damage. A series circuit ensures the actuation of the SMA louvers every time the fog lamps are turned on. SMAs can also be used for sensor and actuation purposes simultaneously. An application that exploits this behavior is the SMA spring for the continuous variable transmission in the Mercedes A class. The spring acts as a sensor that monitors the temperature and actuates a valve at a specific temperature, which changes the direction of oil flow.

A similar actuation system is incorporated in the Shinkansen bullet train gearbox where the temperature in the gear box is monitored and an SMA spring actuates a valve to adjust the oil level in the gearbox [116]. Other applications developed for trains include the thermally actuated switch for the radiator fan in diesel engines and steam traps for the steam heating system in passenger trains. Both of these applications utilize the shape memory effect.

1.10.4 Other Applications

In addition to the aerospace, transportation and medical industries, there are many other fields and applications that incorporate SMAs. The oil industry has shown extensive interest to use the SMA actuation capabilities in release devices and protection systems for downhole drilling equipment. The high operating conditions have also opened the avenue for the use of HTSMAs in these devices [117]. Everyday applications such as coffee makers and rice cookers have also incorporated SMAs. A rice cooker equipped with an SMA valve has the valve actuate when the cooker reaches a certain temperature and releases the excess steam in the chamber. SMA actuated louvers have also been incorporated in air conditioning vents that can adjust depending on the temperature of the air exiting the vents. The SME is also utilized in shower faucet designs where an SMA spring automatically adjusts the flow of hot and cold water to maintain a preset water temperature [118].

The pseudoelastic behavior has also been used in a wide range of applications. Developers of vibration control devices in civil structures have shown interest in pseudoelastic behavior of NiTi due to its capability to dissipate energy through a large hysteresis [119]. Other applications that employ the pseudoelastic behavior are flexible metallic eyeglasses and headphones, that

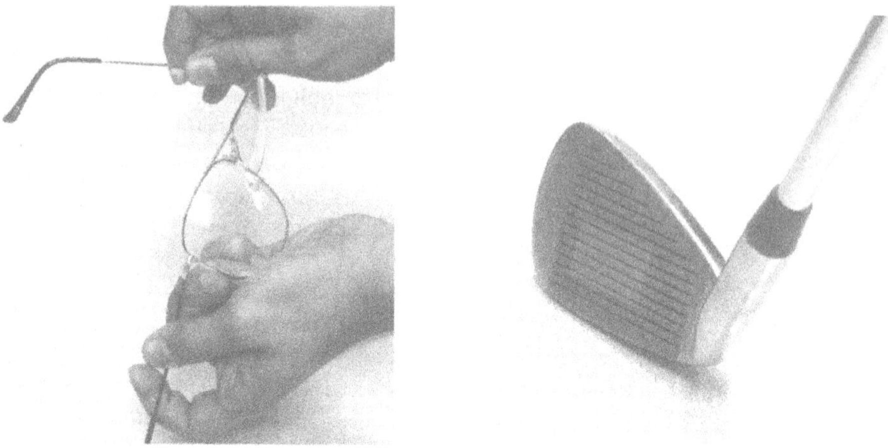

(a) (b)

Fig. 1.27. Commercial pseudoelastic applications of SMAs. **(a)** Pseudoelastic eye glass frames. **(b)** Golf club with embedded pseudoelastic SMA insert.

can be bent without breaking [77]. SMAs have also been used in sporting goods like golf clubs where the SMA embedded in the club absorbs the impact of the strike. Figure 1.27 shows a picture of a pseudoelastic eye glass frame and a golf club with a pseudoelastic SMA embedded in it.

In a recent study, an innovative approach of knitting SMA wires into different patterns has shown to produce complex shape changes such as rolling, spiraling, arching and folding [120]. The ability to generate such unique configurations using SMAs can open prospects for other novel design applications.

1.11 Summary

This chapter introduced the unique characteristic behaviors of shape memory alloys such as the shape memory effect and pseudoelasticity and discussed the underlying microstructural changes associated with such behaviors. The crystallography of martensitic transformation was also discussed, and a brief overview of the effects of alloying on the transformation behavior was presented. Some current implementation of SMAs into industrial and medical applications have also been discussed. As previously mentioned, several other compilations of work performed on SMAs are available. For historical work on SMAs the reader can refer to Perkins [3]. For application considerations, Duerig and coworkers provide a comprehensive summary. For details on microstructural influence on the behavior of SMAs, see Funakubo [4] and Otsuka and Wayman [5]. A more recent compilation of work by Brailovski and coworkers [121] gives an extensive summary on characterization efforts focused on SMAs.

1.12 Problems

1.1. Identify at least one biomedical application for which both the SME and the pseudoelastic effect could provide a solution.

1.2. A space application requires the use of an SMA component for fast actuation. The ambient temperature to which the SMA will be exposed is −30 °C. What properties would govern the choice of a suitable SMA? Suggest an alloy that would be most suited for such an application from the different alloys discussed in Sect. 1.9.

1.3. An SMA wire exhibits 5% recoverable strain upon mechanical unloading. When the wire is cooled, the material expands by 3% and recovers its shape upon heating. What are the different behaviors observed here?

1.4. An industrial process requires a small pressure vessel for storing hot gas. The pressure and the temperature inside the vessel cannot exceed 300 MPa and 80 °C, respectively. Suggest two concepts that implement SMAs in the design of a safety device for the pressure vessel. Suggest an SMA for this application making suitable arguments.

1.5. A particular actuation mechanism in an industry utilizes a temperature sensor that sends a signal to a motor which in turn operates a ratchet-pinion mechanism. The entire setup is fit into a small compartment. The company is searching for a suitable alternative actuation mechanism and is considering the use of SMAs. List the advantages of using SMAs in this particular application.

1.6. A particular application requires multiple cyclic actuation where the actuator needs to expand at high temperature and contract as the temperature decreases. The device assembly or the conditions of operation do not exert any stress on the actuator. How could one use an SMA for such an application?

1.7. A new pair of spectacle frames in the market has the capability to be twisted substantially without failure. Although the frame is made of an SMA, this flexible behavior is not due to a stress induced phase transformation phenomenon. Could this statement be true? Explain.

1.8. Shown in Fig. 1.28 is a simple heat engine. The SMA spring is actuated by resistive heating. A mass of 0.45 kg is suspended on the spring, which has an initial length of 3 cm. The current through and the voltage across the spring are 2 Amps and 6 Volts, respectively. When heated from T$<M_f$ to T$>A_f$, the tip of the spring moves from 12.5 cm (spring length under load) to 5 cm in 5 seconds. With the provided data answer the following questions:

1. What is the amount of work done by the spring?
2. What is the potential energy stored in the spring in the austenitic and martensitic states?

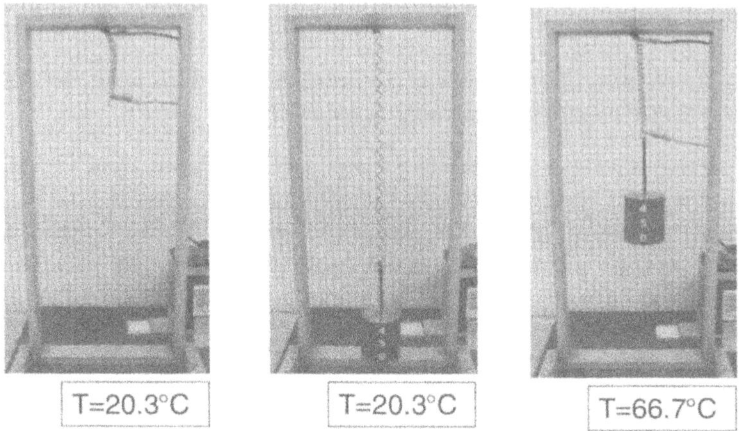

Fig. 1.28. The SMA heat engine.

3. What is the efficiency of the SMA spring?
4. Express the Carnot efficiency of an SMA heat engine, in terms of its transformation temperatures, assuming that the SMA heat engine operates at a nominal stress of 100 MPa, and that the transformation temperatures are shifted by 5 MPa/ °C in the presence of stress. From the SMAs presented in the chapter, select the ones with the highest Carnot efficiency.

1.9. Plot a schematic of the stress-strain response of an SMA specimen when it is thermomechanically loaded along loading paths 1 and 2, shown in the phase

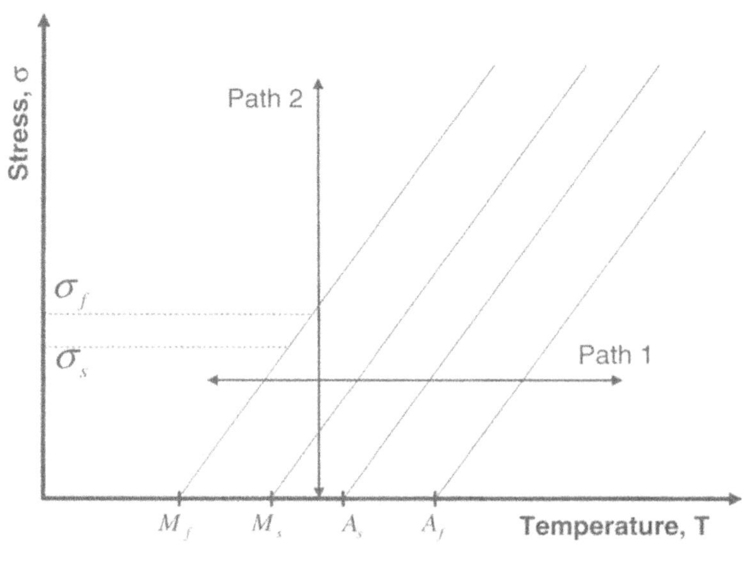

Fig. 1.29. Stress-temperature phase diagram.

diagram in Fig. 1.29. Note that the initial state of the material is austenitic for loading in path 2.

1.10. Compare the shape memory effect exhibited by shape memory alloys with the shape memory effect of shape memory polymers. In particular, compare the steps required to demonstrate the shape memory effect in two coils made of the two different shape memory materials.

1.11. The deployment of solar panels in spacecraft has been attempted by SMA strips (see Fig. 1.23), acting as temperature actuated hinges. Assuming that such hinges are provided in an initial flat shape, describe the process of shape setting for the hinges, so that they can deploy the solar panels upon heating.

References

[1] C. M. Jackson, H. J. Wagner, R. J. Wasilewski, 55-Nitinol—The alloy with a memory: Its physical metallurgy, properties and applications, Tech. Rep. NASA SP-5110, NASA Technology Utilization Office, Washington, D.C. (1972).

[2] T. Duerig, K. Melton, D. Stockel, C. Wayman (Eds.), Engineering Aspects of Shape Memory Alloys, Butterworth-Heinemann, London, 1990.

[3] J. Perkins, Shape Memory Effects in Alloys, Plenum Press, New York, 1975.

[4] H. Funakubo (Ed.), Shape Memory Alloys, Gordon and Breach Science Publishers, 1987.

[5] K. Otsuka, C. M. Wayman (Eds.), Shape Memory Materials, Cambridge University Press, Cambridge, 1999.

[6] G. V. Kurdjumov, L. G. Khandros, First reports of the thermoelastic behaviour of the martensitic phase of Au-Cd alloys, Doklady Akademii Nauk SSSR 66 (1949) 211–213.

[7] W. J. Buehler, J. V. Gilfrich, R. C. Wiley, Effects of low-temperature phase changes on the mechanical properties of alloys near composition TiNi, Journal of Applied Physics 34 (1963) 1475.

[8] F. E. Wang, W. J. Buehler, S. J. Pickart, Crystal structure and a unique "martensitic" transition of TiNi, Journal of Applied Physics 36 (3232-3239).

[9] L. Schetky, Shape-memory alloys, Scientific American 241 (74-82).

[10] M. Wayman, J. Harrison, The origins of the shape memory effect, Journal of Minerals, Metals, and Materials 41 (99) (1989) 26–28.

[11] M. H. Wu, L. M. Schetky, Industrial applications for shape memory alloys, in: Proceedings of the International Conference on Shape Memory and Superelastic Technologies, Pacific Grove, California, 2000.

[12] H. C. Doonkersloot, V. Vucht, Martensitic transformations in Au-Ti, Pd-Ti and Pt-Ti alloys, Journal of Less-Common Metals 20 (1970) 83–91.
[13] K. Melton, O. Mercier, Deformation behavior of NiTi-based alloys, Metallic Tras. 9A (1978) 1487–14488.
[14] S. Miyazaki, K. Mizukoshi, T. Ueki, T. Sakuma, Y. Liu, Fatigue life of Ti-50 at. Science and Engineering (1999) 658–663.
[15] K. Ullakko, J. K. Huang, C. Kantner, R. C. O'Handley, V. V. Kokorin, Large magnetic-field-induced strains in Ni_2MnGa single crystals, Applied Physics Letters 69 (13) (1996) 1966–1968.
[16] H. E. Karaca, I. Karaman, B. Basaran, Y. I. Chumlyakov, H. J. Maier, Magnetic field and stress induced martensite reorientation in NiMnGa ferromagnetic shape memory alloy single crystals, Acta Materialia 54 (1) (2006) 233–245.
[17] L. Contardo, G. Guenin, Training and two way memory effect in Cu-Zn-Al alloy, Acta Metallurgica 38 (7) (1990) 1267–1272.
[18] D. A. Miller, D. C. Lagoudas, Thermo-mechanical characterization of NiTiCu and NiTi SMA actuators: Influence of plastic strains, Smart Materials and Structures 9 (5) (2000) 640–652.
[19] X. Ren, K. Otsuka, Universal symmetry property of point defects in crystals, Physical Review Letters 85 (5) (2000) 1016–1019.
[20] P. Rodriguez, G. Guenin, Stability of the two way memory effect during thermal cycling of a high m_s temperature Cu-Al-Ni alloy, Materials Science Forum 56–58 (2) (1990) 541–546.
[21] D. A. Miller, D. C. Lagoudas, Influence of cold work and heat treatment on the shape memory effect and plastic strain development of NiTi, Material Science & Engineering A 308 (2001) 161–175.
[22] K. N. Melton, O. Mercier, Fatigue of NiTi thermoelastic martensites, Acta Metallurgica 27 (1979) 137–144.
[23] H. Tobushi, T. Hachisuka, S. Yamada, P. H. Lin, Rotating-bending fatigue of a TiNi shape-memory alloy wire, Mechanics of Materials 26 (1997) 35–42.
[24] J. L. McNichols, P. C. Brooks, NiTi fatigue behavior, Journal of Applied Physics 52 (1981) 7442–7444.
[25] H. Tobushi, T. Hachisuka, T. Hashimoto, S. Yamada, Cyclic deformation and fatigue of a TiNi shape memory alloy wire subjected to rotating bending, Journal of Engineering Materials and Technology 120 (1998) 64–70.
[26] D. Miller, Thermomechanical characterization of plastic deformation and transformation fatigue in shape memory alloys, Ph.D. thesis, Texas A&M University (2000).
[27] O. Bertacchini, D. Lagoudas, E. Patoor, Fatigue life characterization of shape memory alloys undergoing thermomechanical cyclic loading, Proceedings of SPIE, Smart Structures and Materials (2003) 612–624.

[28] K. Gall, H. Maier, Cyclic deformation mechanisms in precipitated NiTi shape memory alloys, Acta Materialia 50 (2002) 4643–4657.

[29] G. Eggeler, E. Hornbogen, A. Yawny, A. Heckmann, M. Wagner, Structural and functional fatigue of NiTi shape memory alloys, Materials Science and Engineering A (378) (2004) 24–33.

[30] W. Predki, M. Klönne, A. Knopik, Cyclic torsional loading of pseudoelastic NiTi shape memory alloys: Damping and fatigue failure, Materials Science and Engineering A (417) (2006) 182–189.

[31] K. Madangopal, The self accomodating martensitic microstructure of NiTi shape memory alloys, Acta Materialia 45 (12) (1997) 5347–5365.

[32] W. J. Buehler, R. C. Wiley, Nickel-base alloys, U.S. Patent 3,174,851. (1965).

[33] S. Miyazaki, K. Otsuka, Deformation and transformation behavior associated with the r-phase in Ti-Ni alloys., Metallurgical Transactions A 17A (1986) 53–63.

[34] D. Hartl, B. Volk, D. C. Lagoudas, F. T. Calkins, J. Mabe, Thermomechanical characterization and modeling of Ni60Ti40 SMA for actuated chevrons, in: Proceedings of ASME, International Mechanical Engineering Congress and Exposition (IMECE), 5–10 November, Chicago, IL, 2006, pp. 1–10.

[35] K. Richardson, Nitinol technologies innovator redefining the cutting edge, Outlook 22 (2) (2001) 1–8.

[36] D. J. Clingman, F. T. Calkins, J. P. Smith, Thermomechanical properties of 60-Nitinol, in: Proceedings of the SPIE, Smart Structures and Materials: Active Materials: Behavior and Mechanics, Vol. 5053, 2003, pp. 219–229.

[37] J. Mabe, R. Ruggeri, F. T. Calkins, Characterization of nickel-rich nitinol alloys for actuator development, in: Proceedings of the International Conference on Shape Memory and Superelasticity Technology, 2006.

[38] T. H. Nam, T. Saburi, K. Shimizu, Copper-content dependence of shape memory characteristics in Ti-Ni-Cu alloys, Materials Transactions, JIM 31 (11) (1990) 959–967.

[39] T. Saburi, T. Takagi, S. Nenno, K. Koshino, in: M. Doyama, S. Somlya, R. P. H. Chang (Eds.), Shape Memory Materials, MRS International Meeting on Advanced Materials, Vol. 9, 1989, pp. 147–152.

[40] N. M. Matveeva, V. N. Khachin, V. P. Shivokha, Stable and metastable phase equilibrium in metallic systems, Nauka, Moscow, 1985.

[41] Y. Liu, Mechanical and thermomechanical properties of a $Ti_{0.50}Ni_{0.25}Cu_{0.25}$ melt spun ribbon, Materials Science and Engineering A 354 (2003) 286–291.

[42] Z. L. Xie, J. Van Humbeeck, Y. Liu, L. Delaey, TEM study of Ti50Ni25Cu25 melt spun ribbons, Scripta Materialia 37 (3) (1997) 363–371.

[43] K. N. Melton, J. Simpson, T. W. Duerig, A new wide hysteresis NiTi based shape memory alloy and its applications, in: Proceedings of The

International Conference on Martensitic Transformations, The Japan Institute of Metals, 1986, pp. 1053–1058.
[44] K. N. Melton, J. L. Proft, T. W. Duerig, Wide hysteresis shape memory alloys based on the Ni -Ti -Nb system, MRS 9 (1989) 165–170.
[45] L. C. Zhao, T. W. Duerig, S. Justi, K. N. Melton, J. L. Proft, W. Yu, C. M. Wayman, The study of niobium-rich precipitates in a Ni-Ti-Nb shape memory alloy, Scripta Metallurgica and Materialia 24 (1990) 221–226.
[46] C. S. Zhang, L. C. Zhao, T. W. Duerig, C. M. Wayman, Effects of deformation on the transformation hysteresis and shape memory effect in a Ni47Ti44Nb9 alloy, Scripta Metallurgica and Materialia 2 (1990) 1807–1812.
[47] M. Piao, K. Otsuka, S. Miyazaki, H. Horikawa, Mechanism of the as temperature increase by pre-deformation in thermoelastic alloys, Materials Transactions, JIM 34 (10) (1993) 919–929.
[48] X. M. He, L. J. Rong, D. S. Yan, Y. Y. Li, TiNiNb wide hysteresis shape memory alloy with low niobium content, Materials Science and Engineering A 371 (2004) 193–197.
[49] P. G. Lidquist, C. M. Wayman, Shape memory and transformation behavior of martensitic Ti-Pd-Ni and Ti-Pt-Ni alloys, in: T. W. Duerig, K. N. Melton, D. Stöckel, C. M. Wayman (Eds.), Engineering Aspects of Shape Memory Alloys, Butterworth-Heinemann, London, 1990, pp. 58–68.
[50] P. E. Thoma, J. J. Boehm, Effect of composition on the amount of second phase and transformation temperatures of NixTi90-xHf10 shape memory alloys, Materials Science and Engineering A 273-275 (1999) 385–389.
[51] V. N. Khachin, N. A. Matveeva, V. P. Sivokha, D. V. Chernov, High-temperature shape memory effects in TiNi-TiPd system alloys, Doklady Akademii Nauk SSSR (USSR) 257 (1) (1981) 167–169.
[52] S. Wu, C. Wayman, Martensitic transformations and the shape-memory effect in $Ti_{50}Ni_{10}Au_{40}$ and $Ti_{50}Au_{50}$ alloys, Metallography (1987) 359.
[53] P. K. Kumar, D. C. Lagoudas, K. J. Zanca, M. Z. Lagoudas, Thermomechanical characterization of high temperature SMA actuators, in: Proceedings of SPIE, Vol. 6170, 2006, pp. 306–312.
[54] S. Padula II, G. Bigelow, R. Noebe, D. Gaydosh, A. Garg, Challenges and progress in the development of high-temperature shape memory alloys based on NiTiX compositions for high-force actuator applications, in: Proceedings of the International Conference on Shape Memory and Superelastic Technologies, ASM International, Metals Park, OH, 2006.
[55] Z. Pu, H. Tseng, K. Wu, Martensite transformation and shape-memory effect of NiTi-Zr high- temperature shape-memory alloys, in: SPIE proceedings, Vol. 2441, 1995, p. 171.

[56] B. Strnadel, S. Ohashi, H. Ohtsuka, T. Ishihara, S. Miyazaki, Cyclic stress-strain characteristics of Ti-Ni and Ti-Ni-Cu shape memory alloys, Material Science & Engineering A 202 (1995) 148–156.

[57] P. Potapov, A. Shelyakov, A. Gulyaev, E. Svistunova, N. Matveeva, D. Hodgson, Effect of hf on the structure of Ni-Ti martensitic alloys, Materials Letters 32 (4) (1997) 247–250.

[58] S. Hsieh, S. Wu, Room-temperature phases observed in $Ti_{53-x}Ni_{47}Zr_x$ high-temperature shape memory alloys, Journal of Alloys and Compounds 226 (1998) 276–282.

[59] K. Enami, S. Menno, Metal transformation 2 (1971).

[60] R. Kainuma, M. Ise, C. Jia, H. Ohtani, K. Ishida, Phase equilibria and microstructural control in the Ni-Co-Al system, Intermetallics 4 (1996) 151.

[61] D. Schryvers, P. Boullay, P. Potapov, R. Kohn, J. Ball, Microstructures and interfaces in Ni-Al martensite: comparing HRTEM observations with continuum theories, International Journal of Solids Structures 39 (3543).

[62] H. E. Karaca, I. Karaman, D. C. Lagoudas, H. J. Maier, Y. I. Chumlyakov, Recoverable stress-induced martensitic transformation in a ferromagnetic CoNiAl alloy, Scripta Materialia 49 (2003) 831–836.

[63] H. Morito, A. Fujita, R. Kainuma, K. Ishida, K. Oikawa, Magnetocrystalline anisotropy in single-crystal Co-Ni-Al ferromagnetic shape-memory alloy, Applied Physics Letters 81 (9) (2002) 1657–1659.

[64] P. J. Webster, K. R. A. Ziebeck, S. L. Town, M. S. Peak, Magnetic order and phase transformation in Ni_2MnGa, Philosophical Magazine B 49 (3) (1984) 295–310.

[65] S. J. Murray, M. Marioni, S. M. Allen, R. C. O'Handley, 6% magnetic-field-induced strain by twin-boundary motion in ferromagnetic Ni-Mn-Ga, Applied Physics Letters 77 (6) (2000) 886–888.

[66] R. C. O'Handley, S. M. Allen, D. I. Paul, C. P. Henry, M. Marioni, D. Bono, C. Jenkins, A. Banful, R. Wager, Keynote address: Magnetic field-induced strain in single crystal Ni-Mn-Ga, Proceedings of SPIE, Symposium on Smart Structures and Materials 5053 (2003) 200–206.

[67] A. Sozinov, A. A. Likhachev, N. Lanska, O. Söderberg, K. Ullakko, V. K. Lindroos, Effect of crystal structure on magnetic-field-induced strain in Ni-Mn-Ga, Proceedings of SPIE, Symposium on Smart Structures and Materials 5053 (2003) 586–594.

[68] H. Karaca, I. Karaman, B. Basaran, D. Lagoudas, Y. Chumlyakov, H. Maier, On the stress-assisted magnetic-field-induced phase transformation in ni_2mnga ferromagnetic shape memory alloys, Acta Materialia 55 (2007) 4253–4269.

[69] J. Cui, T. W. Shield, R. D. James, Phase transformation and magnetic anisotropy of an iron-palladium ferromagnetic shape-memory alloy, Acta Materialia 52 (2004) 35–47.

[70] R. D. James, M. Wuttig, Magnetostriction of martensite, Philosophical Magazine A 77 (5) (1998) 1273–1299.
[71] T. W. Shield, Magnetomechanical testing machine for ferromagnetic shape-memory alloys, Review of Scientific Instruments 74 (9) (2003) 4077–4088.
[72] T. Yamamoto, M. Taya, Y. Sutou, Y. Liang, T. Wada, L. Sorensen, Magnetic field-induced reversible variant rearrangement in Fe-Pd single crystals, Acta Materialia 52 (17) (2004) 5083–5091.
[73] S. J. Murray, R. Hayashi, M. Marioni, S. M. Allen, R. C. O'Handley, Magnetic and mechanical properties of FeNiCoTi and NiMnGa magnetic shape memory alloys, Proceedings of SPIE 3675 (1999) 204–211.
[74] A. Fujita, K. Fukamichi, F. Gejima, R. Kainuma, K. Ishida, Magnetic properties and large magnetic-field-induced strains in off-stoichiometric Ni-Mn-Al heusler alloys, Applied Physics Letters 77 (19) (2000) 3054–3056.
[75] M. Wuttig, J. Li, C. Craciunescu, A new ferromagnetic shape memory alloy system, Scripta Materialia 44 (2001) 2393–2397.
[76] T. Sakamoto, T. Fukuda, T. Kakeshita, T. Takeuchi, K. Kishio, Magnetic field-induced strain in iron-based ferromagnetic shape memory alloys, Journal of Applied Physics 93 (10) (2003) 8647–8649.
[77] K. R. Melton, General applications of shape memory alloys and smart materials, in: K. Otsuka, C. M. Wayman (Eds.), Shape Memory Materials, Cambridge University Press, Cambridge, 1999, Ch. 10, pp. 220–239.
[78] C. Liang, F. Davidson, L. M. Schetky, F. K. Straub, Applications of torsional shape memory alloys actuators for active rotor blade control — opportunities and limitations, in: Proceedings of SPIE, Smart Structures and Materials: Smart Structures and Integrated Systems, Vol. 2717, 1996, pp. 91–100.
[79] L. J. Garner, L. N. Wilson, D. C. Lagoudas, O. K. Rediniotis, Development of a shape memory alloy actuated biomimetic vehicle, Smart Materials and Structures 9 (2000) 673–683.
[80] A. A. Ilyin, P. G. Sysolyatin, V. E. Gunter, A. P. Dergilev, M. A. Didin, S. P. Sysolyatin, I. A. Makarova, The use of superelastic shape memory implants in temporo-mandibular joint surgery, Proceedings of the First International Symposium on Advanced Biomaterials (ISAB), Montreal, Canada (1997) 177.
[81] V. Brailovski, F. Trochu, Review of shape memory alloys medical applications in Russia, Bio-Medical Materials & Engineering 6 (4) (1996) 291–298.
[82] B. Sanders, R. Crowe, E. Garcia, Defense advanced research projects agency – Smart materials and structures demonstration program overview, Journal of Intelligent Material Systems and Structures 15 (2004) 227–233.
[83] J. Kudva, Overview of the DARPA smart wing project, Journal of Intelligent Material Systems and Structures 15 (2004) 261–267.

[84] J. K. Strelec, D. C. Lagoudas, M. A. Khan, J. Yen, Design and implementation of a shape memory alloy actuated reconfigurable wing, Journal of Intelligent Material Systems and Structures 14 (2003) 257–273.

[85] R. Mani, D. Lagoudas, O. Rediniotis, MEMS based active skin for turbulent drag reduction, in: Proceedings of SPIE, Smart Structures and Materials, Vol. 5056, San Diego, CA, 2003, pp. 9–20.

[86] M. Tawfik, J. Ro, C. Mei, Thermal post-buckling and aeroelastic behaviour of shape memory alloy reinforced plates, Smart Materials and Structures 11 (2002) 297–307.

[87] C. Nam, A. Chattopadhyay, Y. Kim, Application of shape memory alloy (SMA) spars for aircraft maneuver enhancement, in: Proceedings of SPIE, Smart Structures and Materials, Vol. 4701, San Diego, CA, 2002, pp. 226–236.

[88] D. Pitt, J. Dunne, E. White, E. Garcia, SAMPSON smart inlet SMA powered adaptive lip design and static test, Proceedings of the 42nd AIAA Structures, Structrual Dynamics, and Materials Conference, Seattle, WA, 16–20 April 2001 (2001) 1–11.

[89] S. Wax, G. Fischer, R. Sands, The past, present, and future of DARPAs investment strategy in smart materials, Journal of the Minerals, Metals, and Materials Society 55 (12) (2003) 17–23.

[90] J. Mabe, R. Cabell, G. Butler, Design and control of a morphing chevron for takeoff and cruise noise reduction, in: Proceedings of the 26th Annual AIAA Aeroacoustics Conference, Monterey, CA, 2005, pp. 1–15.

[91] T. Turner, R. Buehrle, R. Cano, G. Fleming, Modeling, fabrication, and testing of a SMA hybrid composite jet engine chevron concept, Journal of Intelligent Material Systems and Structures 17 (2006) 483–497.

[92] J. H. Mabe, F. Calkins, G. Butler, Boeing's variable geometry chevron, morphing aerostructure for jet noise reduction, in: 47th AIAA/ ASME / ASCE / AHS / ASC Structures, Structural Dynamics and Materials Conference, Newport, Rhode Island, 2006, pp. 1–19.

[93] V. Birman, Review of mechanics of shape memory alloy structures, Applied Mechanics Reviews 50 (11) (1997) 629–645.

[94] H. Prahlad, I. Chopra, Design of a variable twist tiltrotor blade using shape memmory alloy (SMA) actuators, in: Proceedings of SPIE, Smart Structures and Materials, Vol. 4327, Newport Beach, CA, 2001, pp. 46–59.

[95] A. Jacot, R. Ruggeri, D. Clingman, Shape memory alloy device and control method, U.S. Patent 7,037,076 (2 May 2006).

[96] K. Singh, J. Sirohi, I. Chopra, An improved shape memory alloy actuator for rotor blade tracking, Journal of Intelligent Material Systems and Structures 14 (2003) 767–786.

[97] D. Kennedy, F. Straub, L. Schetky, Z. Chaudhry, R. Roznoy, Development of an SMA actuator for in-flight rotor blade tracking, Journal of Intelligent Material Systems and Structures 15 (2004) 235–248.

[98] R. Loewy, Recent developments in smart structures with aeronautical applications, Smart Materials and Structures 6 (1997) R11–R42.
[99] C. Test, S. Leone, S. Ameduri, A. Concilio, Feasibility study on rotorcraft blade morphing in hovering, in: Proceedings of SPIE, Smart Structures and Materials, Vol. 5764, San Diego, CA, 2005, pp. 171–182.
[100] A. Johnson, Non-explosive seperation device, U.S. Patent 5,119,555 (June 1992).
[101] O. Godard, M. Lagoudas, D. Lagoudas, Design of space systems using shape memory alloys, in: Proceedings of SPIE, Smart Structures and Materials, Vol. 5056, San Diego, CA, 2003, pp. 545–558.
[102] C. Willey, B. Huettl, S. Hill, Design and development of a miniature mechanisms tool-kit for micro spacecraft, in: Proceedings of the 35th Aerospace Mechanisms Symposium, Ames Research Center, 9–11 May, 2001, pp. 1–14.
[103] A. Peffer, K. Denoyer, E. Fossness, D. Sciulli, Development and transition of low-shock spacecraft release devices, in: Proceedings of IEEE Aerospace Conference, Vol. 4, 2000, pp. 277–284.
[104] B. Carpenter, J. Lyons, EO-1 technology validation report: Lightweight flexible solar array experiment, Tech. rep., NASA Godard Space Flight Center, Greenbelt, MD (8 August 2001).
[105] S. Shabalovskaya, Biological aspects of TiNi alloys surfaces, Journal de Physique IV 5 (1995) 1199–1204.
[106] J. Ryhänen, Biocompatibility evolution of nickel-titanium shape memory alloy, Ph.D. thesis, Univeristy of Oulu, Oulu, Finland (1999).
[107] L. Machado, M. Savi, Medical applications of shape memory alloys, Brazilian Journal of Medical and Biological Research 36 (2003) 683–691.
[108] D. Mantovani, Shape memory alloys: Properties and biomedical applications, Journal of the Minerals, Metals and Materials Society 52 (2000) 36–44.
[109] K. Speck, A. Fraker, Anodic polarization behavior of Ti-Ni and Ti-6Al-4V in simulated physiological solutions, J Dent Res 59 (100) (1980) 1590–1595.
[110] G. Andreasen, T. Hilleman, An evaluation of 55 cobalt subsituted nitinol wire for use in orthodontics, Journal of the American Dental Association 82 (1971) 1373–1375.
[111] S. Thompson, An overview of nickel-titanium alloys used in dentistry, International Endodontic Journal 33 (2000) 297–310.
[112] T. Duerig, A. Pelton, D. Stöckel, Superelastic nitinol for medical devices, Medical Plastics and Biomaterials (1997) 31–42.
[113] D. Lagoudas, E. Vandygriff, Processing and characterization of niti porous sma by elevated pressure sintering, Journal of intelligent material system and structures 13 (2002) 837–850.

[114] J. Paine, C. Rogers, High velocity impact response of composites with surface bonded nitinol-SMA hybrid layers, Journal of Intelligent Material Systems and Structures 5 (4) (1994) 530–535.

[115] B. Barnes, D. B. J. Luntz, A. Browne, K. Strom, Panel deployment using ultrafast SMA latches, in: ASME International Mechanical Engineering Congress and Exposition, Chicago, Illinois, USA, 2006.

[116] K. Otsuka, T. Kakeshita, Science and technology of shape-memory alloys: New developments, bulletin (February 2002).

[117] P. Anderson, A., S. Sangesland, Detailed study of shape memory alloys in oil well applications., Sintef petroleum research, Trondheim, Norway (1999).

[118] I. Ohkata, Y. Suzuki, The design of shape memory alloy actuators and their applications, in: K. Otsuka, C. M. Wayman (Eds.), Shape Memory Materials, Cambridge University Press, Cambridge, 1999, Ch. 11, pp. 240–266.

[119] S. Saadat, J. Salichs, M. Noori, Z. Hou, H. Davoodi, I. Bar-on, An overview of vibration and seismic applications of NiTi shape memory alloy, Smart Materials and Structures 11 (2002) 218–229.

[120] J. Evans, D. Brei, J. Luntz, Preliminary experimental study of SMA knitted actuation architectures, in: ASME International Mechanical Engineering Congress and Exposition, 2006.

[121] V. Brailovski, S. Prokoshkin, P. Terriault, F. Trochu (Eds.), Shape Memory Alloys: Fundamentals, Modeling and Applications, University of Quebec, 2003.

2

Thermomechanical Characterization of Shape Memory Alloy Materials

D. J. HARTL AND D. C. LAGOUDAS

Having introduced the fundamentals of shape memory alloy behavior, the characterization of SMA materials is discussed here. Such a discussion fosters a better understanding of the thermomechanical constitutive response of SMAs. Furthermore, proper determination of required material properties is necessary for the development of comprehensive and accurate SMA material models. Finally, it is perhaps the most important step in the practical implementation of SMAs. Whatever methodology one uses to design a given SMA application, a quantitative evaluation of key material properties is required.

2.1 Introduction

The thermomechanical characterization of an SMA provides a qualitative demonstration of the material behavioral characteristics and allows one to derive distinct quantitative material properties. Like other classes of structural materials, shape memory alloys are tested by subjecting specimens to prescribed thermomechanical inputs while monitoring the exhibited material response. However, though the characterization of SMAs is based on the same principles used to test other material systems, unique considerations exist due to the complex constitutive behavior of shape memory alloys. These issues will be this chapter's focus.

SMAs exhibit coupled thermomechanical behavior, requiring the experimentalist to carefully consider the particular inputs, or *loading paths*, applied to the material. To quantify the complex behavior of SMA materials, various loading paths are imposed while phenomena associated with the phase transformation are recorded. These phenomena include the shape memory effect and pseudoelasticity, as discussed in the first chapter. This chapter will focus on the application and measurement of three thermomechanical fields in particular: stress, strain and temperature. As will be explained in the following chapter, either histories of stress and temperature or strain and temperature will be prescribed while the evolution of the third quantity will be measured, usually by appropriate interpretation of traction, displacement, and temperature measurements at the specimen boundary. To consistently characterize an

D.C. Lagoudas (ed.), *Shape Memory Alloys*, DOI: 10.1007/978-0-387-47685-8_2,

SMA material in a manner that suitably determines its properties, it is useful to establish an experimental process during which various loading paths are applied in a carefully considered progression. Such a process will be proposed in this chapter and will be illustrated by considering the actual testing of various SMA specimens.

2.1.1 Review of SMA Characterization Methods

From the time of the first comprehensive studies published on shape memory alloy properties, particularly with respect to NiTi [1], researchers have sought to fully understand the behavior of these materials. Many of the phenomena observed in SMAs can now be explained in terms of underlying microstructural mechanisms and SMA behavior, under a given thermomechanical loading path, can be phenomenologically described as this current level of knowledge has matured over several decades. This has been made possible by carefully planned and executed experimentation.

Years after the initial well-published discovery at the Naval Ordnance Laboratory, NASA scientists completed a comprehensive report on the properties of Nitinol [2]. This report presents information regarding crystallography, processing techniques, physical, mechanical and chemical properties, and potential applications. Since that time, experimental investigation has been ongoing. Researchers further focused on understanding the metallurgical and microstructural aspects of shape memory and pseudoelastic behavior as exhibited by a number of different alloy systems [3–6]. Experimental research then began to focus on the phenomenological aspects of SMA behavior, including increased investigation of the exhibited thermomechanical coupling and the engineering applications of SMAs [7–13].

As shape memory alloy materials have gained acceptance for use in specialized engineering applications, it has become clear that some testing methods should be standardized, especially in the medical field. ASTM International has addressed this need by publishing standard test methods to guide SMA experimentation efforts, especially for research and development of applications. While some standards are restricted to the definitions of terms or the suitable composition of NiTi alloys [14, 15], others address experimental methods directly. Standard test method F-2004 [16] addresses the details of Differential Scanning Calorimeter (DSC) testing. As shown in Chapter 1, this is an important test for observing transformation behavior in SMAs under zero stress (see Sect. 1.8) and is used as a guideline to derive an estimate for all four key transformation temperatures (i.e., M_s, M_f, A_s, and A_f). Note that some special equipment is required and will be shown in later sections.

As an alternative method to estimate austenite start and finish temperatures (A_s and A_f) only, specification F-2082 [17] recommends the "Bend and Free Recovery" test. This specification makes direct use of the shape memory effect (Sect. 1.4) and does not require specialized equipment. According to the specification, a specimen (wire, tube, or strip) is deformed when bending at a low temperature ($T < A_s$), causing regions to form detwinned martensite. This leads to the generation of transformation strains that are not recovered upon unloading due to the test temperature being below A_s. The sample is then slowly heated, and its geometric configuration is monitored. The temperature at which the sample begins to recover its initial shape is declared to be approximately equal to A_s, and the temperature at which recovery is completed is declared to be approximately equal to A_f. In addition to the martensite to austenite ($M \leftrightarrow A$) transformation, this specification also addresses martensite to R-phase to austenite ($M \rightarrow R \rightarrow A$) transformation for materials that exhibit this behavior (Sect. 1.8). The bend/free recovery method is not addressed further here, but should be considered as a qualitative method of measuring two of the four transformation temperatures.

Finally, specification F-2516 [18] addresses the tensile testing of SMA specimens, especially those exhibiting pseudoelasticity at room temperature (see Sect. 1.5). Applying mechanical loading/unloading cycles is important in the testing of many materials, but for the investigation of SMAs, this is a key step. The parameters described in this specification, such as *upper plateau strength* and *lower plateau strength*, will be shown to correspond to the characterization methods presented in this chapter. The scope of the ASTM standards is a basic understanding of certain aspects of SMA behavior, mostly relevant to medical applications, and is not the full thermomechanical characterization of these materials. Furthermore, they are not intended to calibrate phenomenological models such as those described in Chapter 3. These complex efforts require additional considerations as discussed throughout the remainder of this chapter. As new constitutive models are being developed for SMAs, additional tests are necessary for providing material parameter inputs.

Clearly, the methods and processes presented in this chapter derive directly from decades of previous experimental work performed by many skilled individuals and research teams. The contribution of this text is simply to summarize and organize various experimental techniques into a comprehensive method by which one can deduce the material properties of interest. Furthermore, aspects of experimentation that have challenged others in the past will be addressed.

2.1.2 Shape Memory Alloy Specimens

To begin the discussion of thermomechanical SMA material characterization, it is appropriate to provide examples of the actual material specimens designed for testing. The reader is assumed to have some familiarity with the experimental systems used to provide the necessary loads (experimental inputs) and to monitor material responses. Although there are unique experimental considerations imposed by the behavior of shape memory alloys, such discussion is reserved for Sect. 2.4. This introductory material should be familiar to those who are experienced in experimental material thermomechanical testing.

The most common materials testing performed on SMAs involve tensile axial loading. While the potential configuration of SMA specimens is as varied as those of other metallic material systems, the manner in which SMAs are utilized does motivate the commonality of particular specimen forms. For example, many SMA applications use wire components because they provide relatively high tensile forces and displacements in a compact and simple configuration. Therefore, the SMA specimen forms most frequently tested are wire tensile specimens and this form presents the experimentalist with both advantages and challenges. Perhaps the most beneficial aspect of testing wires is that specimens often arrive from the manufacturer in a useable form such as spooled wire as shown in Fig. 2.1. Due to their long and thin configuration, wire specimens provide a simple gauge length over which stresses are homogeneous. However, because the cross section at the ends of a wire specimen is the same as that in the gauge length, prevention of excessive stress concentrations and resulting premature failure at the grips can be challenging. One possible solution to this problem is presented in Fig. 2.1 where a wire specimen and a suitable gripping mechanism are shown. Such a grip configuration allows the testing of a range of wire diameters.

Testing is often performed for wires, however it is also common that SMA components be manufactured in 2-D and 3-D forms for some chosen applications. In such cases, SMA tensile specimens may be fabricated in the common 'dogbone' configuration per standard metal testing methods (i.e., ASTM E8) [19]. This is most easily accomplished when the as-received material is

Fig. 2.1. Example of spooled thick NiTi wire and prepared specimen with appropriate grips.

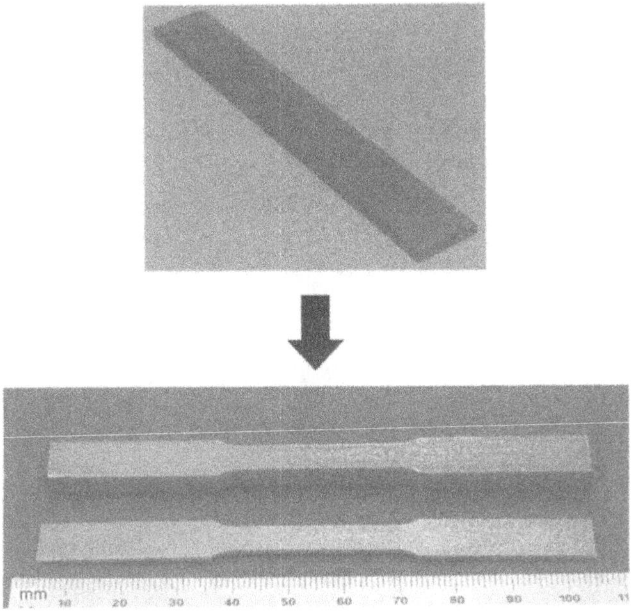

Fig. 2.2. Example of raw Ni60Ti40 (wt.%) plate and subsequent ASTM "dogbone" specimens [19] machined from such plates using EDM.

in plate form. An example of this can be seen in Fig. 2.2 where two NiTi (Ni60Ti40 wt.%) plates of different thickness are shown. It is important to note that due to their thermomechanical behavior, machining SMAs is difficult using current methods. High cutting speeds are necessary for cutting processes that remove material (e.g., drilling and lathing). Methods such as water jet cutting and electrical discharge machining (EDM) are often used as well. EDM is especially useful for machining tensile specimens from different regions throughout properly homogenized bulk material of arbitrary shape. Examples of such a bulk form and the resulting specimens are shown in Fig. 2.3.

Designers and analysts often require information on the compressive behavior of SMAs, which may differ from the complimentary tensile response. For this purpose, compressive specimens are also often required [20, 21]. Though recognized standard protocols (e.g., ASTM standards) for SMA compressive testing do not exist, the general ASTM guidelines for the compressive testing of metals (ASTM E9) provide information on standard specimen configuration and testing methods [22]. Such specimens can also be accurately machined from raw material via EDM, and an example of this is shown in Fig. 2.4.

Finally, it is often necessary to extend the experimental study of an SMA material beyond simple uniaxial loading. To examine the constitutive response under shear or multiaxial loading, SMA specimens in the form of tubes can be

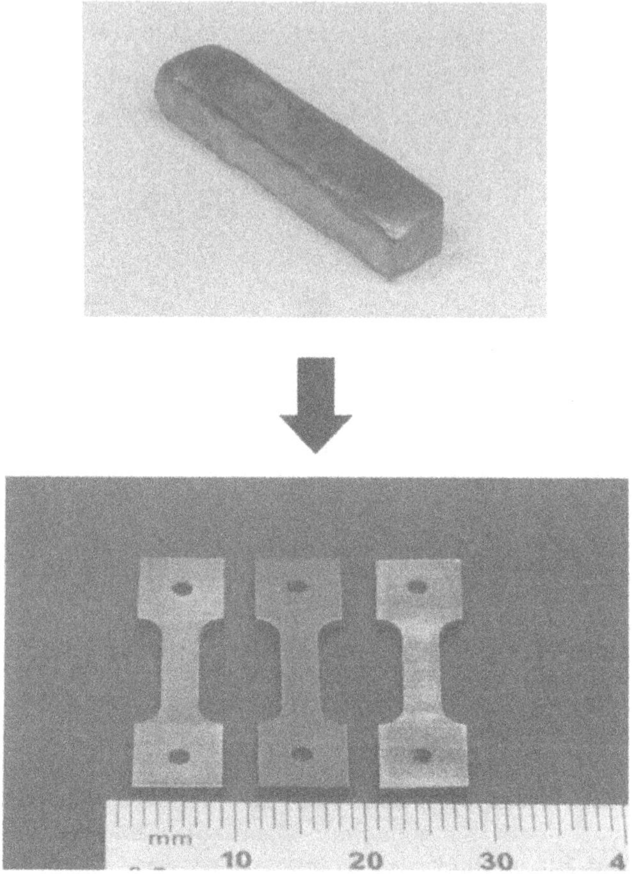

Fig. 2.3. Example of bulk NiTiPd SMA and resulting small SMA "dogbone" specimens machined from its center using EDM.

used. Tubes with sufficiently thin walls (relative to their diameter) provide a region of near-uniform shear stress when subjected to torsion. Response under combined loading can then be assessed if an applied tensile or compressive load is added (compression being applicable for sufficiently short specimens when tested below their buckling load) [23, 24]. Additional application of an internal or external pressure allows consideration of an additional stress state. Example SMA specimens in tube form with varying composition and thickness are shown in Fig. 2.5.

In this section, general aspects of SMA characterization have been discussed. However, to develop an understanding of any material, it is necessary to decide what properties are most useful and then determine under what loading paths these properties are best elucidated. This will be discussed in the following section.

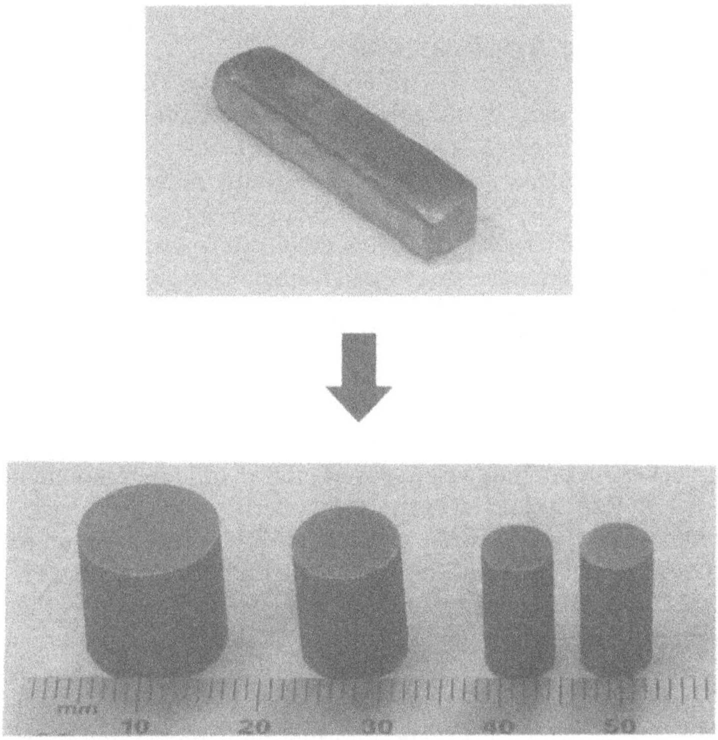

Fig. 2.4. Examples of small NiTiPd SMA compressive specimens EDM machined from the center of bulk raw material. The two specimens on the right follow ASTM E9 specifications [22].

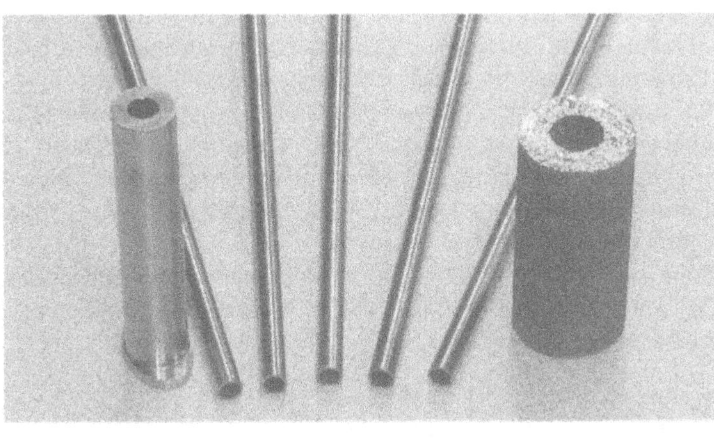

Fig. 2.5. Examples of SMA tube specimens. These include a NiTiPd specimen (left), NiTi high aspect ratio samples (center), and a porous NiTi sample (right).

2.2 Thermomechanical Material Properties of SMAs for Engineering Applications

As it has been discussed in Chapter 1, SMAs represent a specific class of metallic alloys that have two stable solid phases, and by following particular paths in the stress-temperature space, these alloys undergo a transformation from one phase to another. Furthermore, these two phases are distinct in their properties. This transformation can lead to generation and subsequent recovery of strains and cause macroscopic shape changes.

SMAs exhibit nonlinear, hysteretic behavior with a strong thermomechanical coupling. Furthermore, SMAs are highly path-dependent, though the phase transformation itself is not intrinsically loading rate dependent. The thermomechanical coupling, however, can lead to an experimental rate dependence (see Sect. 2.4.3). Each of these aspects leads to experimental complexities, and many of the investigative methods used with other inelastic materials are not completely sufficient to describe the material behavior of an SMA. This section will review the SMA thermomechanical response from an experimental point of view and introduce a general set of material parameters useful in quantitatively describing their constitutive behavior. There are other physical properties that could be experimentally investigated (e.g., electrical conductivity, density, specific heat etc.), but these are not addressed in this chapter. For the interested reader, the comprehensive NASA report provides data on such properties [2], while additional information can be found in more recent compilations of work on SMAs [3, 25].

To examine the thermomechanical engineering properties of shape memory alloys, we first determine what aspects of the material behavior can be parameterized. As an example, we examine the phenomenological response of an SMA specimen undergoing pseudoelastic loading. Recall from Chapter 1 that these experiments are performed by applying prescribed forces (stresses) and temperatures and monitoring exhibited deformations (strains). By considering the unique features of this nonlinear material response (e.g. changes in stiffness, hysteresis, etc.), one can see that a properly chosen set of material parameters is useful in quantitatively describing these most important material behaviors. Note that here the primary interest is in the response of polycrystalline engineering forms of SMA material (e.g. wires, rods, beams etc.), and not the single crystal behavior.

Examination of the new set of experimental pseudoelastic results (uniaxial, tensile) are shown in Fig. 2.6a. The following observations can be made (recalling the discussion in Sect. 1.5):

- During loading:
  - The initial response is nearly linear.
  - At some stress level (σ^{Ms}) the stiffness changes, and a behavior similar to plastic yielding is observed. A 'plateau' is formed.

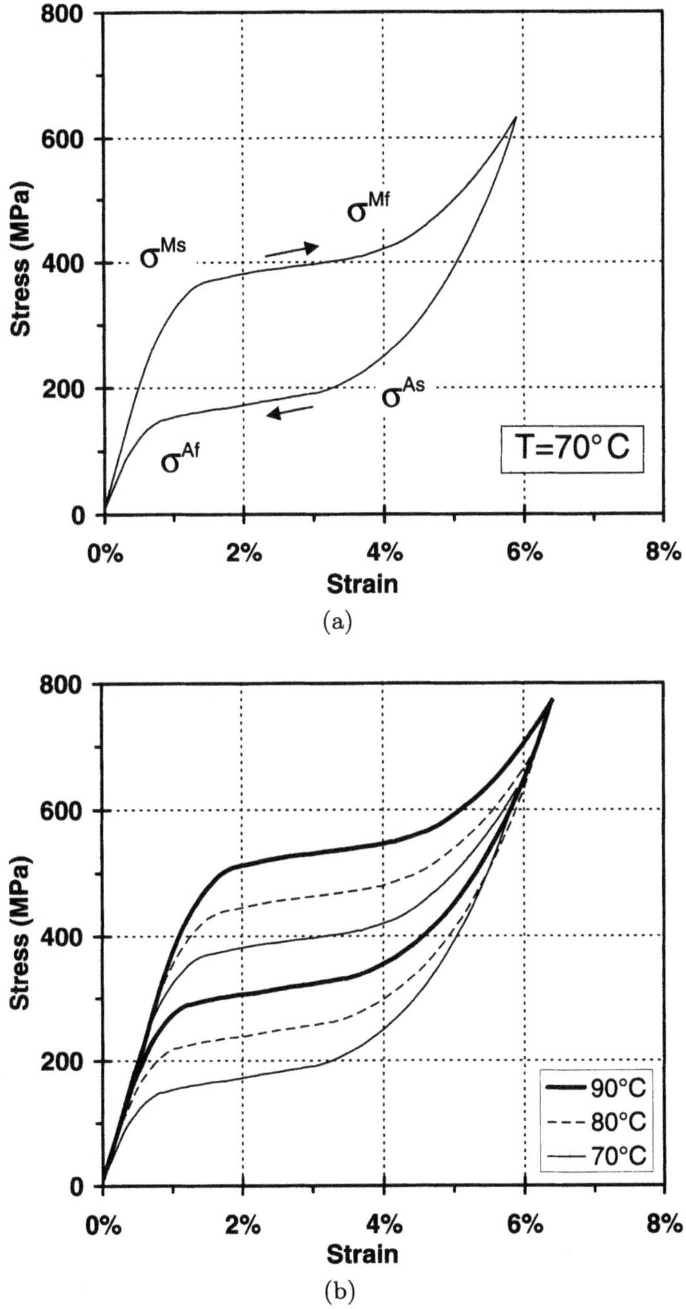

Fig. 2.6. Experimental example of constant temperature phenomenological transformation behavior in NiTi: (a) single temperature of $T = 70\,°\mathrm{C}$, (b) multiple temperatures.

  - As stress increases to a second level (σ^{Mf}), the plateau ends and the response stiffens. A nearly linear response with a slope distinct from the first is observed.
- During unloading:
  - The initial response is nearly linear.
  - A plateau with the same strain length as that observed during loading is formed at a lower stress level (σ^{As}).
  - At the end of the plateau (σ^{Af}), the response stiffens and becomes nearly linear, following the same slope as observed during initial loading.

As explained in Chapter 1, the stiff response at low stress (completely below the loading plateau) corresponds to purely austenitic behavior, while the response at higher stress (completely above the unloading plateau) corresponds to purely martensitic behavior. This indicates the need to consider the independent elastic responses of each phase, and that these responses need not be similar. Such understanding of the elastic response is the first key aspect of SMA characterization.

A second interesting behavior highlights the importance of the current thermomechanical (stress-temperature) state. At the given test temperature, particular stress levels initiate the forward and reverse transformation during loading and unloading, respectively. Furthermore, one can envision a pseudoelastic experiment wherein the same level of force is applied, though at a different temperature (i.e., a different thermomechanical state). Recall that as test temperature increases, the stress needed to initiate each phase transformation (A→M and M→A) increases (see Sect. 1.3). As a result, the response will vary with each change in temperature, as illustrated in Fig. 2.6b.

A third important quantifiable behavior is related to the deformation or strain generated during transformation (plateau region). This represents the amount of macroscopic deformation that can be generated via the underlying microstructural motions and is a clear consequence of the solid-to-solid phase transformation.

The effects discussed above are also observed during thermal transformation at nominally constant stress levels. Engineers are often more interested in the properties of an SMA material as an actuator rather than as a pseudoelastic element. In these cases, constant stress tests are useful in illustrating the three key types of material properties. The strain vs. temperature response during such loading can be seen in the example shown in Fig. 2.7a. Consider first the response at temperatures outside the transformation region (i.e., $T < M_f^\sigma$ and $T > A_f^\sigma$). This behavior is due to thermal expansion, as seen in other metals; this is also a thermoelastic effect (like the linear loading/unloading observed at the beginning and end of pseudoelastic loading).

Second, repetition of such a test at varying stress levels illustrates the relationship between the non-zero stress transformation temperatures and the current stress level, as seen in Fig. 2.7b. This is analogous to varying ambient temperature during pseudoelastic testing.

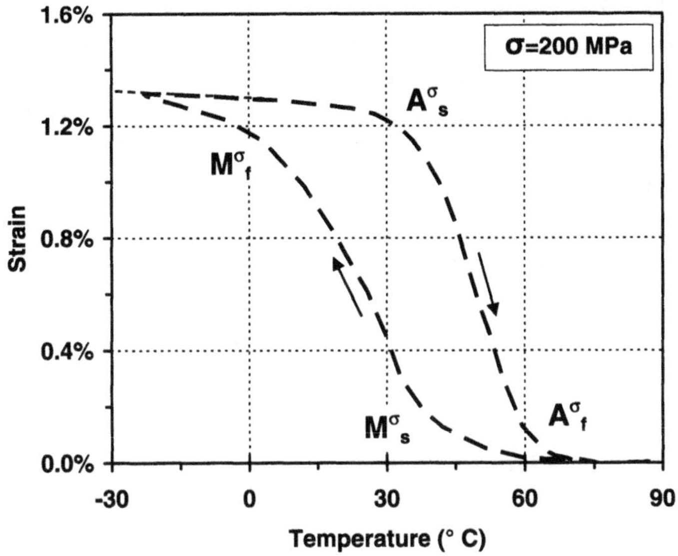

(a)

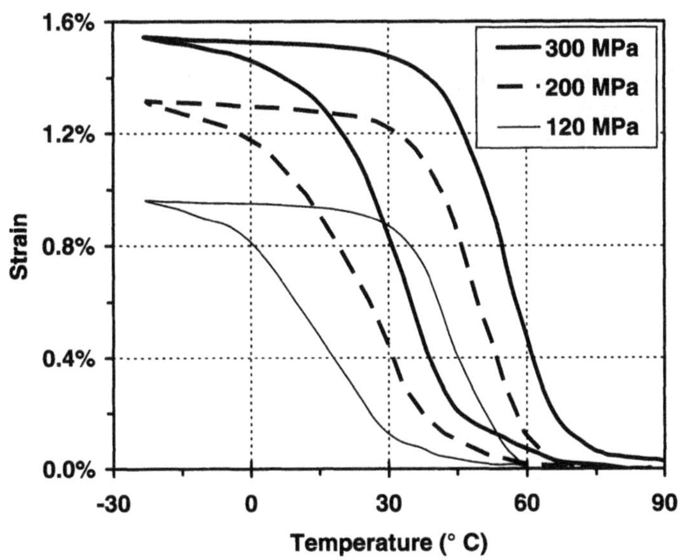

(b)

Fig. 2.7. Experimental example of constant stress phenomenological transformation behavior in NiTi: (a) single stress of 200 MPa, (b) multiple stresses.

Third, note the height of the hysteresis generated during transformation, which represents the amount of inelastic yet recoverable strain being generated and recovered as the transformation evolves.

To review and summarize, we can infer that SMAs require at least three types of material properties to describe three types of behaviors. These types of material properties are:

- *Thermoelastic properties of austenite and martensite* – These parameters, which apply to most structural materials, are necessary to describe the material response when transformation or reorientation is not occurring.
- *Critical stress and temperature states associated with the phase diagram* – These parameters help determine when the process of transformation between phases will begin or end depending on the current thermomechanical state (stress and temperature) and loading history of the material.
- *Transformation strain evolution properties* – These parameters provide a relation between the current state of material transformation (e.g., volume fraction of the various martensitic variants) and the exhibited generation of transformation strain.

2.2.1 Thermoelastic Properties

For each of these three types of properties, one can determine a set of material parameters to sufficiently describe a given shape memory alloy, and we can explore some methods by which they can be found.

Let us first consider the thermoelastic properties. Assuming material isotropy, the elastic stiffness could be represented by the Young's modulus of austenite (E^A), as shown in Fig. 2.8a. The material then transforms into martensite. Once transformation is complete, the fully martensitic material

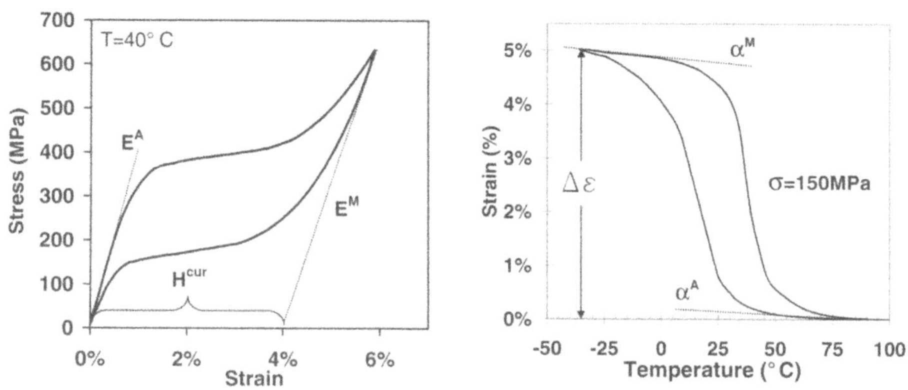

(a) (b)

Fig. 2.8. Experimental examples: (a) the pseudoelastic effect, (b) the shape memory effect.

responds elastically once again. Again assuming isotropy, this can be represented by an elastic modulus of martensite (E^M), which is also schematically shown in Fig. 2.8a. These elastic properties can also be found by simple monotonic loading/unloading of the material in a fully austenitic or martensitic state at stresses sufficiently low to prevent transformation or reorientation.

Other elastic properties such as the Poisson's ratio for each phase can be measured concurrently with the stiffness. The coefficient of thermal expansion is used to predict the thermally induced deformation response of pure austenite and pure martensite. This property can be observed in Fig. 2.8b, where constant stress testing is illustrated. Here, the coefficients for austenite and martensite are shown as the slopes of the strain-temperature plots in the fully austenitic and fully martensitic states, respectively.

So, for each phase elastic properties plus the thermal expansion coefficient is required. For design purposes, it is sometimes also useful to characterize the plastic yield and failure behavior of the material. A potential set of thermoelastic properties is as follows:

- The elastic constants of austenite and martensite, respectively. In the case of isotropy, the Young's Moduli, E^A and E^M, and the Poisson's Ratios, ν^A and ν^M, can be used.
- The coefficients of thermal expansion of austenite and martensite, respectively. In the case of isotropy, only two scalar constants, α^A and α^M, are needed.
- Information on the elastic limit of the material. If the material is isotropic, the yield stress of the material in austenite and martensite (σ_Y^A and σ_Y^M) will suffice.

2.2.2 Critical Stress and Temperature States for Transformation (Phase Diagram)

Finally, we review the properties of the phase diagram, which illustrates the stress and temperature conditions for material transformation. Practically this involves the determination of transformation "surfaces," or boundaries of the transformation regions in a stress-temperature space. These indicate where a given transformation (i.e., austenite to martensite or vice versa) begins or ends.

A schematic example of the phase diagram, previously introduced in Chapter 1, is given in Fig. 2.9. The form of these surfaces might be assumed to be linear, quadratic, or otherwise depending on the behavior of the material. Whatever their general form, these four surfaces are partially described by their intersections with the zero-stress axis (i.e., the zero-stress transformation temperatures). However, these temperatures are insufficient to fully describe the configuration of the phase diagram, and additional parameters are needed. A common set of parameters include the slopes of these surfaces at some stress level (e.g., zero stress). Such slopes are known as "stress influence coefficients."

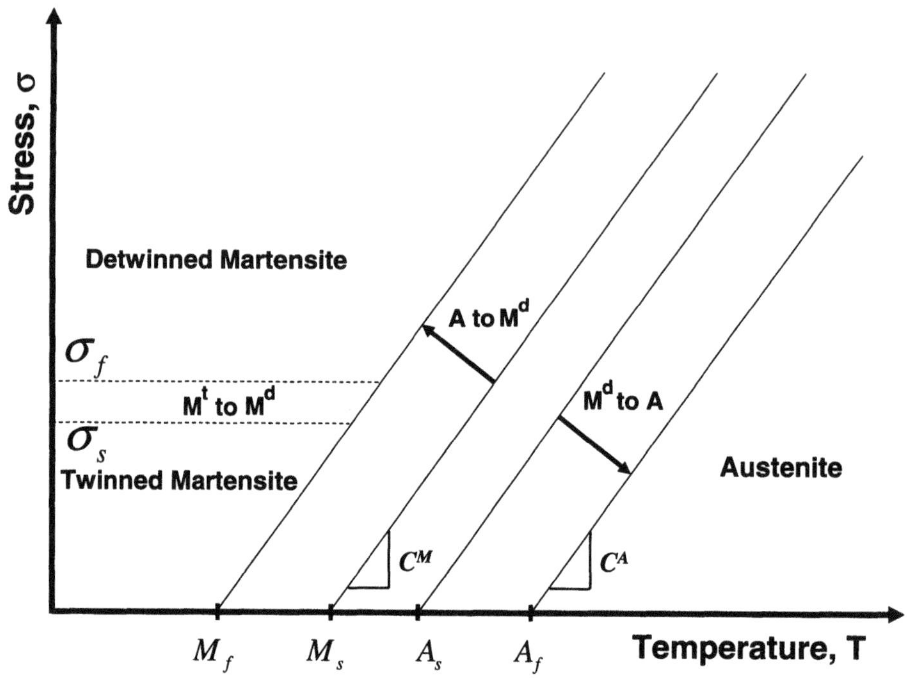

Fig. 2.9. Schematic representation of the phase diagram with possible material properties defined.

Aside from specifying the martensitic phase transformation regions, it is also useful to determine the stresses at which the detwinning of martensite begins and ends. Though this is not technically a transformation but rather a reorientation of martensite, this region can also be plotted on the phase diagram and utilized during design and analysis (see Chapter 6). The parameters required to construct the phase diagram are:

- The initiation and completion temperatures for transformation from austenite to martensite at zero stress (M_s, M_f).
- The initiation and completion temperatures for transformation from martensite to austenite at zero stress (A_s, A_f).
- The *stress influence coefficients* or general slopes of the transformation surfaces. There could be up to four total slopes, though it is often assumed that each pair of surfaces for the two distinct transformations ($A \rightarrow M$ or $M \rightarrow A$) shares a characteristic slope. Thus the zero-stress slopes of transformation regions into austenite (C^A) and into martensite (C^M) are useful parameters, as seen in Fig. 2.9.
- The start and finish stresses for the detwinning of martensite (σ_s, σ_f), which may be temperature-dependent.

2.2.3 Transformation Strain Properties and Hardening

We now continue covering the three key classes of SMA properties by discussing a method to parameterize the generation and recovery of inelastic strains during transformation.

As described in Chapter 1, the formation of martensitic variants from the austenitic parent phase at the microstructural level leads to the observed macroscopic deformation during transformation. However, tracking the detailed configuration of the microstructure (i.e., the evolution of 24 maximum martensitic variants in NiTi) is often beyond the scope of thermomechanical characterization, especially for engineering design and analysis of applications. This would require knowledge of up to 24 internal state variables, for example, which are only observable at length scales much smaller than the engineering scale of thermomechanical testing. Therefore, a more phenomenological approach is often taken whereby the macroscopic deformations are described by a transformation strain field, and the evolution of this field is linked both to the total applied tractions or total applied deformations and to the amount of total martensite present in the material. Theories have been developed wherein the transformation strain dependence on several select variants is considered. Experimentation has supported these theories. Such models will be discussed in Chapter 3 and are addressed in more detail in the literature [26, 27].

When considering the phenomenological behavior of SMAs, one considers the combined effect of microstructural behaviors. In the absence of stresses, local or applied, the martensitic variants form from the parent phase in patterns that lead to negligible macroscopic shape change (i.e., they will self-accommodate). Applied stress results in formation (or reorientation) of preferred martensitic variants, therefore generating observable overall strain due to transformation or detwinning, as discussed in Sect. 1.3. The variant selection is dependent on applied stress, therefore causing the amount of recoverable strain generated to vary with applied stress. Hence, a material parameter to be characterized for SMAs is the maximum strain formed due to transformation at a given stress level and in addition to thermoelastic strains. This strain is the *current maximum transformation strain* and its value depends on several factors, including the crystallography of martensitic transformation, the overall microstructural configuration of grains, texture, configuration and composition of precipitates, in addition to stress level. As a strain measure, the maximum transformation strain could have multiple components. However, for the 1-D experimentation in this chapter, the maximum transformation strain will be considered a scalar quantity given as a function of applied stress for given material conditioning and denoted by $H^{cur}(\sigma)$. If all the martensitic variants are aligned to the maximum extent possible, either due to the effects of applied stress or because of sufficient material training, then the material is said to be in a fully detwinned martensitic state. For a material in such a state, $H^{cur}(\sigma)|_{\sigma \geq \sigma_{align}} = H^{max}$ is the *maximum available transformation strain.* Here the quantity σ_{align} denotes the minimum stress sufficient to fully align the martensitic variants. For sufficiently trained materials, the value of σ_{align} may approach zero. Transformation strain evolution will be further addressed in Sect. 3.3.3.

Other than the value of the maximum transformation strain, it is also important to note the manner in which the strain evolves during loading. Some materials will show a distinct transition from elastic response to transformation and little stress increase during transformation. This is an indication of limited 'hardening' (An example of this will be shown in Sect. 2.5.1). Other SMAs will transition smoothly from elastic to transformation and may show a large increase in stress as the transformation progresses. These materials exhibit more significant hardening. (An example of this will be shown in Sect. 2.5.2). Different material models exist which account for varying kinds of material hardening, and the experimentalist should take note of this behavior. A more detailed discussion of this will be undertaken in Chapter 3.

With each of the aforementioned properties determined, the elastic behavior of each of the two phases is captured. In addition, the evolution of transformation strain is described and the locations of transformation regions in the stress-temperature space are determined. For the models that will be described in Chapter 3, it is required that only the three classes of properties describing these three attributes of SMA behavior be determined. Of course, more complicated models exist that require additional material parameters to account for more general material behavior. Such behaviors include transformation-induced plasticity and the reorientation of martensite. Models incorporating these behaviors will be discussed in Chapter 5 and Chapter 6, respectively.

2.3 Experimental Characterization Process

To experimentally quantify the properties described above for a given SMA material, one could subject a representative specimen to a wide range of uniquely defined thermomechanical loading paths while monitoring the material response. Years of SMA research worldwide have shown that this range can be narrowed to include only a few key tests. The process of characterization described below follows a natural order useful for determining specific material properties as well as for understanding qualitative material behaviors. Furthermore, it allows for an understanding of unstabilized and trained material behavior, an often overlooked distinction in SMA experimentation and application design.

2.3.1 Overview of the General Thermomechanical Characterization Process

In brief, the experimental process proceeds as follows:

- **Experiment 1:** Determination of zero-stress transformation temperatures
- **Experiment 2:** Monotonic loading/monotonic unloading of specimen
  - a) Load at $T < M_f$ to determine martensitic response and zero-stress SME behavior
  - b) Load at $T > A_f$ to determine austenitic response and pseudoelastic behavior
- **Experiment 3:** Determination of SME at non-zero stress level
- **Experiment 4:** Determine effects of cyclic loading, including material response stabilization

The logic behind these steps is straightforward and is based on forming an evolving understanding of the material behavior. Zero-stress transformation temperatures (Experiment 1) are vital to the remainder of the testing process as they allow for experimental system design. Without some knowledge of these temperatures, the proper thermal testing environment cannot be configured.

Monotonic material loading (Experiment 2) is important as a standard test for most materials in determining elastic behavior. It is further important in SMA characterization because such testing above A_f and below M_f provides important information on the key behaviors of pseudoelastic response and stress-free shape memory effect, respectively.

Experiment 3 expands on the SME results of Experiment 2a, testing the capability of the material to provide work output by generating and recovering transformation strain under non-zero stresses.

Finally, the cyclic loading applied during Experiment 4 highlights the change in material response given a particular loading history. After sufficient cycles, the response will often stabilize. This training can significantly modify the material properties and hence requires the repetition of the characterization process (Experiments 1–3).

2.3.2 Illustration of the General Characterization Process

Given the overview, we now demonstrate the general SMA characterization process in more detail. Consider a NiTi wire sample with a diameter of 0.91 mm intended for use in an actuation application. Because the specimen is in wire form, all reported stresses and strains are uniaxial components as obtained in tension. The uniaxial stress is taken as force over a wire cross-sectional area where testing was performed on an MTS 810 system using a 150-pound load cell. This load frame is shown in Fig. 2.10a. An MTS extensometer with a gauge length of one inch was used to record specimen strain levels.

(a) MTS 810 servohydraulic loading frame.

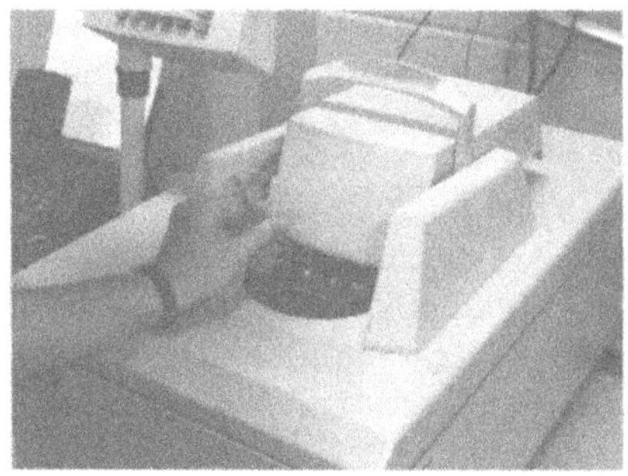

(b) SMA sample loaded into a DSC system.

Fig. 2.10. Experimental equipment used for thermomechanical characterization of shape memory alloys.

Experiment 1: Determination of Zero-Stress Transformation Temperatures

This first experiment is the most fundamental step in the characterization of shape memory alloys. Without knowledge of the stress-free transformation temperatures, it is very difficult to know what phase is present in the alloy at any given stress and temperature state. Recall also that this first experiment is crucial in designing the remainder of the experimental process. To assess these temperatures, a DSC was used (see Sect. 1.8). The small DSC sample is cut from larger bulk material using a low force saw. This method reduces frictional heat and overall material deformation, maximizing the representative integrity of the cut specimen. Note that the use of EDM to cut DSC specimens also leads to accurate DSC results. The specimen is shown being loaded into the DSC machine in Fig. 2.10b, and the test results are shown in Fig. 2.11a. To quantify this heat flow/temperature curve, one may construct lines tangent to the start and finish of each peak and lines tangent to the baseline heat flows. The distinctive intersections of these lines provide one possible measure for the start and finish temperatures of each transformation.

This has been illustrated in Fig. 2.11b, where the transformation temperatures have been noted. It is important to recall, that some materials exhibit the R-phase transformation. For these materials, R-phase transformation indicated in DSC results may be confused with austenitic/martensitic transformations. Given the DSC data alone, it is not always clear if exhibited peaks indicate A→R or A→M, for example. Therefore, the DSC results obtained directly from as-received material should be used as guidelines while zero-stress transformation temperatures derived from other thermomechanical tests are often more practical or applicable in predicting transformation behavior, especially in engineering applications (see the example in Sect. 2.5.1).

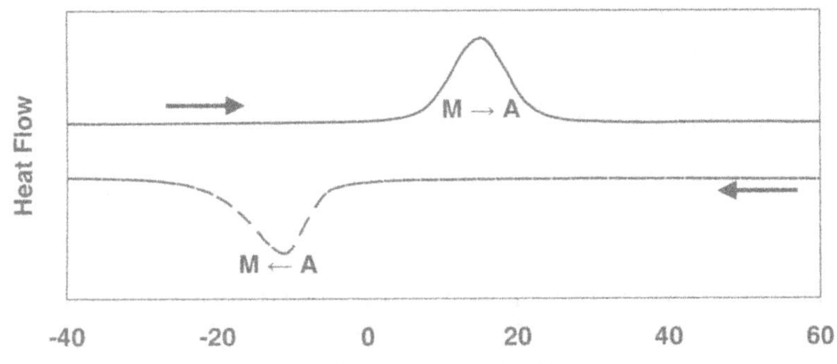

(a) DSC results

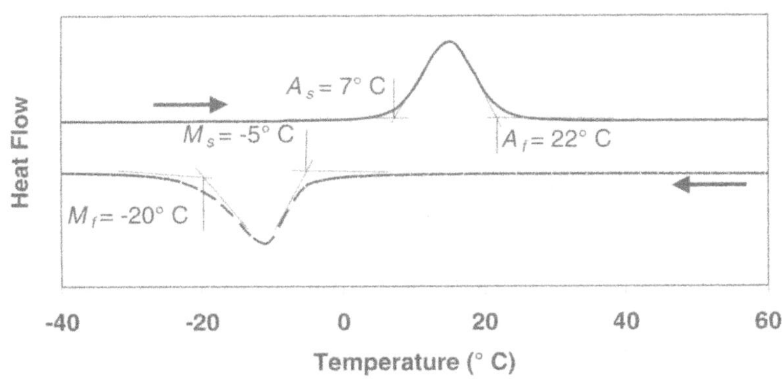

(b) Determination of transformation temperatures using tangent lines

Fig. 2.11. Determination of stress-free transformation temperatures from DSC testing (NiTi, untrained material).

Experiment 2: Monotonic Loading of Specimen

$\mathbf{T < M_f}$:

Monotonic loading below M_f provides information on the elastic properties of twinned (self-accommodated) martensite and determines if the material exhibits crucial shape memory behavior by verifying the capability of the material to exhibit stress-free SME (see Sect. 1.4). Alternatively, additional loading can be applied to failure, to assess the yield and ultimate fracture properties of the particular SMA.

In the current example, data derived from DSC testing is used to determine appropriate testing temperatures. Prior to testing, the sample is first heated above A_f (22°C) to recover any transformation strain and then cooled well below M_f (−20°C) to ensure a fully martensitic material state. This eliminates detwinned martensite which may have formed during inadvertent (and unknown) deformations. The sample is subsequently tested at a temperature below M_f such that martensite is the only stable phase. Testing consists of stressing the material until detwinning completes and to some maximum stress. The maximum stress is not known prior to testing. Rather, it is chosen during the course of testing by noting that detwinning has completed. The sample is then unloaded. The specimen temperature is finally homogeneously increased until it is above A_f, and any strain recovered is monitored.

The results from this test are shown in Fig. 2.12a. Using tangent lines we can approximate the detwinning start and finish stresses as $\sigma_s \approx 140\,\text{MPa}$ and $\sigma_s \approx 170\,\text{MPa}$, respectively. We observe that a significant amount of strain is recovered (6.2%), but some irrecoverable strain remains at the completion of the SME test. This indicates that the material is not responding in a repeatable (stable) manner, and training may be required. Note also that the elastic modulus for martensite is found to be 24 GPa.

$\mathbf{T > A_f}$:

The second monotonic test assesses material behavior at temperatures greater than A_f (i.e., possible pseudoelasticity). Here a loading/unloading cycle is applied to the specimen at $T = 30\,°\text{C}$, and nearly full pseudoelasticity is observed with some residual plastic strain. These results are shown in Fig. 2.12b. The martensitic elastic modulus found here is notably higher than that observed in Fig. 2.12a. This is an important observation and is due to the fact that some remnant austenite may remain at the end of loading. The elastic modulus observed during unloading may then reflect a combination of the moduli of both austenite and martensite. Note that, according to the definitions in ASTM test method F-2516, the *upper plateau stress* or stress at 3% strain during loading is $\simeq 300\,\text{MPa}$. By the same standard, the *lower plateau stress* or stress at 2.5% strain during unloading is $\simeq 100\,\text{MPa}$.

Finally, if device design is a goal, it may be useful to determine the material plastic yield and ultimate failure values. Such a test was not performed on the current specimen, but can be easily accomplished. With respect to the yield

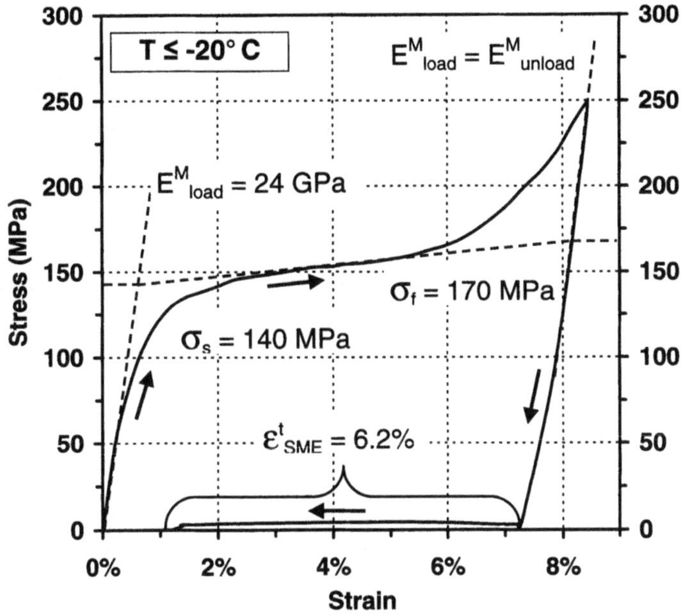

(a) $T < M_f$

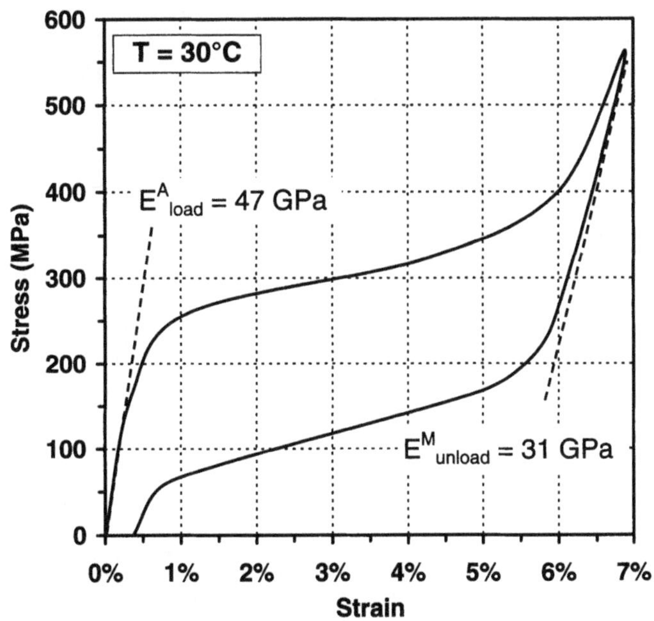

(b) $T > A_f$

Fig. 2.12. Results of monotonic loading/unloading.

and failure properties of martensite, one may employ a used SME specimen to perform these tests. The critical stress for plastic yield is often notably higher than the start and finish stresses for detwinning and should be easily identified. Regarding determination of the austenitic yield limit, note again that such a test must be performed at a temperature sufficiently above A_f to keep the material in the austenitic state (i.e., hot enough to prevent forward transformation into martensite). The phase diagram (estimated or exact) can be used to determine such an appropriate temperature. An additional material property not yet considered, though commonly discussed, is the maximum temperature at which austenite can be transformed to martensite via the application of stress without first plastically yielding. This temperature is often denoted M_d.

Experiment 3: Determination of SME at Non-zero Stress Level

For actuators, it is important to observe not only that the zero-stress shape memory effect is exhibited, but also that the material is able to perform work by providing displacement while under some load. A simple test of this behavior is a constant stress actuation test. In the case of uniaxial loading, this is sometimes referred to as *isobaric* testing. To perform this test for the current example, the material sample is heated well beyond A_f (22 °C) and then stressed to 200 MPa. This load is held constant while the temperature is slowly and homogeneously lowered until forward transformation into martensite is completed. Finally, the temperature is slowly increased until reverse transformation is completed. Throughout this test, the strain is monitored and recorded. The results of this experiment are shown in Fig. 2.13a.

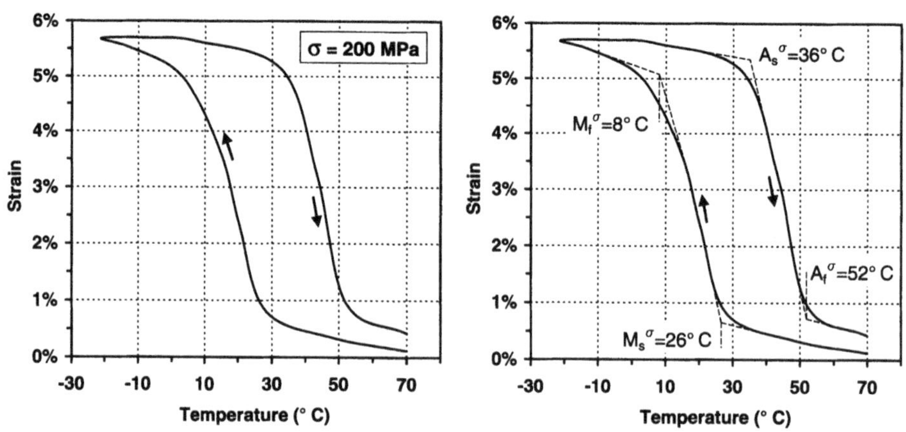

(a) (b)

Fig. 2.13. Results of constant stress thermal cycling (first cycle): (a) strain-temperature loading curve, (b) determination of transformation temperatures at stress.

Before continuing with the next experiment in the characterization process, it is appropriate to examine the results observed up to this point. This allows proper planning of additional experiments.

Investigation of Transformation Strain Generation and the Phase Diagram

After completion of the first three experiments, results can be examined and the performance of the material can be assessed. Recalling the three proposed classes of material properties (elastic, transformation criteria, and transformation strain properties), determination of any elastic parameters from the experimental data is often quite straightforward. Properties from the other two classes are addressed at this point.

The ability of the material to generate and then recover transformation strain must be evaluated. It was observed that 6.2% transformation strain was exhibited during stress-free SME testing (Experiment 2a). While this value is important, it is usually more valuable to examine the amount of strain recovered under load. To do so accurately requires not only measuring the total strain generated during transformation, but also considering the contributions of thermal and elastic strains. It is therefore necessary to review the relationship between these quantities.

For the magnitude of strains conventionally experienced by SMAs, the total strain can be additively decomposed into an elastic component, a transformation component, and a thermal component ($\varepsilon = \varepsilon^e + \varepsilon^t + \varepsilon^{th}$). Because the characterization of SMA samples is often 1-D in nature, simple relations can be used to describe the elastic and thermal response at any point in the loading path. In one dimension, the thermal strain may be simply given as $\varepsilon^{th} = \alpha(T - T_0)$ where α denotes the current thermal expansion coefficient. This leads to the following common 1-D elastic relation:

$$\sigma = E[\varepsilon - \varepsilon^t - \alpha(T - T_0)]. \tag{2.3.1}$$

Here the T_0 is assumed to be the reference testing temperature. We can now consider the stress found in the wire when it has been fully transformed into one phase or the other. When the material is purely martensitic, it is fully detwinned such that $\varepsilon^t = H^{cur}(\sigma)$. In this pure martensitic state and at the reference temperature ($(T - T_0) = 0$), this yields the following for the stress in terms of the current total strain:

$$\sigma = E^M[\varepsilon - H^{cur}(\sigma)]. \tag{2.3.2}$$

It is further assumed that when the material is purely austenitic, all transformation strain has been recovered ($\varepsilon^t = 0$). For materials in such an austenitic state, this yields the following for the austenitic stress in terms of the total strain:

$$\sigma = E^A[\varepsilon - \alpha^A (T - T_0)]. \tag{2.3.3}$$

Therefore, given one isobaric test at one given constant stress level (see Fig. 2.13b), one can derive the maximum transformation strain by solving both (2.3.2) and (2.3.3) for the current strain in each of the pure phases and assuming a constant testing stress, σ. This yields:

$$\varepsilon^M = \frac{\sigma}{E^M} + H^{cur}(\sigma), \tag{2.3.4}$$

$$\varepsilon^A = \frac{\sigma}{E^A} + \alpha^A (T - T_0), \tag{2.3.5}$$

where ε^M and ε^A denote strain in a purely martensitic and purely austenitic state, respectively. Observing Fig. 2.8b, the height of a given hysteresis is $\Delta\varepsilon = \varepsilon^M - \varepsilon^A$, which, considering (2.3.4) and (2.3.5), leads to the following relation for the current maximum transformation strain at a given constant stress test level:

$$H^{cur}(\sigma_i) = \Delta\varepsilon + \alpha^A (T - T_0) + \sigma_i \frac{E_M - E_A}{E_M E_A} \tag{2.3.6}$$

Therefore, having determined the elastic moduli of martensite and austenite, and using an appropriate common value for the austenitic coefficient of thermal expansion (10^{-6} [3]), one can derive the maximum transformation strain, $H^{cur}(\sigma_i)$, exhibited under the application of each constant stress σ_i. For the single test performed thus far (Fig. 2.13b) where $\sigma_i = 200$ MPa, this is found to be $H^{cur}(200) = 5.1\%$.

It is also important to determine at what stress and temperature states the material can be expected to transform. The transformation criteria are best understood via construction of the phase diagram, which describes the transformation behavior in stress-temperature space. In relation to the progress of the current example, it is now important to have an approximate understanding of the transformation behavior for the purpose of planning a suitable cyclic loading experiment. To construct such an estimation requires two data points per transformation surface, assuming the transformation boundaries are approximately linear. The first set of points can sometimes be estimated via DSC testing (Experiment 1) and indicate at what temperatures the transformations begin and end under zero stress.

The second set of points can be determined from the constant stress actuation cycle (Experiment 3), or from the constant temperature pseudoelastic cycle (Experiment 2b), as applicable. Since the current example involves material intended for actuation, here we determine the discrete transformation temperatures at 200 MPa constant applied stress (Fig. 2.13a). For the purpose of illustration, tangent lines are constructed and their intersections are used to provide discrete values for these non-zero stress transformation temperatures. This is illustrated in Fig. 2.13b. Here M_s^σ and M_f^σ represent the non-zero stress forward transformation start and finish temperatures, respectively. Likewise, A_s^σ and A_f^σ represent the *non-zero stress* reverse transformation

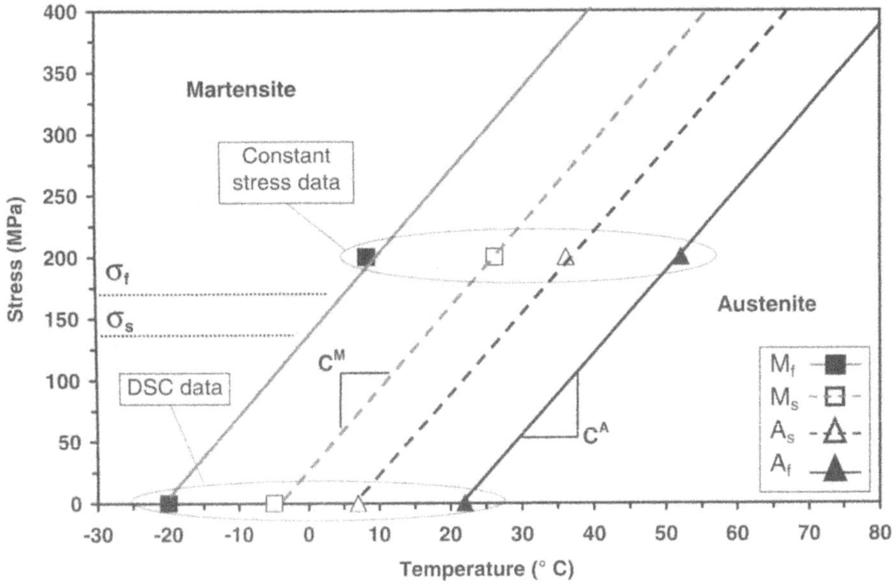

Fig. 2.14. Preliminary phase diagram, as-received material.

start and finish temperatures, respectively. Plotting the stress-free transformation temperatures and the 200 MPa transformation temperatures in the same stress-temperature space, an approximate phase diagram is constructed, as shown in Fig. 2.14. While this approximate phase diagram is insufficient to calibrate a model that accurately captures material behavior at all thermomechanical states, it is useful to estimate at what states the material will be fully austenitic and fully martensitic. It is also useful in planning the training process, as previously mentioned.

Experiment 4: Determination of Cyclic Loading Effects (Training)

To determine the effects of cyclic loading on the thermomechanical response of an SMA, the material is subjected to multiple transformation cycles. In the case of a material intended for use as an actuator, a straightforward method consists of applying many thermal transformation cycles under constant load to the specimen. This is often referred to as *training*. If the goal of cyclic loading is eventual stability of response, training loads (thermal and mechanical) which exceed those expected in the application should be applied (see also Sect. 1.6). For the current example, it will be seen that all further experimentation is performed at stress levels of 200 MPa and below. A constant stress of 200 MPa is therefore chosen for cyclic loading. A total of 80 thermal cycles are applied and the results are shown in Fig. 2.15. Note that while the initial strain response of the material evolves with each cycle, it eventually stabilizes.

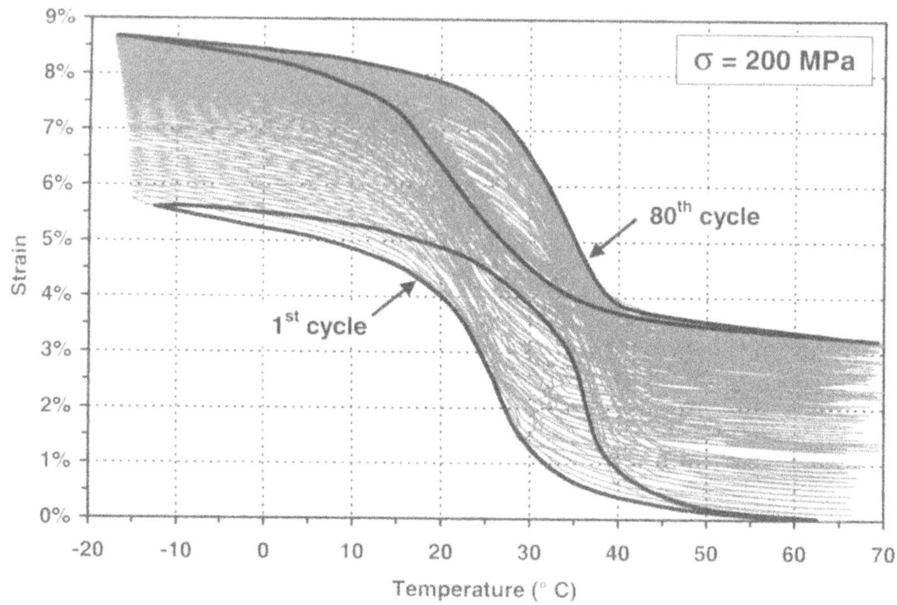

Fig. 2.15. Results from isobaric training at 200 MPa.

As an alternative to constant temperature or constant stress cyclic loading, application-based cycling may be used wherein an SMA specimen is stabilized in the same manner in which it is used (e.g., a smart structure incorporating active SMA elements is repeatedly actuated). The SMA components can then be removed from the application context and tested in the standard manner. Whatever stabilization or training method is used, it is important to remember that the particular choice of loading path used to train a specimen can substantially influence the final material properties and should be carefully chosen.

Evaluation of Stabilized or Trained Material

When a material has been cyclicly loaded and its response has evolved and stabilized, it is necessary to repeat the most important tests of the characterization process to attain the new trained material properties. The results of such repeated testing will be summarized below.

The first interesting and important result involves the effect of training on the DSC test results. During training of polycrystalline SMAs, widespread permanent dislocations are generated at the micro-scale within the material. This results in a heterogeneous microscopic stress state, which, per the behavior characterized by the phase diagram, results in transformation temperatures that vary from locale to locale. Because this occurs throughout the specimen, a distribution of localized transformation temperatures is expected.

This "smoothing" diminishes the usefulness of the DSC results in ascertaining the overall zero-stress transformation temperatures by effectively eliminating the peaks observed during heating and cooling. Therefore, other methods of determining these temperatures are often required after cyclic loading.

Recalling that the intended use of the material in the current example is actuation under some non-zero load, it is clear that a repetition of the stress-free shape recovery testing, or SME testing, is not necessary. Observation of the stabilized constant stress actuation behavior, however, is not only important but necessary for the determination of several material parameters. Determination of the temperatures at which the phase transformations begin and end for different constant stresses allows construction of the final phase diagram.

In addition, by examining the amount of strain generated during each isobaric cycle, one may derive a functional relationship between applied stress and current maximum transformation strain for this stabilized material. To this end, thermal cycles at constant stress are applied to the specimen. Stresses ranged from $\sim$ 2.5 MPa to 200 MPa in $\sim$ 50 MPa increments. The low stress test is important in not only determining the zero-stress transformation temperatures, but also in indicating the presence of transformation strain generation at zero-stress (the so-called "two-way shape memory effect," see Sect. 1.6). The results of these constant stress tests are shown in Fig. 2.16.

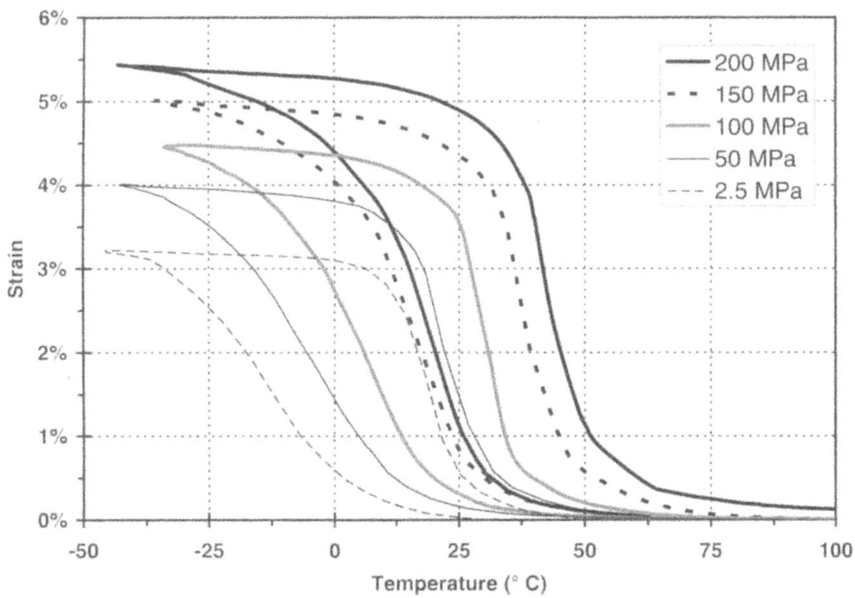

Fig. 2.16. Results of isobaric thermal cycling at various stress levels, trained material.

By examining the constant stress uniaxial test results, the amount of uniaxial transformation strain generated by the material for a given constant stress can be determined. This again requires that 2.3.6 be used to derive the current maximum transformation strain $H^{cur}(\sigma_i)$ exhibited under the application of each constant stress level σ_i. The experimental plot of this relationship is shown in Fig. 2.17. The dashed fitted curve represents the proposed functional form of $H^{cur}(\sigma)$ and highlights the tendency of the current maximum transformation strain to saturate with increasing stress.

Finally, the phase diagram for the stabilized material is derived. To construct the phase diagram (i.e., to determine the transformation temperatures at various constant stress levels), tangent lines are again used as shown in Fig. 2.18. Here the transformation temperatures are investigated for the material under an applied stress of 150 MPa. The experimental phase diagram is derived from a whole series of such tests and is shown in Fig. 2.19. Simple linear regressions are used to schematically represent each of the continuous transformation surfaces. One key change from the untrained, estimated phase diagram is the obvious broadening of the transformation regions (see Fig. 2.14). The mechanisms behind this behavior are the same as those previously discussed concerning the reduced usefulness of the DSC results. However, extrapolation of the transformation surfaces supersedes the DSC in providing an accurate indication of the zero-stress transformation temperatures.

The derived material properties for the experimental study summarized in this example are given in Table 2.1. Most properties were determined for the stabilized material, with the exception of elastic moduli.

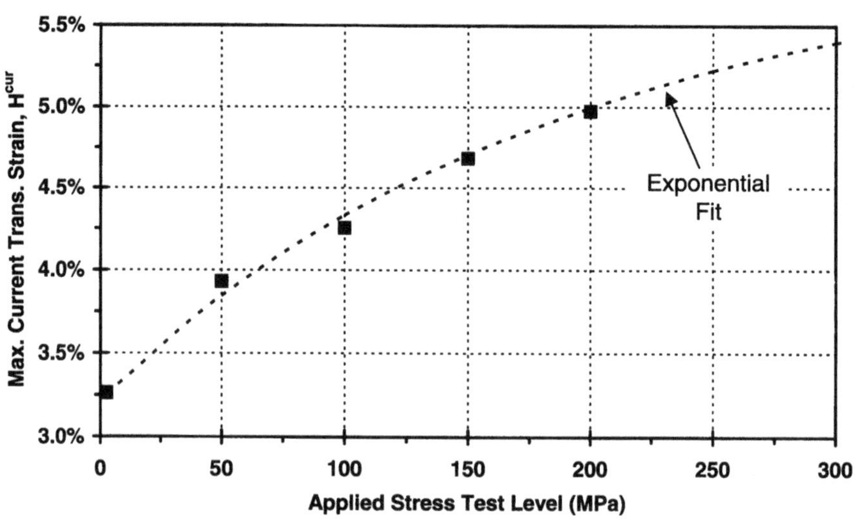

Fig. 2.17. Determination of exhibited maximum transformation strain as a function of applied constant stress, trained material.

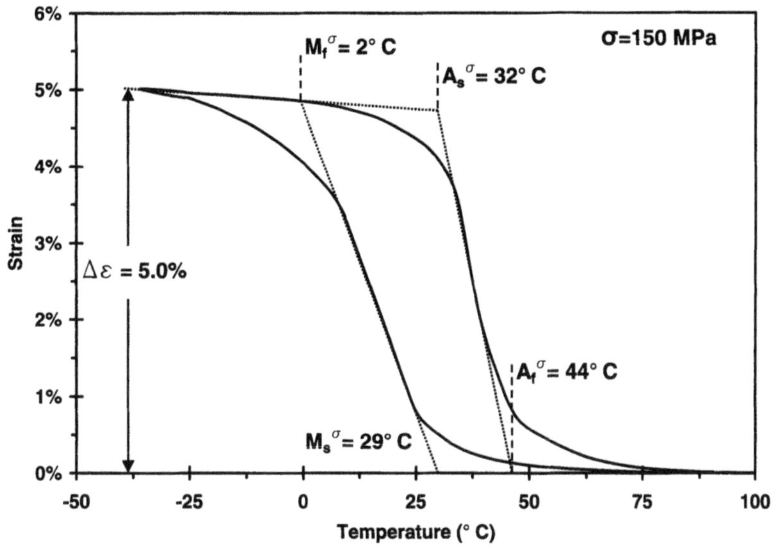

Fig. 2.18. Determination of transformation temperatures using tangent lines at a constant stress of 150 MPa, trained material.

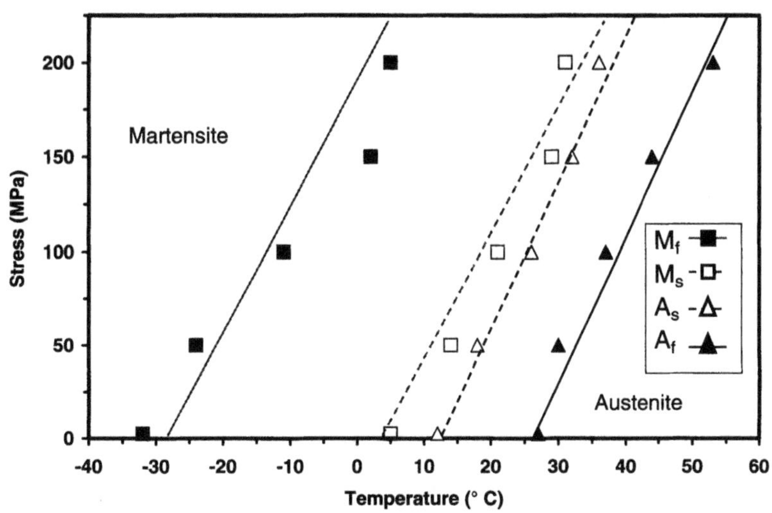

Fig. 2.19. Experimentally derived phase diagram, trained material.

This concludes the general characterization process for a particular example of SMA material. While interesting material behaviors such as fatigue, plastic hardening, transformation-induced plasticity, or others could also be studied, these will not be covered at this time. The intent of the above example was an illustrated overview of the most common experiments. However, while the characterization process for SMA specimens is conceptually straightforward,

Table 2.1. Experimentally derived material parameters; trained material.

Material Parameter	Value
E^A	47 GPa
E^M	24 GPa
σ_s	140 MPa
σ_f	170 MPa
M_s	3 °C
M_f	−29 °C
A_s	12 °C
A_f	26 °C
C^A	8.3 MPa/ °C
C^M	6.7 MPa/ °C
$H^{cur}(\sigma)$	$= H^{cur}(0) + (H^{max} - H^{cur}(0))[1 - \exp(-k_H\sigma/E^A)]$
	$= 0.032 + (0.06 - 0.032)[1 - \exp(-235\sigma/E^A)]$

there are many unique challenges that the SMA experimentalist must address. Such considerations will now be discussed.

2.4 Experimental Considerations Unique to SMA Thermomechanical Characterization

In the previous section, a description of a characterization procedure for SMAs was given and instructions for the completion of each experimental step were provided. However, the unique properties of SMAs are eventually manifested as a set of important experimental details and challenges that must also be discussed. These topics include consideration of loading rates, material statistical variation, material history, etc., and will be described with others in more detail below.

2.4.1 Influence of Total Material History on Shape Memory Behavior

As with other metals, the constitutive properties of shape memory alloys are strongly dependent on several factors, including the exact alloy composition, the particular heat treatments previously applied, prior history, and cold-working. However, because SMAs undergo important microstructural changes not attainable in other metallic systems, the material sensitivity to these and other historical occurrences can be much higher in thermomechanical response. The designer must be aware of these effects during the material selection phase, and it is important that the experimentalist keeps these details in mind to ensure accurate and representative testing.

The most influential characteristic of a material specimen is its alloy composition. As previously reviewed (see Sec. 1.9), several alloy systems are known to exhibit shape memory behavior (NiTi, NiTiCu, NiMnGa, etc.), and each has significantly different properties. It is reasonable that a change in elemental constituents will cause a change in exhibited material behavior. However, SMAs are also highly sensitive to the atomic balance when considering even just one alloy system (i.e., NiTi). A very small change in the balance between nickel and titanium will noticeably affect properties, especially with respect to the transformation temperatures. Furthermore, two material specimens of nominally identical composition will often behave dissimilarly. In this way, the sensitivity of shape memory alloys to composition surpasses that of aluminum, iron and other more conventional metals. For SMAs, one cannot often define a required set of material properties and then accurately choose a material composition that will provide them (though it is possible to estimate an appropriate composition).

It has also been shown that the exact thermomechanical response of an SMA is highly sensitive to heat treatments. This is because imposed thermal manipulations, such as high temperature soaks and rapid quenches, greatly affect a metal at the microscopic scale. Internal stresses can be generated or relaxed, and precipitates can be formed or dissolved. As with other metals, alterations in the elastic, plastic and ultimate failure properties will occur. In addition, transformation temperatures will shift, and even the ability of the alloy to exhibit some effects will improve or degrade (i.e., pseudoelasticity). For example, pseudoelasticity is rarely observed in some NiTi systems that have been fully annealed and is only possible if sufficient precipitates are formed via aging or other such treatments [25].

Like heat treatments, cold-working and hot-working performed during material processing can significantly alter material behavior. Examples of such working important to the SMA experimentalist include the drawing of raw material into wires or the rolling of the material into plates. Such permanent deformations can induce significant internal stresses while also altering the configuration of the microstructure. The density of dislocations increases and the configuration of precipitates formed during initial fabrication and heat treatment can be altered (see Sect. 1.9.1).

Perhaps a more fundamental effect of such processes is the alteration of the grain structure. Grains are often refined, but can also be notably reoriented, inducing material anisotropy and texturing effects. Such material alterations continue to occur during specimen preparation and testing. The thermomechanical conditions imposed during material specimen training are especially influential on the final properties and should be chosen carefully. As with heat treatments, microstructural changes incurred during working will alter both the conventional and especially the shape memory properties of SMAs.

These and other considerations lead to an important conclusion, especially when design of some application is the goal: to accurately characterize a material, one must ensure that the configuration and history of the SMA test specimen matches that of the SMA component used to the intended application component.

2.4.2 Comparison of Test Specimen to Intended Application Component

For designers and analysts, the end goal of material characterization is often the eventual ability to predict the response of some application, and carefully planned and executed characterization is required for such correct model calibration. The material composition and loads chosen for characterization must be highly representative of those found in the final application system. Between the material intended for use in an application and the complimentary material used for characterization, a one-to-one correspondence is required for the following attributes, several of which have been previously discussed:

1. The exact alloy composition.
2. The methodology of fabrication of the alloy component, including hot/cold working and heat treatments.
3. The loading history of the material prior to use (i.e., thermomechanically stabilized or untrained).
4. The stress state applied (e.g., axial, shear, or multicomponent).
5. The thermomechanical loading path applied, especially in relation to the phase diagram.

The first two of these items are common to the characterization and application of all classes of materials. It is necessary to characterize the same material that is to be used in the application. This includes materials with the same composition, same microstructural texture (as a result of any cold-working), and same heat treatment history.

The third of these items also pertains to other material classes in some degree. However, it is especially important for materials such as SMAs that undergo large deformations and may, over some number of cycles, accumulate significant irrecoverable plastic strain. The generation of this strain results from significant changes in the microstructure of the material, indicating that other material properties may have been altered. As an example, SMA specimens that have been cycled many times and have generated significant plastic strain will often exhibit less maximum transformation strain than was shown before these cycles were applied. It is also common for the transformation regions to broaden (i.e., the difference between M_s and M_f increases). Each of these effects were observed in Section 2.3.1. In the case of a multi-use actuator application, a stabilized material specimen should be characterized. If an application is to be used very few times, or only once, as in the case of some release mechanisms (see Sect. 1.10), then as-received, non-stabilized specimens of identical composition should be characterized. Experimentation itself necessarily imposes loading cycles and that material property evolution is strongest during the first repetitions of loading. Thus, if as-received (non-stabilized) behavior is required, used specimens should be replaced often with new, untested specimens.

It is also important to consider the applied stress state used to derive the material constants. The similarity between the stress state applied during characterization and that experienced by a component in an application has been shown to affect the material parameter calibration accuracy. Specifically, this concerns the distinction between extension vs. shear. Tension/compression asymmetry is also common, and this stress state should be considered as well. If an SMA component undergoes loading consisting mainly of axial tension and/or compression (i.e., wire or rod applications), uniaxial testing should be used. For shear-based applications (i.e., torque tubes, etc.), characterization in shear will often lead to more accurate modeling. In this way G^A and G^M, the shear moduli for austenite and martensite, respectively, may be directly and accurately determined. Furthermore, material for use in tensile applications should be characterized in tension, and similarly for compression.

Finally, let's consider the thermomechanical (stress or temperature) loading path applied to the SMA component of interest during common use. Because the micromechanical consequences of repeated constant stress loading may differ from those of repeated isothermal loading, material properties measured in each of these two ways have been shown to differ significantly. If a design requires an SMA component to undergo nominally isothermal loading (i.e., vibration isolation and other pseudoelastic applications), then isothermal material characterization will be the most accurate. The same requirement applies for components undergoing thermal actuation cycles.

2.4.3 Importance of Mechanical and Thermal Loading Rates

Recall that the forward transformation into martensite is exothermic while the reverse transformation into austenite is endothermic. This introduction of the latent heat phenomenon and resulting possible temperature changes during the characterization process imposes additional constraints on the loading rates applied to SMA specimens. The challenge is particularly applicable to pseudoelastic testing. Such experiments are often performed at constant temperature to simplify interpretation of the transformation behavior. During loading, however, thermal energy can be added to the specimen if the heat generated during the exothermic forward transformation is not dissipated. If loading is performed slow enough, convective and conductive processes will remove this additional heat without noticeably raising the specimen temperature. If loading or unloading progresses too rapidly, the temperature of the specimen will rise during loading and fall during unloading, violating any isothermal assumptions.

Several studies have been performed to assess the influence of loading rates on material response. Many of these address pseudoelastic loading [12, 28, 29] and have found the thermal effects present during this exothermic/endothermic cycle to be significant, especially at higher rates. Throughout the various research efforts in the area of shape memory alloys, many displacement and thermal loading rates have been suggested. For isothermal loading at any temperature, rates should not exceed $\sim$ 0.01-0.05% strain/sec [12, 18]. This helps to ensure quasi-static testing. Others have also explored the martensitic reorientation (detwinning) rate dependency in various stress configurations and in various alloys [24, 30], and have shown that the detwinning process is generally independent of the strain rate.

Finally, dynamic impact testing has also been performed in several studies [31, 32] and phenomenological model predications and experimental results have been compared while others [33] have investigated martensitic transformation mechanisms under these dynamic loading conditions.

The use of reasonable strain rates is straightforward in strain-driven or deformation-driven experimentation, but not all thermomechanical experiments on metals are performed using prescribed strain or displacement inputs. Most testing systems also provide the option to use force control, which may present advantages in some situations. For example, using force control during pseudoelastic unloading allows the experimentalist to set an exact ending force value, such as 0 MPa or 7 MPa for ASTM standard testing [18]. However, for SMAs exhibiting "flat" pseudoelastic loading/unloading plateaus, force control can lead to problems. A constant force rate that is suitable during the elastic portion of loading quickly leads to an unacceptably high strain rate during stress-induced transformation as the testing system seeks to provide the same constant force rate. An example of this is shown in Fig. 2.20.

Here the main result involves a specimen undergoing two pseudoelastic loading cycles at an ambient temperature of 37 °C (nominal). For the first test, a strain rate of $0.05\%/s$ was applied during both loading and unloading. For the second test, a strain rate of $0.05\%/s$ was applied during loading, but a 0.08 N/s force rate was applied during unloading. In Fig. 2.20a the forward loading (forward transformation) paths for both tests are nearly identical while the unloading (reverse transformation) curves are not. During unloading, the use of force control in the second test caused high strain rates as the material response reached the lower plateau stress. This in turn caused a reduction in specimen temperature from 37 °C to 35 °C as seen in Fig. 2.20b. This temperature change causes the unloading curve to follow a different isothermal load path (i.e., the result for a test temperature of 35 °C) at the end of reverse transformation.

Whatever the deformation rate imposed, the measurement of deformations and strains requires careful attention. In some specimen configurations, the stress-induced strains that result from martensitic transformation or reorientation can initiate at discrete locations and then propagate in a wave-like manner along the length of the test section [34]. For this reason, devices such as strain gauges with short gauge lengths are not particularly effective at measuring the macroscopic strain. Instead, they have been shown to indicate "jumps" in strain as transformation or reorientation is initiated in the local region where the strain gauge is attached [12]. Extensometers, which generally have larger gauge lengths, have been more effective for measuring

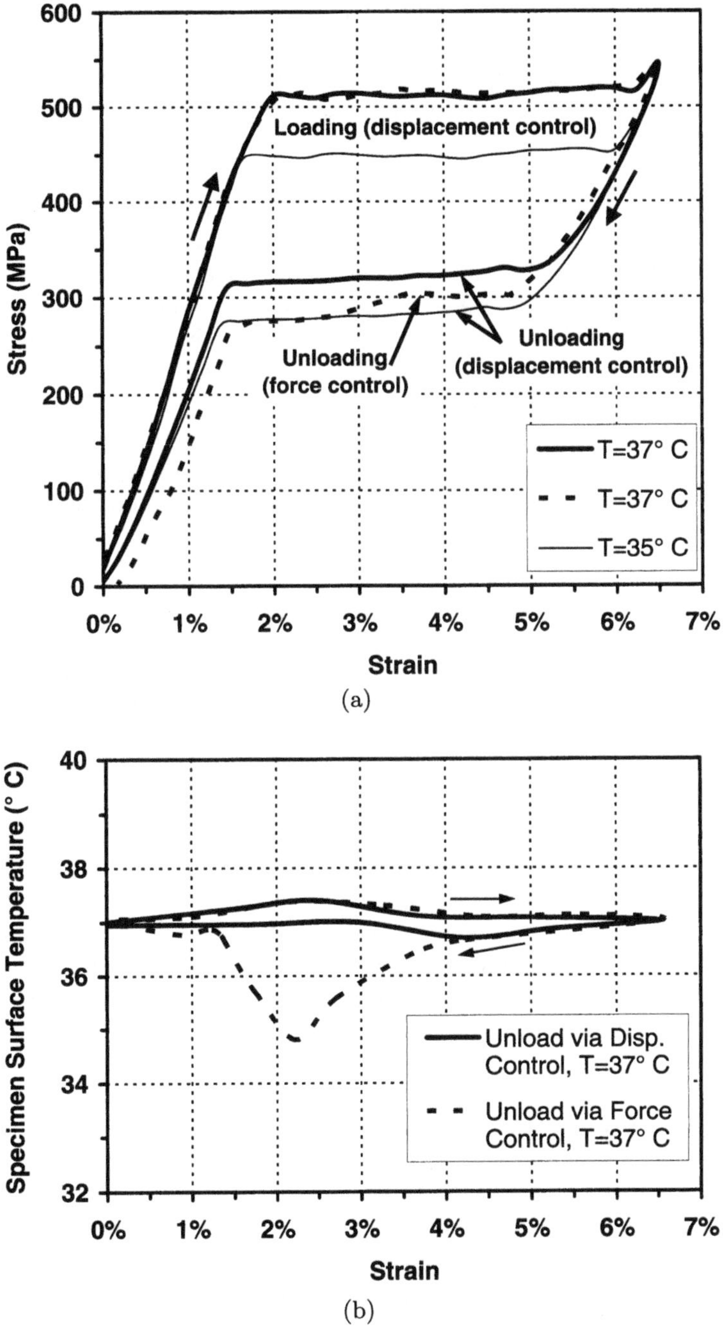

Fig. 2.20. Comparison of unloading control methods for pseudoelastic testing: (a) stress-strain response, (b) specimen surface temperature response.

phenomenological response. Furthermore, the initiation of phase transformation or martensite reorientation often occurs near the grips and this can lead to inconsistencies when comparing strain measurement acquired over the entire length of a specimen (i.e., via crosshead displacement) with those acquired in a region of homogeneous stress (i.e., via extensometry).

It is also important to impose reasonable heating and cooling rates when prescribed thermal variations are imposed. For specimens such as wires, direct resistive Joule heating is common, while in other cases load frame-mounted furnaces are required. To cool the specimen, conventional liquid baths such as chilled water [12] or laboratory coolant [35] have been used, especially for the testing of wires. For temperatures substantially below room temperature, cooling can be provided by the use of liquid nitrogen [7]. While the mechanical strain rates described above for isothermal testing are generally agreed upon, appropriate heating and cooling rates are often more dependent on the experimental system employed. Such rates must be carefully considered and should be adjusted as necessary to ensure slow, homogenous heating/cooling of the material sample test section.

2.4.4 Stochastic Variation in Material Response

One must also consider stochastic or statistical variation across different material samples. The thermomechanical behavior from specimen to specimen and test to test will often deviate noticeably from some nominal response. To correctly perform material characterization, several material samples for each thermomechanical loading path are often required. Statistical variation may be minimized by ensuring that samples are prepared from homogeneous bulk material in a repeatable manner. However, despite such efforts, some variation will occur and this must be accounted for during subsequent interpretation of results, calibration of any models, design, and analysis. An example is provided in the ASTM standard for DSC testing (ASTM F-2004), where the repeatability across multiple samples is briefly discussed [16]. This issue will be further exemplified in Example 2 in the following section.

2.5 Examples of SMA Characterization

Having expanded on various details of SMA characterization, this section highlights how different considerations affect both the planning and execution of material characterization. Here, three examples are presented. The first demonstrates SMA wire specimens for use in a pseudoelastic application. The second addresses pseudoelastic testing once more, though the focus is the stochastic variation observed when several specimens are taken from the same source material. Finally, the third example illustrates the derivation of material parameters used to model an actuation application in which the material used is not the common equiatomic NiTi.

2.5.1 Example 1. Characterization of NiTi Wire Intended for Pseudoelastic Application

For the first example of characterization, consider conventional, commercially available NiTi wire, intended for use in an application requiring wire components and stabilized material response. Because the material raw form (wire) is the same as the form needed for the application, specimen preparation is minimal. The application is used for vibration isolation research and thus requires an SMA that displays pseudoelasticity at room temperature. The as-received spooled wire fulfills this requirement.

As outlined above, the first step in the characterization process is the estimation of the zero-stress transformation temperatures for the as-received material. DSC testing was again used, and the results from this test are shown in Fig. 2.21.

The next pertinent step is monotonic mechanical testing of the material. However, in this example the response of the material below M_f was of little interest, and the shape memory behavior of the material was not assessed. To plan the pseudoelastic testing process, the experimentalist noted that the A_f temperature estimated from the DSC was $\approx 20\,°\mathrm{C}$, below room temperature. This indicates that the first loading path can be applied at room temperature. However, to ensure that the wire temperature remains constant, a heating and cooling system was installed to negate the effects of self-heating and self-cooling during the exothermic and endothermic transformations (2.4.3). This precaution was indeed a necessity as the applied strain rate of 0.13% strain/sec used in this case exceeded the recommendations of Sect. 2.4.3.

The final decision left to the experimentalist was the maximum stress level applied. It is common to decide on this value during testing, while monitoring the stress *in situ*. The maximum stress should clearly exceed the end of forward transformation, but should not be so high as to initiate permanent plastic yielding.

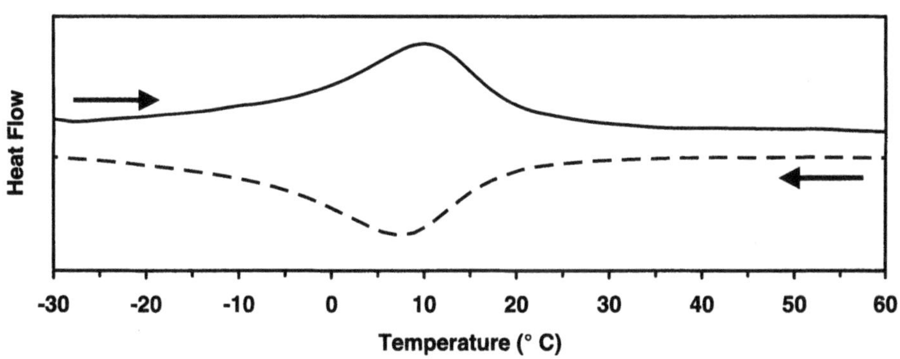

Fig. 2.21. Estimation of zero-stress transformation temperature using a DSC; Example 1, as-received material.

After an initial isothermal loading cycle was performed, the experimentalist noted acceptable results. Thus cyclic loading was commenced to stabilize the material. Recall also that the intended application will be used at room temperature. This further supports 21 °C as an optimal training temperature. Training was performed by applying 15 sequential loading/unloading cycles with a maximum stress of 700 MPa. The results for all applied cycles are shown in Fig. 2.22, with the final cycle highlighted. No significant plastic strain is generated at the end of this last cycle. This material exhibits relatively low hardening with very distinct transitions from elastic to transformation.

With the material satisfactorily stabilized, the experimentalist repeats the characterization process with the goal of deriving final material parameters. Continuing, isothermal loading is applied to the trained wire specimens at different constant temperatures (25, 35, 45, and 55 °C). By noting where transformations begin and end, a new, detailed, phase diagram can be constructed. By measuring different shape parameters found in each pseudoelastic curve, elastic properties and the maximum transformation strain can be derived as shown below. Figure 2.23 shows the pseudoelastic curves for all four constant temperatures at which the SMA wire was tested. To provide a more detailed illustration of material parameter derivation, refer to Fig. 2.24, which shows the material response during testing at 25 °C. Here E^A and E^M are measured in a straightforward manner. Although there appear to be two possible martensitic elastic slopes in this figure (loading and unloading), the slope during loading is influenced by very minor continuing transformation in some

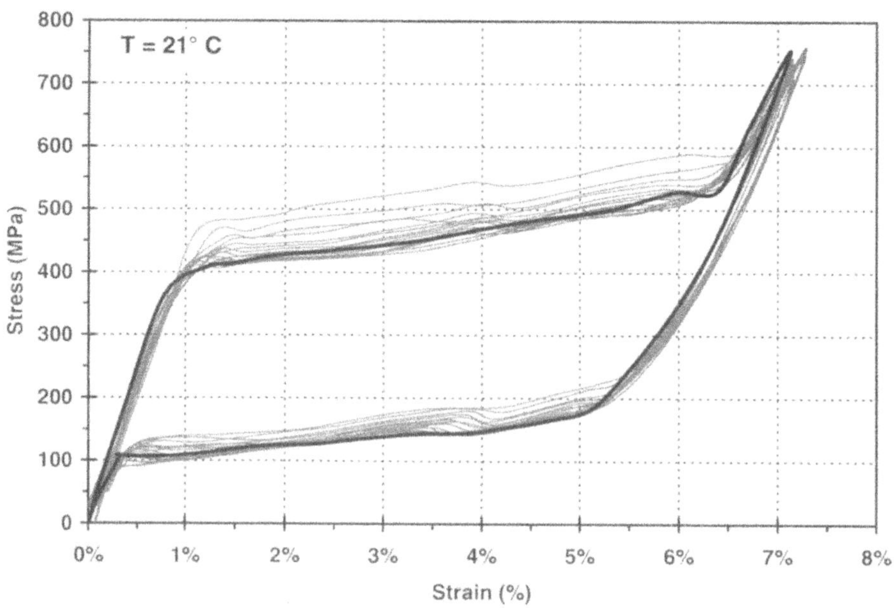

Fig. 2.22. Pseudoelastic stabilization of material at $T = 21\,°\mathrm{C}$. 15 cycles applied, last cycle darkened.

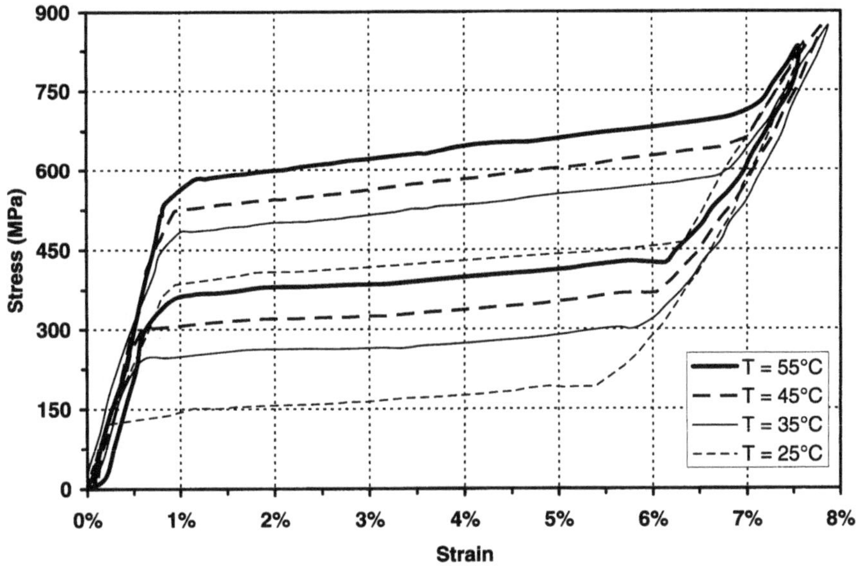

Fig. 2.23. Pseudoelastic loading/unloading results for various testing temperatures.

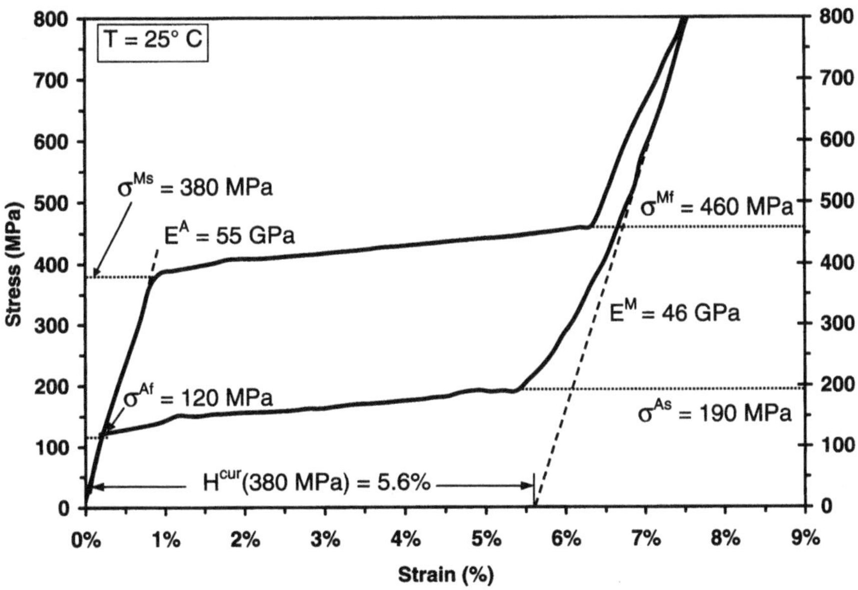

Fig. 2.24. Pseudoelastic loading/unloading results and derivation of parameters for various testing temperatures of 25 °C.

parts of the specimen. The slope during unloading is therefore more representative of truly elastic response. Since the analysis and testing are for 1-D case, ν^A and ν^M are not needed.

The next step is to determine the critical stresses for initiation and completion of phase transformation. At a known constant temperature, these stresses are denoted as σ^{Ms} and σ^{Mf} for martensite and σ^{As} and σ^{Af} for austenite. The values for these stresses at a test temperature of 25 °C are shown in Fig. 2.24. By examining the four isothermal pseudoelastic tests performed (Fig. 2.23), one can determine the stresses for the initiation and completion of both transformations at four distinct temperatures. Construction of the phase diagram proceeds using this experimental data, and results for the current case are shown in Fig. 2.25. The various material parameters found during this experimental study are summarized in Table 2.2.

In addition to the determination of the final phase diagram, the maximum transformation strain, H^{cur}, was also derived. This value is found by considering again (2.3.2) and solving it for H^{cur}. This yields the following relation:

$$H^{cur} = \varepsilon - \frac{\sigma}{E^M} \tag{2.5.7}$$

where the stress used in this relation to determine the current maximum transformation strain is σ^{Ms}. Therefore, given a set of pseudoelastic test results such as those shown in Fig. 2.24, and considering an imagined martensitic stress of zero, it can be inferred that the maximum transformation strain, H^{cur}, is equivalent to the amount of strain indicated when the martensitic elastic stress response is extrapolated to the zero-stress axis. This has been shown in Fig. 2.24.

One key feature of these results is the incompatibility between the phase diagram (Fig. 2.25) and the DSC results (Fig. 2.21). The DSC was performed on the as-received specimen without any preparatory heat treatment (see ASTM F-2004 [16]). The results indicate transformation temperatures at zero-stress that do not agree with those derived from the phase diagram via extrapolation. As mentioned previously, this apparent contradiction is not uncommon in materials exhibiting R-phase transformations (e.g., nickel-rich NiTi alloys).

Table 2.2. Experimentally derived material parameters; Example 1, trained material.

Material Parameter	Value
E^A	55 GPa
E^M	46 GPa
M_s	−28 °C
M_f	−43 °C
A_s	−3 °C
A_f	7 °C
$C^A = C^M$	7.4 MPa/ °C
$H^{cur}(\sigma) = H^{max}$	0.056

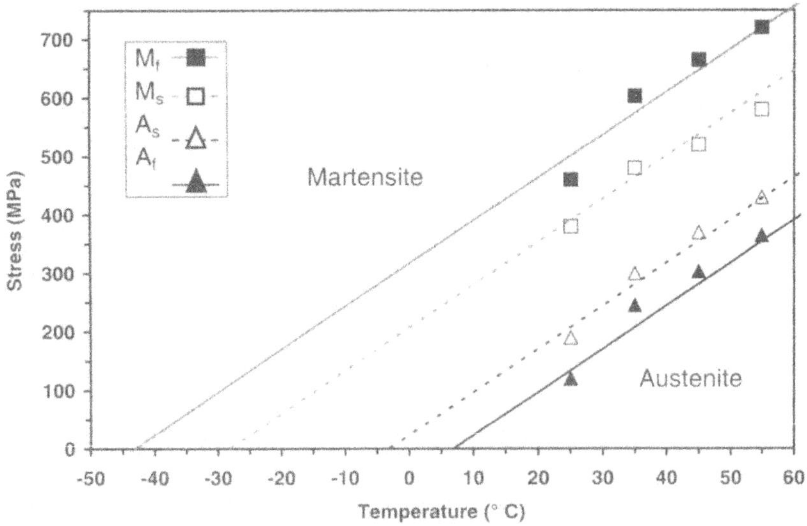

Fig. 2.25. Experimental estimation of phase diagram using pseudoelastic experiments; Example 1, trained material.

2.5.2 Example 2. Characterization of NiTi Wire for Determination of Stochastic Variation

This example is intended to illustrate only the effects of stochastic variation across samples (see Sect. 2.3) where the effect of interest was pseudoelasticity. Four NiTi (50.3 Ni, at.%) wire samples with a diameter of 2.16 mm were used, each taken from the same original roll and prepared (i.e., heat treated) in a consistent manner.

The zero-stress transformation temperatures of the material were first determined via DSC testing. The heat flow curves for the heating portion of the tests for each of the first three wires are shown in Fig. 2.26. Note the slight variation in the results for wire specimens from the same source and prepared in the same manner.

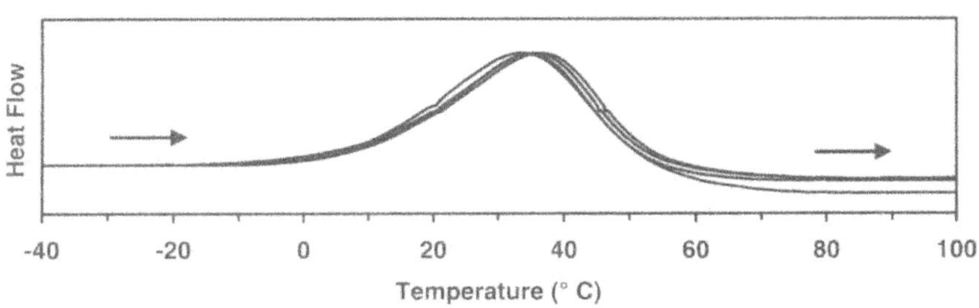

Fig. 2.26. DSC results (heating cycle) showing slight variation across four wires; Example 2, as-received material.

Based on the results of the DSC, it was determined that 80°C was an appropriate constant temperature at which to test the pseudoelastic characteristics of the wire samples. The first of the four specimens was loaded to $\approx 800\,\mathrm{MPa}$ and then unloaded. Such a load was repeated 20 times to stabilize the material. The results of this pseudoelastic testing can be seen below in Fig. 2.27. Note the substantial reduction in the pseudoelastic hysteresis caused by repeated application and removal of the load.

To assess the stochastic variation across multiple samples, this same loading scheme was applied to each of the remaining three wire specimens. Stress/strain results for the first cycle and the last cycle were then plotted and compared. These results can be seen in Fig. 2.28. Here it is observed that the statistical variation in the response of the specimens is not negligible, and is more noticeable than the variation in their DSC results (Fig. 2.26). While the qualitative behavior exhibited by the first cycle and the stabilized 20th cycle is similar for all samples, there is a marked variation in each response. Perhaps most notable is the inconsistency in the material hardening observed in each wire during the final loading cycle. Such a result reiterates the need for designers and analysts to always account for some statistical error when deriving model parameters from experimental data. The material properties averaged across the four samples are given in Table 2.3. Because tests were not performed at varying temperatures, no accurate phase diagram parameters were derived.

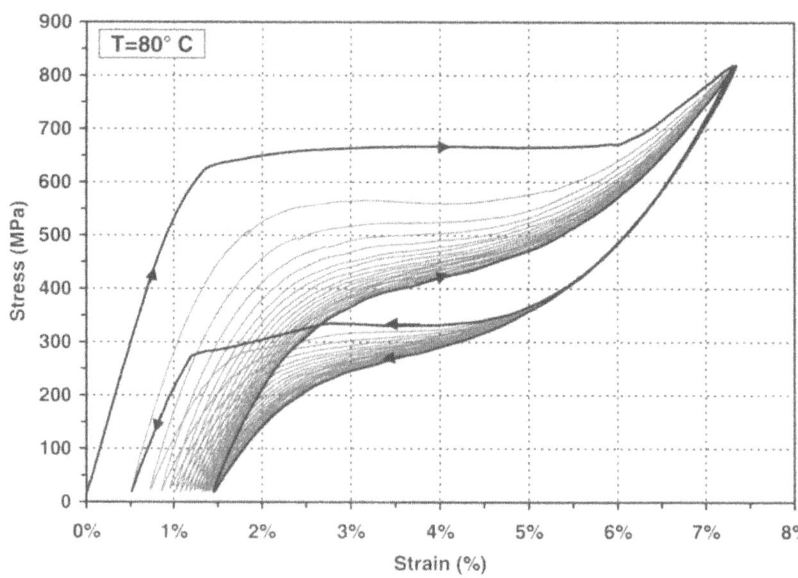

Fig. 2.27. Results of repeated pseudoelastic testing of large diameter NiTi wires; Example 2, training of material.

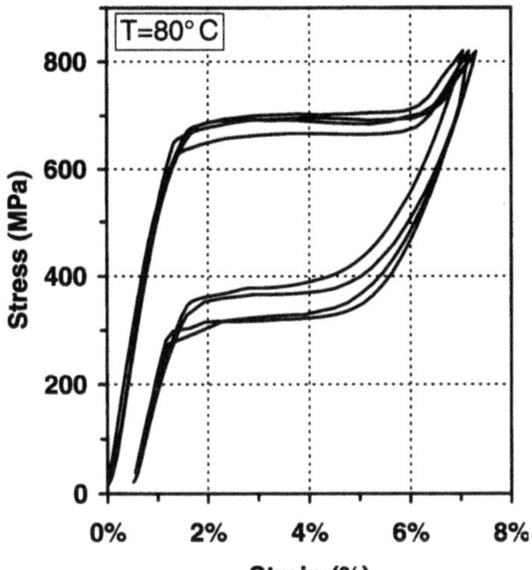

(a) Four wires, 1st cycle (as-received)

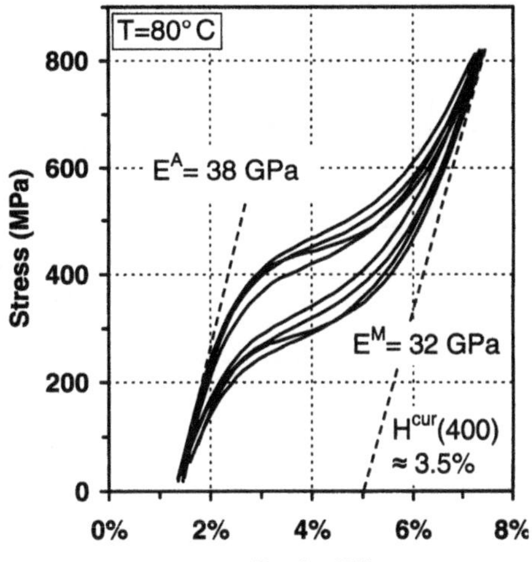

(b) Four wires, 20th cycle

Fig. 2.28. Results of pseudoelastic testing of multiple large diameter wire specimens from same bulk NiTi material; Example 2, as-received and trained material.

Table 2.3. Experimentally derived material parameters; Example 2, trained material.

Material Parameter	Value
E^A	38 GPa
E^M	32 GPa
$H^{cur}(\sigma) = H^{max}$	0.035

2.5.3 Example 3. Characterization of Ni60Ti40 (wt%) Plate Intended for Actuation Application

This example pertains to the characterization of a NiTi alloy intended for use in an actuation application. The material received for testing was in the form of plates 267 mm long, 38 mm wide, and 1.8 mm thick. The SMA components as utilized in the application were in a beam configuration, providing a bending moment to the aerostructure on which they were mounted. This implies that the most prominent stresses within the SMA component would be axial and would vary throughout the component. This experimental study provided an opportunity to showcase how simple 1-D characterization could be used to predict the performance of a 3-D SMA component experiencing complex internal stresses. Thus, careful and accurate model calibration was required. Also, the application is intended to be used repeatedly. Stable material response is then required.

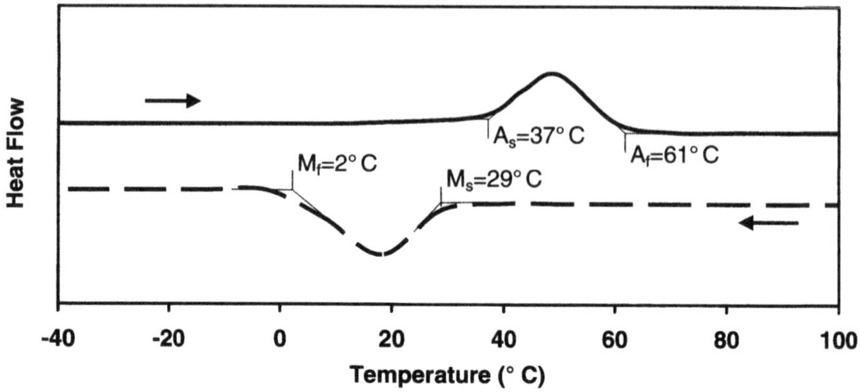

Fig. 2.29. Determination of zero-stress transformation temperatures using a DSC; Example 3, as-received material.

Here, as before, the first step was to determine the zero-stress transformation temperatures via DSC testing. For this purpose, small portions ($\sim$ 20 mg) were cut from the received plate with a low force saw. The results of DSC testing are are shown in Fig. 2.29.

The next step was to subject the material to monotonic loading. The first test was performed at $T < M_f$, and the results are given in Fig. 2.30a. As is often the case, the maximum stress was not known beforehand but was chosen during testing such that the majority of detwinning was completed, yet obvious plastic yielding did not begin to occur. For some SMAs, such as the material in this example, the detwinning start stress may be clearly observed during testing while the detwinning finish stress may be less clear (or completely obscured) due to significant material hardening and plastic strain generation. In this case, an accurate determination of the detwinning start stress is useful, and $\sigma_s \approx 140\,\text{MPa}$ was calculated using tangent lines. Upon unloading, the specimen was heated to above A_f, and the recoverable transformation strain was recorded. This material does not exhibit recoverable strains of the same magnitude as those seen in equiatomic NiTi materials (Sect. 1.9.1).

Following monotonic testing below M_f, such loading was repeated at $T > A_f$. A moderate load was applied and the elastic modulus of austenite was determined. Because of the intended use of the material in an actuation application, pseudoelasticity was not of direct interest and the specimen was not loaded sufficiently to induce this effect. The results for $T > A_f$ are shown in Fig. 2.30b.

The material specimen was then subjected to isobaric thermal cycles. Based on knowledge of the stresses to be expected in the final application, a moderate constant loading level of 150 MPa was chosen. The results of this test are shown in Fig. 2.31. Note the significant amount of plastic strain remaining at the end of the test, this indicates a certain need for material

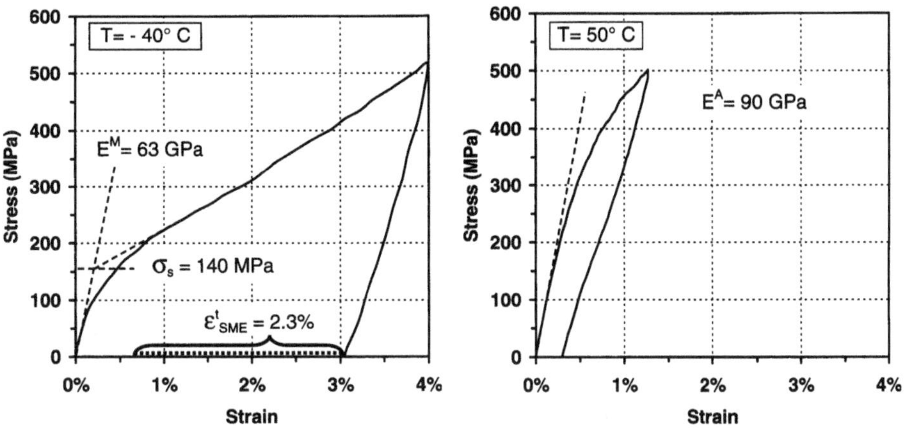

(a) $T < M_f$ and SME recovery (b) $T > A_f$

Fig. 2.30. Monotonic loading of Ni60Ti40 (wt. %) material in martensite and austenite; Example 3, as-received material.

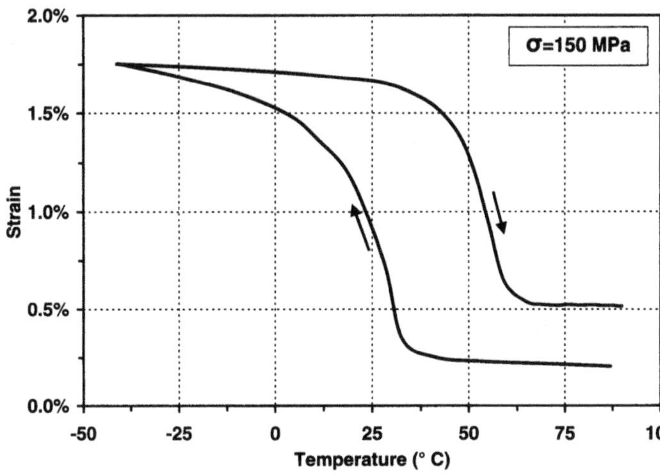

Fig. 2.31. Results of first isobaric test at 150MPa; Example 3, as-received material.

training. Plotting both the zero-stress transformation temperatures and the newfound 150 MPa isobaric transformation temperatures concurrently allows for an initial estimation of the phase diagram. This can be seen in Fig. 2.32.

With an estimate of the phase diagram constructed, the material stabilization procedure and apparatus were designed. Further review of the intended application revealed that the maximum stresses in the beam actuators would not exceed 300 MPa. Therefore, this was chosen as an appropriate stabiliza-

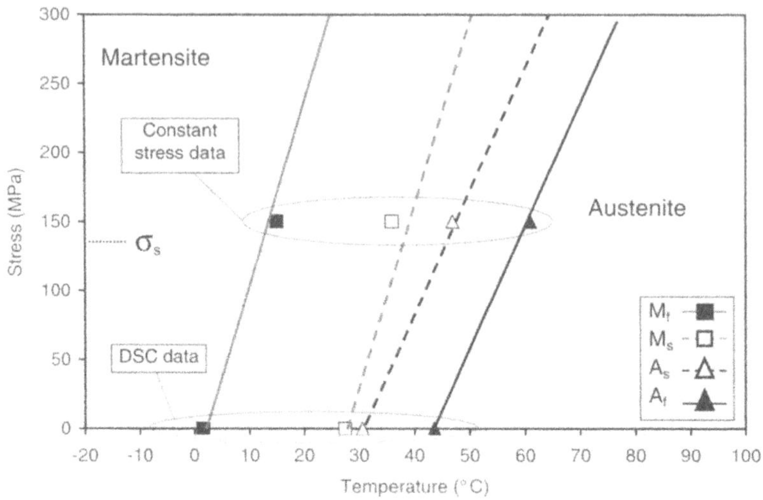

Fig. 2.32. Initial estimate of the phase diagram; Example 3, as-received material.

tion constant stress level. To complete experimental design, the approximate phase diagram was consulted, and it was determined that, for full reverse transformation into austenite at this stress level, the entirety of the specimen gauge length should be heated to $\sim$ 100 °C. A total of 100 thermal cycles at a constant stress of 300 MPa were applied while the strain response was monitored. In this case, training was performed at an accelerated rate; therefore, correct measurement of the specimen temperature was not guaranteed. However, the end goal of training is the application of multiple cycles, not the careful determination of specimen behavior. The training results are shown in Fig. 2.33a. Note that the final cycle has been darkened. For each cycle, the maximum strain represents the deformation in the martensitic state while the minimum strain represents the austenitic response. The difference can be used to derive the current maximum transformation strain per (2.3.6). Each of these three strain measures was monitored with each cycle, and their evolution is shown in Fig. 2.33b. Note how the material begins to stabilize rapidly in the first $\sim$ 20 cycles, and then more slowly after this.

After material training, the careful characterization process was repeated. As the DSC is destructive (i.e., it requires a small portion to be cut from the sample), it was not appropriate to perform this test at this time. Also, exhibition of the pseudoelastic effect is inconsequential considering the final application. Therefore, the next goal was the determination of the functional form of $H^{cur}(\sigma)$. Multiple isobaric tests were performed at various stress levels, from 300 MPa down to 90 MPa and the results are shown in Fig. 2.34. The hysteresis height $\Delta\varepsilon$ was measured for each curve, and from these measurements $H^{cur}(\sigma_i)$ is derived via (2.3.6) (where $\sigma_1 = 300$ MPa, $\sigma_2 = 250$ MPa, etc). The

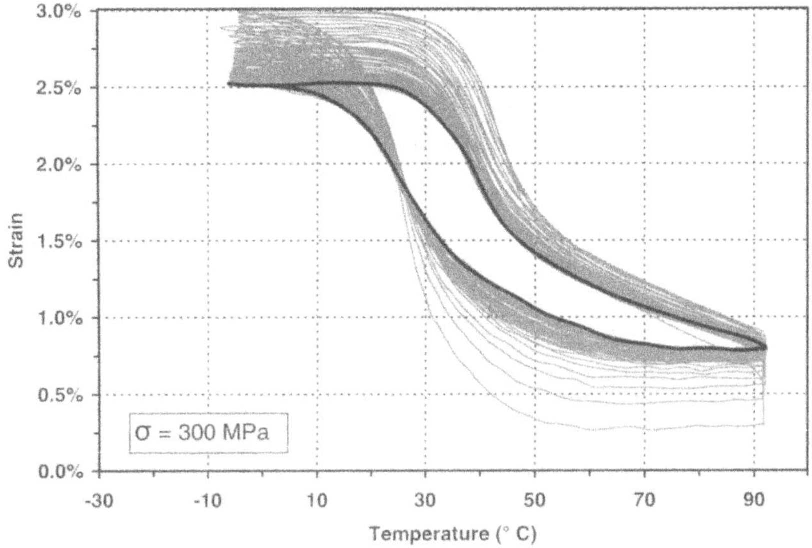

(a) Strain/temperature training cycles.

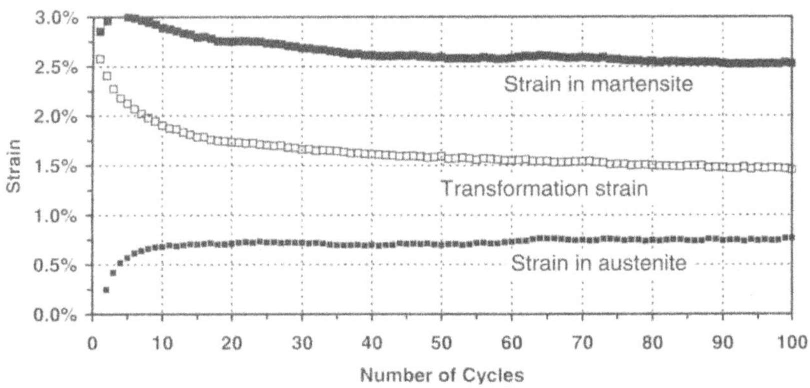

(b) Stabilization of response.

Fig. 2.33. Results of 100 applied constant stress cycles at a stress of 300 MPa; Example 3.

results for the current experimental study are shown below in Fig. 2.35 along with an analytical fit which suggests that an exponential form of $H^{cur}(\sigma)$ is appropriate.

Finally, an accurate phase diagram was determined. The transition temperatures exhibited at each constant stress applied during isobaric testing were plotted concurrently with an estimated transformation surface as shown in Fig. 2.36. The material properties determined from these experiments are

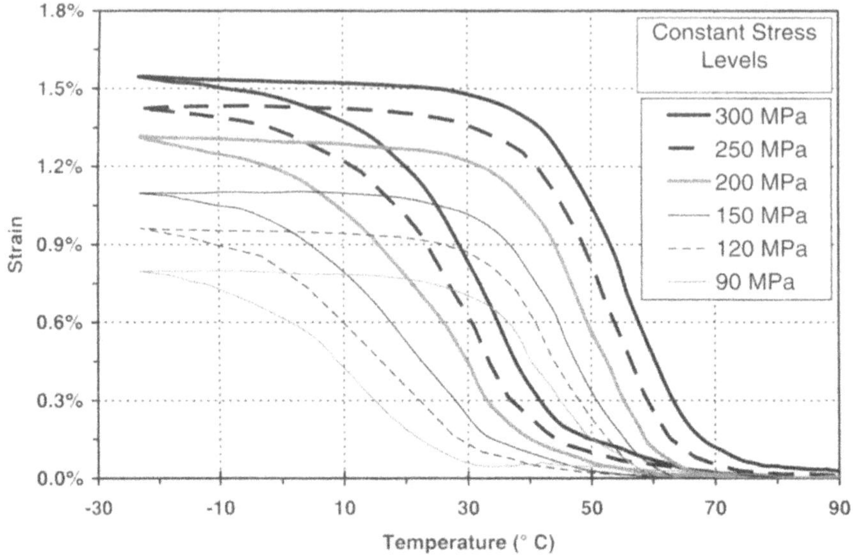

Fig. 2.34. Isobaric thermal cycling results for six applied stress levels; Example 3, trained material.

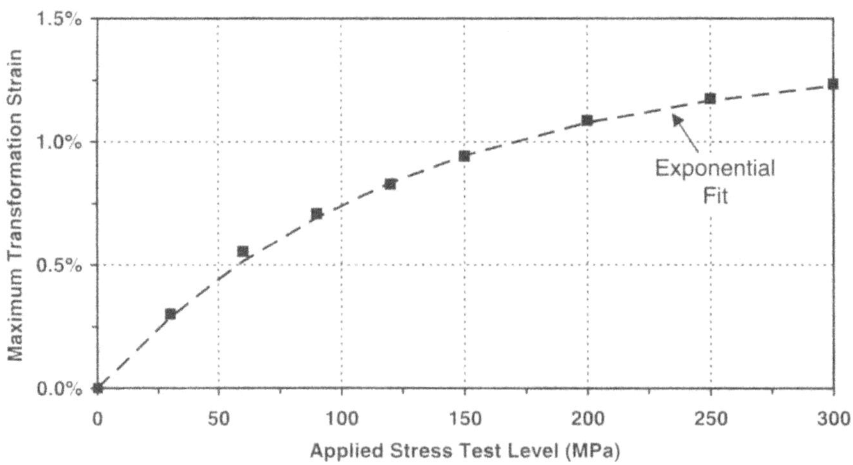

Fig. 2.35. Variation of maximum transformation strain with applied stress level; Example 3, trained material.

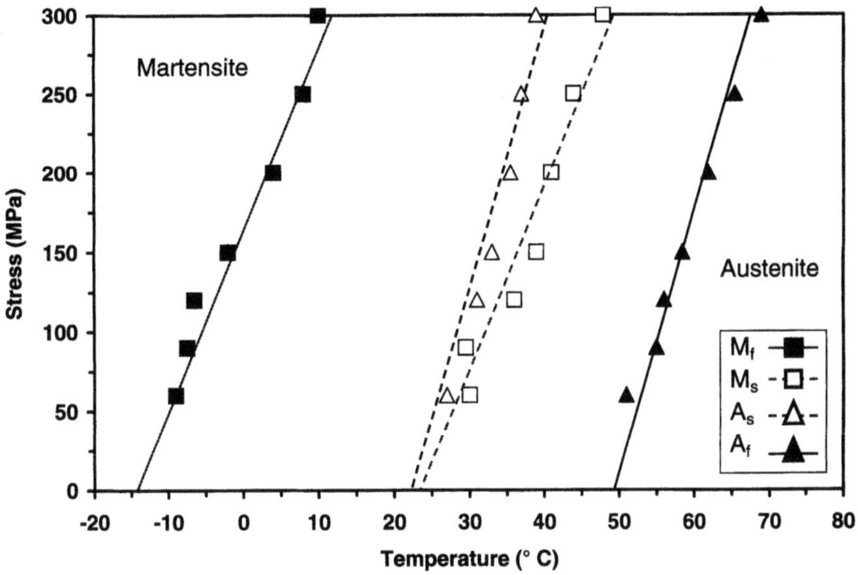

Fig. 2.36. Final experimental phase diagram derived from constant stress loading paths; Example 3, trained material.

given in Table 2.4. To graphically summarize a portion of the characterization process described in this example, Fig. 2.37 is also provided. Here is shown how bulk material (raw SMA plate) is used to construct appropriate testing specimens which are then subject to thermomechanical loading paths (cooling via liquid nitrogen spray under constant load). This subsequently leads to the generation of useful data which can be interpreted according to the methods discussed in this chapter.

Table 2.4. Experimentally derived material parameters; Example 3, trained material.

Material Parameter	Value
E^A	90 GPa
E^M	63 GPa
M_s	23 °C
M_f	−14 °C
A_s	22 °C
A_f	49 °C
C^A	16.0 MPa/ °C
C^M	11.4 MPa/ °C
$H^{cur}(\sigma)$	$= 0.0135[1 - \exp(-720\sigma/E^A)]$

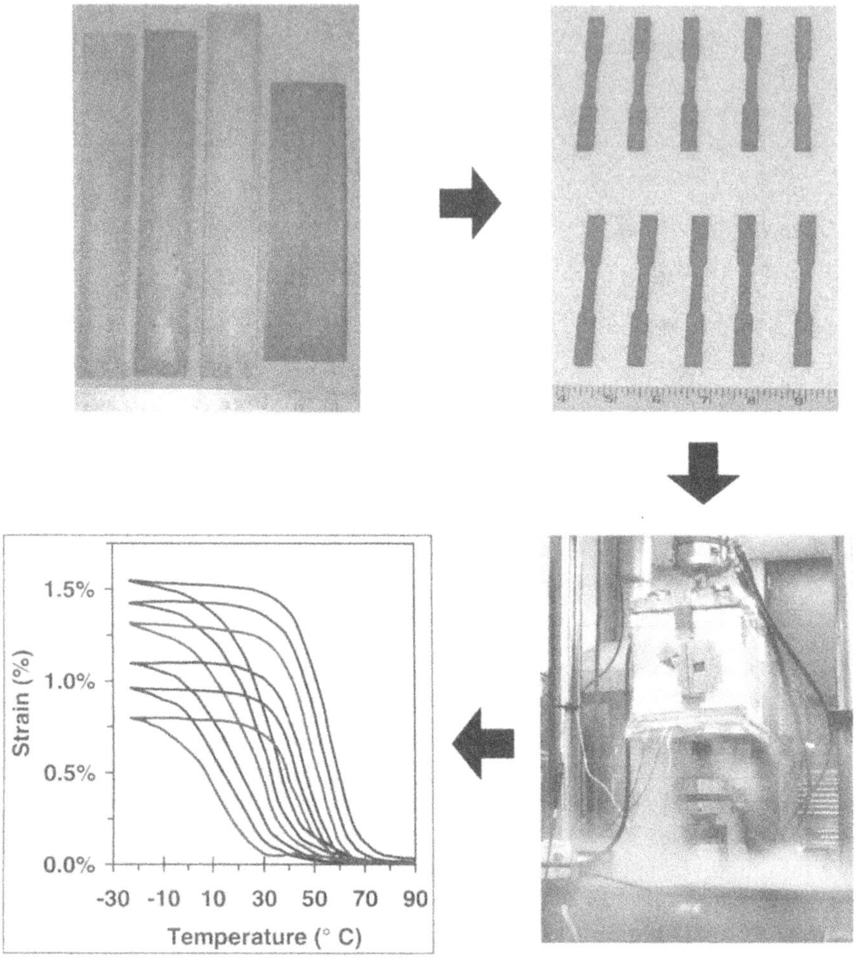

Fig. 2.37. Graphical summary of the thermomechanical characterization process as used in Example 3.

2.6 Simple SMA Application Design and Empirical 1-D Analysis

To begin the initial design of an SMA component for use in an engineering application, one must be able to choose a suitable alloy composition and then estimate the overall system response. To choose an alloy, knowledge of various required design parameters is required, and this will be discussed below. For the purposes of initial response estimation, a simple 1-D empirical model is also described in this section.

2.6.1 Application Design Considerations

The first and most fundamental step in the design of an SMA application is the determination of the operational limits. From the discussion of these first two chapters, it should be clear that important SMA behaviors are based on the current stress-temperature state and loading history. Therefore, the phase diagram represents the key design space. Using this tool, one can estimate upper and lower bounds for the operating temperature and the required stress range.

Furthermore, the limits on the maximum transformation strain for a given material should also be considered. Each of these three limit determinations are valuable in helping a designer choose a particular shape memory alloy and subsequent thermomechanical treatment for a particular set of material properties. Specifically, operation stress and temperature allow selection of the transformation temperatures, and this knowledge, combined with estimates of required actuation strain, further allow an alloy to be chosen.

Often, the most easily determined system parameter is the required operating temperature range of an active SMA component. It is estimated by considering the ambient temperature, the ability to heat the SMA element (power availability), and the ability to cool the element (via active cooling or heat conduction/convection). The operating stress range is defined by considering the SMA component loading path, especially in stress-temperature space. For example, pseudoelastic components usually undergo large variations in stress over time while some actuators may experience large temperature variations with little stress deviation.

With a stress-operating temperature envelope defined, one can now estimate the zero-stress transformation temperatures required of an SMA material for a given application. Knowledge of these temperatures is key to choosing a particular shape memory alloy. An example of such an operational envelope is given in Fig. 2.38. On a stress-temperature phase diagram, one should first identify the points $(T^{\min}, \sigma^{\min})$ and $(T^{\max}, \sigma^{\max})$. These are the estimated minimum and maximum stress and temperature states that the application is expected to apply to the installed SMA component of interest. For complete phase transformation, the maximum temperature must exceed the non-zero stress austenitic finish temperature A_f^σ as computed at maximum stress (i.e. $T^{\max} \geq A_f^\sigma,\ \sigma = \sigma^{\max}$) and likewise, during cooling $(T^{\min} \leq M_f^\sigma,\ \ \sigma = \sigma^{\min})$. Assuming a simple linear relationship between stress and temperature on the transformation lines in the phase diagram, one may then select reasonable slopes (stress influence coefficients). Values for C^A and C^M from 5 MPa/K to 10 MPa/K are generally the most reasonable. Extrapolating down to the zero-stress axis via relations (2.6.8), A_f and M_f can be found (assuming complete actuation is required).

$$A_f \leq T^{\max} - \frac{\sigma^{\max}}{C^A} \qquad M_f \geq T^{\min} - \frac{\sigma^{\min}}{C^M} \tag{2.6.8}$$

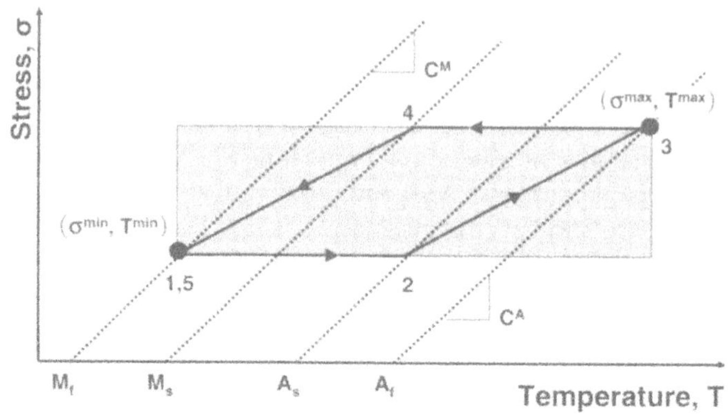

Fig. 2.38. Example of SMA application design space.

The shaded area in Fig. 2.38 designates the operating regime of the SMA actuator. A possible actuation path is shown as 1-2-3-4, point 1 being the starting point, 2 being the temperature for transformation initiation under σ^{min}, 3 the temperature for actuation completion under σ^{max}, and 4 being the temperature for the initiation of transformation back to the original configuration or state, 5.

The determination of the required actuation strain is the final step in determining a suitable alloy. This can be derived from the amount of mechanical motion required of a given actuator and is thus application-specific. For one-time use elements, one can effectively utilize actuation strains that are near the maximum attainable by a given alloy, i.e., up to 8% for NiTi (Sect. 1.9.1). However, for elements that are used repeatedly, lower transformation strains are favorable as they increase fatigue life and decrease plastic strain development. Considering the known operating temperature, stress and strain, it is now possible to choose an appropriate shape memory alloy. For many conventional applications, NiTi will suffice. It is both relatively affordable and widely available. However, for specialty applications, other alloys should be researched. A summary of approximate material parameters for various common SMAs, including NiTiPd for use at high temperatures, is given in Table 2.5. Recall that representative transformation temperatures were provided in Table 1.1.

For many applications using simple SMA components subject to uniform or near-uniform stress states (i.e., wires, torque tubes), one can also consider the *end states* of operation. These represent the device configurations or states at the end of forward and reverse transformation, or when the component has fully transformed in martensite and austenite. To do so requires only the use of (2.3.2) and (2.3.3), coupled with additional relations describing the operation of the device (e.g., external forces, biasing spring loads, etc.). These above equations require only the distinctive properties of pure austenite and pure martensite. We must develop a more complete material model to predict the

Table 2.5. Approximate material properties for various SMA material systems

Property	NiTi	NiTiCu	NiTiPd	CuAlNi
E^A	70 GPa	50 GPa	15 GPa	90 GPa
E^M	30 GPa	25 GPa	25 GPa	80 GPa
ν	0.3	0.3	0.3	0.3
α	$10 \cdot 10^{-6}/\,^{\circ}\mathrm{C}$	$10 \cdot 10^{-6}/\,^{\circ}\mathrm{C}$	$10 \cdot 10^{-6}/\,^{\circ}\mathrm{C}$	$15 \cdot 10^{-6}/\,^{\circ}\mathrm{C}$
C^A	7 MPa/ °C	10 MPa/ °C	5 MPa/ °C	–
C^M	7 MPa/ °C	10 MPa/ °C	5 MPa/ °C	–
H^{max}	6%	5%	3%	4%
ρ	6500 kg/m^3	6500 kg/m^3	8200 kg/m^3	7500 kg/m^3
σ^y	700 MPa	600 MPa	400 MPa	300 MPa

system response during phase transformation. A simple empirical 1-D example is derived next, and full 3-D modeling accounting for more complex material behaviors will be introduced throughout the remainder of this book.

2.6.2 Experimentally-Based 1-D Material Model

Based on the experimental results presented throughout this chapter and in Chapter 1, we can derive a simple empirical 1-D material model to capture the overall material behavior. Furthermore, we can use the material parameters discussed in this chapter and derived in the examples to calibrate such a model. Of course, for complex engineering applications of SMAs using 3-D components, such a model is not appropriate. However, it is useful for the design and analysis of applications based on SMA components undergoing homogeneous stress with only one component (i.e., uniaxial or shear stress). For simplicity, the model derived below does not account for partial transformation, which is the subject of Problem 2.10 at the end of this chapter. Additional empirical assumptions will be described as the model is derived.

Relations for the stable, fully transformed material state have already been given in (2.3.3) and (2.3.2), which follow easily from (2.3.1). Here we seek to account for 1-D material response during transformation as well. The model is derived based on the same "strength of materials" understanding of the constitutive behavior as was previously employed, and uses the *total martensitic volume fraction*, ξ, to track the progression of the phase transformation. Specifically, $\xi = 0$ when the material is fully austenitic, or in the parent phase, and $\xi = 1$ when the material is fully martensitic.

Recall Sect. 2.2.3 regarding $H^{cur}(\sigma)$ and H^{max}: a material transforming into pure detwinned martensite will exhibit the maximum attainable transformation strain, H^{max}. However, a material transforming into martensite of multiple variants will generate a recoverable strain that depends on stress, $H^{cur}(\sigma)$. In this model, and in all models presented in Chapters 3–6, ξ will denote the *total martensitic volume fraction*, which may include multiple variants (fully twinned, fully detwinned, or some combination). Because H^{max} is exhibited when only pure detwinned martensite is present, we can also define a relation that gives the *detwinned martensitic volume fraction* as:

$$\xi^d = \frac{H^{cur}(\sigma)}{H^{max}}\xi \tag{2.6.9}$$

Of course it is straightforward to see that we could also give the *twinned martensitic volume fraction* as:

$$\xi^t = \xi - \xi^d = \frac{H^{max} - H^{cur}(\sigma)}{H^{max}}\xi \tag{2.6.10}$$

To continue, only ξ is considered and the following three assumptions are made:

- **Assumption 1:** The elastic stiffness, unique for each phase, varies linearly with varying martensitic volume fraction. This is consistent with a *rule of mixtures* approach to accounting for austenitic and martensitic elastic stiffness. In more complex models, other relations employing the methods of micromechanics are sometimes used.

$$E = E^A + \xi(E^M - E^A) \tag{2.6.11}$$

- **Assumption 2:** The coefficient of thermal expansion (α) is a constant.
- **Assumption 3:** The 1-D transformation strain varies linearly with varying total martensitic volume fraction such that:

$$\varepsilon^t = \xi H^{cur}(\sigma) \tag{2.6.12}$$

Note that the sign of H^{cur} is dependent on the particular stress applied (i.e., $H^{cur} \geq 0$ for tensile stress, $H^{cur} \leq 0$ for compressive stress). Substituting these two relations into (2.3.1) yields:

$$\sigma = [E^A + \xi(E^M - E^A)][\varepsilon - \xi H^{cur}(\sigma) - \alpha(T - T_0)] \tag{2.6.13}$$

To further capture transformation behavior, it is necessary to use the phase diagram that captures the critical 1-D stress and temperature states at which transformation is induced and completed. From observation of the numerous phase diagram examples shown throughout the chapter, the following assumption is inferred:

- **Assumption 4:** The transformation start and finish temperatures at a given stress (e.g., M_s^σ and A_f^σ) are linearly related to the applied stress. This implies the following relations:

$$M_s^\sigma = M_s + \frac{\sigma}{C^M} \tag{2.6.14}$$

$$M_f^\sigma = M_f + \frac{\sigma}{C^M} \tag{2.6.15}$$

$$A_s^\sigma = A_s + \frac{\sigma}{C^A} \tag{2.6.16}$$

$$A_f^\sigma = A_f + \frac{\sigma}{C^A} \tag{2.6.17}$$

To complete the 1-D empirical model, the evolution of martensitic volume fraction with changing stress-temperature state must be addressed. A final assumption is then required.

- **Assumption 5:**
  - During *forward transformation*, the martensitic volume fraction, ξ, linearly increases with decreasing temperature from M_s^σ to M_f^σ. Because $M_s^\sigma - M_f^\sigma = M_s - M_f$, this can be written as:

$$\xi = \frac{M_s^\sigma - T}{M_s - M_f} \tag{2.6.18}$$

  - During *reverse transformation*, the martensitic volume fraction, ξ, linearly decreases with increasing temperature from A_s^σ to A_f^σ. Because $A_f^\sigma - A_s^\sigma = A_f - A_s$, this can be written as:

$$\xi = \frac{A_f^\sigma - T}{A_f - A_s} \tag{2.6.19}$$

This is schematically illustrated in Fig. 2.39 where forward transformation is shown in Fig. 2.39a and reverse transformation in Fig. 2.39b. Note that this assumption on the evolution of ξ is applicable regardless of the arbitrary loading path experienced by the SMA material. Whenever the stress-temperature state is within the transformation region, the current transformation temperatures at given (non-zero) stress can always be calculated via (2.6.14)–(2.6.17).

Assumption 5 combined with previous assumptions allows us to define the martensitic volume fraction for all points on the phase diagram when

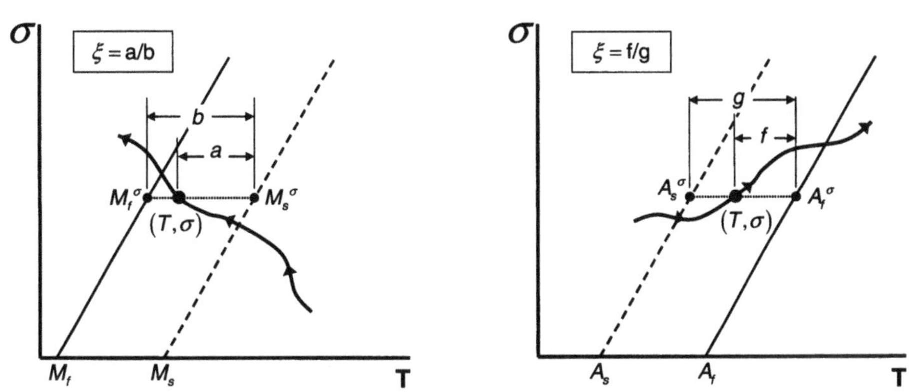

(a) Forward transformation. (b) Reverse transformation.

Fig. 2.39. Schematic illustrating the assumed transformation behavior for the evolution of martensitic volume fraction.

cooling/loading into martensite and when heating/loading into austenite. The relations for ξ are given below where (2.6.14) and (2.6.17) are substituted into (2.6.18) and (2.6.19), respectively.

For *loading/cooling into martensite*, the martensitic volume fraction is given as:

$$\xi = \begin{cases} 0, & T \geq M_s^{\sigma}, \\ \dfrac{M_s + \dfrac{\sigma}{C^M} - T}{M_s - M_f}, & M_f^{\sigma} < T < M_s^{\sigma}, \\ 1, & T \leq M_f^{\sigma}, \end{cases} \tag{2.6.20}$$

while for *loading/heating into austenite* is given as:

$$\xi = \begin{cases} 1, & T \leq A_s^{\sigma}, \\ \dfrac{A_f + \dfrac{\sigma}{C^A} - T}{A_f - A_s}, & A_s^{\sigma} < T < A_f^{\sigma}, \\ 0, & T \geq A_f^{\sigma}. \end{cases} \tag{2.6.21}$$

Using (2.6.13) combined with either (2.6.20) or (2.6.21), as appropriate, there are sufficient relations to describe the 1-D SMA behavior during both transformation ($0 < \xi < 1$) and thermoelastic loading ($\xi = 0$ or $\xi = 1$). Recall that this model in its current form does not account for partial transformation.

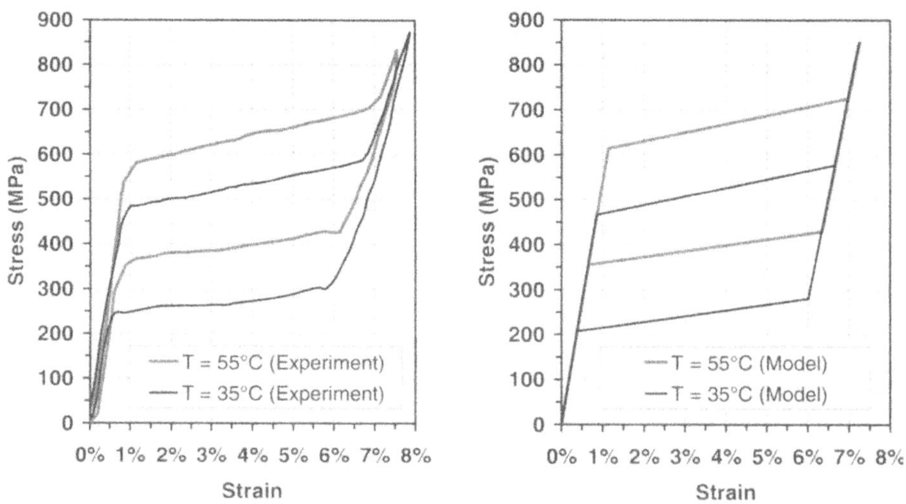

(a) Experiment (b) Empirical model

Fig. 2.40. Comparison of experimental pseudoelastic results and 1-D empirical model predictions (cf. Example 1, Sect. 2.5.1).

This model has been used to present the pseudoelastic results shown in the first example in Sect. 2.5.1. The material properties used to calibrate the model are given in Table 2.2. Here (Fig. 2.40) the isothermal responses at 35 °C and 55 °C are shown, and the model accurately captures the material behavior.

Finally, note that the 1-D relations given above can be re-written for homogeneous shear. Several engineering applications of SMAs utilize shear deformation, and many of these include torque tube components. For sufficiently thin SMA torque tubes, the variation in stress and strain through the wall thickness is small enough to be negligible. For the design of such applications, it is advantageous to utilize an analytical model such as that given above to approximate the actuation behavior. This requires that the relations above be rewritten for shear, where, for example τ and γ denote shear stress and strain, respectively, and $\gamma = 2\epsilon_{12}$. For such a model, material calibration should be formed using experimental data taken from shear loading.

2.7 Summary

Understanding the experimental characterization of shape memory alloys is important not only for those who plan to quantify material properties, but also for those analysts and designers who must form a solid understanding of SMA behavior. As with characterization of any material, known inputs are applied and exhibited outputs are monitored and evaluated. As one becomes more familiar with the uncommon properties of shape memory alloys, an intuition into the material response to a given thermomechanical load is formed. An understanding of the simple 1-D phenomenological response allows the construction of empirical material models, as exemplified above. As this understanding grows, the formulation and refinement of more advanced theoretical models that explain or even predict more complicated behavior becomes possible.

Furthermore, once a model has been proposed, numerical implementation is necessary to convert a theory into a useable tool for the design and analysis of engineering applications. The formulation and calibration of various models and their numerical implementation will be the topic of the remainder of this book.

2.8 Problems

2.1. Recall that the determination of the stress-free transformation temperatures A_s and A_f can be performed using either DSC or free bend recovery. How might the temperatures determined by these two methods differ? What would cause such a difference?

2.2. Describe two other methods by which all four stress-free transformation temperatures could be experimentally determined and which does not involve the measurement of heat flow (i.e., DSC) or deformations (i.e., bend and free recovery).

2.3. In the experimental work performed by Shaw and Kyriakides [12], a fluid bath was employed for most isothermal tests. Why was such an experimental setup used? Explain the consequences of using an ambient air environment instead.

2.4. Consider further the experimental work performed by Shaw and Kyriakides [12]. Explain the mechanisms that lead to the significant variation in "strain" as measured via crosshead displacement ($\Delta l/l_0$) vs. measurement by locally-mounted extensometers (ε) (see "Figure 12d" in the referenced work).

2.5. Consider the isothermal stress-strain experimental results shown in Fig. 2.6b. From these results, determine appropriate material parameters. Assume material isotropy.

a) Determine the elastic stiffnesses of the material.
b) Construct the phase diagram for this material configuration.
c) Determine a suitable maximum transformation strain. Does the amount of transformation strain generated during forward transformation vary significantly with increasing upper plateau stress level?

2.6. Consider the constant stress experimental results shown in Fig. 2.41. From these results, determine the material parameters assuming material isotropy.

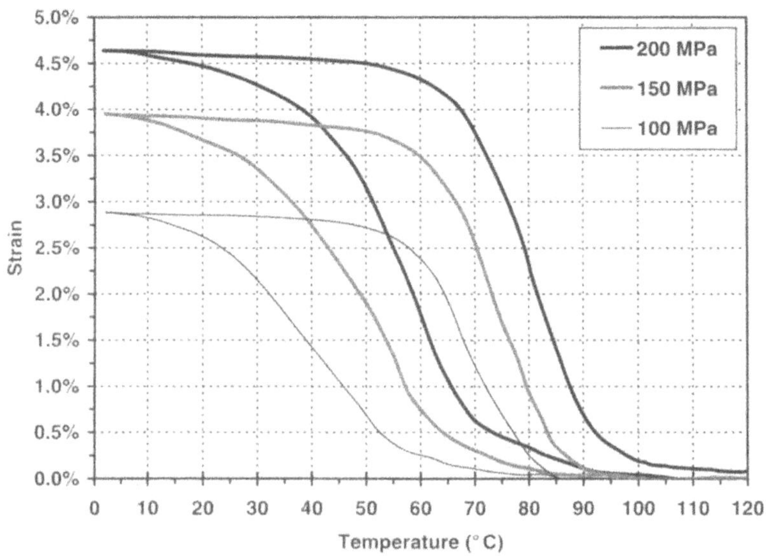

Fig. 2.41. Example of isobaric strain-temperature experimental results for near-equiatomic NiTi.

a) Use the elastic properties from Problem 2.5.
b) Construct the phase diagram for this material configuration.
c) Determine a suitable function to describe the maximum transformation strain at all effective stress levels.

2.7. Consider the experimental work performed and reported by Miyazaki et al. [7].

a) Using the data in "Fig. 1" from this reference, reconstruct and complete the phase diagram shown in "Fig. 2" (i.e., add the transformation *finish* surfaces).
b) Based on this phase diagram, approximate three distinct strain-temperature cyclic responses if this material is subjected to full thermal cycles at three distinct constant stresses.

2.8. Determine the three thermomechanical states (i.e., stress-strain-temperature states) of an SMA wire with a length of 0.5 m and a diameter of 0.25 mm if it is initially unstressed at $T = -10\,°\mathrm{C}$ (Point 1), then subjected to a constant load of 15 N at −10 °C (Point 2), then heated to 150 °C (Point 3). Sketch these points and the connecting paths on the phase diagram, on a stress-strain plot, and on a strain-temperature plot.

a) Use the material properties from Problem 2.5.
b) Use the material properties from Problem 2.6.

2.9. Determine the three equilibrium states of the SMA wire from Problem 2.8 if it is initially unstressed, then placed in opposition to an elastic spring with a spring constant of 4 N/mm with a prestress of 75 MPa at −10 °C, then heated to 150 °C. Sketch these paths on the phase diagram and on a stress-strain plot.

a) Use the material properties from Problem 2.5.
b) Use the material properties from Problem 2.6.

2.10. Repeat Problem 2.8 using the simple 1-D model derived in Sect. 2.6.2. Specifically, derive and plot the continuous analytical solution of the wire stress-strain response. (*Hint: The solution must be determined incrementally to account for the beginnings/endings of transformation.*)

2.11. Consider the case in which two SMA rods are arranged as shown in Fig. 2.42 where both rods exhibit the material properties from Problem 2.5. The cool wire is assumed to be initially fully detwinned and the initial tensile stress level in both rods is given.

a) If rod 1 is heated from T_0 to T_f, and then cooled back to T_0, determine the thermomechanical states (stress, strain, and temperature) at the end of heating and at the end of cooling. Plot these two points on an approximate quantitative phase diagram.
b) Repeat the case where the temperature of rod 2 is a constant 100 °C.

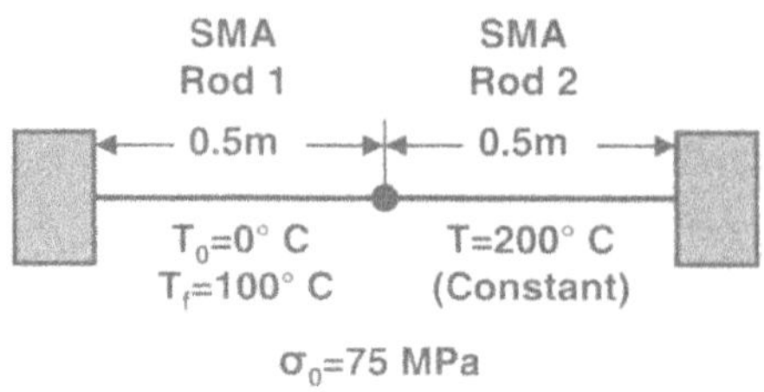

Fig. 2.42. SMA rods in an "antagonistic" configuration.

(*Hint: The solution must be determined incrementally using the model in Sect. 2.6.2.*)

2.12. For the simple 1-D model presented in Sect. 2.6.2, add the necessary relations to account for partial transformation. Model a case of isothermal loading at $T = 40\,°\mathrm{C}$ using the material parameters given in Table 2.2. Load the material such that ξ reaches a value of 0.5, and then unload. At what stress does ξ reach 0.5? At what stress state should reverse transformation begin? (*Hint: Consider reformulating 2.6.21 considering that ξ must be continuous.*)

2.13. An SMA wire has length $2L$, diameter D, and maximum transformation strain $H^{cur} = H^{max}$. It is initially in a twinned martensitic state and is installed between two rigid supports in a stress-free, straight configuration. A weight W sufficient to induce and complete detwinning is then hung on the wire. This configuration is shown in Fig. 2.43. Assuming that $E^A = E^M$, determine the initial angle θ_i just after the weight is added. Furthermore, determine the elevation, e, of the weight when the wire becomes fully austenitic upon heating. For the case of $E^A \neq E^M$, find the new elevation and compare its percent difference with respect to the previous case. For this problem you may neglect any influence of thermal expansion and the weight of the wire itself.

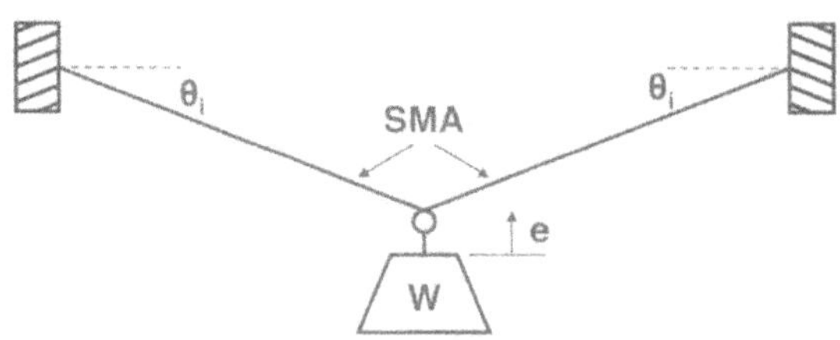

Fig. 2.43. Weight suspended from an SMA wire.

2.14. Consider a composite cylinder as shown in Fig. 2.44, formed by first expanding a martensitic SMA cylinder to an inner radius R through a detwinning process which causes a tangential transformation strain of $\varepsilon_\theta^t = 0.03$. The

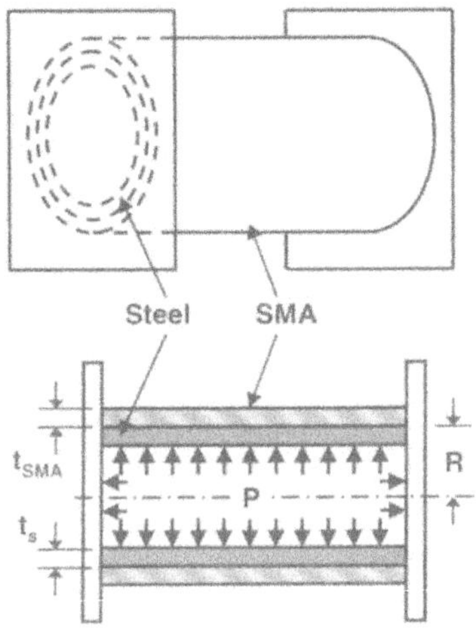

Fig. 2.44. SMA/steel composite cylinder of radius R and subjected to internal pressure P.

SMA cylinder is then brought over a steel cylinder of outer radius R. Assume that the thicknesses of both cylinders are very small compared to their radius (i.e., $t_{SMA} \ll R$ and $t_s \ll R$). The composite cylinder is pressurized by an internal pressure, P, while it is held between two rigid surfaces that allow only radial expansion or contraction but not any axial deformation. The composite cylinder is then subjected to a uniform temperature change, sufficient to cause 90% phase transformation of the SMA from detwinned martensite to austenite, while small enough to neglect any thermal expansion mismatch effects. Determine the stress in both the steel and the SMA cylinders for the case $t_s = t_{SMA} = 0.03R$. Assume also that $E_s = 200\,\text{GPa}$, $\nu_s = 0.3$; $E_{SMA} = 30\,\text{GPa}$, $\nu_{SMA} = 0.3$ (Both austenite and martensite elastic properties are assumed to be the same).

2.15. Describe the feasibility of the following SMA/SMA antagonistic actuator composed of two concentric cylinders. Assume that the inner SMA cylinder is prestrained axially until $\varepsilon^t = 0.056$ and then relaxed at $T < A_s$, at which time it is equal in length to the outer SMA cylinder. The outer cylinder, also at $T < A_s$, remains in a twinned martensite state ($\varepsilon^t = 0$). The actuator is assembled at a temperature below A_s and this assembly is shown below in Fig. 2.45 in the stress-free configuration. The actuator length $L = 150\,\text{mm}$ and the cross-sectional area is $50\,\text{mm}^2$ for both SMA cylinders. Use SMA properties from Table 2.2. Assume that when the inner SMA cylinder is heated above A_s, the outer remains at a temperature below A_s.

a) By assuming a uniaxial stress state in the cylinder assembly, find the maximum actuation force that the actuator can generate (actuator "blocking force") as the temperature of the inner cylinder is raised to A_f.

b) Determine the stroke of the actuator for actuation force of 10% of the blocking force.

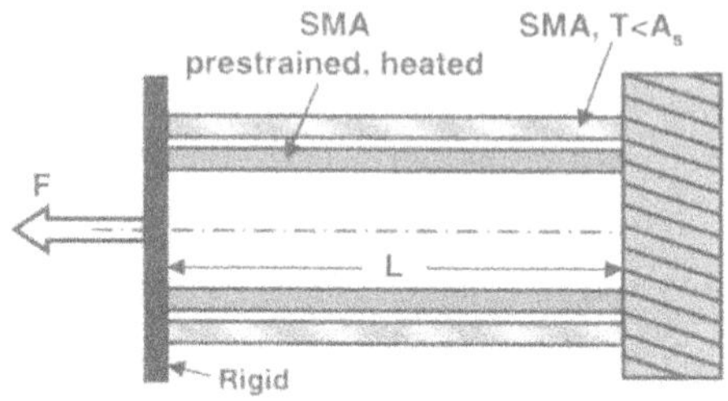

Fig. 2.45. SMA/SMA concentric cylinder antagonistic actuator.

2.16. Consider the cyclic response of an SMA-linear antagonistic actuator, shown in Fig. 2.46, under temperature variations. Use SMA properties for NiTi from Table 2.5. The length of both SMA bars is 150 mm and their cross-section is 25 mm^2. The initial state of the SMA actuators is martensitic with the left SMA prestrained to $\varepsilon^t = 0.06$. The initial assembly of the antagonistic actuation system is such that each SMA component is stress free. An SMA thermal history will be considered whereby the temperature of a component is heated from its initial temperature $T_0 < A_s$ to a maximum temperature of $T_{max} = A_s + 50\,°\text{C}$ and then cooled to $T_{min} = M_s - 50\,°\text{C}$.

a) Determine the actuator system blocking force F_{block} by finding the maximum force when point A is constrained (zero displacement) while the left SMA actuator is heated to T_{max}.
b) Consider the following heat/cool cycle: the right actuator, which is initially in a unstressed twinned martensite state, remains at T_0 while the left SMA actuator is heated to T_{max} and then cooled to T_{min}, after which the right actuator is heated to T_{max} and then cooled to T_{min}. This heating/cooling cycle is alternately applied to the right and left actuators until the system response stabilizes. Assuming $F = 0.5F_{block}$, determine the cyclic response of the actuator by plotting the displacement of point A with respect to temperature for several temperature cycles until a repeatable actuation path is reached.

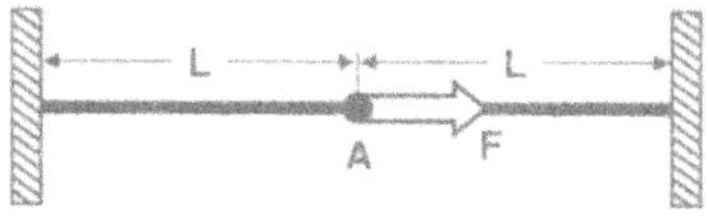

Fig. 2.46. SMA linear antagonistic actuator.

c) Discuss the advantages and disadvantages of this form of antagonistic actuator when compared to the configuration described in Prob. 2.15.

2.17. Figure 2.47 shows an SMA-powered automobile designed and constructed by freshmen engineering students at Texas A&M University. This vehicle uses a "10-cylinder" NiTi SMA engine as opposed to a conventional internal combustion engine.

The SMA wires have material properties as given for NiTi in Table 2.5. The 10 wires are each prestrained to $\varepsilon^t = 0.06$ prior to installation on the automobile and the bias springs, once attached, apply a nearly constant force resulting in a stress of 150 MPa. The diameter of the wires is 0.15 mm and their length is 80 mm. The repeating process of "firing" all 10 SMA pistons is assumed to take 10 seconds per cycle, with each wire heated independently and for an equal amount of time to $T_{max} = 80\,°C > A_f^\sigma$. Cooling is convective and $T_{min} = 30\,°C < M_f^\sigma$. The duty cycle for each wire is 10% heating, 90% cooling. The electrical resistivity of the SMA material is 80 $\mu\Omega$-cm, the latent heat of transformation is 20 J-g^{-1}, and the heat capacity is 0.32J °C^{-1}g^{-1}. The car is powered by a 9 V battery.

a) Find the horsepower of this SMA automobile engine. Estimate its efficiency as a "green" electric car.
b) What would be the required cross-sectional area of 10 SMA wires (or rods), each with a length of 1.0 m, to design a 100 HP engine? How much electric power would be required to operate such an SMA engine?

Fig. 2.47. SMA-powered 10 "cylinder" automobile (SMA wires accentuated for clarity).

2.18. Consider the SMA-actuated grippers shown in Fig. 2.48 (where all dimensions are given in mm). A single NiTiCu SMA wire with a diameter of 0.6 mm and properties as given in Table 2.5 is attached to the gripper at two locations, one on each jaw. The attachment points are shared with a bias spring that has a spring constant of $k = 0.5\,\mathrm{N/mm}$. The SMA wire also passes over three pulleys: two are symmetric and adjustable and the third is located on the gripper axis of symmetry. Before installation in a stress free state (grips closed), the SMA wire is prestrained to 5% transformation strain. The preload in the spring when the grips are closed is 25 N.

Moving outward from the centerline, the first adjustable pulley position is located 40 mm from the centerline of the grippers. Five additional possible positions are spaced 10 mm apart along a line perpendicular to the centerline (see Fig. 2.48).

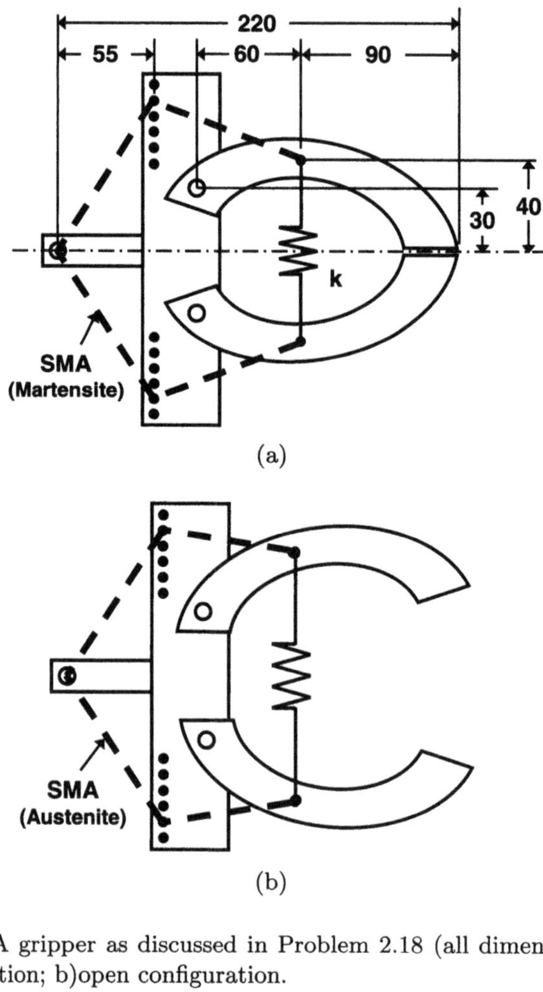

Fig. 2.48. SMA gripper as discussed in Problem 2.18 (all dimensions in mm); a) closed configuration; b)open configuration.

Determine the optimal symmetric pulley position such that the opening motion of the grips is maximized when the SMA is heated into full austenite. Do not allow the SMA stress to exceed 70% of the yield stress. Repeat for the case where the spring constant is doubled to 1.0 N/mm. Repeat once more if the preload on the spring is 50 N. How could this design be improved if there were no constraints on the location of the symmetric pulleys?

References

[1] W. Buehler, R. Wiley, The properties of TiNi and associated phases, Tech. rep., U.S. Naval Ordnance Laboratory (1961).

[2] C. M. Jackson, H. J. Wagner, R. J. Wasilewski, 55-Nitinol—The alloy with a memory: Its physical metallurgy, properties and applications, Tech. Rep. NASA SP-5110, NASA Technology Utilization Office, Washington, D.C. (1972).

[3] K. Otsuka, C. M. Wayman (Eds.), Shape Memory Materials, Cambridge University Press, Cambridge, 1999.

[4] J. Perkins, Shape Memory Effects in Alloys, Plenum Press, New York, 1975.

[5] H. Funakubo (Ed.), Shape Memory Alloys, Gordon and Breach Science Publishers, 1987.

[6] X. Ren, K. Otsuka, Universal symmetry property of point defects in crystals, Physical Review Letters 85 (5) (2000) 1016–1019.

[7] S. Miyazaki, K. Otsuka, Y. Suzuki, Transformation pseudoelasticity and deformation behavior in a Ti-50.6at%Ni alloy, Scripta Materialia 15 (1981) 287–292.

[8] Z. Bo, D. C. Lagoudas, Thermomechanical modeling of polycrystalline SMAs under cyclic loading, Part I: Theoretical Derivations, International Journal of Engineering Science 37 (1999) 1089–1140.

[9] D. C. Lagoudas, Z. Bo, Thermomechanical modeling of polycrystalline SMAs under cyclic loading, Part II: Material characterization and experimental results for a stable transformation cycle, International Journal of Engineering Science 37 (1999) 1141–1173.

[10] Z. Bo, D. C. Lagoudas, Thermomechanical modeling of polycrystalline SMAs under cyclic loading, Part III: Evolution of plastic strains and two-way shape memory effect, International Journal of Engineering Science 37 (1999) 1175–1203.

[11] Z. Bo, D. C. Lagoudas, Thermomechanical modeling of polycrystalline SMAs under cyclic loading, Part IV: Modeling of minor hysteresis loops, International Journal of Engineering Science 37 (1999) 1205–1249.

[12] J. Shaw, S. Kyriakides, Thermomechanical aspects of NiTi, Journal of the Mechanics and Physics of Solids 43 (8) (1995) 1243–1281.

[13] T. Duerig, K. Melton, D. Stockel, C. Wayman (Eds.), Engineering Aspects of Shape Memory Alloys, Butterworth-Heinemann, London, 1990.

[14] ASTM International, Standard Terminology for Nickel-Titanium Shape Memory Alloys (2005).

[15] ASTM International, Standard Specification for Wrought Nickel-Titanium Shape Memory Alloys for Medical Devices and Surgical Implants (2005).

[16] ASTM International, Standard Test Method for Transformation Temperature of Nickel-Titanium Alloys by Thermal Analysis (2005).

[17] ASTM International, Standard Test Method for Determination of Transformation Temperature of Nickel-Titanium Shape Memory Alloys by Bend and Free Recovery (2003).

[18] ASTM International, Standard Test Method for Tension Testing of Nickel-Titanium Superelastic Materials (2006).

[19] ASTM International, Standard Test Method for Tension Testing of Metallic Materials (2004).

[20] H. Sehitoglu, I. Karaman, R. Anderson, X. Zhang, K. Gall, H. J. Maier, Y. Chumlyakov, Compressive response of NiTi single crystals, Acta Materialia 48 (13) (2000) 3311–3326.

[21] H. Sehitoglu, I. Karaman, X. Y. Zhang, H. Kim, Y. I. Chumlyakov, H. J. Hans Maier, I. Kireeva, Deformation of NiTiCu single crystals in compression, Metallurgical and Material Transactions A 32 (2001) 477–489.

[22] ASTM International, Standard Test Methods of Compression Testing of Metallic Materials at Room Temperature (2000).

[23] J. T. Lim, D. L. McDowell, Mechanical behavior of a Ni-Ti shape memory alloy under axial-torsional proportional and nonproportional loading, Journal of Engineering Materials and Technology 121 (1999) 9–18.

[24] A. Keefe, G. Carman, Thermo-mechanical characterization of shape memory alloy torque tube actuators, Smart Materials and Structures 9 (2000) 665–672.

[25] K. Otsuka, X. Ren, Physical metallurgy of Ti–Ni-based shape memory alloys, Progress in Materials Science 50 (2005) 511–678.

[26] E. Patoor, D. C. Lagoudas, P. B. Entchev, L. C. Brinson, X. Gao, Shape memory alloys, Part I: General properties and modeling of single crystals, Mechanics of Materials 38 (5–6) (2006) 391–429.

[27] D. C. Lagoudas, P. B. Entchev, P. Popov, E. Patoor, L. C. Brinson, X. Gao, Shape memory alloys, Part II: Modeling of polycrystals, Mechanics of Materials 38 (5–6) (2006) 430–462.

[28] B. Chang, J. Shaw, M. Iadicola, Thermodynamics of shape memory alloy wire: Modeling, experiments, and application, Continuum Mechanics and Thermodynamics 18 (1–2) (2006) 83–118.

[29] C. Lexcellent, J. Rejzner, Modeling of the strain rate effect, creep, and relaxation of a Ni-Ti shape memory alloy under tension (compression)-

torsional proportional loading in the pseudoelastic range, Smart Materials and Structures 9 (2000) 613–621.

[30] Y. Liu, Y. Li, K. Ramesh, Rate dependence of deformation mechanisms in a shape memory alloy, Philosophical Magazine A 82 (12) (2002) 2461–2473.

[31] P. Popov, K. Ravi-Chandar, D. Lagoudas, Dynamic loading of polycrystalline shape memory alloy rods, Mechanics of Materials 35 (7) (2003) 689–716.

[32] J. Nemat-Nasser, W. Choi, G. Guo, J. Isaacs, Very high strain-rate response of a NiTi shape-memory alloy, Mechanics of Materials 37 (2–3) (2005) 287–298.

[33] J. Escobar, R. Clifton, On pressure-shear plate impact for studying the kinetics of stress-induced phase transformations, Material Science & Engineering A 170 (1993) 125–142.

[34] P. Feng, Q. Sun, Experimental investigation on macroscopic domain formation and evolution in polycrystalline NiTi microtubing under mechanical force, Journal of the Mechanics and Physics of Solids 54 (8) (2006) 1568–1603.

[35] D. A. Miller, D. C. Lagoudas, Thermo-mechanical characterization of NiTiCu and NiTi SMA actuators: Influence of plastic strains, Smart Materials and Structures 9 (5) (2000) 640–652.

3

Thermomechanical Constitutive Modeling of SMAs

L. G. MACHADO AND D. C. LAGOUDAS

With the fundamental concepts of SMA behavior introduced and the experimental results of this behavior illustrated, one can now begin to investigate methods by which SMA behavior might be accurately predicted. This chapter introduces the basic ideas and concepts behind the thermomechanical constitutive modeling of polycrystalline SMAs. From a review of first principles through the consideration of key assumptions, a procedure for obtaining constitutive equations for SMAs using thermodynamic potentials is discussed. The reader will be guided through examples which illustrate how such a model can be simplified as needed. Specific examples involving different thermomechanical loading paths are also presented. Finally, this chapter provides an overview of different constitutive modeling methodologies currently used by SMA researchers throughout the field.

3.1 Introduction

The fundamentals of shape memory alloy behavior and methods for obtaining material properties have been introduced in the first two chapters. This chapter introduces the basic elements underlying the modeling of SMA thermomechanical behavior.

We have seen in Chapter 1 that SMAs are metallic alloys that have the ability to recover from significant deformation, thereby regaining a previous shape, when subjected to specific thermomechanical loads. In addition, we have seen that the martensitic phase transformations occurring in SMAs are diffusionless, solid-solid phase transformations, where the parent (austenite) and product (martensite) phases can coexist during the phase transformation, which is largely influenced by the state of stress and temperature.

Chapter 2 presented concepts related to experimental tests that are useful in characterizing shape memory alloys. In the current chapter, we will discuss the thermomechanical modeling of polycrystalline shape memory alloys. A brief review of the conservation laws and continuum mechanics is also presented in the beginning of this chapter. A methodology for obtaining the thermomechanical constitutive equations for SMAs by enforcing the conservation

D.C. Lagoudas (ed.), *Shape Memory Alloys*, DOI: 10.1007/978-0-387-47685-8_3,

laws and basic principles of continuum thermodynamics in the manner first exemplified by Coleman, Noll, and Gurtin [1, 2] will then be discussed in detail.

3.2 Brief Review of Continuum Mechanics

The three basic components of continuum mechanics are kinematics, conservation (balance) laws and constitutive equations. The kinematics component describes the geometry of motion and deformation of a continuum body, without considering the cause of motion or deformation. The conservation laws express how external effects influence the motion of a continuum body. Finally, constitutive equations mathematically describe the main characteristics of material behavior that can only be understood and/or predicted through an understanding of experimental observations, such as described in Chapter 2 for shape memory alloys. In the coming sections a brief review of continuum mechanics is given, following standard texts such as Truesdell and Noll [3], Malvern [4] and Gurtin [5] in addition to relatively more recent texts such as Lai et al. [6], Slaughter [7], and Batra [8].

3.2.1 Kinematics of SMAs

Let us assume that an SMA material body in its deformed configuration occupies a region Ω, at time t, with boundary surface $\partial\Omega$. The position of an SMA material point in the deformed configuration in relation to the position of the same material point in the reference configuration is given by the displacement vector field, $\mathbf{u}(\mathbf{X},t)$, where $\mathbf{X}$ is the reference position of the material point. One can define a measure of deformation of neighboring material points in terms of the *deformation gradient tensor*, $\mathbf{F}$, given by

$$\mathbf{F} = \boldsymbol{\nabla}\mathbf{u} + \mathbf{1} \tag{3.2.1}$$

where $\mathbf{1}$ is the second-order identity tensor and $\boldsymbol{\nabla}\mathbf{u}$ is the *displacement gradient.* The symmetric second-order *Green-Lagrange strain tensor*, which is invariant under rigid body rotation and translation in the deformed configuration, can be defined in terms of the deformation gradient as

$$\mathbf{E} = \frac{1}{2}\left[\mathbf{F}^T\mathbf{F} - \mathbf{1}\right]. \tag{3.2.2}$$

We can rewrite the expression of the Green-Lagrange strain tensor with respect to $\boldsymbol{\nabla}\mathbf{u}$ as follows:

$$\mathbf{E} = \frac{1}{2}\left[(\boldsymbol{\nabla}\mathbf{u}) + (\boldsymbol{\nabla}\mathbf{u})^T + \left((\boldsymbol{\nabla}\mathbf{u})^T(\boldsymbol{\nabla}\mathbf{u})\right)\right]. \tag{3.2.3}$$

The notion of choosing a reference configuration for SMAs, with respect to which deformations can be described under applied loads, is more complicated than for other conventional materials. This is because there are two natural reference configurations to choose from, i.e., the stress-free configurations of the austenitic and martensitic phases. In this work, we will select the austenitic parent phase as the reference configuration and the transition from austenite to martensite will be accounted for by an internal state variable to be introduced in Sect. 3.3. When the deformation gradient tensor, $\mathbf{F}$, approaches the identity tensor, $\mathbf{1}$, or equivalently, for infinitesimal displacement gradients, the quadratic term in (3.2.3) can be neglected. As a result, the difference between the reference and the deformed configuration becomes negligible. This assumption, which is realistic for most applications of polycrystalline SMAs, leads to the *infinitesimal strain tensor* of the following form:

$$\boldsymbol{\varepsilon} = \frac{1}{2}\left[(\boldsymbol{\nabla}\mathbf{u}) + (\boldsymbol{\nabla}\mathbf{u})^T\right]. \tag{3.2.4}$$

The displacement gradient used in the infinitesimal strain above can be thought of as defined over the deformed configuration of the body, since the difference between the two configurations is a higher order effect. The infinitesimal strain tensor, $\boldsymbol{\varepsilon}$, will be used in this chapter and the remainder of the book. Appendix A gives a construction of thermomechanical constitutive equations for SMAs where the full Green-Lagrange strain tensor is used.

3.2.2 Conservation (Balance) Laws

The basic conservation laws of continuum mechanics are:

1. Conservation of mass
2. Conservation of linear momentum
3. Conservation of angular momentum
4. Conservation of energy

In continuum thermodynamics , the conservation of energy is also called the first law of thermodynamics, while the second law corresponds to the entropy inequality principle.

Conservation of Mass

The law of conservation of mass states that the total mass of a continuum body cannot change with time or deformation. Considering the mass of an SMA body to be related to the density, ρ, by

$$M = \int_{\Omega} \rho \mathrm{dV}, \tag{3.2.5}$$

the law of conservation of mass can be written as

$$\frac{DM}{Dt} = \frac{D}{Dt}\left(\int_{\Omega} \rho \mathrm{dV}\right) = 0, \tag{3.2.6}$$

where $\frac{D}{Dt}$ is the material time derivative. Following standard derivations in continuum mechanics, one can obtain the local form of the conservation of mass as

$$\frac{\partial \rho}{\partial t} + \operatorname{div}(\rho \mathbf{v}) = 0, \tag{3.2.7}$$

where $\mathbf{v}$ is the velocity of a material point and "div" denotes the divergence operator acting on a vector field in the deformed configuration.

Conservation of Linear Momentum

The law of conservation of linear momentum states that the rate of change of linear momentum of a continuum body is equal to the total sum of surface and body forces applied to it. The integral form of the conservation of linear momentum is given by

$$\frac{D}{Dt}\left(\int_{\Omega} \rho \mathbf{v} \mathrm{dV}\right) = \int_{\partial\Omega} \mathbf{t} \mathrm{dS} + \int_{\Omega} \mathbf{b} \mathrm{dV}, \tag{3.2.8}$$

where $\mathbf{t}$ is the surface traction vector, and $\mathbf{b}$ is the body force vector. The local form of the conservation of linear momentum is given by

$$\operatorname{div}(\boldsymbol{\sigma}) + \mathbf{b} = \rho \dot{\mathbf{v}}, \tag{3.2.9}$$

where $\boldsymbol{\sigma}$ is the Cauchy stress tensor and "div" here denotes the divergence operator acting on the stress tensor. The accelaration of the material point is $\dot{\mathbf{v}}$ where $\dot{()}$ indicates the material time derivative operation.

Conservation of Angular Momentum

The law of conservation of angular momentum states that the rate of change of angular momentum of a continuum body is equal to the sum of the moments applied by the surface and body forces in addition to distributed body couples. In the absence of distributed body couples, the integral form of the conservation of angular momentum is given by

$$\frac{D}{Dt}\left(\int_{\Omega} \mathbf{r} \times \rho \mathbf{v}\right) \mathrm{dV} = \int_{\partial\Omega} \mathbf{r} \times \mathbf{t} \mathrm{dS} + \int_{\Omega} \mathbf{r} \times \mathbf{b} \mathrm{dV}, \tag{3.2.10}$$

where $\mathbf{r}$ is the position vector of a material point and $\times$ denotes the cross product between two vectors.

The local form of the conservation of angular momentum states that the Cauchy stress tensor is symmetric, i.e.,

$$\boldsymbol{\sigma} = \boldsymbol{\sigma}^T. \tag{3.2.11}$$

Conservation of Energy - First Law of Thermodynamics

The law of conservation of energy states that the time rate of change of the total energy (kinetic plus internal energy) of a continuum body is equal to the rate at which external mechanical work is done to that body by surface tractions and body forces plus the rate at which thermal energy is added by heat flux, $\mathbf{q}$ and heat sources, r. The integral form of the conservation of energy is given by the following equation:

$$\begin{aligned} \frac{D}{Dt}\left(\int_{\Omega} \frac{1}{2}\rho \mathbf{v}\cdot\mathbf{v}\mathrm{dV} + \int_{\Omega}\rho u \mathrm{dV}\right) = \int_{\partial\Omega} \mathbf{t}\cdot\mathbf{v}\mathrm{dS} + \int_{\Omega}\mathbf{b}\cdot\mathbf{v}\mathrm{dV} + \\ \int_{\partial\Omega} -\mathbf{q}\cdot\mathbf{n}\mathrm{dS} + \int_{\Omega}\rho r \mathrm{dV}, \end{aligned} \tag{3.2.12}$$

where u is the *specific internal energy* (per unit mass). The local form of conservation of energy is given by

$$\rho\dot{u} = \boldsymbol{\sigma} : \dot{\boldsymbol{\varepsilon}} - \operatorname{div}(\mathbf{q}) + \rho r, \tag{3.2.13}$$

where the sign (:) indicates double dot product operation between two tensors and $\dot{u}$ and $\dot{\boldsymbol{\varepsilon}}$ are the material time derivatives of the specific internal energy and infinitesimal strain tensor, respectively.

Entropy Inequality Principle - Second Law of Thermodynamics

The entropy inequality principle expresses the second law of thermodynamics and it states that the internal entropy production is always greater than or equal to zero. The second law of thermodynamics can mathematically be expressed by the Clausius-Duhem inequality as

$$\frac{D}{Dt}\left(\int_{\Omega}\rho s \mathrm{dV}\right) + \int_{\partial\Omega}\frac{\mathbf{q}}{T}\cdot\mathbf{n}\mathrm{dS} - \int_{\Omega}\frac{\rho r}{T}\mathrm{dV} \geq 0 \tag{3.2.14}$$

where s is the *specific entropy* (per unit mass). The local form of the Clausius-Duhem inequality is given by

$$\rho\dot{s} + \frac{1}{T}\operatorname{div}(\mathbf{q}) - \frac{1}{T^2}\mathbf{q}\cdot\boldsymbol{\nabla}T - \frac{\rho r}{T} \geq 0. \tag{3.2.15}$$

Based on experimental observations that heat only flows spontaneously from a hotter material point to a colder one, we can assume that the term

$-\frac{1}{T^2}(\mathbf{q} \cdot \nabla T)$ is always greater than or equal to zero. The strong form of the second law then reduces to

$$\rho\dot{s} + \frac{1}{T}\text{div}(\mathbf{q}) - \frac{\rho r}{T} \geq 0, \tag{3.2.16}$$

which is also known as Clausius-Planck inequality.

Need for Constitutive Equations

So far the conservation laws of mass, linear momentum, and angular momentum, as well as the first and second laws of thermodynamics have been presented. At this point it is necessary to count the number of unknown field variables and available equations to verify that we have a well-posed system with the same number of equations and unknowns. Thus, starting at the stress and strain tensors, we have six components from the symmetric stress tensor, $\boldsymbol{\sigma}$, and six from the symmetric infinitesimal strain tensor, $\boldsymbol{\varepsilon}$. We also have three unknowns from the components of the displacement vector, $\mathbf{u}$, three from the heat flux vector, $\mathbf{q}$, and three additional unknowns from temperature, density and internal energy, which are scalar functions. Therefore, we have a total of 21 unknown field variables that need to be determined throughout Ω as functions of $\mathbf{x}$ and t, where $\mathbf{x}$ denotes the position of material points in the deformed configuration.

A count of the available field equations totals 11, as shown in Table 3.1. We have one equation from the conservation of mass, (3.2.7), three equations from the conservation of linear momentum, (3.2.9) and one equation from the conservation of energy, (3.2.13). In addition, we have six equations from kinematics (3.2.4). Therefore, we have a total of 21 unknowns, but only 11 equations, as summarized in Table 3.1. Note that three equations from the conservation of angular momentum, (3.2.11), have already been accounted for by the symmetry of the stress tensor. Consequently, we require 10 more equations to construct a system with unique solutions. These can be found by introducing appropriate constitutive equations. This topic will be discussed in the next section, emphasizing materials with evolving microstructure.

3.2.3 Constitutive Equations in the Presence of Internal State Variables

Constitutive equations are mathematical models intended to describe the principal features of a material behavior in an idealized form. SMAs are materials that undergo a phase transformation and therefore are characterized by a sequence of thermodynamic states that can be described by the introduction of additional internal state variables such as the martensitic volume fraction as discussed in Chapter 2, Sect. 2.6.2. In this section we present a procedure for obtaining constitutive equations in the presence of internal state variables.

Table 3.1. Unknown field variables and available field equations in continuum mechanics.

Field variables			Field equations	
Name	Symbol	Number of Unknowns	Name	Number of Equations
Displacement (vector)	$\mathbf{u}$	3	Conservation of Linear Momentum	3 (3.2.9)
Strain (second-order symmetric tensor)	$\boldsymbol{\varepsilon}$	6	Kinematics	6 (3.2.4)
Stress (second-order symmetric tensor by 3.2.11)	$\boldsymbol{\sigma}$	6		
Density (scalar)	ρ	1	Conservation of Mass	1 (3.2.7)
Internal Energy (scalar)	u	1		
Temperature (scalar)	T	1	Conservation of Energy	1 (3.2.13)
Heat Flux (vector)	$\mathbf{q}$	3		
Number of Unknowns:		21	Number of Equations	11

Thermodynamic state variables are those that represent all quantities that characterize a material body at a certain state (Coleman and Noll [1], Coleman and Gurtin [2]). If these variables can be observed, they are called (*external*) *state variables*, otherwise they are called *internal state variables*. The thermodynamic state of an SMA can be fully determined by a combination of external and internal state variables.

A *thermodynamic potential* is a function that characterizes a certain thermodynamic state of a material and it depends on state variables, both external and internal (Holzapfel [9]). Four thermodynamic potentials are commonly defined, according to a certain choice of the independent state variables. These four thermodynamic potentials are the *specific internal energy*, u, the *specific Helmholtz free energy*, ψ, the *specific enthalpy*, h, and the *specific Gibbs free energy*, G, as specific quantities all defined per unit mass. The internal energy, u, can be defined as a measure of kinetic and potential energy of the material points within the material system. The Helmholtz free energy, ψ, is defined to be the portion of the internal energy available for doing work at constant temperature, whereas enthalpy is the portion of internal energy that can be released as heat at constant applied stress. The Gibbs free energy, G, is finally the portion of enthalpy available for doing work at constant temperature.[1]

The first natural choice for a thermodynamic potential when deriving constitutive equations can be the internal energy. However, the internal energy

[1] All specific quantities defined per unit mass, such as internal energy, enthalpy and Helmholtz free energy, are represented by lower case letters. The only exception is the Gibbs free energy that is represented by the capital letter G, following standard notation in the literature of SMAs.

is not often used because it depends on the entropy which is not an easily measured quantity. For the same reason, enthalpy is not commonly used. The two remaining choices for thermodynamic potentials are the Helmholtz and the Gibbs free energies. The choice between Gibbs and Helmholtz free energies is based on the state variable one can control during experimental measurement, i.e., stress or strain. Table 3.2 presents the four thermodynamic potentials and their relation to the specific internal energy, obtained through a Legendre transformation. In the list of internal variables, a set of internal state variables $\boldsymbol{\zeta}$ is included.

Constitutive Assumptions for Materials with Internal State Variables

The first step in developing constitutive equations for a specific material is to properly select the independent and dependent state variables that will characterize the behavior of the material. At first, we discuss how to obtain constitutive equations by utilizing the Helmholtz free energy as the thermodynamic potential. Later, an example of how to obtain constitutive equations for a thermoelastic material using the Gibbs free energy as the thermodynamic potential is discussed. The derivation of constitutive equations for a polycrystalline SMA using the Gibbs free energy is discussed in the next section.

A simple material is one for which the stress at each material point can be determined from a function of the local configurational history of the point (Coleman and Noll [10], Truesdell and Noll [3]). Later, Coleman [11] included the history of the temperature and temperature gradient and added the heat flux, internal energy and entropy to the dependent variables. The constitutive behavior of a material point for a special class of simple materials is characterized by four response functions, ψ, $\boldsymbol{\sigma}$, s and $\mathbf{q}$, which depend on $\boldsymbol{\varepsilon}$, T, $\mathbf{g} = \nabla T$, and a set of internal state variables $\boldsymbol{\zeta}$ to account for the loading path history dependence. The response functions therefore can be written as

Table 3.2. Thermodynamic potentials and their relationships.

Specific Thermodynamic Potentials	Symbol	Relation to internal energy u (Legendre Transformation)	Independent Variables
Internal energy	u	u	s, $\boldsymbol{\varepsilon}$, $\boldsymbol{\zeta}$
Enthalpy	h	$h = u - \frac{1}{\rho}\boldsymbol{\sigma} : \boldsymbol{\varepsilon}$	s, $\boldsymbol{\sigma}$, $\boldsymbol{\zeta}$
Helmholtz free energy	ψ	$\psi = u - sT$	T, $\boldsymbol{\varepsilon}$, $\boldsymbol{\zeta}$
Gibbs free energy	G	$G = u - \frac{1}{\rho}\boldsymbol{\sigma} : \boldsymbol{\varepsilon} - sT$	T, $\boldsymbol{\sigma}$, $\boldsymbol{\zeta}$

$$\boldsymbol{\sigma} = \boldsymbol{\sigma}(\boldsymbol{\varepsilon}, T, \mathbf{g}, \boldsymbol{\zeta}); \tag{3.2.17}$$

$$\mathbf{q} = \mathbf{q}(\boldsymbol{\varepsilon}, T, \mathbf{g}, \boldsymbol{\zeta}); \tag{3.2.18}$$

$$s = s(\boldsymbol{\varepsilon}, T, \mathbf{g}, \boldsymbol{\zeta}); \tag{3.2.19}$$

$$\psi = \psi(\boldsymbol{\varepsilon}, T, \mathbf{g}, \boldsymbol{\zeta}). \tag{3.2.20}$$

Note that in the above response functions it is assumed that there is dependence on the gradient of T, namely $\mathbf{g}$, only. There is no dependence on the gradient of $\boldsymbol{\varepsilon}$ and $\boldsymbol{\zeta}$. It should also be noted that the above response functions are the simplification of the corresponding functions when the Green-Lagrange strain tensor, $\mathbf{E}$, is reduced to the infinitesimal strain tensor, $\boldsymbol{\varepsilon}$, and $\mathbf{g}$ and $\boldsymbol{\zeta}$ are assumed to have the same evaluation in both reference and deformed configurations. Moreover, the assumptions in (3.2.17-3.2.20) are based on Truesdell's *principle of equipresence*, which states that a quantity presented as independent variable in one constitutive equation should be also present in all, unless its presence contradicts some law of physics or the assumed symmetry of the material.

The key point in the process of obtaining constitutive equations is to consider that every admissible thermodynamic process in the body must obey the entropy inequality at each time t and for all material points in the body. Also it is assumed that, for a fixed material point at a given time t, the variables $\boldsymbol{\varepsilon}$, $\dot{\boldsymbol{\varepsilon}}$, T, $\dot{T}$, $\mathbf{g}$, $\dot{\mathbf{g}}$, $\boldsymbol{\zeta}$ and $\dot{\boldsymbol{\zeta}}$ can all be varied independently, and $\dot{T}$, $\dot{\mathbf{g}}$ and $\dot{\boldsymbol{\zeta}}$ are not arguments in the response functions. Substituting the first law of thermodynamics (3.2.13), and the time derivative of the Legendre transformation as expressed in terms of the Helmholtz free energy given by Table 3.2, into the Clausius-Planck inequality, (3.2.16), we obtain the following form of the second law:

$$\boldsymbol{\sigma} : \dot{\boldsymbol{\varepsilon}} - \rho\dot{\psi} - \rho s\dot{T} \geq 0. \tag{3.2.21}$$

Assuming that ψ is a continuous function, we can use the chain rule of differentiation to obtain the time derivative of the Helmholtz free energy as

$$\dot{\psi} = \frac{\partial\psi}{\partial\boldsymbol{\varepsilon}} : \dot{\boldsymbol{\varepsilon}} + \frac{\partial\psi}{\partial T}\dot{T} + \frac{\partial\psi}{\partial\mathbf{g}} \cdot \dot{\mathbf{g}} + \frac{\partial\psi}{\partial\boldsymbol{\zeta}} \cdot \dot{\boldsymbol{\zeta}}. \tag{3.2.22}$$

After substituting (3.2.22) into (3.2.21) we obtain

$$\boldsymbol{\sigma} : \dot{\boldsymbol{\varepsilon}} - \rho\left[\frac{\partial\psi}{\partial\boldsymbol{\varepsilon}} : \dot{\boldsymbol{\varepsilon}} + \frac{\partial\psi}{\partial T}\dot{T} + \frac{\partial\psi}{\partial\mathbf{g}} \cdot \dot{\mathbf{g}} + \frac{\partial\psi}{\partial\boldsymbol{\zeta}} \cdot \dot{\boldsymbol{\zeta}}\right] - \rho s\dot{T} \geq 0. \tag{3.2.23}$$

We can now fix all variables, i.e., $\boldsymbol{\varepsilon}$, $\dot{\boldsymbol{\varepsilon}}$, T, $\dot{T}$, $\mathbf{g}$, $\boldsymbol{\zeta}$, $\dot{\boldsymbol{\zeta}}$, but let $\dot{\mathbf{g}}$ vary arbitrarily. Since $\dot{\mathbf{g}}$ can assume either positive or negative values, (3.2.23) can only be satisfied through the requirement that $\partial\psi/\partial\mathbf{g} \equiv 0$. So, the Helmholtz free energy is not a function of $\mathbf{g}$, and consequently, no other state variable depends on $\mathbf{g}$ according to the principle of equipresence. Next, we fix all independent variables and their increments, except for $\dot{T}$, which results in the requirement that

$$s = -\frac{\partial \psi}{\partial T}. \tag{3.2.24}$$

Following the same idea, the constitutive equation for stress can be obtained as

$$\boldsymbol{\sigma} = \rho \frac{\partial \psi}{\partial \boldsymbol{\varepsilon}}. \tag{3.2.25}$$

After substituting the above constitutive equations for entropy and stress, into (3.2.23) the Clausius-Planck inequality results in the following form:

$$-\rho \frac{\partial \psi}{\partial \boldsymbol{\zeta}} \cdot \dot{\boldsymbol{\zeta}} \geq 0. \tag{3.2.26}$$

Depending on the specific form of the Helmholtz free energy, the response functions for stress and entropy can thus be determined through (3.2.25) and (3.2.24), respectively. To complete the list of response functions, i.e., (3.2.17-3.2.20), $\mathbf{q}$ has to be determined so that it satisfies the constraint imposed by the strong form of the second law, $\mathbf{q} \cdot \mathbf{g} \geq 0$, while $\dot{\boldsymbol{\zeta}}$ should satisfy (3.2.26). Therefore, the seven equations (3.2.24) and (3.2.25) and three equations for $\mathbf{q}$ will constitute the required 10 equations necessary to complete the number of equations as summarized in Table 3.1.

Example 3.1. Derivation of constitutive equations for a thermoelastic material

As an example, we derive constitutive equations for a thermoelastic material for which there is no dependence on the set of internal state variables $\boldsymbol{\zeta}$. This time, instead of using the Helmholtz free energy as one of the thermodynamic potentials, we use the Gibbs free energy. As a result, we need to determine the following response functions:

$$\boldsymbol{\varepsilon} = \boldsymbol{\varepsilon}\left(\boldsymbol{\sigma}, T, \mathbf{g}\right); \tag{3.2.27}$$

$$\mathbf{q} = \mathbf{q}\left(\boldsymbol{\sigma}, T, \mathbf{g}\right); \tag{3.2.28}$$

$$s = s\left(\boldsymbol{\sigma}, T, \mathbf{g}\right); \tag{3.2.29}$$

$$G = G\left(\boldsymbol{\sigma}, T, \mathbf{g}\right). \tag{3.2.30}$$

Following the same procedure of combining the first law with the second law of thermodynamics, and then substituting the time derivative of the Legendre transformation from Table 3.2, we obtain

$$-\rho \dot{G} - \dot{\boldsymbol{\sigma}} : \boldsymbol{\varepsilon} - \rho s \dot{T} \geq 0. \tag{3.2.31}$$

The time derivative of the Gibbs free energy is given by

$$\dot{G} = \frac{\partial G}{\partial \boldsymbol{\sigma}} : \dot{\boldsymbol{\sigma}} + \frac{\partial G}{\partial T} \dot{T} + \frac{\partial G}{\partial \mathbf{g}} \cdot \dot{\mathbf{g}}. \tag{3.2.32}$$

After substituting (3.2.32) into (3.2.31) we get the following expression for the Clausius-Planck inequality:

$$-\rho\left[\frac{\partial G}{\partial \boldsymbol{\sigma}}:\dot{\boldsymbol{\sigma}}+\frac{\partial G}{\partial T}\dot{T}+\frac{\partial G}{\partial \mathbf{g}}\cdot\dot{\mathbf{g}}\right]-\boldsymbol{\varepsilon}:\dot{\boldsymbol{\sigma}}-\rho s\dot{T}\geq 0. \tag{3.2.33}$$

By applying the same procedure of fixing all but one variable, we find that $\partial G/\partial \mathbf{g}\equiv 0$, and we arrive at the following constitutive relations:

$$s=-\frac{\partial G}{\partial T}, \tag{3.2.34}$$

$$\boldsymbol{\varepsilon}=-\rho\frac{\partial G}{\partial \boldsymbol{\sigma}}. \tag{3.2.35}$$

If we choose a quadratic polynomial form of the Gibbs free energy given by

$$\begin{aligned} G(\boldsymbol{\sigma},T)=&-\frac{1}{2\rho}\boldsymbol{\sigma}:\boldsymbol{\mathcal{S}}:\boldsymbol{\sigma}-\frac{1}{\rho}\boldsymbol{\sigma}:\boldsymbol{\alpha}\left(T-T_0\right)+\\ &c\left[\left(T-T_0\right)-T\ln\left(\frac{T}{T_0}\right)\right]-s_0T+u_0, \end{aligned} \tag{3.2.36}$$

we can obtain the following response functions for the entropy, s, and strain, $\boldsymbol{\varepsilon}$, respectively, valid for a linear thermoelastic material:

$$s=\frac{1}{\rho}\boldsymbol{\sigma}:\boldsymbol{\alpha}+c\ln\left(\frac{T}{T_0}\right)+s_0, \tag{3.2.37}$$

$$\boldsymbol{\varepsilon}=\boldsymbol{\mathcal{S}}:\boldsymbol{\sigma}+\boldsymbol{\alpha}\left(T-T_0\right). \tag{3.2.38}$$

Notice that after substituting the expressions for entropy,(3.2.37) and strain, (3.2.38), into (3.2.33), the Clausius-Planck entropy inequality is identically satisfied.

3.3 Constitutive Modeling of SMAs

In this section we present the derivation of a constitutive model for polycrystalline shape memory alloys. The constitutive equations will be determined by following the procedure of thermodynamics considering internal state variables presented in the previous sections. We will use the Gibbs free energy as the thermodynamic potential, instead of the Helmholtz free energy, because it is customary to define thermomechanical loading path for SMAs in the stress-temperature space, often using the phase diagram introduced in Chapter 1. The constitutive model presented here follows the formulation proposed by Boyd and Lagoudas [12].

3.3.1 Choice of Internal State Variables

The first step to obtain constitutive equations for SMAs is the selection of the internal state variables that will best represent the SMA thermomechanical response corresponding to change in the microstructure. We have seen in Chapter 1 that the martensitic phase itself can exist in two different forms: the twinned martensite M^t, and the detwinned martensite M^d. Recall that twinned martensite is induced by cooling and it is formed by different variants of martensite configured in an energetically favorable manner such that no macroscopic deformation is observed during transformation. Detwinned martensite, on the other hand, is induced by stress and it is formed predominantly by a single variant of martensite. We also saw in Chapter 1 that the martensitic phase transformation into detwinned martensite in the presence of stress leads to a macroscopic shape change. This transformation induced deformation is associated with the transformation strain, $\boldsymbol{\varepsilon}^t$, introduced already in the empirical formulation of SMAs in Chapter 2. The recovery of the transformation strain, either upon mechanical loading or heating results in shape recovery as the SMA returns to its parent austenitic phase.

The constitutive model considers the martensitic volume fraction, ξ, and the transformation strain, $\boldsymbol{\varepsilon}^t$, to be the internal state variables since both play an important role in characterizing the phase transformation and the observable thermomechanical response of SMAs. The martensitic volume fraction is assumed to be a scalar quantity, and it incorporates the contribution of all the different martensitic variants present in the material. An extended SMA constitutive model that distinguishes the twinned from the detwinned martensite will be presented in Chapter 6.

3.3.2 Kinematic Assumptions

Experimental observations have shown that polycrystalline SMAs can achieve recoverable transformation strains of about 6%. Therefore, one can use small strain formulation, as mentioned earlier in Section 3.2, to describe their deformations. With the assumption of infinitesimal strains, there is no longer a distinction between reference and current configurations. As a consequence, the total strain tensor can be given by the displacement gradient from (3.2.4) which is appropriate for small strains. We also assume that the total strain can be decomposed additively into two parts, a thermoelastic part $\boldsymbol{\varepsilon}^{th}$, and an inelastic part $\boldsymbol{\varepsilon}^{in}$ as follows:

$$\boldsymbol{\varepsilon} = \boldsymbol{\varepsilon}^{th} + \boldsymbol{\varepsilon}^{in}. \tag{3.3.39}$$

The inelastic strain could further be decomposed into additional components to account for various phenomena, including the transformation strain, $\boldsymbol{\varepsilon}^t$ and the formation of plastic strains, $\boldsymbol{\varepsilon}^p$, either directly by yielding or by transformation-induced plasticity. However, in this chapter we assume that the inelastic component of strain is associated with only the transformation strain

$$\boldsymbol{\varepsilon}^{in} = \boldsymbol{\varepsilon}^{t}, \tag{3.3.40}$$

while the transformation-induced plastic strain will be discussed in Chapter 5 and the reorientation of martensitic variants induced strain will be introduced in Chapter 6.

3.3.3 Thermomechanical Constitutive Assumptions for SMAs

Now that we have selected the set of internal state variables, ξ and $\boldsymbol{\varepsilon}^t$, the next step is to define the thermodynamic potential that will be used. In this model, the Gibbs energy is selected to be the thermodynamic potential as mentioned in the beginning of Section 3.3. Note that the Gibbs free energy, G, is a function of the independent state variables stress, $\boldsymbol{\sigma}$, and temperature, T, which can be more suitable when comparing numerical with experimental results, and the state variables, ξ and $\boldsymbol{\varepsilon}$. The explicit form of the Gibbs free energy is given by (Boyd and Lagoudas [12], and Qidwai and Lagoudas [13])

$$\begin{aligned} G\left(\boldsymbol{\sigma}, T, \xi, \boldsymbol{\varepsilon}^t\right) = & -\frac{1}{2\rho}\boldsymbol{\sigma} : \boldsymbol{\mathcal{S}} : \boldsymbol{\sigma} - \frac{1}{\rho}\boldsymbol{\sigma} : \left[\boldsymbol{\alpha}\left(T - T_0\right) + \boldsymbol{\varepsilon}^t\right] + \\ & c\left[\left(T - T_0\right) - T\ln\left(\frac{T}{T_0}\right)\right] - s_0 T + u_0 + \frac{1}{\rho}f(\xi), \end{aligned} \tag{3.3.41}$$

where T_0 is a reference temperature. The material parameters $\boldsymbol{\mathcal{S}}$, $\boldsymbol{\alpha}$, c, s_0, and u_0 are the fourth-order effective compliance tensor, the second-order effective thermal expansion tensor, the effective specific heat, the effective specific entropy at the reference state, and the effective specific internal energy at the reference state, respectively. The function $f(\xi)$ is a transformation hardening function and will be defined in the following sections. The above expression for the Gibbs free energy extends the concept of a linear thermoelastic material by combining two such materials representing austenite and martensite into one through ξ.

The effective material properties can be determined in terms of the properties for the pure phases and the martensitic volume fraction ξ via the following expressions:

$$\boldsymbol{\mathcal{S}}(\xi) = \boldsymbol{\mathcal{S}}^A + \xi\left(\boldsymbol{\mathcal{S}}^M - \boldsymbol{\mathcal{S}}^A\right) = \boldsymbol{\mathcal{S}}^A + \xi\Delta\boldsymbol{\mathcal{S}} \tag{3.3.42}$$

$$\boldsymbol{\alpha}(\xi) = \boldsymbol{\alpha}^A + \xi\left(\boldsymbol{\alpha}^M - \boldsymbol{\alpha}^A\right) = \boldsymbol{\alpha}^A + \xi\Delta\boldsymbol{\alpha} \tag{3.3.43}$$

$$c(\xi) = c^A + \xi\left(c^M - c^A\right) = c^A + \xi\Delta c \tag{3.3.44}$$

$$s_0(\xi) = s_0^A + \xi\left(s_0^M - s_0^A\right) = s_0^A + \xi\Delta s_0 \tag{3.3.45}$$

$$u_0(\xi) = u_0^A + \xi\left(u_0^M - u_0^A\right) = u_0^A + \xi\Delta u_0 \tag{3.3.46}$$

where the superscripts A and M denote the austenitic and martensitic phases, respectively. Note that the above approximation of the effective compliance tensor is different from the approximation of effective stiffnesses introduced in Chapter 2 for the empirical model. One can use micromechanics methods to estimate more accurate approximations for the effective compliance and thermal expansion tensors, if knowledge of the microstructure is available. For example, the self-consistent or Mori-Tanaka methods can be implemented if the shape of the martensitic variants forming in the austenitic phase within a grain is known and the orientation distribution and shape of the grains forming the polycrystalline SMA is also available. Further details can be found in the review articles by Patoor, Lagoudas, and others [14, 15] and some references mentioned therein. For the purposes of this chapter and to simplify identification of material parameters from simple experiments, we will adopt the above approximations.

Constitutive relations are obtained following a similar procedure presented in Sect. 3.2.3. The Gibbs free energy, G, and the internal energy, u, which are related to each other through the strong form of the Legendre transformation as shown in Table 3.2, are substituted into the first and second law of thermodynamics as expressed in the Clausius-Planck inequality (3.2.16). After following the Coleman and Noll procedure described in Section 3.2.3, the total infinitesimal strain tensor and the entropy are given by

$$\boldsymbol{\varepsilon} = -\rho \frac{\partial G}{\partial \boldsymbol{\sigma}} = \boldsymbol{\mathcal{S}} : \boldsymbol{\sigma} + \boldsymbol{\alpha} (T - T_0) + \boldsymbol{\varepsilon}^t, \tag{3.3.47}$$

$$s = -\frac{\partial G}{\partial T} = \frac{1}{\rho} \boldsymbol{\sigma} : \boldsymbol{\alpha} + c \ln \left(\frac{T}{T_0} \right) + s_0. \tag{3.3.48}$$

with the identification of $\boldsymbol{\zeta}$ with $(\boldsymbol{\varepsilon}^t, \xi)$ as the set of internal state variables, the Clausius-Planck inequality, (3.2.16), assumes the following form:

$$\left(-\rho \frac{\partial G}{\partial \boldsymbol{\varepsilon}^t} \right) : \dot{\boldsymbol{\varepsilon}}^t + \left(-\rho \frac{\partial G}{\partial \xi} \right) \dot{\xi} \geq 0. \tag{3.3.49}$$

If the specific form of the Gibbs free energy is substituted from (3.3.41) into (3.3.49) the following result is obtained:

$$\boldsymbol{\sigma} : \dot{\boldsymbol{\varepsilon}}^t + \left(-\rho \frac{\partial G}{\partial \xi} \right) \dot{\xi} \geq 0. \tag{3.3.50}$$

Next, we will propose evolution equations for the internal state variables, ξ and $\boldsymbol{\varepsilon}^t$.

Evolution of Internal State Variables and Kuhn-Tucker Conditions

Now that the expressions for the total strain, entropy and the local form of the entropy inequality have been defined, we need to determine the evolution equations for the internal state variables, $\boldsymbol{\varepsilon}^t$ and ξ. One key assumption underlies this class of models wherein the martensitic phase transformation occurs without consideration for martensitic variant reorientation: any change in the current microstructural state of the material is strictly a result of a change in the martensitic volume fraction (Boyd and Lagoudas [12]). Given this assumption, a relation between the evolution of the transformation strain and the evolution of the martensitic volume fraction during forward and reverse transformation transformation (flow rule) can be postulated as

$$\dot{\boldsymbol{\varepsilon}}^t = \boldsymbol{\Lambda}\dot{\xi}, \tag{3.3.51}$$

where $\boldsymbol{\Lambda}$ is the *transformation tensor*, which determines the transformation strain direction, and is assumed to have the following form (Boyd and Lagoudas [12]):

$$\boldsymbol{\Lambda} = \begin{cases} \dfrac{3}{2} H^{max} \dfrac{\boldsymbol{\sigma}'}{\bar{\sigma}'} \; ; & \dot{\xi} > 0 \\ H^{max} \dfrac{\boldsymbol{\varepsilon}^{t-r}}{\bar{\varepsilon}^{t-r}} \; ; & \dot{\xi} < 0 \end{cases} \tag{3.3.52}$$

where H^{max} is a material parameter associated with the maximum transformation strain. Note that H^{max} could also be defined as a function of the applied stress ($H^{cur}(\bar{\sigma}')$), as presented in Chapter 2. For the remainder of this chapter and for simplicity, H^{max} will be written simply as H. The above form of the transformation tensor is motivated by the assumption that for the forward phase transformation from austenite to martensite ($\dot{\xi} > 0$) the transformation strain will form in the direction of the deviatoric stress. During the reverse transformation from martensite to the parent austenitic phase ($\dot{\xi} < 0$), the transformation strain will be recovered proportionally to the existing transformation strain at the reversal point from forward to the reverse transformation.

The *deviatoric stress tensor* $\boldsymbol{\sigma}'$ is defined by

$$\boldsymbol{\sigma}' = \boldsymbol{\sigma} - \frac{1}{3}\mathrm{tr}\left(\boldsymbol{\sigma}\right)\mathbf{1}, \tag{3.3.53}$$

while the *effective (von Mises equivalent) stress* is given by (3.3.54) below:

$$\bar{\sigma}' = \sqrt{\frac{3}{2}||\boldsymbol{\sigma}'||^2}. \tag{3.3.54}$$

The *transformation strain at the reversal point* is denoted by $\boldsymbol{\varepsilon}^{t-r}$ and the *effective transformation strain at the reversal of the phase transformation* is given by the following equation:

$$\bar{\varepsilon}^{t-r} = \sqrt{\frac{2}{3}||\varepsilon^{t-r}||^2}. \tag{3.3.55}$$

For both of the above equations, $||\cdot||^2 = (\cdot : \cdot)$ denotes the inner product of the enclosed quantity.

By substituting the flow rule, (3.3.51) into (3.3.49), the Clausius-Planck inequality can be written as

$$\left(\boldsymbol{\sigma} : \boldsymbol{\Lambda} - \rho \frac{\partial G}{\partial \xi}\right) \dot{\xi} = \pi \dot{\xi} \geq 0, \tag{3.3.56}$$

where π is defined by the above equation and is the general thermodynamic force conjugated to ξ. For the Gibbs free energy proposed by (3.3.41) the explicit evaluation of π is

$$\begin{aligned}\pi(\boldsymbol{\sigma}, T, \xi) &= \boldsymbol{\sigma} : \boldsymbol{\Lambda} + \frac{1}{2}\boldsymbol{\sigma} : \Delta\boldsymbol{\mathcal{S}} : \boldsymbol{\sigma} + \boldsymbol{\sigma} : \Delta\boldsymbol{\alpha}\,(T - T_0) + \\ &- \rho\Delta c\left[(T - T_0) - T\ln\left(\frac{T}{T_0}\right)\right] + \rho\Delta s_0 T - \rho\Delta u_0 - \frac{\partial f}{\partial \xi}.\end{aligned} \tag{3.3.57}$$

Equation (3.3.51) connects the evolution equation of the transformation strain to the evolution of the martensitic volume fraction. As a result, the transformation strain is no longer an independent state variable. Therefore, it is sufficient to define an evolution equation for only the martensitic volume fraction, ξ, compatible with the entropy inequality (3.3.56).

The conditions for the occurrence of the martensitic phase transformations, forward and reverse, need to be defined now. The constitutive model presented in this chapter and throughout the book assumes that the martensitic phase transformation will take place whenever the thermodynamic force π reaches a critical value. This assumption must be implemented in such a way that the Clausius-Planck inequality is satisfied for all possible thermomechanical paths. The key points related to forward and reverse phase transformation are summarized below.

- When the forward martensitic transformation is taking place, $\dot{\xi}$ assumes positive values since austenite is being transformed into martensite. Therefore, the only way that the Clausius-Planck inequality, (3.3.56), can be satisfied is for π to assume a positive value. Consequently, for the forward martensitic transformation, $\dot{\xi} > 0$, the function π assumes the threshold value of $\pi = Y$.
- When the reverse martensitic transformation is taking place, $\dot{\xi}$ assumes negative values since martensite is transforming back into austenite. Therefore, the only way that the Clausius-Planck inequality, (3.3.56), can be satisfied is for π to assume a negative threshold value. Consequently, for the reverse martensitic transformation, $\dot{\xi} < 0$, the function π assumes the value of $\pi = -Y$.
- Finally, when the state of stress and temperature of the SMA is such that no phase transformation is taking place the value of ξ remains constant and $\dot{\xi} = 0$. Therefore, the Clausius-Planck inequality is satisfied regardless of the value of π, because $\pi\dot{\xi} = 0$.

The above assumptions for the forward and the reverse martensitic transformation can be captured by introducing a transformation function , Φ, such that

$$\Phi = \begin{cases} \pi - Y & \text{whenever} \quad \dot{\xi} > 0 \quad (A \rightarrow M) \\ -\pi - Y & \text{whenever} \quad \dot{\xi} < 0 \quad (M \rightarrow A) \end{cases} \tag{3.3.58}$$

The transformation function, Φ, satisfies the condition of

$$\Phi = 0, \tag{3.3.59}$$

during both forward and reverse phase transformations. The transformation function represents two families of transformation surfaces for $0 \leq \xi \leq 1$, and the two surfaces for $\xi = 0$ and $\xi = 1$ represent the boundaries of each transformation surface family, when the material is fully austenite and fully martensite, respectively.

The constraints on the evolution of the martensitic volume fraction presented above can be expressed in terms of the so-called Kuhn-Tucker conditions, which are given below for both the forward and reverse phase transformations:

$$\dot{\xi} \geq 0; \qquad \Phi\left(\boldsymbol{\sigma}, T, \xi\right) = \pi - Y \leq 0; \qquad \Phi\dot{\xi} = 0; \tag{3.3.60}$$

$$\dot{\xi} \leq 0; \qquad \Phi\left(\boldsymbol{\sigma}, T, \xi\right) = -\pi - Y \leq 0; \qquad \Phi\dot{\xi} = 0. \tag{3.3.61}$$

Appendix A discusses the Kuhn-Tucker conditions as part of a constraint minimization process, following the principle of maximum dissipation for the evolution of the internal state variables. Equation (3.3.60) states that during the thermoelastic loading of austenite, $\dot{\xi} = 0$ and $\Phi < 0$, until the transformation condition $\Phi = 0$ is reached and forward transformation begins ($\dot{\xi} > 0$). Likewise, (3.3.61) states that during the thermoelastic loading of martensite, $\dot{\xi} = 0$ and $\Phi < 0$, until the reverse transformation condition $\Phi = 0$ is reached and reverse transformation begins ($\dot{\xi} < 0$).

During phase transformation, the stress and temperature should remain on the transformation surface (Qidwai and Lagoudas [16]). This condition is mathematically expressed by the consistency condition (Simo and Hughes [17]) which states that $\dot{\Phi} = 0$ and this is explicitly given by the following expression:

$$\dot{\Phi} = \frac{\partial \Phi}{\partial \boldsymbol{\sigma}} : \dot{\boldsymbol{\sigma}} + \frac{\partial \Phi}{\partial T}\dot{T} + \frac{\partial \Phi}{\partial \xi}\dot{\xi} = 0. \tag{3.3.62}$$

The above formulation is consistent with the assumption of rate-independent behavior of SMAs. As discussed in Chapters 1 and 2, the martensitic transformation is a rate-independent process due to its diffusionless nature.

The last step in the constitutive formulation is to select a hardening function, $f(\xi)$. The hardening function is used to account for the interactions between the austenitic phase and the martensitic phase, and also among the martensitic variants themselves. So far, we have not considered any particular form of the hardening function, which appears in (3.3.41) and is fundamental to the construction of (3.3.57). Here we consider a second-order polynomial representation of the transformation-hardening function (Boyd and Lagoudas [12]). The hardening function $f(\xi)$ assumes the following form during the forward and reverse phase transformation:

$$f(\xi) = \begin{cases} \frac{1}{2}\rho b^M \xi^2 + (\mu_1 + \mu_2)\,\xi; & \dot{\xi} > 0 \\ \frac{1}{2}\rho b^A \xi^2 + (\mu_1 - \mu_2)\,\xi; & \dot{\xi} < 0 \end{cases} \tag{3.3.63}$$

where ρb^M, ρb^A, μ_1, μ_2 represent model parameters to be determined experimentally or from micromechanics evaluations.

It will be demonstrated later in the chapter that different constitutive models can be gathered under the same thermodynamic framework presented for the current model, by considering an appropriate form of the hardening function. For example, the model presented here assumes a second-order polynomial hardening function, as shown in (3.3.63), whereas a model proposed by Tanaka [18] assumes an exponential hardening function. A discussion on different hardening functions will be presented in Section 3.4.

Material Parameters

In deriving this model, it was necessary to introduce various material parameters which cannot be derived directly from experimental data using the methods of Chapter 2. To use the model for the purposes of engineering design and analysis, it is necessary to identify these material parameters such that the model can accurately predict the experimental behavior of the SMAs. This model calibration can be done in a straightforward manner if some of the model parameters (b^A, b^M, μ_1, μ_2 and Y) are cast in terms of directly observable material parameters, such as transformation temperatures and stress influence coefficients. The remaining material parameters are thermoelastic constants for the austenitic and martensitic phases and the maximum transformation strain, which can be measured directly. Note that the model parameter, Δs_0, must also be identified, but it can be shown (see Problem 3.3) that a direct relationship exists between $\rho\Delta s_0$ and the stress influence coefficients and maximum transformation strain. The reference entropy difference (Δs_0) can thus be written directly in terms of experimentally derived material parameters. Writing these above model parameters in terms of the experimental parameters requires the use of five independent equations containing these quantities. The first four of these equations are taken from applying the transformation criterion (3.3.59) at four different thermomechanical states. The four states and their complimentary equations are:

1. Beginning of forward phase transformation at zero stress:

$$\pi(\boldsymbol{\sigma}, T, \xi) = Y \quad \text{at} \quad \boldsymbol{\sigma} = \mathbf{0}, \quad T = M_s, \quad \xi = 0. \tag{3.3.64}$$

2. Ending of forward phase transformation at zero stress:

$$\pi(\boldsymbol{\sigma}, T, \xi) = Y \quad \text{at} \quad \boldsymbol{\sigma} = \mathbf{0}, \quad T = M_f, \quad \xi = 1. \tag{3.3.65}$$

3. Beginning of reverse phase transformation at zero stress:

$$\pi(\boldsymbol{\sigma}, T, \xi) = -Y \quad \text{at} \quad \boldsymbol{\sigma} = \mathbf{0}, \quad T = A_s, \quad \xi = 1. \tag{3.3.66}$$

4. Ending of reverse phase transformation at zero stress:

$$\pi(\boldsymbol{\sigma}, T, \xi) = -Y \quad \text{at} \quad \boldsymbol{\sigma} = \mathbf{0}, \quad T = A_f, \quad \xi = 0. \tag{3.3.67}$$

The fifth equation derives from the requirement of continuity of the Gibbs free energy and thus the continuity of $f(\xi)$ at the point of martensitic reversal ($\xi = 1$). Note that the continuity of $f(\xi)$ at the austenitic reversal point ($\xi = 0$) is automatically satisfied. The fifth equation then has the following mathematical form:

5. Continuity of Gibbs free energy:

$$f(\xi = 1)|_{\dot{\xi}>0} = f(\xi = 1)|_{\dot{\xi}<0}. \tag{3.3.68}$$

These five relations lead to a system of five independent algebraic equations which can be solved to express the five model parameters (b^A, b^M, μ_1, μ_2 and Y) in terms of experimentally observable material parameters. The final result is shown in Table 3.3. Note that Δu_0 is not explicitly needed in the implementation of the quadratic hardening function since it is eliminated from the final evaluation of the thermodynamic force, π, in (3.3.57).

Example 3.2. Determination of transformation surfaces in 2-D stress- temperature space

Here we determine the transformation surfaces assuming that the stress state of the SMA material is given by

Table 3.3. Model parameters for the quadratic polynomial hardening function model (Boyd and Lagoudas [12]).

$$\begin{aligned}
Y &= \tfrac{1}{4}\rho\Delta s_0 \left(M_s + M_f - A_f - A_s\right) \\
b^A &= -\Delta s_0 \left(A_f - A_s\right) \\
b^M &= -\Delta s_0 \left(M_s - M_f\right) \\
\mu_1 &= \tfrac{1}{2}\rho\Delta s_0 \left(M_s + A_f\right) - \rho\Delta u_0 \\
\mu_2 &= \tfrac{1}{4}\rho\Delta s_0 \left(A_s - A_f - M_f + M_s\right)
\end{aligned}$$

$$[\sigma] = \begin{bmatrix} \sigma_{11} & 0 & 0 \\ 0 & \sigma_{22} & 0 \\ 0 & 0 & 0 \end{bmatrix}.$$

The same material constants used in Chapter 2, Sect. 2.6.2, are used in this example, and are given in Table 3.4.

To determine the transformation surface boundaries related to the forward phase transformation, we need to use the transformation function (3.3.58) for the case of $\Phi = 0$. Then, by noting that $\Delta c = 0$ and $\Delta \boldsymbol{\alpha} = 0$, from Table 3.4, we can write (3.3.59) as

$$\Phi = \boldsymbol{\sigma} : \boldsymbol{\Lambda} + \frac{1}{2}\boldsymbol{\sigma} : \Delta\boldsymbol{\mathcal{S}} : \boldsymbol{\sigma} + \rho\Delta s_0 T - \rho\Delta u_0 - \frac{\partial f}{\partial \xi} - Y = 0. \tag{3.3.69}$$

The term $\boldsymbol{\sigma} : \boldsymbol{\Lambda}$ in (3.3.69) is computed by using the stress state given above, together with the form of the transformation tensor in (3.3.52). For the forward transformation, the transformation tensor is a function of the deviatoric stress tensor, given in (3.3.53). After some algebraic manipulations, we obtain

$$\boldsymbol{\sigma} : \boldsymbol{\Lambda} = H\left(\sigma_{11}^2 - \sigma_{11}\sigma_{22} + \sigma_{22}^2\right)^{1/2}. \tag{3.3.70}$$

After carrying out the evaluation of the term $\boldsymbol{\sigma} : \Delta\boldsymbol{\mathcal{S}} : \boldsymbol{\sigma}$, the forward phase transformation condition (3.3.69) has the explicit evaluation

$$\begin{aligned}\Phi =& \frac{1}{2}\left(\sigma_{11}^2\Delta\mathcal{S}_{1111} + \sigma_{22}^2\Delta\mathcal{S}_{2222} + \sigma_{22}\Delta\mathcal{S}_{2211}\sigma_{11} + \sigma_{11}\Delta\mathcal{S}_{1122}\sigma_{22}\right) + \\ & H\left(\sigma_{11}^2 - \sigma_{11}\sigma_{22} + \sigma_{22}^2\right)^{1/2} + \rho\Delta s_0 T - \rho\Delta u_0 \\ & - \left(\frac{1}{2}\rho b^M \xi + \mu_1 + \mu_2\right) - Y = 0,\end{aligned} \tag{3.3.71}$$

where the coefficients $\Delta\mathcal{S}_{1111} = \Delta\mathcal{S}_{2222} = 1/E^M - 1/E^A$ and $\Delta\mathcal{S}_{1122} = \Delta\mathcal{S}_{1122} = (-\nu^M/E^M) - (-\nu^A/E^A)$, with the engineering modulus for the elastic austenite and martensite given in Table 3.4. Note that the elastic compliance tensor for both austenite and martensite is assumed to correspond to

Table 3.4. Values of the SMA material parameters used in Example 3.2.

$E^A = 55$ GPa	$E^M = 46$ GPa
$\alpha^A = 22 \cdot 10^{-6}$ /K	$\alpha^M = 22 \cdot 10^{-6}$ /K
$\nu = \nu^A = \nu^M = 0.33$	$\rho\Delta c = 0.0$
$M_f = 230$ K	$M_s = 245$ K
$A_s = 270$ K	$A_f = 280$ K
$H = 0.056$	$C^A = C^M = 7.4$ MPa/K

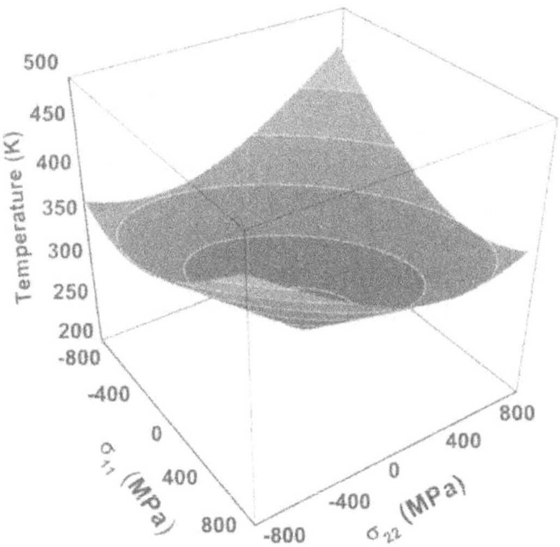

(a) Transformation surface for the onset of forward phase transformation ($\xi = 0$).

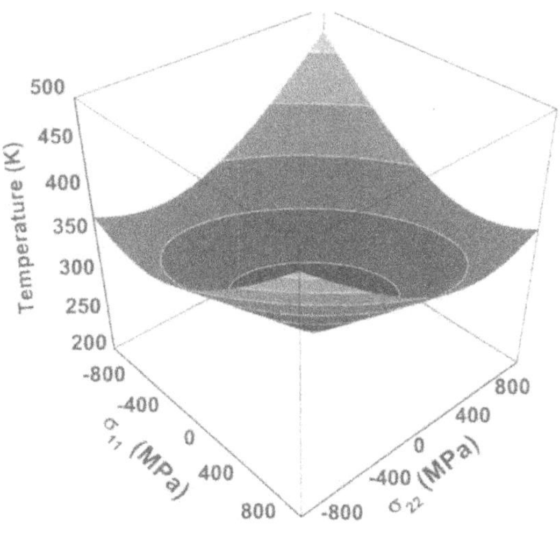

(b) Transformation surface for the end of forward phase transformation ($\xi = 1$).

Fig. 3.1. Transformation surfaces for the forward phase transformation in a three-dimensional stress-temperature space.

that of an isotropic material, consistent with the elastic behavior of a polycrystalline SMA.

Figure 3.1 shows the three-dimensional plot of the transformation surfaces for the forward phase transformation in the $\sigma_{11} - \sigma_{22} - T$ space for the cases of $\xi = 0$, and $\xi = 1$ using (3.3.71).

We can take a slice of the transformation surface boundaries of Fig. 3.1, considering a constant temperature $T = 308\mathrm{K}$, to obtain the two-dimensional plot in stress space $(\sigma_{11}, \sigma_{22})$, as given in Fig. 3.2, both for the forward and the reverse phase transformations. The equation for the reverse phase transformation condition can be derived in a similar fashion (see Problem 3.2). Similarly, considering $\sigma_{22} = 0$ we obtain the boundaries of the transformation surfaces in the uniaxial stress-temperature space, as the intersection of the three-dimensional plot with the plane $\sigma_{22} = 0$. The uniaxial stress-temperature plot, shown in Fig. 3.3 is also called the phase diagram, as introduced in Chapter 1.

3.3.4 Thermomechanical Coupling in SMAs

It has been experimentally observed that thermomechanical coupling is very strong in SMAs. Since latent heat is produced/absorbed during the forward/reverse martensitic phase transformation, temperature variations can occur in the material, influencing its mechanical behavior. Therefore, the thermomechanical coupling can be a key factor to be considered in the modeling of SMAs, especially if phase transformations are assumed to occur over relatively

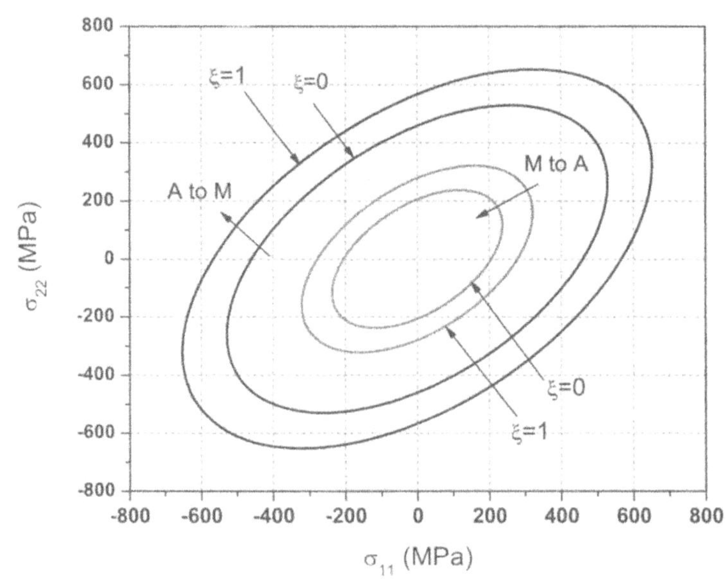

Fig. 3.2. Transformation surface plot in a two-dimensional stress space for forward and reverse phase transformation at a given fixed temperature ($T = 308\mathrm{K}$).

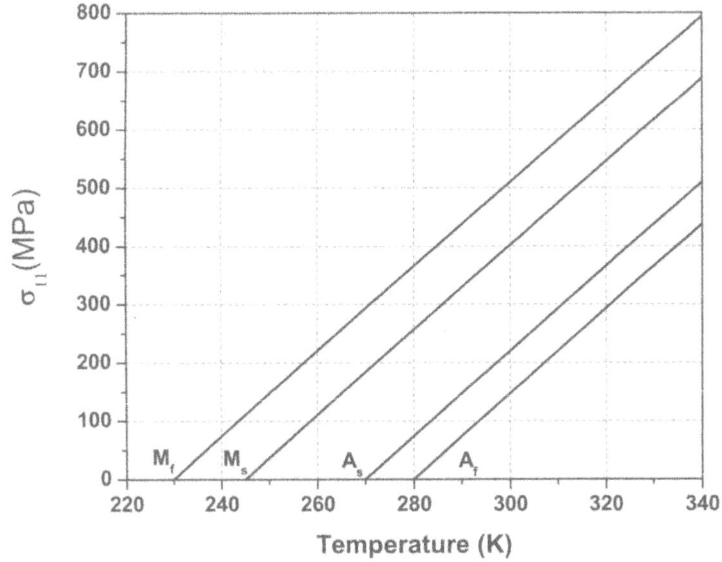

Fig. 3.3. Phase transformation diagram in the σ_{11} - temperature plane.

short time spans. The amount of latent heat produced is strongly related to the amount of martensitic phase transformation.

The first step in modeling the thermomechanical coupling is to start from the first law of thermodynamics, (3.2.13), which expresses the conservation of energy. We can rewrite (3.2.13), after combining it with the second law of thermodynamics, (3.2.16), and the Legendre transformation, as follows:

$$\rho T \dot{s} = \pi \dot{\xi} - \mathrm{div}\,(\mathbf{q}) + \rho r. \tag{3.3.72}$$

The expression for the time derivative of the entropy $\dot{s}$ is obtained from (3.2.16), as

$$\dot{s} = -\frac{\partial \dot{G}}{\partial T} = -\frac{\partial^2 G}{\partial T \partial \boldsymbol{\sigma}} : \dot{\boldsymbol{\sigma}} - \frac{\partial^2 G}{\partial T^2}\dot{T} + \frac{1}{\rho}\frac{\partial \pi}{\partial T}\dot{\xi}. \tag{3.3.73}$$

The evaluation of the different terms in (3.3.73) results in

$$\frac{\partial^2 G}{\partial \boldsymbol{\sigma} \partial T} = -\frac{\boldsymbol{\alpha}}{\rho}, \tag{3.3.74}$$

$$\frac{\partial^2 G}{\partial T^2} = -\frac{T}{c}, \tag{3.3.75}$$

and the partial derivative of π with respect to the temperature, derived from (3.3.57) is given by

$$\frac{1}{\rho}\frac{\partial \pi}{\partial T} = \frac{1}{\rho}\Delta\boldsymbol{\alpha} : \boldsymbol{\sigma} - \Delta c \ln\left(\frac{T}{T_0}\right) + \Delta s_0. \tag{3.3.76}$$

When (3.3.74), (3.3.75) and (3.3.76) are substituted into (3.3.73), the entropy rate is reduced to the following explicit evaluation:

$$\dot{s} = \frac{1}{\rho}\boldsymbol{\alpha} : \dot{\boldsymbol{\sigma}} + \frac{c}{T}\dot{T} + \left(\frac{1}{\rho}\Delta\boldsymbol{\alpha} : \boldsymbol{\sigma} - \Delta c \ln\left(\frac{T}{T_0}\right) + \Delta s_0\right)\dot{\xi}. \tag{3.3.77}$$

Finally, by substituting (3.3.77) into (3.3.72), one gets the following relation:

$$T\boldsymbol{\alpha} : \dot{\boldsymbol{\sigma}} + \rho c\dot{T} + \left(-\pi + T\Delta\boldsymbol{\alpha} : \boldsymbol{\sigma} - \rho\Delta c T \ln\left(\frac{T}{T_0}\right) + \rho\Delta s_0 T\right)\dot{\xi} = -\mathrm{div}\,(\mathbf{q}) + \rho r, \tag{3.3.78}$$

which is the three-dimensional form of the fully thermomechanical coupled energy balance equation for shape memory alloys. The first term of the left-hand side of (3.3.78) expresses how the temperature changes due to a change in the stress state of the material, while the second term of the left-hand side is related to the heat capacity. The third term of the left-hand side expresses how the temperature of the SMA changes due to a variation of the martensitic volume fraction, associated with the latent heat of phase transformation. The first and second terms of the right-hand side of (3.3.78) are related to the heat transfer processes by the heat flux, $\mathbf{q}$, and heat sources, ρr.

As a special case, let us consider adiabatic heat transfer conditions, which can be simulated by eliminating the right-hand side of the energy balance (heat) equation, (3.3.78), i.e., $-\mathrm{div}\,(\mathbf{q})+\rho r = 0$. Moreover, by assuming $\Delta\boldsymbol{\alpha} = 0$, and $\Delta c = 0$, the heat equation assumes the form

$$T\boldsymbol{\alpha} : \dot{\boldsymbol{\sigma}} + \rho c\dot{T} + (-\pi + \rho\Delta s_0 T)\,\dot{\xi} = 0. \tag{3.3.79}$$

Recall that the consistency condition is given by (3.3.62), as follows:

$$\dot{\Phi} = \frac{\partial \Phi}{\partial \boldsymbol{\sigma}} : \dot{\boldsymbol{\sigma}} + \frac{\partial \Phi}{\partial T}\dot{T} + \frac{\partial \Phi}{\partial \xi}\dot{\xi} = 0.$$

For forward transformation, $\dot{\xi} > 0$, the evaluation of the terms appearing in (3.3.80) is given by

$$\frac{\partial \Phi}{\partial \boldsymbol{\sigma}} = \Delta\boldsymbol{S} : \boldsymbol{\sigma} + \boldsymbol{\Lambda}, \tag{3.3.80}$$

$$\frac{\partial \Phi}{\partial T} = \Delta\boldsymbol{\alpha} : \boldsymbol{\sigma} + \rho\Delta s_0, \tag{3.3.81}$$

$$\frac{\partial \Phi}{\partial \xi} = -\frac{\partial^2 f}{\partial \xi^2} = -\rho b^M \xi - \mu_1 - \mu_2, \tag{3.3.82}$$

while for the case of reverse phase transformation, $\dot{\xi} < 0$, we have

$$\frac{\partial \Phi}{\partial \boldsymbol{\sigma}} = -\Delta \boldsymbol{\mathcal{S}} : \boldsymbol{\sigma} - \boldsymbol{\Lambda}, \tag{3.3.83}$$

$$\frac{\partial \Phi}{\partial T} = -\Delta \boldsymbol{\alpha} : \boldsymbol{\sigma} - \rho \Delta s_0, \tag{3.3.84}$$

$$\frac{\partial \Phi}{\partial \xi} = \frac{\partial^2 f}{\partial \xi^2} = -\rho b^A \xi - \mu_1 + \mu_2. \tag{3.3.85}$$

The evolution of martensitic volume fraction can be evaluated from the consistency condition (3.3.80) as

$$\dot{\xi} = -\frac{(\boldsymbol{\Lambda} + \Delta \boldsymbol{\mathcal{S}} : \boldsymbol{\sigma}) : \dot{\boldsymbol{\sigma}} + (\rho \Delta s_0) \dot{T}}{\rho \Delta s_0 \left(M_s - M_f \right)}, \tag{3.3.86}$$

for the forward phase transformation and as

$$\dot{\xi} = -\frac{(\boldsymbol{\Lambda} + \Delta \boldsymbol{\mathcal{S}} : \boldsymbol{\sigma}) : \dot{\boldsymbol{\sigma}} + (\rho \Delta s_0) \dot{T}}{\rho \Delta s_0 \left(A_s - A_f \right)}, \tag{3.3.87}$$

for the reverse phase transformation. Now, the temperature evolution equation for forward phase transformation can be derived by substituting the equation (3.3.86) into the energy balance equation for adiabatic conditions (3.3.79). After solving for $\dot{T}$, one can get the following expression:

$$\dot{T} = -\frac{T\boldsymbol{\alpha} - \boldsymbol{\chi}}{(\rho c - \delta)} : \dot{\boldsymbol{\sigma}}, \tag{3.3.88}$$

where $\boldsymbol{\chi}$ is given by

$$\boldsymbol{\chi} = \frac{(\boldsymbol{\Lambda} + \Delta \boldsymbol{\mathcal{S}} : \boldsymbol{\sigma})}{\rho \Delta s_0 \left(M_s - M_f \right)} \left(-Y + \rho \Delta s_0 T \right), \tag{3.3.89}$$

and δ is given by

$$\delta = \frac{\rho \Delta s_0}{\rho \Delta s_0 \left(M_s - M_f \right)} \left(-Y + \rho \Delta s_0 T \right). \tag{3.3.90}$$

A similar expression for $\dot{T}$ can be derived for the reverse phase transformation and the details are left as an exercise for the interested reader.

3.4 Unification of Different SMA Constitutive Models

Different constitutive models are distinguished intrinsically by a different selection of internal state variables and their respective evolution equations.

If the internal state variables of different SMA constitutive models are the same, then these models can possibly be unified under the same thermodynamic framework through the selection of appropriate forms of the hardening function f. In this section, we present an approach to the unification of the model derived above with two well-known constitutive models for SMAs: the exponential model proposed by Tanaka and coworkers [18, 19], and the cosine model proposed by Liang and Rogers [20].

Exponential (Tanaka) Model

This model [18, 19] can be assimilated into the present thermodynamic framework if the hardening function f is selected as

$$f(\xi) = \begin{cases} \dfrac{\rho \Delta s_0}{a_e^M}\left[(1-\xi)\ln(1-\xi) - \xi\right] + (\mu_1^e + \mu_2^e)\,\xi; & \dot{\xi} > 0 \\ \dfrac{\rho \Delta s_0}{a_e^A}\xi\left[\ln(\xi) - 1\right] + (\mu_1^e - \mu_2^e)\,\xi; & \dot{\xi} < 0 \end{cases} \tag{3.4.91}$$

where a_e^M, a_e^A, μ_1^e, μ_2^e are transformation strain hardening model parameters. These model parameters used in the exponential model are defined in Table 3.5, where they are given in terms of the transformation temperatures and the reference entropy difference which is related to the stress influence coefficients (cf. Problem 3.3). Note that Y^e is the critical value for the thermodynamic force π to cause phase transformation. The remaining terms in the Gibbs free energy and other key relations remain unchanged from those discussed in previous sections. The exponential model is intended to capture the effects of a phase transformation process that corresponds to nucleation of the martensitic phase and growth from the parent austenitic phase. Additional details can be found in Lagoudas et al. [21].

Table 3.5. Parameters for the exponential model.

$$Y^e = -\frac{\rho \Delta s_0}{2}(A_s - M_s) + \frac{\rho \Delta s_0}{2\ln(0.01)}(M_s - M_f + A_f - A_s)$$

$$a_e^A = -\frac{\ln(0.01)}{(A_s - A_f)}$$

$$a_e^M = \frac{\ln(0.01)}{(M_s - M_f)}$$

$$\mu_1^e = \frac{1}{2}\rho \Delta s_0 (M_s + 2A_f - A_s) - \rho \Delta u_0$$

$$\mu_2^e = \frac{1}{2}\rho \Delta s_0 \left(\frac{1}{a_e^A} - \frac{1}{a_e^M}\right)$$

Cosine (Liang and Rogers) Model

This model has been proposed by Liang and Rogers [20] and it assumes the following hardening function, within the present thermodynamic framework:

$$f(\xi) = \begin{cases} \int\limits_0^{\xi} -\frac{\rho\Delta s_0}{a_c^M}\left[\pi - \cos^{-1}(2\xi - 1)\right] d\xi + (\mu_1^c + \mu_2^c)\,\xi; & \dot{\xi} > 0 \\ \int\limits_0^{\xi} -\frac{\rho\Delta s_0}{a_c^A}\left[\pi - \cos^{-1}(2\xi - 1)\right] d\xi + (\mu_1^c - \mu_2^c)\,\xi; & \dot{\xi} < 0 \end{cases} \tag{3.4.92}$$

In the above expression a_c^M, a_c^A, μ_1^c, μ_2^c are transformation strain hardening model parameters. Again, the remaining terms in the Gibbs free energy and other key relations remain unchanged from those discussed in previous sections. These model parameters are presented in Table 3.6 where π represents the universal constant 3.14... and is not the thermodynamical force conjugated to ξ. The model parameters here are given in terms of the transformation temperatures and the reference entropy difference (related to the stress influence coefficients, cf. Problem 3.3). The critical value for the thermodynamic force to initiate transformation for the cosine model is denoted by Y^c.

The cosine model was first introduced to better capture experimental response observed in the testing of SMA actuators. Any of the three models presented here (polynomial, exponential, cosine) under the unified thermodynamic framework is an acceptable choice if it is shown to fit experimental data well. The comparison of the three models is given in Fig 3.4 where results were generated using standard SMA properties to be discussed in the following chapter. It should also be noted that the polynomial (quadratic) hardening model presented in this chapter can be extended to a higher order polynomial hardening function if necessary, based on experimental observations.

Table 3.6. Parameters for the cosine model.

$$Y^c = -\frac{\rho\Delta s_0}{2}(A_s - M_s) + \frac{\rho\Delta s_0}{4}(M_s - M_f + A_f - A_s)$$

$$a_c^A = -\frac{\pi}{(A_s - Af)}$$

$$a_c^M = \frac{\pi}{(M_s - M_f)}$$

$$\mu_1^c = \frac{1}{4}\rho\Delta s_0 (M_s + A_f) - \rho\Delta u_0$$

$$\mu_2^c = \frac{\pi}{4}\rho\Delta s_0\left(\frac{1}{a_c^M} - \frac{1}{a_c^A}\right)$$

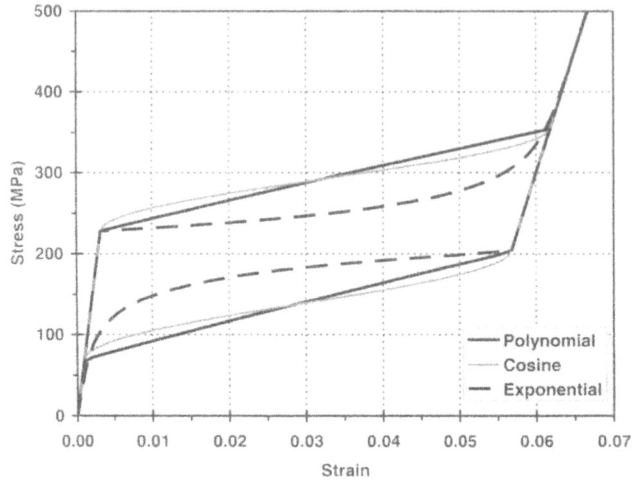

Fig. 3.4. Comparison of pseudoelastic stress vs. strain response as predicted by three distinct hardening functions.

Smooth Transformation Hardening Model

Next, we introduce a hardening function that is a more general polynomial function in terms of ξ that allows smooth transitions between the elastic and transformation regimes. This hardening function is constructed in such a way that it has continuous derivatives. The proposed hardening function has the following form:

$$f(\xi) = \begin{cases} \frac{1}{2}\rho b^M \left(\xi + \frac{\xi^{n_1+1}}{(n_1+1)} + \frac{(1-\xi)^{n_2+1}}{(n_2+1)}\right) ; & \dot{\xi} > 0 \\ \frac{1}{2}\rho b^A \left(\xi + \frac{\xi^{n_3+1}}{(n_3+1)} + \frac{(1-\xi)^{n_4+1}}{(n_4+1)}\right) ; & \dot{\xi} < 0, \end{cases} \tag{3.4.93}$$

while the first derivative of f with respect to ξ is given by

$$\frac{\partial f(\xi)}{\partial \xi} = \begin{cases} \frac{1}{2}\rho b^M \left(1 + \xi^{n_1} - (1-\xi)^{n_2}\right) ; & \dot{\xi} > 0 \\ \frac{1}{2}\rho b^A \left(1 + \xi^{n_3} - (1-\xi)^{n_4}\right) ; & \dot{\xi} < 0, \end{cases} \tag{3.4.94}$$

where ρb^M and ρb^A are given in Table 3.3. The exponents n_1, n_2, n_3 and n_4 assume real values, and are determined from experimental test data.

To compare the proposed smooth transformation hardening model with the model presented in Sect. 3.3, the isothermal pseudoelastic SMA response as predicted by each of these two models has been plotted and these are compared to experimental results. It should be mentioned that the material parameters used in these simulations had to be adjusted slightly when a different model was used. This improved the accuracy of each model prediction. Specifically, the transformation temperatures for the quadratic hardening

Table 3.7. Material parameters for the smooth transformation hardening model.

$E^A = 32.5 \cdot 10^9\text{Pa}$	$E^M = 23.0 \cdot 10^9\text{Pa}$
$\alpha^A = 22.0 \cdot 10^{-6}\text{K}$	$\alpha^M = 22.0 \cdot 10^{-6}\text{K}$
$c^A = 400.0\frac{\text{J}}{\text{kgK}}$	$c^M = 400.0\frac{\text{J}}{\text{kgK}}$
$H = 0.033$	$\rho = 6500\frac{\text{kg}}{\text{m}^3}$
$T_0 = 313\text{K}$	$\rho\Delta s_0 = -11.55 \cdot 10^4\frac{\text{J}}{\text{m}^3\text{K}}$
$A_s = 217\text{K}$	$M_s = 264\text{K}$
$A_f = 290\text{K}$	$M_f = 160\text{K}$
$n_1 = 0.17, n_2 = 0.27$	$n_3 = 0.25, n_4 = 0.35$

model were selected to be $M_f = 194\text{K}$, $M_s = 226\text{K}$, $A_s = 241\text{K}$, $A_f = 290\text{K}$. These transformation temperatures give a better prediction for this model. For more details the reader is referred to additional work by Machado [22]. The material parameters used by the smooth transformation hardening model are presented in Table 3.7 and an example comparing the smooth model with two alternatives is given in Fig. 3.5.

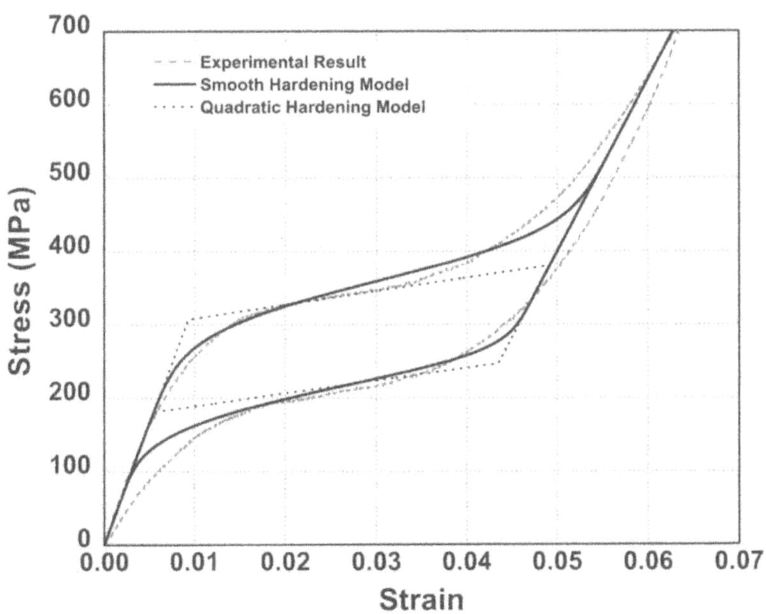

Fig. 3.5. Comparison of experimental results with the polynomial (quadratic) and smooth transformation hardening model results, for $T = 313\text{K}$.

3.5 Analytical Solutions and 1-D Examples

Next, we present examples that use the analytical solutions discussed to predict the behavior of an SMA wire under different thermomechanical loading paths. We first reduce the unified model presented above to a 1-D form. Following this, example 3.3 shows how we can obtain the phase diagram, σ vs. T, using the analytical solutions. Example 3.4 shows how to derive isothermal loading paths for different temperatures, whereas Example 3.5 demonstrates how to derive constant stress loading paths for different stress levels, according to the same phase diagram of Example 3.3. Example 3.6 compares isothermal and adiabatic heat conditions of a pseudoelastic loading path. Example 3.7 explores a more diverse loading paths where isothermal and constant stress loading paths are in sequence.

3.5.1 1-D Reduction of the SMA Constitutive Model

The model for SMAs presented here can be reduced for the 1-D case by assuming a uniaxial loading of an SMA prismatic bar along its long axis, assumed to coincide with the x_1-direction. The stress tensor has only one non-zero component, that is

$$\sigma_{11} = \sigma \neq 0 \tag{3.5.95}$$

The transformation strain components are given by

$$\varepsilon^t_{11} = \varepsilon^t; \quad \varepsilon^t_{22} = \varepsilon^t_{33} = -\frac{1}{2}\varepsilon^t; \quad \varepsilon_{12} = \varepsilon_{23} = \varepsilon_{31} = 0, \tag{3.5.96}$$

where $\boldsymbol{\varepsilon}^t$ is the axial transformation strain assuming that it results in isochoric deformations due to phase transformation.

Due to the fact that the stress tensor has one non-zero component and the transformation strain tensor is of the form presented above, the double dot product between tensor quantities of the equations presented in previous sections will be reduced to a simple scalar multiplication. As a result, the fourth-order compliance tensors $\mathcal{S}^A$ and $\mathcal{S}^M$ reduce to the scalars S^A and S^M, respectively. The second-order thermal expansion tensors $\boldsymbol{\alpha}^A$ and $\boldsymbol{\alpha}^M$ reduce to the scalars α^A and α^M, whereas the transformation tensor $\boldsymbol{\Lambda}$ reduces to H or $H^{cur}(\sigma)$ in the one-dimensional case. The equation for the evolution of transformation strain during forward transformation becomes

$$\dot{\varepsilon}^t = H\text{sgn}(\sigma)\dot{\xi}, \tag{3.5.97}$$

where the sign function is defined as

$$\text{sgn}(\sigma) = \begin{cases} 1 & \text{if } \sigma \geq 0 \\ -1 & \text{if } \sigma < 0 \end{cases} \tag{3.5.98}$$

It is left to the reader to derive the form for the reverse transformation (see Problem 3.5). The above equation (3.5.97) can be integrated, resulting in the following expression:

$$\varepsilon^t = H\text{sgn}(\sigma)\xi. \tag{3.5.99}$$

Therefore, the one-dimensional form of the total strain and the entropy are given by

$$\varepsilon = S\sigma + \alpha\left(T - T_0\right) + \varepsilon^t, \tag{3.5.100}$$

$$s = \frac{1}{\rho}\sigma\alpha + c\ln\left(\frac{T}{T_0}\right) + s_0, \tag{3.5.101}$$

where the expressions for S and α are defined in terms of (3.3.42) and (3.3.43), as

$$S\left(\xi\right) = S^A + \xi\left(S^M - S^A\right) = S^A + \xi\Delta S, \tag{3.5.102}$$

$$\alpha\left(\xi\right) = \alpha^A + \xi\left(\alpha^M - \alpha^A\right) = \alpha^A + \xi\Delta\alpha. \tag{3.5.103}$$

The one-dimensional form of the thermodynamic force π is given by

$$\pi = \left|\sigma\right| H + \frac{1}{2}\Delta S\sigma^2 + \Delta\alpha\left(T - T_0\right) - \rho\Delta c\left[\left(T - T_0\right) - T\ln\left(\frac{T}{T_0}\right)\right] + \rho\Delta s_0 T - \rho\Delta u_0 - \frac{\partial f}{\partial\xi}. \tag{3.5.104}$$

For the quadratic hardening function and forward transformation (3.5.104) reduces to

$$\pi = \left|\sigma\right| H + \frac{1}{2}\Delta S\sigma^2 + \rho\Delta s_0 T - \rho\Delta u_0 - \left[\rho b^M\xi + \left(\mu_1 + \mu_2\right)\right] = Y. \tag{3.5.105}$$

Solving (3.5.105), we can get the explicit expression for ξ

$$\xi = \frac{1}{\rho b^M}\left[\left|\sigma\right| H + \frac{1}{2}\Delta S\sigma^2 + \rho\Delta s_0\left(T - M_s\right)\right]. \tag{3.5.106}$$

Substituting (3.5.105) into (3.5.100) for the case of forward martensitic transformation, the total strain becomes

$$\varepsilon = S\sigma + \alpha\left(T - T_0\right) + \frac{H\text{sgn}\left(\sigma\right)}{\rho b^M}\left[\left|\sigma\right| H + \frac{1}{2}\Delta S\sigma^2 + \rho\Delta s_0\left(T - M_s\right)\right], \tag{3.5.107}$$

where

$$S = S^A + \xi\left(S^M - S^A\right); \qquad \alpha = \alpha^A + \xi\left(\alpha^M - \alpha^A\right).$$

Next, consider the case for reverse phase transformation. The thermodynamic force π is given as

$$\pi = |\sigma| H + \frac{1}{2}\Delta S\sigma^2 + \rho\Delta s_0 T - \rho\Delta u_0 - \left[\rho b^A \xi + (\mu_1 - \mu_2)\right] = -Y, \quad (3.5.108)$$

which can be solved for ξ as

$$\xi = \frac{1}{\rho b^A}\left[Y + |\sigma| H + \frac{1}{2}\Delta S\sigma^2 + \rho\Delta s_0\left(T - A_f\right)\right]. \quad (3.5.109)$$

Substituting (3.5.108) into (3.5.100) we get an expression for the total strain for the case of reverse martensitic transformation, given by

$$\varepsilon = S\sigma + \alpha\left(T - T_0\right) + \frac{H\operatorname{sgn}(\sigma)}{\rho b^A}\left[|\sigma| H + \frac{1}{2}\Delta S\sigma^2 + \rho\Delta s_0\left(T - A_f\right)\right]. \quad (3.5.110)$$

We summarize the above formulas in Table 3.8.

3.5.2 Example Solutions for Various Thermomechanical Loading Paths

Example 3.3. Plotting of analytical transformation surfaces (1-D stress state)

Using the same material parameters of Example 3.2, we show how to derive the stress vs. temperature phase diagram using analytical solutions for the stress interval of $0 < \sigma < 800\text{MPa}$ and temperature interval of $224K < T < 336K$, assuming $\Delta\alpha = 0 = \Delta c = 0$.

Table 3.8. Summary of SMA constitutive model - One-dimensional formulation.

Thermoelastic response:

$$\varepsilon = S^A\sigma + \alpha^A\left(T - T_0\right) \qquad \text{or} \qquad \varepsilon = S^M\sigma + \alpha^M\left(T - T_0\right)$$

Forward phase transformation:

$$\pi = |\sigma| H + \tfrac{1}{2}\Delta S\sigma^2 + \rho\Delta s_0 T - \rho\Delta u_0 - \left[\rho b^M \xi + (\mu_1 + \mu_2)\right] = Y$$

$$\varepsilon = S\sigma + \alpha\left(T - T_0\right) + \frac{H\operatorname{sgn}(\sigma)}{\rho b^M}\left[|\sigma| H + \frac{1}{2}\Delta S\sigma^2 + \rho\Delta s_0\left(T - M_s\right)\right]$$

Reverse phase transformation:

$$\pi = |\sigma| H + \tfrac{1}{2}\Delta S\sigma^2 + \rho\Delta s_0 T - \rho\Delta u_0 - \left[\rho b^A \xi + (\mu_1 - \mu_2)\right] = -Y$$

$$\varepsilon = S\sigma + \alpha\left(T - T_0\right) + \frac{H\operatorname{sgn}(\sigma)}{\rho b^A}\left[|\sigma| H + \frac{1}{2}\Delta S\sigma^2 + \rho\Delta s_0\left(T - A_f\right)\right]$$

The first step to find the transformation surface at the beginning of the forward phase transformation is to substitute $\xi = 0$ into (3.5.105). We obtain

$$|\sigma| H + \frac{1}{2}\Delta S\sigma^2 + \rho\Delta s_0 (T - M_s) = 0. \tag{3.5.111}$$

Similarly, we substitute $\xi = 1$ into (3.5.105) to find out the transformation surface at the end of the forward phase transformation, with the following result:

$$|\sigma| H + \frac{1}{2}\Delta S\sigma^2 + \rho\Delta s_0 (T - M_s) - \rho\Delta s_0 \left(M_f - M_s\right) = 0. \tag{3.5.112}$$

To find the transformation surface at the beginning of the reverse phase transformation, we substitute $\xi = 1$ into (3.5.108) to get

$$|\sigma| H + \frac{1}{2}\Delta S\sigma^2 + \rho\Delta s_0 \left(T - A_f\right) - \rho\Delta s_0 \left(A_s - A_f\right) = 0. \tag{3.5.113}$$

Finally, for the transformation surface pertaining to the end of the reverse phase transformation, we substitute $\xi = 0$ into (3.5.108), which yields

$$|\sigma| H + \frac{1}{2}\Delta S\sigma^2 + \rho\Delta s_0 \left(T - A_f\right) = 0. \tag{3.5.114}$$

Now, using the known material parameters, we can plot the phase diagram, as shown in Fig. 3.6. Note that the transformation surfaces as derived using this model and the given material properties exhibit only minor nonlinearity. Examination of (3.5.111)–(3.5.114) indicates that the difference in compliance between austenite and martensite is the only term contributing to the curvature of the transformation surfaces. This compliments the discussion of Chapter 2 in which such linearity was often assumed (see also Sect. 2.6.2).

Example 3.4. Determination of 1-D solutions for isothermal loading paths

In this example, we show how to generate the stress vs. temperature and stress vs. strain expressions for isothermal loading paths with different initial temperatures greater than A_f. Because the initial temperature of the material is $T > A_f$ and the initial stress will be zero, the reference state of the material can be considered austenitic and the configuration of the body assuming zero stress and zero transformation strain can be thought of as the reference configuration. The selected temperatures for isothermal loading are $T_1 = 328$K, $T_2 = 308$K, $T_3 = 276$K and $T_4 = 260$K. The same material parameters will be used as used in Example 3.2, and these are given in Table 3.4.

Following the same procedure for computing the transformation surfaces as described in Example 3.3, we can plot the isothermal loading path in the stress vs. temperature plot for the four different initial temperatures in Fig. 3.7.

We need to derive the isothermal stress vs. strain curves for all the four temperatures. First, we need to compute the values of σ^{Ms}, σ^{Mf}, σ^{As} and

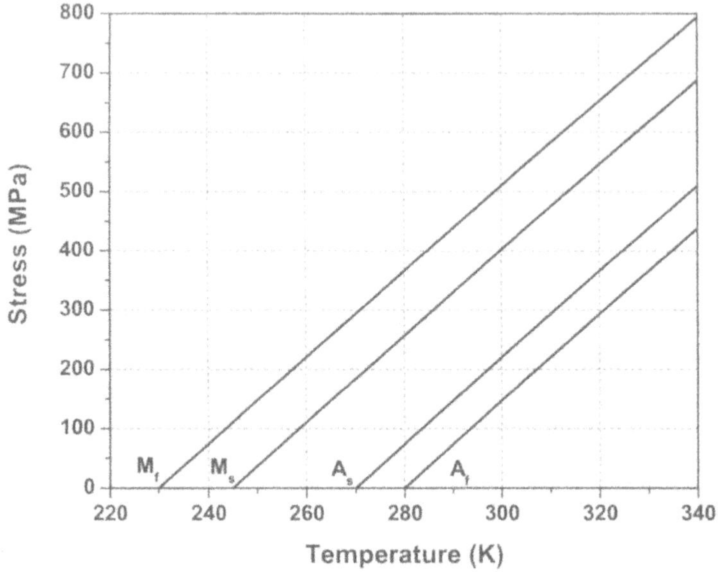

Fig. 3.6. Uniaxial stress-temperature phase diagram for Example 3.3.

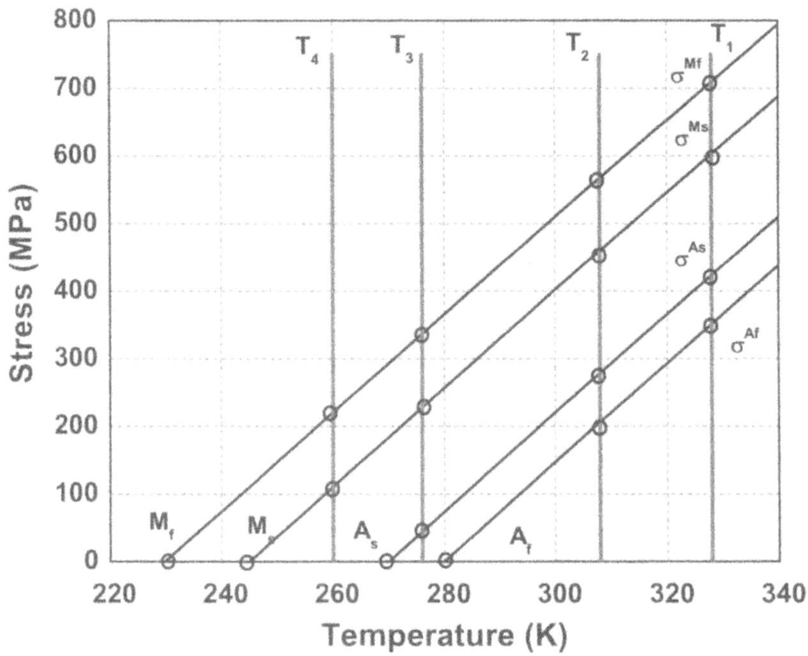

Fig. 3.7. Phase diagram - isothermal loading path.

σ^{Af} for each corresponding temperature. These values of stress represent the beginning and end of both forward and reverse transformation. A general procedure for obtaining different stress levels at the beginning and at the end of the phase transformation follows.

We find the value of σ^{Ms} by substituting $\xi = 0$ and $\sigma = \sigma^{Ms}$ in (3.5.105), which gives

$$\frac{1}{2}\Delta S(\sigma^{Ms})^2 + \sigma^{Ms} H + \rho\Delta s_0 \left(T - M_s\right) = 0. \tag{3.5.115}$$

Similarly, the values of σ^{Mf}, σ^{As}, σ^{Af} can be obtained by substituting $\xi = 1$ into (3.5.105) and (3.5.108), and substituting $\xi = 0$ into (3.5.108), which yields the following evaluations:

$$\begin{aligned}
&\frac{1}{2}\Delta S(\sigma^{Mf})^2 + \sigma^{Mf} H + \rho\Delta s_0 \left(T - M_f\right) - \rho b^M = 0;\\
&\frac{1}{2}\Delta S(\sigma^{Af})^2 + \sigma^{Af} H + \rho\Delta s_0 \left(T - A_f\right) - \rho b^A = 0;\\
&\frac{1}{2}\Delta S(\sigma^{As})^2 + \sigma^{As} H + \rho\Delta s_0 \left(T - A_s\right) = 0.
\end{aligned} \tag{3.5.116}$$

Since the value of the stress levels of the forward and reverse phase transformation are computed from the equations above, we now need to define the equations for the elastic and transformation regimes. Let us consider the region $0 \leq \sigma \leq \sigma^{Ms}$, the interval where the elastic austenitic phase occurs. In this interval of stress, no phase transformation takes place. Therefore, we can calculate the response of the material using the following set of equations:

$$\begin{aligned}
&\xi = 0; \quad S = S^A; \quad \alpha = \alpha^A;\\
&\varepsilon = S^A\sigma + \alpha^A \left(T - T_0\right)
\end{aligned} \tag{3.5.117}$$

The forward martensitic transformation occurs when the stress level is in the interval $\sigma^{Ms} < \sigma < \sigma^{Mf}$. Therefore, before calculating the strains for this interval, we need to first find the martensitic volume fraction related to each stress and temperature of the SMA. After the martensitic volume fraction is obtained, we can compute the material properties and then find the strain. For this purpose, we use the following set of equations:

$$\begin{aligned}
&\xi = \frac{1}{\rho b^M}\left[|\sigma|\, H + \frac{1}{2}\Delta S\sigma^2 + \rho\Delta s_0 \left(T - M_s\right)\right];\\
&S = S^A + \xi\left(S^M - S^A\right);\\
&\alpha = \alpha^A + \xi\left(\alpha^M - \alpha^A\right);\\
&\varepsilon = S\sigma + \alpha\left(T - T_0\right) + H\mathrm{sgn}(\sigma)\xi.
\end{aligned} \tag{3.5.118}$$

The forward martensitic transformation ends when the stress level reaches σ^{Mf}. Above σ^{Mf}, a second elastic response related to the martensitic phase will be exhibited until the maximum value of the stress $\sigma^{\max}$ for the loading is reached. Upon unloading, the martensitic phase responds elastically until the stress level reaches σ^{As}. For this martensitic elastic regime in the interval $\sigma^{Mf} \leq \sigma \leq \sigma^{max}$ (loading) and $\sigma^{max} \geq \sigma \geq \sigma^{As}$ (unloading), we have the following set of equations:

$$
\begin{aligned}
&\xi = 1; \quad S = S^{M}; \quad \alpha = \alpha^{M}; \\
&\varepsilon = S^{M}\sigma + \alpha^{M}\left(T - T_0\right) + H\mathrm{sgn}(\sigma).
\end{aligned} \tag{3.5.119}
$$

The reverse martensitic transformation occurs when the stress level reaches σ^{As}. Therefore, in the interval of $\sigma^{As} > \sigma > \sigma^{Af}$, we have the following set of equations with which to compute the strains:

$$
\begin{aligned}
&\xi = \frac{1}{\rho b^{A}}\left[\left|\sigma\right| H + \frac{1}{2}\Delta S\sigma^2 + \rho\Delta s_0\left(T - A_f\right)\right]; \\
&S = S^{A} + \xi\left(S^{M} - S^{A}\right); \\
&\alpha = \alpha^{A} + \xi\left(\alpha^{M} - \alpha^{A}\right); \\
&\varepsilon = S\sigma + \alpha\left(T - T_0\right) + H\mathrm{sgn}(\sigma)\xi.
\end{aligned} \tag{3.5.120}
$$

Finally, for the stress interval $\sigma^{Af} \geq \sigma \geq 0$, we return to the elastic response of the austenitic phase. Then, we can use the following set of equations:

$$
\begin{aligned}
&\xi = 0; \quad S = S^{A}; \quad \alpha = \alpha^{A}; \\
&\varepsilon = S^{A}\sigma + \alpha^{A}\left(T - T_0\right).
\end{aligned} \tag{3.5.121}
$$

Substituting the value of the material parameters into the equations presented above, one can plot the pseudoelastic stress vs. strain curves for isothermal conditions as shown in Fig. 3.8. Note that any effects due to thermal expansion have been neglected.

Example 3.5. Determination of 1-D solutions for constant stress loading paths

In this example, we show how to obtain the constitutive response related to constant stress loading paths using the same material parameters as the previous example. Note that in the case of a uniaxial stress state, constant stress paths are often known as "isobaric" paths, and this terminology will be used throughout the remainder of this chapter. Specifically, we derive the necessary equations for three isobaric paths for stress levels of $\sigma = 100$ MPa, $\sigma = 150$ MPa, and $\sigma = 200$ MPa, and also plot strain vs. temperature for these three stress levels.

Figure 3.9 shows the stress-temperature phase diagram with the thermomechanical loading paths given by straight horizontal lines. The first step is to

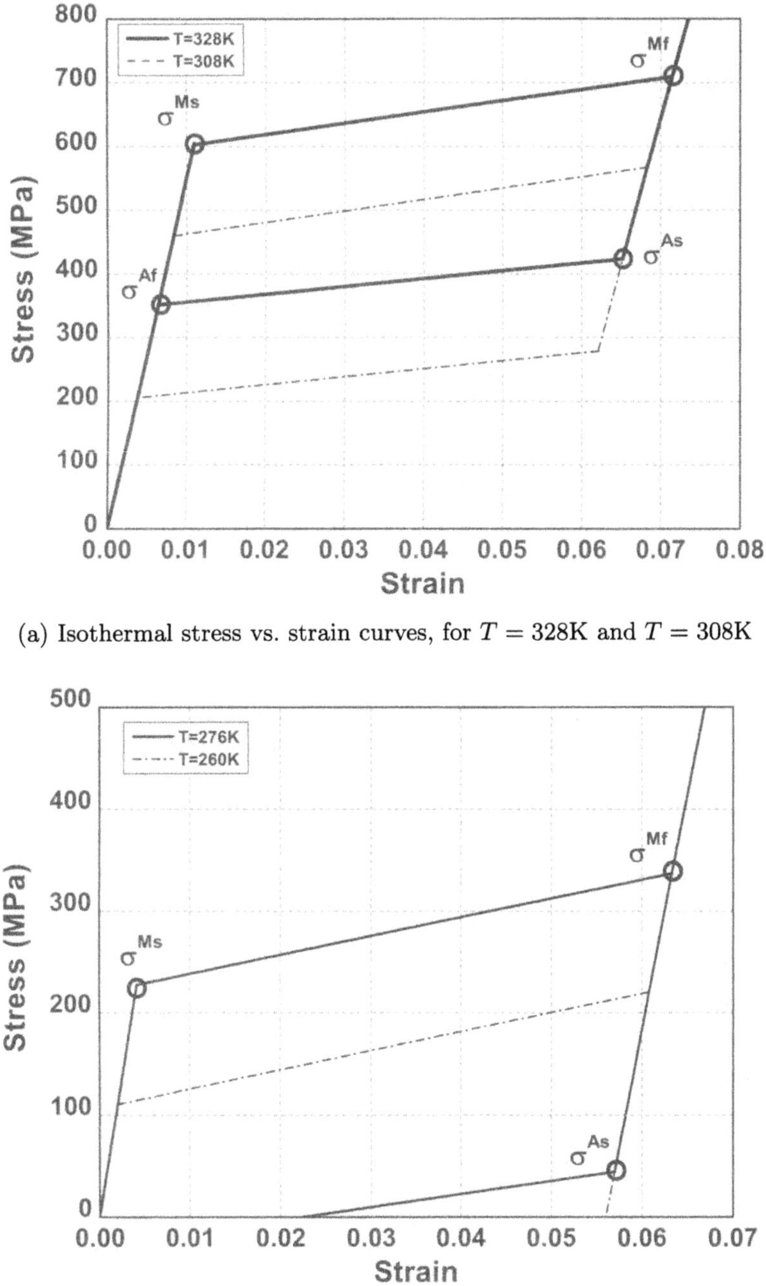

(a) Isothermal stress vs. strain curves, for $T = 328$K and $T = 308$K

(b) Isothermal stress vs. strain curves, for $T = 276$K and $T = 260$K

Fig. 3.8. Isothermal stress vs. strain curves for different temperatures.

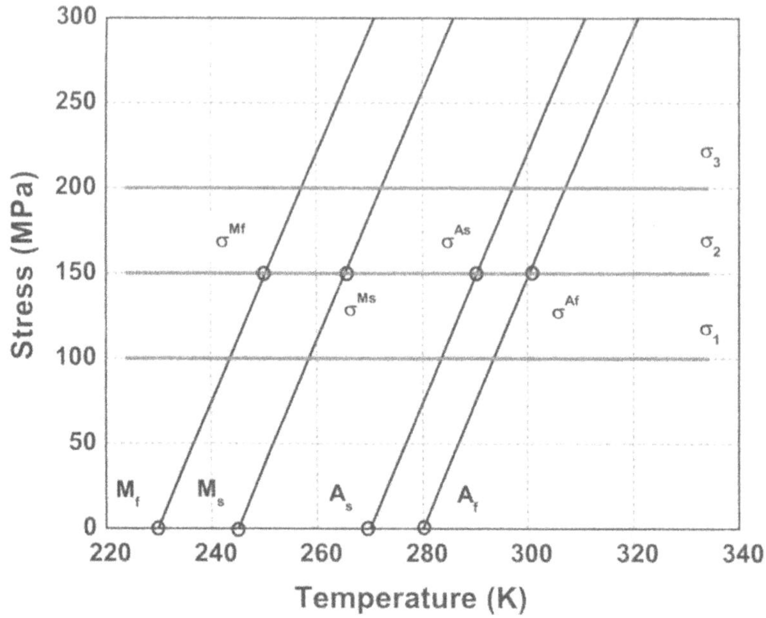

Fig. 3.9. Phase diagram - isobaric loading path.

compute the transformation temperatures for each given stress level. For this purpose, we use (3.5.105) to compute the values of the non-zero stress transformation temperatures M_s^σ and M_f^σ for the forward phase transformation, and we use (3.5.108) to compute A_f^σ and A_s^σ for the reverse phase transformation. Therefore, by substituting $\xi = 0$ and $T = M_s^\sigma$, into (3.5.105), we can solve for M_s^σ. The temperature M_f^σ can be obtained in the same manner, in this case by substituting $\xi = 1$ and $T = M_f^\sigma$ into (3.5.105). Similarly, A_s^σ and A_f^σ can be obtained by using (3.5.108). We then obtain the following set of equations:

$$
\begin{aligned}
&\frac{1}{2}\Delta S\sigma^2 + \sigma H + \rho\Delta s_0\left(M_s^\sigma - M_s\right) = 0;\\
&\frac{1}{2}\Delta S\sigma^2 + \sigma H + \rho\Delta s_0\left(M_f^\sigma - M_f\right) - \rho b^M = 0;\\
&\frac{1}{2}\Delta S\sigma^2 + \sigma H + \rho\Delta s_0\left(A_f^\sigma - A_f\right) - \rho b^A = 0;\\
&\frac{1}{2}\Delta S\sigma^2 + \sigma H + \rho\Delta s_0\left(A_s^\sigma - A_s\right) = 0.
\end{aligned} \tag{3.5.122}
$$

Let the SMA material be subjected to a given stress level at a starting temperature $T > A_f^\sigma$ (i.e., $T = T^{max}$). Assume also that $T_0 = T^{max}$. This will generate a small elastic strain in the austenitic phase of the SMA. Next, we gradually start reducing the temperature while the stress level remains constant. Therefore, in the interval of temperature $M_s^\sigma < T < T^{max}$, we still have $\xi = 0$, and no phase transformation is taking place. Therefore, we compute the thermomechanical response using the following set of equations:

$$\begin{aligned} &\xi = 0; \quad S = S^A; \quad \alpha = \alpha^A; \\ &\varepsilon = S^A\sigma + \alpha^A (T - T_0). \end{aligned} \tag{3.5.123}$$

At the temperature M_s^σ, the forward phase transformation starts and it continues until the temperature of the SMA reaches M_f^σ. As the SMA cools through this range, an increase in the total strain due to the generation of transformation strain is observed. For the interval of temperature $M_s^\sigma < T \leq M_f^\sigma$, we have the following set of equations:

$$\begin{aligned} \xi &= \frac{1}{\rho b^M}\left[|\sigma|\, H + \frac{1}{2}\Delta S\sigma^2 + \rho\Delta s_0 \left(T - M_s\right)\right]; \\ S &= S^A + \xi\left(S^M - S^A\right); \\ \alpha &= \alpha^A + \xi\left(\alpha^M - \alpha^A\right); \\ \varepsilon &= S(\xi)\sigma + \alpha(\xi)\left(T - T_0\right) + H\mathrm{sgn}\left(\sigma\right)\xi. \end{aligned} \tag{3.5.124}$$

If we continue to cool the SMA to a temperature T_{min} less than M_f^σ, no more phase transformation will occur, and the material behaves thermoelastically in the martensitic phase. For the interval of temperature $T_{min} < T \leq M_f^\sigma$ we then have

$$\begin{aligned} &\xi = 1; \quad S = S^M; \quad \alpha = \alpha^M; \\ &\varepsilon = S^M\sigma + \alpha^M (T - T_0) + H\mathrm{sgn}(\sigma). \end{aligned} \tag{3.5.125}$$

After the cooling process ends, we heat the material again. The material initially behaves elastically, and (3.5.125) applies until the temperature of the SMA reaches A_s^σ. In the interval $A_s^\sigma < T \leq A_f^\sigma$, the reverse transformation takes place and the material recovers part of its transformation strain. During the reverse transformation, we have

$$\begin{aligned} \xi &= \frac{1}{\rho b^A}\left[|\sigma|\, H + \frac{1}{2}\Delta S\sigma^2 + \rho\Delta s_0 \left(T - A_f\right)\right]; \\ S &= S^A + \xi\left(S^M - S^A\right); \\ \alpha &= \alpha^A + \xi\left(\alpha^M - \alpha^A\right); \\ \varepsilon &= S(\xi)\sigma + \alpha(\xi)\left(T - T_0\right) + H\mathrm{sgn}\left(\sigma\right)\xi. \end{aligned} \tag{3.5.126}$$

Finally, after the reverse transformation, the SMA behaves thermoelastically again. The temperature range for this elastic region is $A_f^\sigma < T < T^{max}$. We then have the following constitutive response:

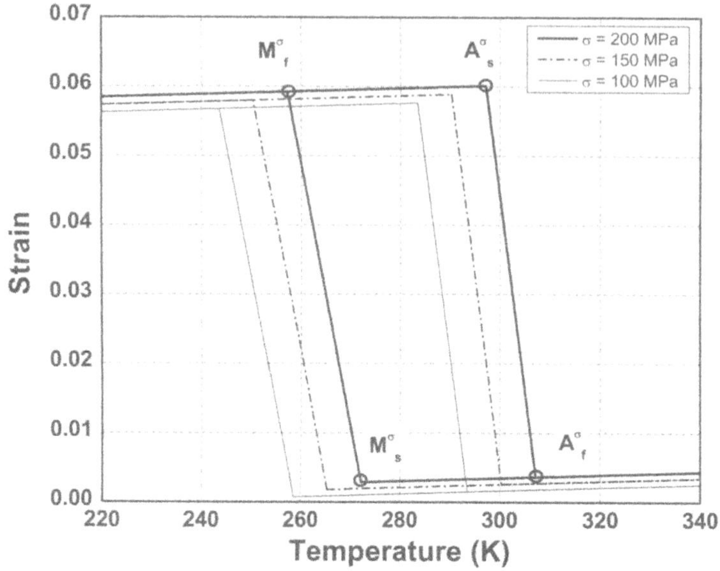

Fig. 3.10. Strain vs. temperature plot.

$$\xi = 0; \quad S = S^A; \quad \alpha = \alpha^A;$$
$$\varepsilon = S^A\sigma + \alpha^A (T - T_0). \tag{3.5.127}$$

Figure 3.10 shows the strain vs. temperature plots for the isobaric loading-unloading conditions, for three different stress levels of 100 MPa, 150 MPa and 200 MPa.

Example 3.6. Determination of 1-D solutions for adiabatic pseudoelastic loading paths

In this example, we compare the isothermal pseudoelastic loading-unloading path with an adiabatic one. We again plot the phase diagram (stress vs. temperature plot) and show how to derive and and the stress vs. strain behavior for both isothermal and adiabatic conditions using the material parameters of Example 3.3.

The solution of the isothermal pseudoelastic loading-unloading path is exactly the same as presented in Example 3.4 and will not be presented here. However, the solution of the adiabatic loading path needs some further consideration since the temperature changes during phase transformation. For the elastic loading in the austenitic phase in the interval of $0 < \sigma \leq \sigma^{Ms}$, we have the following set of equations:

$$\xi = 0; \quad S = S^A; \quad \alpha = \alpha^A;$$
$$\varepsilon = S^A\sigma + \alpha^A (T - T_0). \tag{3.5.128}$$

Next, we consider the interval at which phase transformation takes place, $\sigma^{Ms} < \sigma \leq \sigma^{Mf}$. The adiabatic assumption for the SMA material implies complete thermal insulation from the environment; there is no heat exchange. Since the SMA cannot exchange heat with the surrounding environment, its temperature increases during the exothermic forward phase transformation. We compute the increment of temperature, ΔT, for a given small increment of stress, $\Delta\sigma$, integrating (3.3.88), and assuming that all variables remain constant during the increment of stress. This results in the following relation:

$$\Delta T = -\frac{T\alpha - \left[\dfrac{(H\mathrm{sgn}(\sigma) + \Delta S\sigma)}{\rho\Delta s_0\left(M_s - M_f\right)}(-Y + \rho\Delta s_0 T)\right]}{\left[\rho c - \left(\dfrac{\rho\Delta s_0}{\rho\Delta s_0\left(M_s - M_f\right)}(-Y + \rho\Delta s_0 T)\right)\right]}\Delta\sigma,$$

The temperature is then updated. The stress and temperature (updated values) are then used to calculate ξ and, consequently, the updated value of strain:

$$\begin{aligned}
\xi &= \frac{1}{\rho b^M}\left[|\sigma| H + \frac{1}{2}\Delta S\sigma^2 + \rho\Delta s_0 (T - M_s)\right];\\
S &= S^A + \xi\left(S^M - S^A\right);\\
\alpha &= \alpha^A + \xi\left(\alpha^M - \alpha^A\right);\\
\varepsilon &= S(\xi)\sigma + \alpha(\xi)(T - T_0) + H\mathrm{sgn}(\sigma)\xi.
\end{aligned} \tag{3.5.129}$$

For values of $\sigma > \sigma^{Mf}$, in the interval of $\sigma^{Mf} < \sigma \leq \sigma^{max}$, we have only elastic loading in the martensitic phase, and no phase transformation occurs. Therefore, we can compute the constitutive response using the following set of equations:

$$\begin{aligned}
&\xi = 1; \quad S = S^M; \quad \alpha = \alpha^M;\\
&\varepsilon = S^M\sigma + \alpha^M (T - T_0) + H\mathrm{sgn}(\sigma)\xi.
\end{aligned} \tag{3.5.130}$$

The procedure for computing the unloading adiabatic path follows from the same discussion, and is thus not explicitly described. Figure 3.11 presents the results of the stress vs. temperature, while Fig. 3.12 and stress vs. strain curves for the complete adiabatic loading path. Figure 3.12 provides a comparison of the isothermal response (dashed curve) and the adiabatic response (black solid curve). The results reiterate the conclusions addressed in Chapter 2, Sect. 2.4.3 about the importance of loading rate. The adiabatic loading path is representative of very fast cycling during which loading and unloading complete in such a short time span that heat transfer mechanisms have not yet become effective.

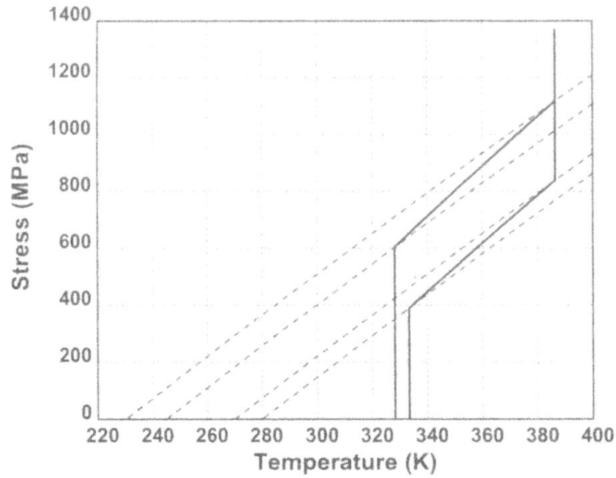

Fig. 3.11. Stress vs. temperature response assuming adiabatic conditions.

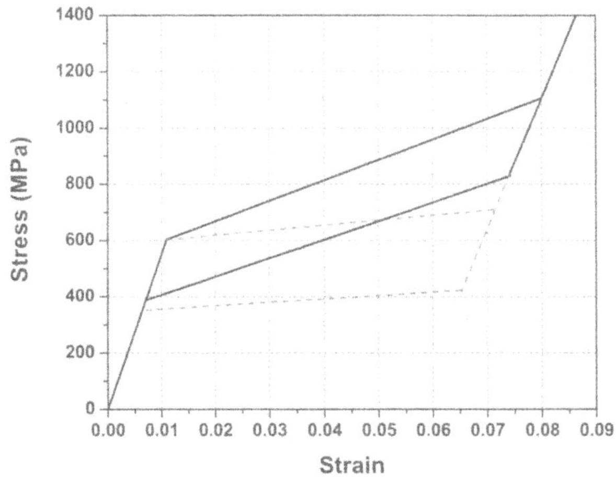

Fig. 3.12. Stress vs. strain plots: isothermal conditions (dashed) and adiabatic conditions (solid).

Example 3.7. Determination of 1-D solutions for sequential isothermal-constant stress loading paths

In this example, we consider a sequence of isothermal and isobaric loading paths. Consider an SMA bar with the material properties given in Table 3.9. Note that Problem 3.3 addresses the relation between the stress influence coefficients C^A and C^M and the material parameter $\rho\Delta s_0$ as used in the derivation of the constitutive model.

Table 3.9. Material properties for Example 3.7.

$E^A = 33$ GPa	$E^M = 15$ GPa
$\alpha^A = 22 \cdot 10^{-6}$ /K	$\alpha^M = 22 \cdot 10^{-6}$ /K
$H = 0.023$	$\rho \Delta c = 0.0$
$M_f = 227$ K	$M_s = 242$ K
$A_s = 261$ K	$A_f = 270$ K
$C^A = 4.5\,\mathrm{MPa/K}$	$C^M = 4.5\,\mathrm{MPa/K}$

Assuming the reference temperature to be $T_0 = 298\,\mathrm{K}$ and the reference stress to be zero, we seek to predict the SMA behavior when it is subjected to the following loading path:

1. Cooled at zero stress to $T = 255\mathrm{K}$.
2. Loaded isothermally to $\sigma = 200\,\mathrm{MPa}$ and then unloaded isothermally.
3. Loaded once more isothermally to $100\,\mathrm{MPa}$.
4. Heated at constant stress ($\sigma = 100\,\mathrm{MPa}$) to $T = 298\,\mathrm{K}$.
5. Unloaded isothermally to a zero stress level.

For such a stress-temperature loading path, an appropriately constructed phase diagram provides a graphical aid as we seek to predict the material response at different points in the loading path. We plot the stress-temperature phase diagram for the following interval: $0 \leq \sigma \leq 200$ MPa and $220 \leq T \leq 325\mathrm{K}$. The procedure to construct the phase diagram is discussed in Example 3.3, and it is not repeated here. The phase diagram is shown in Fig. 3.13.

The solution method to plot the stress-strain curve, stress-temperature curve, and strain-temperature curve for these loading steps is as follows:

- The process of cooling under zero load is a purely thermoelastic step as the thermomechanical state of the material does not enter the forward transformation region. Therefore, we have:

$$\varepsilon = S^A \sigma + \alpha^A \left(T - T_0\right), \tag{3.5.131}$$

  with $\xi = 0$, $S = S^A$ and $\alpha = \alpha^A$. Equation (3.5.131) corresponds to the path from point 0 to 1 shown in Fig. 3.13.
- At temperature $T = 255$ K, we load isothermally. Recall that the material is initially in austenite. It is important to compute the level of stress where the forward martensitic phase transformation starts and finishes (i.e., σ^{Ms} and σ^{Mf}). By following the approach presented in Example 3.4, we find that $\sigma^{Ms} = 56\,\mathrm{MPa}$ and $\sigma^{Mf} = 115\,\mathrm{MPa}$ as shown in Fig. 3.13. Therefore, for the interval $0 \leq \sigma \leq 56$ MPa the material behaves elastically. Then, we have the following constitutive response:

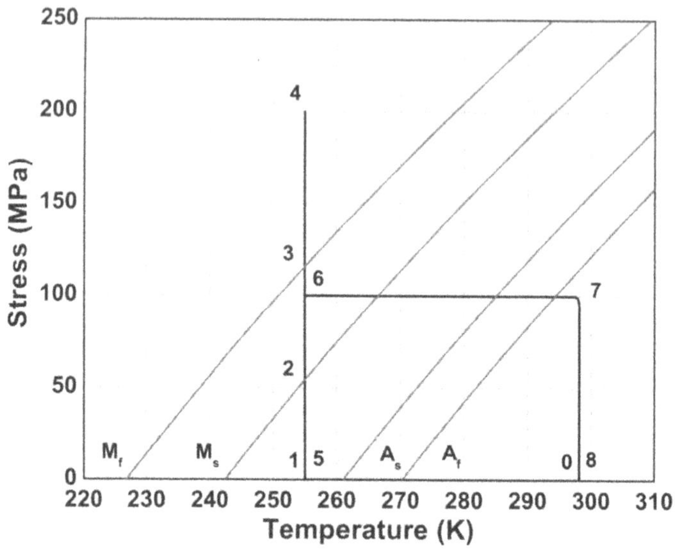

Fig. 3.13. Stress vs. temperature phase diagram with loading path described in Example 3.7.

$$\varepsilon = S^A\sigma + \alpha^A (T - T_0) \tag{3.5.132}$$

with $\xi = 0$, $S = S^A$ and $\alpha = \alpha^a$. The equation (3.5.132) corresponds to the path from point 1 to 2 shown in Fig. 3.13.

The martensitic phase transformation occurs during the interval of applied stress $56 < \sigma < 115$ MPa. Therefore, we have the following set of equations, for the martensitic volume fraction, compliance coefficient, thermal expansion coefficient and constitutive response

$$\begin{aligned}
\xi &= \frac{1}{\rho b^M}\left[|\sigma| H + \frac{1}{2}\Delta S\sigma^2 + \rho\Delta s_0 (T - M_s)\right];\\
S &= S^A + \xi\left(S^M - S^A\right);\\
\alpha &= \alpha^A + \xi\left(\alpha^M - \alpha^A\right);\\
\varepsilon &= S\sigma + \alpha (T - T_0) + H\mathrm{sgn}(\sigma)\xi.
\end{aligned} \tag{3.5.133}$$

The set of equations (3.5.133) correspond to the path from point 2 to 3 shown in Fig. 3.13.

For the stress interval $115 \leq \sigma \leq 200$ MPa, no phase transformation takes place. The material behaves elastically in the martensitic phase. The same behavior occurs when unloading from $\sigma = 200$ MPa to $\sigma = 0$ MPa. Therefore, in the interval $0 \leq \sigma \leq 200$ MPa upon unloading, we have the following constitutive response:

$$\varepsilon = S^M\sigma + \alpha^M (T - T_0), \tag{3.5.134}$$

with $\xi = 1$, $S = S^M$ and $\alpha = \alpha^M$. Equation (3.5.134) corresponds to the path 3-4-5 as shown in Fig. 3.13.

- From the previous results, we have seen that the material is in martensite after unloading. Therefore, by loading the material again until $\sigma = 100$ MPa does not induce any phase change, and the same set of equations (3.5.134) for elastic martensitic phase holds.
- For the next thermomechanical loading step (a constant stress heating at $\sigma = 100\,\text{MPa}$) we first compute the values of temperature where the forward and reverse martensitic phase transformation starts and finishes (M_s^σ, M_f^σ, A_s^σ, A_f^σ). The necessary equations to compute the temperatures at which the reverse phase transformations start and finish for a constant stress heating path were already presented in Example 3.5. Using the same methodology, we calculate $M_s^\sigma = 266$ K, $M_f^\sigma = 251$ K, $A_s^\sigma = 285$ K and $A_s^\sigma = 294$ K (see Fig. 3.13).
Therefore, for the interval of $255 \leq T \leq 285$ K at $\sigma = 100$ MPa, the material behaves elastically. We have the following constitutive response:

$$\varepsilon = S^M \sigma + \alpha^M (T - T_0), \tag{3.5.135}$$

with $\xi = 1$, $S = S^M$ and $\alpha = \alpha^M$.
The reverse phase transformation takes place in the interval $285 \leq T \leq 294$ K. For this temperature interval, we have the following set of equations:

$$\begin{aligned} \xi &= \frac{1}{\rho b^A} \left[|\sigma| H + \frac{1}{2} \Delta S \sigma^2 + \rho \Delta s_0 \left(T - A_f \right) \right]; \\ S &= S^A + \xi \left(S^M - S^A \right); \\ \alpha &= \alpha^A + \xi \left(\alpha^M - \alpha^A \right); \\ \varepsilon &= S(\xi)\sigma + \alpha(\xi)(T - T_0) + H \operatorname{sgn}(\sigma)\, \xi. \end{aligned} \tag{3.5.136}$$

The material behaves elastically again in the temperature interval $294 < T < 298$ K at the stress level of $\sigma = 100$ MPa. Then we have the equations:

$$\varepsilon = S^A \sigma + \alpha^A (T - T_0), \tag{3.5.137}$$

with $\xi = 0$, $S = S^A$ and $\alpha = \alpha^A$. The set of equations (3.5.135), (3.5.136) and (3.5.137) correspond to the total path from point 6 to 7 as shown in Fig. 3.13 and Fig. 3.14.
- At the temperature of $T = 298$ K, we perform the final unloading to $\sigma = 0$. By examining the constructed phase diagram we can see that no phase transformation is expected to take place during this step. Therefore, we can use the same constitutive equation in (3.5.137), which will also correspond to the path from point 7 to 8 shown in Fig. 3.13.

Fig. 3.14 presents the stress vs. strain for the isothermal loading path described above, while Fig. 3.15 shows the strain vs. temperature plot. We also

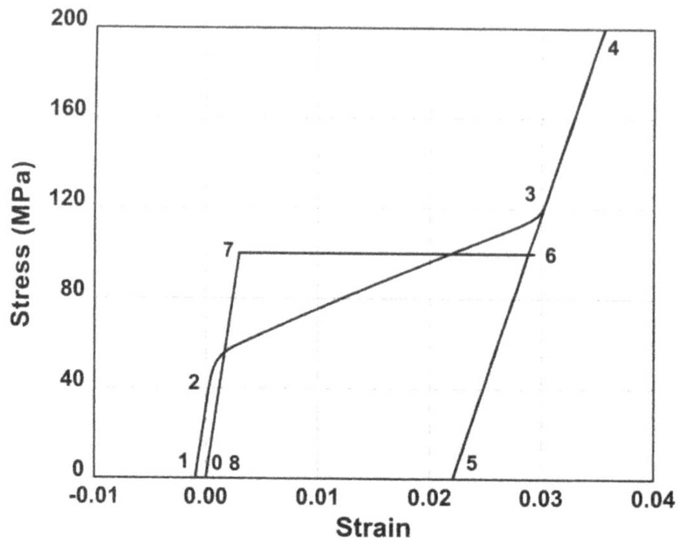

Fig. 3.14. Stress vs. strain response at corresponding to the thermomechanical loading path of Fig. 3.13.

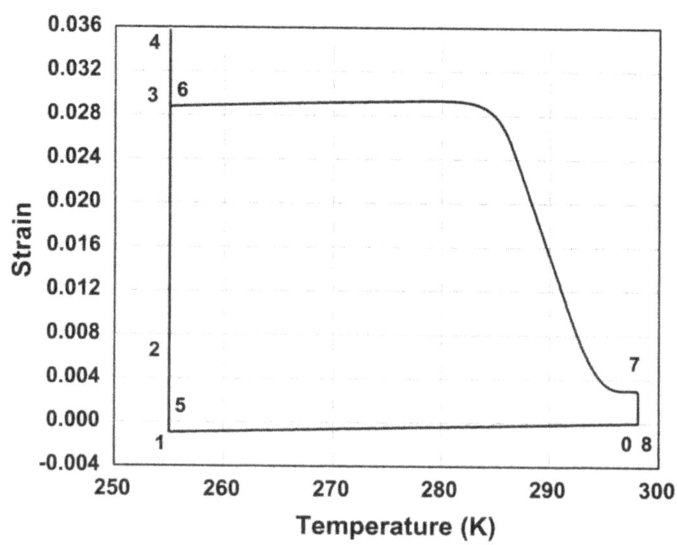

Fig. 3.15. Strain vs. temperature response corresponding to the thermomechanical loading path of Fig. 3.13.

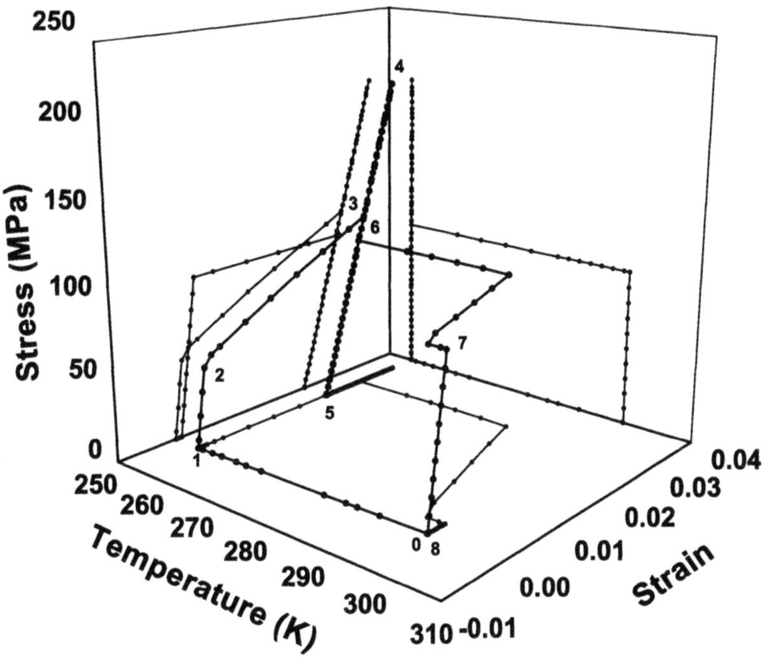

Fig. 3.16. Stress vs. strain vs. temperature response of the SMA corresponding to the loading path of Fig. 3.13.

present a three-dimensional plot of the loading-unloading path in Fig. 3.16. Note that the path labeled with the thermomechanical state points (0–8) represents the total stress-strain-temperature path experienced by the material element and that two projections onto the stress-strain and stress-temperature planes are also shown.

3.5.3 Application of the Smooth Hardening Model to a Nonlinear Oscillator

The hysteretic behavior of pseudoelastic SMAs results in a high dissipation capacity that can be used to attenuate undesired vibrations of a mechanical system or structure. Although the evolving thermomechanical properties and high dissipation capacity are very interesting characteristics observed in the dynamic response of SMAs, they can also lead to a very complex dynamical response, in some cases leading to chaotic vibration. Chaotic response implies that two very close but different dynamical orbits can diverge as time progresses, and consequently, chaos is related to long-term unpredictability. Therefore, it is important to study the nonlinear dynamical response of SMA systems. In this example we investigate the nonlinear dynamical response of a one-degree of freedom SMA oscillator (Fig. 3.17), which consists of a mass m attached to an SMA element, assumed to be a prismatic SMA bar of length

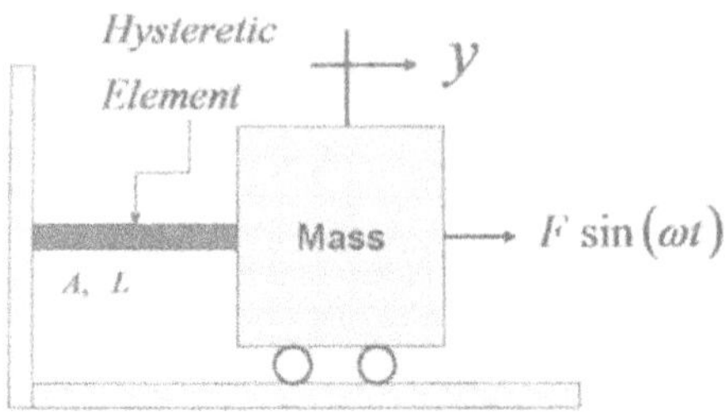

(a) One-degree of freedom SMA oscillator

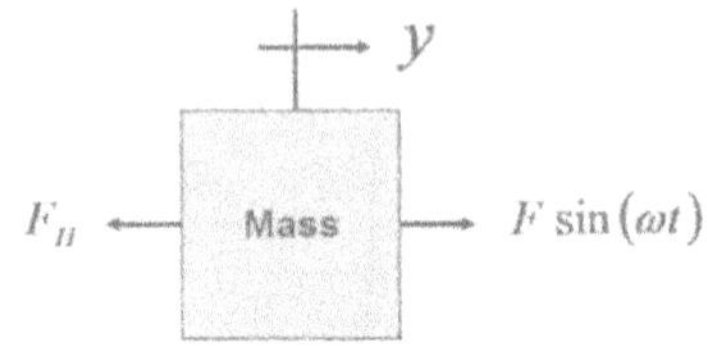

(b) Free body diagram of the SMA oscillator

Fig. 3.17. Single-degree of freedom SMA oscillator.

L and cross-section area A. For more details about the SMA oscillator, the interested reader is referred to [22]. The system is harmonically excited by a force $F \sin(\omega t)$.

The equation of motion of the oscillator is given by

$$m\ddot{y} + F_H = F \sin(\omega t), \tag{3.5.138}$$

where y is the mass displacement from its reference position, relative to an inertial frame, ω is the forcing frequency, F is the amplitude of the excitation force and F_H is the force exerted by the hysteretic SMA element on the mass.

A non-dimensional version of Eqn. 3.5.138, can be obtained by assuming that the SMA element restitution force is equally distributed at all points of the SMA bar. We can then assume $\sigma = F_H/A$, where σ represents the nominal uniaxial stress in the SMA bar, and $\varepsilon = y/L$, where ε corresponds to the axial strain of the SMA element. The equation of motion of the oscillator, Eqn. 3.5.138 then results in the following form:

$$\ddot{\varepsilon} + \frac{\sigma A}{mL} = \frac{F}{mL} \sin(\omega t). \tag{3.5.139}$$

Let us now define the following non-dimensional variables:

$$\begin{gathered} \omega_0 = \sqrt{\frac{E^A A}{mL}}; \qquad \hat{F} = \frac{F}{mL\omega_0^2}; \\ \hat{t} = \omega_0 t; \qquad \hat{\omega} = \frac{\omega}{\omega_0}; \qquad \hat{\sigma} = \frac{\sigma}{E^A} \end{gathered} \tag{3.5.140}$$

where $E^A = 1/S^A$ is the Young's modulus of an SMA bar in fully austenitic phase and ω_0 is related to the natural frequency of the system (e.g., identified as the natural frequency of the system when the SMA element is in the austenitic phase). With the above definitions, and after introducing the derivative with respect to non-dimensional time, $\varepsilon' = \partial\varepsilon/\partial\hat{t}$, the equation of motion, Eqn. 3.5.139, can be re-written in a non-dimensional form as

$$\varepsilon'' + \hat{\sigma} = \hat{F} \sin\left(\hat{\omega}\hat{t}\right) \tag{3.5.141}$$

where the normalized uniaxial stress in the SMA bar is given by

$$\hat{\sigma} = \frac{1}{\hat{S}}\left[\varepsilon - \hat{\alpha}\left(\hat{T} - \hat{T}_0\right) - \varepsilon^t\right]. \tag{3.5.142}$$

The non-dimensional effective material properties can be defined in terms of (3.3.42)-(3.3.46), as follows:

$$\hat{S} = \hat{S}^A + \xi\left(\hat{S}^M - \hat{S}^A\right) \tag{3.5.143}$$

$$\hat{\alpha} = \hat{\alpha}^A + \xi\left(\hat{\alpha}^M - \hat{\alpha}^A\right) \tag{3.5.144}$$

$$\hat{c} = \hat{c}^A + \xi\left(\hat{c}^M - \hat{c}^A\right) \tag{3.5.145}$$

$$\hat{s}_0 = \hat{s}_0^A + \xi\left(\hat{s}_0^M - \hat{s}_0^A\right) \tag{3.5.146}$$

$$\hat{u}_0 = \hat{u}_0^A + \xi\left(\hat{u}_0^M - \hat{u}_0^A\right) \tag{3.5.147}$$

where

$$\hat{S}^A = S^A E^A; \quad \hat{S}^M = S^M E^A; \quad \hat{\alpha}^A = \alpha^A A_s; \quad \hat{\alpha}^M = \alpha^M A_s;$$

$$\hat{T} = \frac{T}{A_s}; \quad \hat{T}_0 = \frac{T_0}{A_s}; \quad \hat{c}^A = \frac{\rho}{E^A} A_s c^A; \quad \hat{c}^M = \frac{\rho}{E^A} A_s c^M;$$

$$\hat{s}_0^A = \frac{\rho}{E^A} A_s s_0^A; \quad \hat{s}_0^M = \frac{\rho}{E^A} A_s s_0^M; \quad \hat{u}_0^A = \frac{\rho}{E^A} u_0^A; \quad \hat{u}_0^M = \frac{\rho}{E^A} u_0^M.$$

Free vibrations are first considered by eliminating the forcing term from the right-hand side of (3.5.138), and by giving appropriate initial conditions to the oscillator. Figure 3.18 shows results related to the free vibration of the isothermal SMA oscillator. Results are presented in the form of stress vs. strain and phase space curves (where the "phase space" represents a kind of speed vs. position space). For a high energy initial condition $(\varepsilon(0), \varepsilon'(0)) = (0.0, 0.04)$, and $\hat{T} = 1.258$, the system initially dissipates energy due the hysteresis loop. The level of energy dissipated per cycle is equivalent to the area of the hysteresis loop, defined by the amount of phase transformation that the SMA underwent. However, over the course of time, as the SMA dissipates energy, the system converges to the elastic regime. Since there is no phase transformation in the elastic regime, no energy dissipation occurs, and as a result, the oscillator motion converges to a periodic orbit. Results for non-isothermal conditions can be found in an alternate work by Machado [22]. In the same work the details for the numerical integration of the dynamical system combine with the constitutive model are also presented.

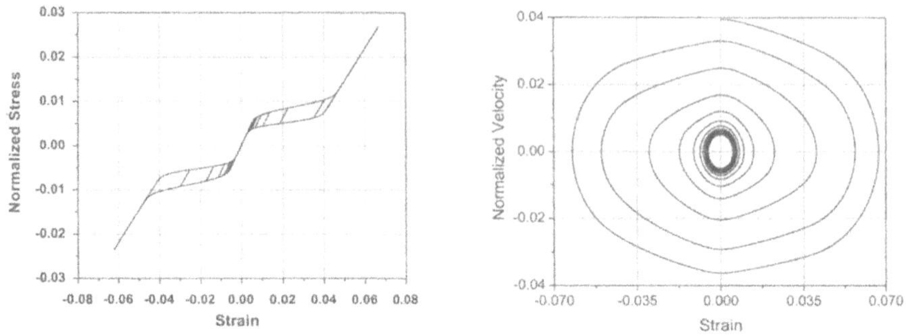

(a) $\hat{\sigma}$ vs. ε (b) ε' vs. ε

Fig. 3.18. Free response of the SMA oscillator: Stress vs. strain and phase portrait curves.

Forced vibrations of the pseudoelastic SMA oscillator are investigated by considering a fixed amplitude of the excitation force and different values of the excitation frequency. Since, at first, we are assuming isothermal conditions, the temperature of the SMA element is fixed at $\hat{T} = 1.258$. In addition, the amplitude of the excitation force is selected to be $\hat{F} = 0.008$ for all simulations.

Figure 3.19a presents the bifurcation diagram of the SMA oscillator subjected to isothermal conditions, for the range of frequencies of $0.24 < \hat{\omega} < 0.76$. One can observe that Fig. 3.19 contains regions of clouds of points separated by regions with lines. Usually the regions of point clouds are associated with chaotic vibration, and the regions of lines are related to periodic motion. Figure 3.19b shows an enlargement of Fig. 3.19a for the interval of $0.35 < \hat{\omega} < 0.55$, where the chaotic regimes can more clearly be identified.

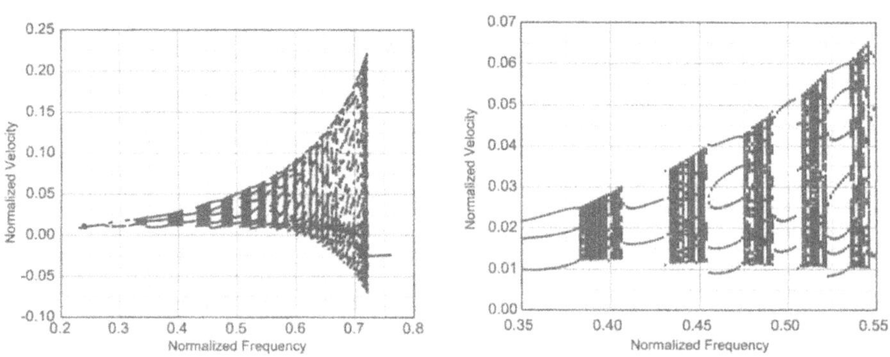

(a) Interval: $0.2 < \hat{\omega} < 0.8$ (b) Interval: $0.35 < \hat{\omega} < 0.55$

Fig. 3.19. Bifurcation diagram for the response of a single degree of freedom SMA oscillator assuming isothermal conditions.

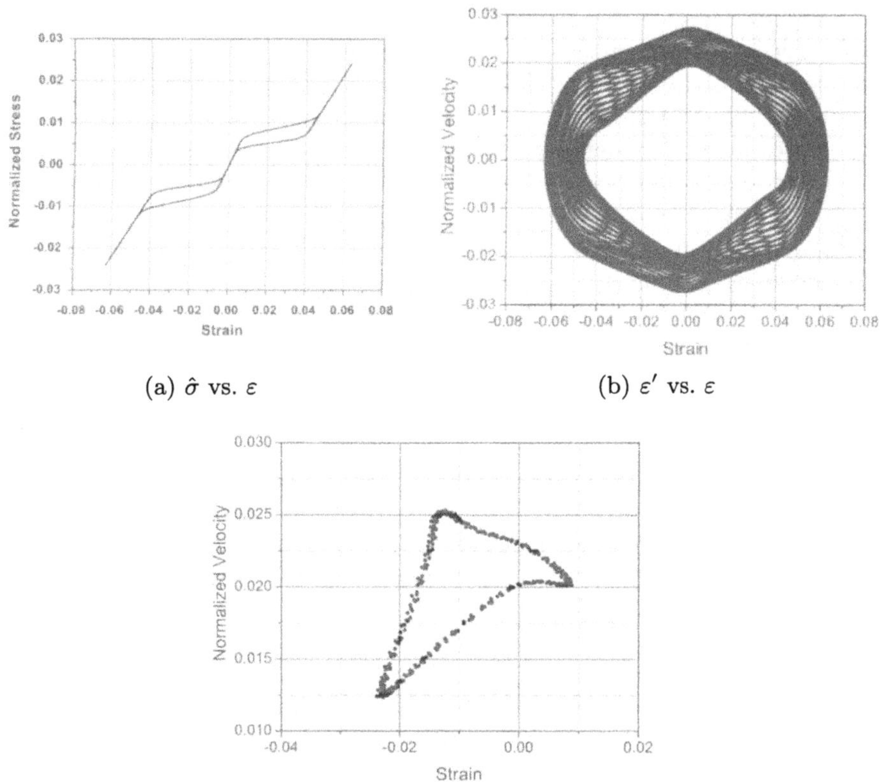

(a) $\hat{\sigma}$ vs. ε

(b) ε' vs. ε

(c) Poincaré Map: ε' vs. ε

Fig. 3.20. Forced response of the SMA oscillator for $\hat{F} = 0.008$ and $\hat{\omega} = 0.397$Hz under isothermal conditions.

The next analysis is concerned with the motion of the oscillator when the excitation frequency is $\hat{\omega} = 0.397$ which is potentially chaotic based on Fig. 3.19b. Figure 3.20a presents the stress vs. strain plot, while Fig. 3.20b presents the phase space curves. The so-called Poincaré map is shown in Fig. 3.20c and presents a cloud of points that can be associated with chaotic motion. However, only after the evaluation of the Lyapunov exponents can one know with certainty that the response of the system is truly chaotic. The Lyapunov exponents for this case have been found to be $(\lambda_1, \lambda_2 = +0.021, -0.074)$ and the positive value of λ_1 is suggesting chaotic response [22].

3.6 Brief Overview of Other Thermomechanical Constitutive Models for SMAs

Throughout the history of investigation into the complex thermomechanical response of shape memory alloys, many researchers have proposed a variety of different constitutive models with the goal of predicting material behavior. This section presents a summary of several of these models.

Describing the complex characteristics in the phase transitions in polycrystalline SMAs has been a significant challenge to researchers. These characteristics include modeling the hardening during phase transformation, the asymmetric response that SMAs exhibit in tension and compression, the modeling of detwinning of martensite, two-way shape memory effect and the effects of reorientation and the accumulation of plastic strains during cyclic loading (Birman [23], Patoor et al. [14], Lagoudas et al. [15]).

The topic of transformation surfaces (see Appendix A) and tension-compression asymmetry of the SMA response has been investigated by Raniecki and Lexcellent[24] who presented a model for pseudoelasticity of SMAs. A distinct feature of the model is its capability to take into account the difference between the tension and compression loading. This is accomplished by using a $J_2 - J_3$ transformation surface. The model uses exponential hardening functions. It was used in a later work by Raniecki et al. [25] to study bending of SMA beams undergoing pseudoelastic loading. In this work, the tension-compression difference was not addressed. The authors were able to determine the distribution of the martensitic volume fraction along the thickness of the beam during both loading and unloading. Additional results included plots of the beam curvature versus the applied moment. Rejzner et al. [26] have further extended the work on pseudoelastic beams, by including the effect of tension-compression asymmetry in the analysis and comparing the results with experimental data. It was found, however, that the tension-compression asymmetry does not have a significant influence on the macroscopic beam response.

The comprehensive study of Qidwai and Lagoudas [16] focused on the choice of different transformation functions and their effect on the material response. In particular, the asymmetry of the material behavior under tension and compression, as well as the volumetric transformation strain, can be modeled by choosing an appropriate functional form of transformation function. Qidwai and Lagoudas [16] proposed a transformation function, based on the J_2, J_3 and I_1 stress invariants, which can account for the observed asymmetry. The subject of the form of the transformation function has been revisited in a recent paper by Lexcellent et al. [27]. Multiaxial experiments on polycrystalline SMAs were performed to determine the initial transformation surface. The experiments have revealed tension-compression asymmetry, consistent with the results found in the literature. Motivated by the experimental results, Lexcellent et al. [27] proposed an analytical expression for the transformation function, based on the J_2 and J_3 stress invariants.

Another important aspect of the SMA response is the detwinning of martensite. A 1-D model that separates the martensitic volume fraction into two parts: thermally-induced (self-accommodated) and stress-induced (detwinned) has been presented by Brinson [28]. Leclercq and Lexcellent [29] have presented a similar model formulated in a 3-D framework; however, only

1-D implementation and numerical results have been provided. Comparisons are made with experimental data for both pseudoelastic mechanical loading as well as for isobaric thermally-induced transformation. The results are in reasonably good agreement, with the largest discrepancies observed for the case of isobaric thermally-induced transformation. In another study, Lagoudas and Shu [30] have proposed a 3-D model with three internal variables but again with only 1-D numerical implementation and results. The main difference with the earlier model by Leclercq and Lexcellent [29] is in the type of transformation hardening function. The extended SMA modeling presented in Chapter 6 further addresses some of the modeling issues associated with detwinning of martensite, based on the recent 3-D work by Popov and Laogudas [31].

One of the important problems recently addressed by researchers is the behavior of SMAs under cycling loading (Lexcellent and Bourbon [32], Fischer [33], Lexcellent et al. [34], Abeyaratne and Kim [35], Bo and Lagoudas [36–39], Lagoudas and Entchev [40]). During cyclic phase transformation, a substantial amount of plastic strains is accumulated. In addition, the transformation loop evolves with the number of cycles and TWSME is developed. Based on the experimental observations, researchers have attempted to create models able to capture the effects of cycling loading. One-dimensional models for the behavior of SMA wires under cycling loading have been presented by Tanaka et al. [41], Lexcellent and Bourbon [32], Lexcellent et al. [34], and Abeyaratne and Kim [35], among others.

A series of papers by Bo and Lagoudas [36–39] studies the cyclic behavior of SMA wires in one dimension. The work focuses on the modeling of stress-induced transformation, where both transformation and plastic strains occur simultaneously as a result of the applied stress. The resulting model is able to account for simultaneous development of transformation and plastic strains during phase transformation under applied loads. In addition to the plastic strain, the changes in the material response are also modeled by introducing evolution equations for the material parameters. Finally, minor hysteresis loops are also modeled by Bo and Lagoudas [38]. This is accomplished by modifying the transformation criterion and the hardening parameters during a minor loop. All of the above-mentioned features of the model have been demonstrated and the results have been compared with experimental data for NiTi SMA actuators. The results are in very good agreement.

A three-dimensional model for transformation induced plasticity has been presented by Fischer [33]. In contrast to the work by Lagoudas and Bo [37],a separate phase transformation condition and plasticity yield condition are used by Fischer [33]. The theory is presented in general terms, but the identification of the material parameters and the implementation are not discussed.

Many other approaches to modeling SMAs have been proposed. The work of Brocca et al. [42] presents a three-dimensional model for SMAs which is based on the microplane model by Bazant [43]. The main idea of the model is to deduce the macroscopic constitutive behavior of an SMA by describing the response of the SMA along planes of different orientations, called *microplanes*. The SMA constitutive behavior on the microplanes is described by a one-dimensional model. First, the normal and shear components of the stress on each microplane are defined in terms of the unit normal and tangential vectors of the plane and the macroscopic stress tensor. Next, the normal and shear components of the strain are calculated based on the constitutive model for the microplane. Finally, the components of the macroscopic strain tensor are calculated from the normal and shear strain components for a set of microplanes using the principal of virtual work. The particular SMA constitutive model on the microplane implemented in the work by Brocca et al. [42] is the one presented by Bekker and Brinson [44]. The effect of the hydrostatic pressure and the tension-compression asymmetry are also taken into account by modifying the critical stress values for phase transformation and the transformation temperatures. Various results demonstrating the capabilities of the microplane model are presented and compared with experimental data. The recent model of Anand and Gurtin [45] also captures this asymmetry and is formulated assuming finite deformations whereby the deformation gradient is multiplicatively decomposed into an elastic term and a separate term accounting for deformation due to the formation of martensitic variants.

One of the latest one-dimensional models for SMA wire actuator is presented by Shaw [46]. The model is capable of simulating both the pseudoelastic behavior and the shape memory effect. The new development of the model by Shaw [46] is the accounting for the instabilities during martensitic phase transformation by including strain gradient effects and by allowing a softening stress-strain response. Recent 1-D SMA modeling by Paiva, Savi and coworkers [47] has taken a different path, addressing also tension/compression asymmetry and the capability to capture the generation of plastic strain at stress levels exceeding the yield stress.

Different implementation issues for SMA constitutive models have been presented and discussed in a series of papers by Govindjee and coworkers [48–50]. The model presented by Govindjee and Kasper [48] is based on the stress-temperature phase diagram for SMAs. The approach taken in that work is similar to the one presented by Bekker and Brinson [51]. It is assumed that during the martensitic phase transformation, two different martensitic variants may form, depending on the sign of the applied loading. In addition, if the material is cooled in a stress-free condition, both variants form simultaneously, resulting in zero macroscopic transformation strain. Further mechanical loading will result in growth of one martensitic variant at the expense of the other, producing observable macroscopic transformation strain in the direction of the loading. In a later work Govindjee and Hall [52] present a constitutive model based on statistical physics. Similar to their previous work Govindjee and Kasper [48] the evolution of two martensitic variants is prescribed. A numerical implementation of the model and results for different loading conditions, including stress-induced and thermally-induced transformation are presented. A three-dimensional model for SMAs has been presented by Govindjee and Miehe [49] and Hall and Govindjee [50], where the number of martensitic variants formed during transformation is not restricted to two. The general equations of the model have been presented as well as its numerical implementation.

In conclusion, a historical summary of some commonly used phenomenological SMA models and their features is presented in the tables which conclude this chapter. For a thorough review of the many micromechanical models developed over the years, please see the text by K. Bhattacharya [54].

1-D Model	**Formulation**	**Features**
Tanaka [18]		Uses an exponential hardening rule for the phase transformation. Material properties remain constant during phase transformation. (1986)
Achenbach [55]	Potential energy wells	Considers three phases, austenite (A), tensile martensite (M_+), and compressive martensite (M_-). The method of potential energy wells is used to determine probabilistically when transformation will occur. (1989)
Liang and Rogers [20]	Helmholtz free energy	Uses a cosine hardening rule for the phase transformation. Material properties remain constant during phase transformation. (1990)
Brinson [28]	Helmholtz free energy	Introduces two internal variables allowing modeling of both detwinned and self accommodated martensite. Uses a cosine hardening law and variable elastic stiffness during phase transformation. (1993)
Abeyaratne and Knowles [56]	Helmholtz free energy	Considers the initiation and propagation of phase transformation in the context of a 1-D bar. Volume fraction of martensite not explicitly considered; instead, strain present in each phase is directly employed. (1993)
Ivshin and Pence [57]	Phase fraction as state function	Models the transformation between austenite and a single favored variant of martensite. Phase fraction of austenite directly postulated to have a given functional form where cases of hysteretic and non-hysteretic phase transformations are considered. Hyperbolic tangent function is used to model the transformation hysteresis. (1994)

1-D Model	**Formulation**	**Features**
Auricchio and Sacco [58]	Phase diagram based	Addresses isothermal pseudoelasticity. Critical transformation stresses at a given temperature are used to calibrate this plasticity-based model. Various micromechanical schemes are investigated to account for the change in elastic properties with phase change. A linear hardening law is used. (1997)
Bekker and Brinson [44, 51]	Phase diagram based	The model is formulated in terms of possible thermomechanical paths in stress-temperature space. Cosine hardening function is used for the phase transformation. (1997)
Govindjee and Kasper [48]	Phase diagram based	A distinction is made for martensitic variants in tension M^+ and compression M^-. Constant stiffness and thermal expansion coefficients are assumed. The model also includes kinematic hardening plasticity. The plastic strains influence the total amount of martensite that can be produced through an exponential relation. (1999)
Rajagopal and Srinivasa [59]	Helmholtz free energy	Transformation from austenite to one preferred variant of martensite is modeled by considering two distinct reference configurations, one being the austenitic (parent) phase, and another being the martensitic (product) phase. (1999)
Seelecke and Müller [60]	Potential energy wells	Expands on the work of Achenbach [55], considering three phases: austenite (A), tensile martensite (M_+), and compressive martensite (M_-). Here the thermomechanical coupling is emphasized. (2004)
Savi and Paiva [47, 53]	Helmholtz free energy	Considers plastic strain effects and plastic-phase transformation coupling, which makes it possible to describe TWSME. Tensile-compressive asymmetry and internal subloops are also considered. (2005)

3-D Model	**Formulation**	**Features**
Patoor, Eberhardt and Berveiller [61]	Gibbs free energy	Transformation of single crystals with multiple martensitic variants is first addressed. The methods are then extended to polycrystalline materials where Drucker-Prager-type transformation surfaces are assumed. (1987)
Liang and Rogers [62]	Helmholtz free energy	This is a three dimensional extension of Liang and Rogers [20] based on a J_2-type transformation surface. (1992)
Sun and Hwang [63, 64]	Gibbs free energy	Uses micromechanics to formulate a model at the macroscale. The internal variables are the total martensitic volume fraction and reoriented martensitic volume fraction. (1993)
Boyd and Lagoudas [12]	Gibbs free energy	Uses the volume fraction of martensite as internal variable and a transformation evolution equation to connect it with the transformation strain. A J_2-type transformation surface is used for the forward phase transformation. The model accounts for non-proportional loading paths by using a non-associative evolution equation during reverse transformation. A polynomial hardening function is introduced. (1996)
Auricchio, Taylor, and Lubliner [65–67]	Phase diagram based	This set of works on isothermal pseudoelasticity precedes that of Auricchio and Sacco [58], and considers three-dimensional response using a Drucker-Prager-type surface to describe the critical stresses for pseudoelastic transformation. An exponential hardening law is used, and focus is placed on finite element implementation of the model. (1996)
Lagoudas Bo, and Qidwai [21]	Gibbs free energy	Provides a unified framework for the earlier models of Tanaka [18], Liang and Rogers [20], Boyd and Lagoudas [12]. (1996)

3-D Model	Formulation	Features
Leclercq and Lexcellent [29]	Helmholtz free energy	Uses two internal variables to allow modeling of both detwinned and self-accommodated martensite. The transformation surfaces are J_2-type. Exponential hardening is used for both the transformation and the reorientation processes. (1996)
Raniecki and Lexcellent [24]	Gibbs free energy	Models tension-compression asymmetry using a $(J_2 - J_3)$-type transformation surface. Exponential transformation hardening functions are used. (1998)
Bo and Lagoudas [36–39]	Gibbs free energy	Extends the work of Boyd and Lagoudas [12] to include modeling of transformation induced plasticity. This is done by introducing the following state variables: transformation strain, plastic strain, drag stress, back stress and connecting their evolution to the evolution of the martensitic volume fraction ξ. (1999)
Qidwai and Lagoudas [16]	Gibbs free energy	Comprehensive modeling of tension-compression asymmetry. An extension of the work by Boyd and Lagoudas [12]. Several different transformation surfaces based on I_1-, J_2- and J_3-type invariants are proposed. (2000)
Govindjee, Hall, and Miehe [49, 50]	Helmholtz/Gibbs free energy	This work generalizes earlier phenomenological models by considering an arbitrary number of martensitic variants. The focus of the paper is the numerical implementation of the model via a nonlinear optimization method. (2001)
Brocca, Brinson, and Bažant [42]	Microplane based	The constitutive model on each microplane is the model by Brinson [28]. The model is able to account for nonproportional loading paths. (2002)
Juhasz et al. [68]	Helmholtz free energy	The effects of reorientation are taken into account by taking the entire transformation strain ε^t as an internal variable instead of just the detwinned martensitic volume fraction. (2002)

3-D Model	Formulation	Features
Lagoudas and Entchev [40]	Gibbs free energy	The model accounts for the simultaneous development of transformation and plastic strains during stress-induced phase transformation. The work extends and implements in 3-D the model presented in Bo and Lagoudas [37]. (2003)
Helm and Haupt [69]	Helmholtz free energy	Both temperature-induced and stress-induced martensite are considered. This work approaches the forward and reverse transformation using a single transformation function with kinematic hardening. (2003)
Anand and Gurtin [45]	Helmholtz free energy	Derived in a finite deformation framework whereby the deformation gradient is multiplicatively decomposed into an elastic term and an additional term accounting for deformations caused by the martensitic transformation. Tension-compression asymmetry in single crystals is considered, as well as the influence of texture in polycrystals. A three-dimensional implementation is performed. (2003)
Popov and Lagoudas [31]	Gibbs free energy	The polycrystalline SMA is considered a mixture of three species-self accommodated martensite M^t, detwinned martensite M^d and austenite A extending the work of Lagoudas and Shu [70]. The internal variables describe the three possible transitions between the different species. In addition, the model presents a three-dimensional implementation. (2007)
Reese and Christ [71]	Helmholtz free energy	The focus of this model is large deformation finite element analysis considering only isothermal pseudoelasticity. Large rotations and moderate strains are considered. In the style of Helm and Haupt [69], a single transformation function and kinematic hardening are used to describe the critical stresses for transformation. (2008)

3.7 Summary

This chapter discussed the basic elements of thermomechanical modeling of polycrystalline SMAs. A methodology for obtaining constitutive equations for SMAs by enforcing the basic principles of continuum thermodynamics was discussed in detail. The derivation of the SMA constitutive model proposed by Boyd and Lagoudas was reviewed in detail and different thermomechanical loading paths were considered. Various examples were presented throughout the chapter to illustrate the capability of the model to describe different aspects of the SMA behavior. This chapter also presented the unification of different constitutive models for SMAs under the same thermodynamic framework, and summarized other SMA constitutive models available in the literature.

3.8 Problems

3.1. The procedure for obtaining constitutive assumptions for dissipative materials was presented in Sec. 3.2.3. You are required to obtain the form of Helmholtz free energy, ψ, of a thermoelastic material that is a function only of strain, $\boldsymbol{\varepsilon}$, and temperature, T, by performing a Taylor series expansion of $\psi(\boldsymbol{\varepsilon}, T)$ around a reference state. Moreover, obtain the constitutive relations for $\boldsymbol{\sigma}$ and s. Hint: See Batra [8].

3.2. Derive the equation for the reverse phase transformation condition which is complimentary to the equation for the forward transformation given in (3.3.71).

3.3. Starting with the consistency condition as given in (3.3.62), derive a relation between the difference in reference entropy, $\rho\Delta s_0$, and the stress influence coefficients, C^A and C^M (as discussed in Chapter 2) at zero stress. Note that the martensitic volume fraction is constant at the transformation surface boundaries, for forward and reverse transformations ($\xi = 0$ or $\xi = 1$).

3.4. In Example 3.2 the transformation surfaces for the forward martensitic phase transformation, for $\xi = 0$ and $\xi = 1$ were plotted. Plot the transformation surfaces for the reverse phase transformation for $\xi = 0$ and $\xi = 1$, using the same stress state of the material given in Example 3.2.

3.5. Show that the transformation tensor $\boldsymbol{\Lambda}$ reduces to $H^{cur}(\sigma)$ in the one dimensional case for the case of forward phase transformation. Also give the one-dimensional form of $\boldsymbol{\Lambda}$ for the reverse phase transformation.

3.6. In Sect. 3.5.1 we presented the analytical solutions for the Boyd-Lagoudas model. Derive analytical solutions using the hardening functions given in Sect. 3.4 for the exponential and cosine models.

3.7. Rework Example 3.4, where different isothermal pseudoelastic loading paths were plotted, but using the hardening functions given in Sect. 3.4. Compare the stress vs. strain curves of the exponential and cosine model with the polynomial model.

3.8. Rework Example 3.5, where different isothermal loading paths were plotted, but using the hardening functions given in Sect. 3.4. Compare the strain vs. temperature curves of the exponential and cosine model with the polynomial model.

3.9. Rework Example 3.6, where the cases of isothermal and adiabatic pseudoelastic loading paths were compared, but using the hardening functions given in Sect. 3.4. Compare the stress vs. strain curves of the exponential and cosine model with the polynomial model for the isothermal and adiabatic conditions.

3.10. Rework Example 3.7, using the smooth transformation hardening function given in Sect. 3.4.

3.11. Use the 1-D reduction of the polynomial constitutive model presented in this chapter to rework Problem 2.15 which addresses the cyclic thermomechanical response of a differential SMA actuator.

3.12. Write the transformation criteria for the onset and end of martensitic to austenitic phase transformation, as well as the forward phase transformation from austenite to martensite using principal stresses. Plot the transformation surfaces on the π-plane, which is the plane perpendicular to the hydrostatic axis, for $T > A_f$.

3.13. Derive the appropriate equations describing minor hysteresis loops according to the polynomial model presented in this chapter. As a demonstration, plot the uniaxial stress vs. strain curves for pseudoelastic condition at $T = A_s + 288K$, unloading at values of the martensitic volume fraction of $\xi = 0.1, 0.3, 0.5, 0.7$ and 0.9. Use the material properties used in Example 3.2.

Discuss the relation of the above minor hysteresis loops with respect to the major hysteresis loop, which derived for the full forward and reverse phase transformation.

3.14. A shape memory alloy can also be thought of as two materials combined into one. This idea leads to the assumption that SMAs have two reference configurations, one being the austenitic phase (parent phase) and the second one being the crystallographically different martensitic phase. The approach take in this chapter is to assume the parent phase to be the main reference (stress-free) configuration and the transition to the second stress-free configuration of martensite is achieved through the internal state variable ξ and the corresponding transformation strain. How can one reformulate the approach presented in this chapter by choosing martensite to be the parent phase?

3.15. In the paper of Bo and Lagoudas [39], minor hysteresis loops are introduced to account for experimental observations, which are different from the minor hysteresis loops described by the models of this chapter. Explain the key idea of the minor hysteresis loops model in Bo and Lagoudas [39].

3.16. Explain the key ideas and assumptions in the minor hysteresis loops model of Ivshin and Pence [57].

3.17. Study two cases, one isothermal and one adiabatic, of the chaotic response of one-degree of freedom oscillators.

3.18. Derive the equations of a one-degree of freedom oscillator in the case of heat convection.

3.19. The constitutive models presented in this chapter are all rate-independent (i.e., they are described by differential equations homogeneous in time) and real time can be replaced by any convenient loading parameter that captures the loading path dependence. Describe a way of introducing real time dependence in the evolution equations for the internal state variables for the thermodynamic constitutive models of this chapter.

3.20. Model a uniaxial pseudoelastic experiment subject to the loading path shown in Fig. 3.21 where loading is assumed to be under adiabatic conditions while unloading is assumed to be under isothermal conditions.

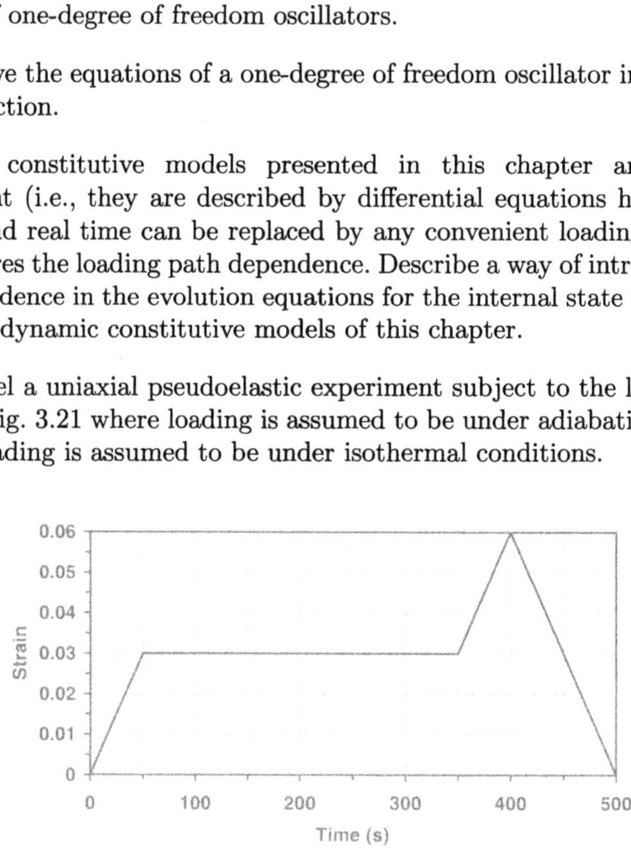

Fig. 3.21. Strain loading path for consideration in Problem 3.20.

References

[1] B. D. Coleman, W. Noll, The thermodynamics of elastic materials with heat conduction and viscosity, Archive for Rational Mechanics and Analysis 13 (1963) 167–178.

[2] B. D. Coleman, M. E. Gurtin, Thermodynamics with internal state variables, The Journal of Chemical Physics 47 (2) (1967) 597–613.

[3] C. Truesdell, W. Noll, The Non-Linear Field Theories of Mechanics, Springer, 1965.

[4] L. E. Malvern, Introduction to the Mechanics of a Continuous Medium, Series in Engineering of the Physical Sciences, Prentice Hall, Upper Saddle River, NJ, 1969.

[5] M. E. Gurtin, An Introduction to Continuum Mechanics, Vol. 158 of Mathematics in Science and Engineering, Academic Press, San Diego, 1981.

[6] W. M. Lai, D. Rubin, E. Krempl, Introduction to Continuum Mechanics, 3rd Edition, Butterworth-Heinemann, Woburn, MA, 1993.

[7] W. S. Slaughter, The Linearized Theory of Elasticity, Birkhäuser, Boston, 2002.

[8] R. Batra, Elements of Continuum Mechancis, AIAA, Reston, VA, 2006.

[9] G. Holzapfel, Nonlinear Solid Mechanics: A Continuum Approach for Engineering, Wiley Chichester, UK, 2000.

[10] B. D. Coleman, W. Noll, Material symmetry and thermostatic inequalities in finite elastic deformations, Archive for Rational Mechanics and Analysis 15 (1964) 87–111.

[11] B. D. Coleman, Thermodynamics of materials with memory, Archive for Rational Mechanics and Analysis 17 (1964) 1–46.

[12] J. G. Boyd, D. C. Lagoudas, A thermodynamical constitutive model for shape memory materials. Part I. The monolithic shape memory alloy, International Journal of Plasticity 12 (6) (1996) 805–842.

[13] M. A. Qidwai, D. C. Lagoudas, Numerical implementation of a shape memory alloy thermomechanical constitutive model using return mapping algorithms, International Journal for Numerical Methods in Engineering 47 (2000) 1123–1168.

[14] E. Patoor, D. C. Lagoudas, P. B. Entchev, L. C. Brinson, X. Gao, Shape memory alloys, Part I: General properties and modeling of single crystals, Mechanics of Materials 38 (5–6) (2006) 391–429.

[15] D. C. Lagoudas, P. B. Entchev, P. Popov, E. Patoor, L. C. Brinson, X. Gao, Shape memory alloys, Part II: Modeling of polycrystals, Mechanics of Materials 38 (5–6) (2006) 430–462.

[16] M. A. Qidwai, D. C. Lagoudas, On the thermodynamics and transformation surfaces of polycrystalline NiTi shape memory alloy material, International Journal of Plasticity 16 (2000) 1309–1343.

[17] J. C. Simo, T. J. R. Hughes, Computational Inelasticity, Vol. 7 of Interdisciplinary Applied Mathematics, Springer-Verlag, New York, 1998.

[18] K. Tanaka, A thermomechanical sketch of shape memory effect: One-dimensional tensile behavior, Res Mechanica 18 (1986) 251–263.

[19] K. Tanaka, S. Kobayashi, Y. Sato, Thermomechanics of transformation pseudoelasticity and shape memory effect in alloys, International Journal of Plasticity 2 (1986) 59–72.

[20] C. Liang, C. A. Rogers, One-dimensional thermomechanical constitutive relations for shape memory materials, Journal of Intelligent Material Systems and Structures 1 (1990) 207–234.

[21] D. C. Lagoudas, Z. Bo, M. A. Qidwai, A unified thermodynamic constitutive model for SMA and finite element analysis of active metal matrix composites, Mechanics of Composite Materials and Structures 3 (1996) 153–179.

[22] L. Machado, Shape memory alloys for vibration isolation and damping, Ph.D. thesis, Texas A&M University, College Station, TX (2007).

[23] V. Birman, Review of mechanics of shape memory alloy structures, Applied Mechanics Reviews 50 (11) (1997) 629–645.

[24] B. Raniecki, C. Lexcellent, Thermodynamics of isotropic pseudoelasticity in shape memory alloys, European Journal of Mechanics - A/Solids 17 (2) (1998) 185–205.

[25] B. Raniecki, J. Rejzner, C. Lexcellent, Anatomization of hysteresis loops in pure bending of ideal pseudoelastic SMA beams, International Journal of Mechanical Sciences 43 (5) (2001) 1339–1368.

[26] J. Rejzner, C. Lexcellent, B. Raniecki, Pseudoelastic behaviour of shape memory alloy beams under pure bending: experiments and modelling, International Journal of Mechanical Sciences 44 (4) (2002) 665–686.

[27] C. Lexcellent, A. Vivet, C. Bouvet, S. Calloch, P. Blanc, Experimental and numerical determinations of the initial surface of phase transformations under biaxial loading in some polycrystalline shape-memory alloys, Journal of the Mechanics and Physics of Solids 50 (2002) 2717–2735.

[28] L. C. Brinson, One-dimensional constitutive behavior of shape memory alloys: Thermomechanical derivation with non-constant material functions and redefined martensite internal variable, Journal of Intelligent Material Systems and Structures 4 (1993) 229–242.

[29] S. Leclercq, C. Lexcellent, A general macroscopic description of the thermomechanical behavior of shape memory alloys, JMPS 44 (6) (1996) 953–980.

[30] D. C. Lagoudas, S. G. Shu, Residual deformation of active structures with SMA actuators, International Journal of Mechanical Sciences 41 (1999) 595–619.

[31] P. Popov, D. C. Lagoudas, A 3-D constitutive model for shape memory alloys incorporating pseudoelasticity and detwinning of self-accommodated martensite, International Journal of Plasticity 23 (10–11) (2007) 1679–1720.

[32] C. Lexcellent, G. Bourbon, Thermodynamical model of cyclic behavior of Ti-Ni and Cu-Zn-A— shape memory alloys under isothermal undulated tensile tests, Mechanics of Materials 24 (1996) 59–73.

[33] F. Fischer, E. Oberaigner, K. Tanaka, F. Nishimura, Transformation induced plasticity revised an updated formulation, International Journal of Solids and Structures 35 (18) (1998) 2209–2227.

[34] C. Lexcellent, S. Leclerq, B. Gabry, G. Bourbon, The two way shape memory effect of shape memory alloys: An experimental study and a phenomenological model, International Journal of Plasticity 16 (2000) 1155–1168.

[35] R. Abeyaratne, S. Kim, Cyclic effects in shape-memory alloys: a one-dimensional continuum model, International Journal of Solids and Structures 34 (25) (1997) 3273–3289.

[36] Z. Bo, D. C. Lagoudas, Thermomechanical modeling of polycrystalline SMAs under cyclic loading, Part I: Theoretical Derivations, International Journal of Engineering Science 37 (1999) 1089–1140.

[37] D. Lagoudas, Z. Bo, Thermomechanical modeling of polycrystalline SMAs under cyclic loading, part II: Material characterization and experimental results for a stable transformation cycle, International Journal of Engineering Science 37 (9) (1999) 1141–1173.

[38] Z. Bo, D. C. Lagoudas, Thermomechanical modeling of polycrystalline SMAs under cyclic loading, Part III: Evolution of plastic strains and two-way shape memory effect, International Journal of Engineering Science 37 (1999) 1175–1203.

[39] Z. Bo, D. C. Lagoudas, Thermomechanical modeling of polycrystalline SMAs under cyclic loading, Part IV: Modeling of minor hysteresis loops, International Journal of Engineering Science 37 (1999) 1205–1249.

[40] D. C. Lagoudas, P. B. Entchev, Modeling of transformation-induced plasticity and its effect on the behavior of porous shape memory alloys. Part I: Constitutive model for fully dense SMAs, Mechanics of Materials 36 (9) (2004) 865–892.

[41] K. Tanaka, F. Nishimura, T. Hayashi, H. Tobushi, C. Lexcellent, Phenomenological analysis on subloops and cyclic behavior in shape memory alloys under mechanical and/or thermal loads, Mechanics of Materials 19 (1995) 281–292.

[42] M. Brocca, L. Brinson, Z. Bažant, Three-dimensional constitutive model for shape memory alloys based on microplane model, Journal of the Mechanics and Physics of Solids 50 (5) (2002) 1051–1077.

[43] Z. Bazant, Microplane model for strain controlled inelastic behavior, Mechanics of engineering materials (1984) 45–59.

[44] A. Bekker, L. C. Brinson, Phase diagram based description of the hysteresis behavior of shape memory alloys, Acta Materialia 46 (10) (1998) 3649–3665.

[45] L. Anand, M. E. Gurtin, Thermal effects in the superelasticity of crystalline shape-memory materials, Journal of the Mechanics and Physics of Solids 51 (2003) 1015–1058.

[46] J. Shaw, A thermomechanical model for a 1-D shape memory alloy wire with propagating instabilities, International Journal of Solids and Structures 39 (5) (2002) 1275–1305.

[47] A. Paiva, M. A. Savi, A. M. B. Bragra, P. M. C. L. Pacheco, A constitutive model for shape memory alloys considering tensile-compressive

asymmetry and plasticity, International Journal of Solids and Structures 42 (11–12) (2005) 3439–3457.

[48] S. Govindjee, E. P. Kasper, Computational aspects of one-dimensional shape memory alloy modeling with phase diagrams, Computer Methods in Applied Mechanics and Engineering 171 (1999) 309–326.

[49] S. Govindjee, C. Miehe, A multi-variant martensitic phase transformation model: formulation and numerical implementation, Computer Methods in Applied Mechanics and Engineering 191 (2001) 215–238.

[50] G. Hall, S. Govindjee, Application of a partially relaxed shape memory free energy function to estimate the phase diagram and predict global microstructure evolution, Journal of the Mechanics and Physics of Solids 50 (3) (2002) 501–530.

[51] A. Bekker, L. C. Brinson, Temperature-induced phase transformation in a shape memory alloy: Phase diagram based kinetics approach, Journal of the Mechanics and Physics of Solids 45 (6) (1997) 949–988.

[52] S. Govindjee, G. J. Hall, A computational model for shape memory alloys, International Journal of Solids and Structures 37 (5) (2000) 735–760.

[53] M. A. Savi, A. Paiva, Describing internal subloops due to incomplete phase transformations in shape memory alloys, Archive of Applied Mechanics 74 (9) (2005) 637–647.

[54] K. Bhattacharya, Microstructure of Martensite: Why it Forms and How it Gives Rise to the Shape-Memory Effect, Oxford University Press, Oxford, 2003.

[55] M. Achenbach, A model for an alloy with shape memory, International Journal of Plasticity 5 (1989) 371–395.

[56] R. Abeyaratne, J. K. Knowles, Continuum model of a thermoelastic solid capable of undergoing phase transformation, Journal of the Mechanics and Physics of Solids 41(3) (1993) 541–571.

[57] I. Ivshin, T. J. Pence, A thermodynamical model for one-variant shape memory alloy material, Journal of Intelligent Material Systems and Structures 5 (1994) 455–473.

[58] F. Auricchio, E. Sacco, A one-dimensional model for superelastic shape-memory alloys with different elastic properties between austenite and martensite, International Journal of Non-Linear Mechanics 32 (6) (1997) 1101–1114.

[59] K. Rajagopal, A. R. Srinivasa, On the thermomechanics of shape memory wires, Z. angew. Math. Phys. 50 (1999) 459–496.

[60] S. Seelecke, I. Müller, Shape memory alloy actuators in smart structures: Modeling and simulation, Applied Mechanics Reviews 57 (1) (2004) 23–46.

[61] E. Patoor, A. Eberhardt, M. Berveiller, Potential pseudoelastique et plasticité de transformation martensitique dans les mono- et polycristaux metalliques, Acta Materialia 35 (11) (1987) 2779–2789.

[62] C. Liang, C. A. Rogers, A multi-dimensional constitutive model for shape memory alloys, Journal of Engineering Mathematics 26 (1992) 429–443.

[63] Q. P. Sun, K. C. Hwang, Micromechanics modelling for the constitutive behavior of polycrystalline shape memory alloys — I. Derivation of general relations, Journal of the Mechanics and Physics of Solids 41(1) (1983) 1–17.

[64] Q. P. Sun, K.C. Hwang, Micromechanics modelling for the constitutive behavior of polycrystalline shape memory alloys — II. Study of individual phenomena, Journal of the Mechanics and Physics of Solids 41 (1) (1983) 19–33.

[65] J. Lubliner, F. Auricchio, Generalized plasticity and shape memory alloys, International Journal of Solids and Structures 33 (7) (1996) 991–1003.

[66] F. Auriccho, R. L. Taylor, Shape-memory alloys: Modelling and numerical simulation of the finite-strain superelastic behavior., Computer Methods in Applied Mechanics and Engineering 143 (1–2) (1997) 175–194.

[67] F. Auricchio, R. L. Taylor, J. Lubliner, Shape-memory alloys: Macromodelling and numerical simulations of the superelastic behavior, Computer Methods in Applied Mechanics and Engineering 146 (1997) 281–312.

[68] L. Juhász, E. Schnack, O. Hesebeck, H. Andrä, Macroscopic modeling of shape memory alloys under non-proportional themo-mechanical loadings, Journal of Intelligent Material Systems and Structures 13 (2002) 825–836.

[69] D. Helm, P. Haupt, Shape memory behavior: Modelling within continuum thermomechanics, International Journal of Solids and Structures 40 (2003) 827–849.

[70] D. C. Lagoudas, S. G. Shu, Residual deformation of active structures with SMA actuators, International Journal of Mechanical Sciences 41 (1999) 595–619.

[71] S. Reese, D. Christ, Finite deformation pseudo-elasticity of shape memory alloys – Constitutive modelling and finite element implementation, International Journal of Plasticity 24 (2008) 455–482.

4

Numerical Implementation of an SMA Thermomechanical Constitutive Model Using Return Mapping Algorithms

M. A. Siddiq Qidwai, D. J. Hartl, and D. C. Lagoudas

In the previous chapter we described the derivation of a 3-D SMA thermomechanical constitutive model. We now address the numerical implementation of this model and the development of numerical tools to support the process of designing SMA devices for use in load bearing 3-D structures. In this chapter, the numerical implementation of SMA thermomechanical constitutive response is presented using return mapping algorithms appropriate for rate-independent inelastic constitutive models. The closest point projection return mapping algorithm and the convex cutting plane return mapping algorithm are discussed, and finite element analysis examples are provided.

4.1 Introduction

Recall the many SMA applications introduced in Chapter 1. Many of these devices were designed and built without the use of modern tools of design and analysis. Throughout all industrial sectors, most SMA systems are the result of repeated design/build/test cycles. One reason for this is that reliable models that accurately account for the complex thermomechanical behavior of SMA components under general loading conditions were slow to be developed and integrated into finite element analysis software. In addition, the legacy finite element codes commonly used by industry did not easily incorporate new advances in material constitutive modeling. Some nonlinear commercial packages have begun including SMAs as material options; however, the constitutive models included are more suitable for loading paths such as pseudoelastic loading, rather than thermally activated paths over multiple cycles which are of greater interest (as in actuators for aerospace systems).

The *finite element method* (FEM) is a tool for the design and analysis of engineering structures. To be used in the analysis and design of shape memory alloy applications, FEM must be integrated with the numerical implementation of SMA constitutive models in 3-D. Such FEM implementations are a relatively recent development made possible by the increasing fidelity of 3-D SMA constitutive models. These improved models are themselves made

D.C. Lagoudas (ed.), *Shape Memory Alloys*, DOI: 10.1007/978-0-387-47685-8_4,

possible by the expanding availability of experimental data. Each of these developments is driven by the needs of the application design community and requires accurate models as well as robust numerical algorithms.

As research on the modeling of SMA materials began in the late 1980s and early 1990s, the focus was on 1-D models such as those derived at the end of Chapter 2. Soon, such models based on 1-D experimental results were being generalized to 3-D: the exponential model [1], the cosine model [2], the polynomial model [3], the modified cosine model [4], and the unified thermomechanical model [5] among others. These and other models have been discussed in more detail in the previous chapter. For the purpose of numerical implementation in this chapter, the unified thermomechanical model is chosen. Recall that it unifies the aforementioned SMA constitutive models (exponential, polynomial, and cosine) under a consistent thermodynamic framework; however, while the implementation of the unified model is described here, the methods outlined are suitable for any similar phenomenological model.

The reader will recall that the unified model implemented in this chapter was formulated under the assumption of infinite strains, which is most applicable for the analysis of structures undergoing small strains and moderate displacements. Recent work has addressed the more general assumption of finite strains, including the work by Anand and Gurtin [6] and, more recently, the work by Reese and Christ [7], where both groups address finite element implementations.

Return mapping algorithms have been developed to numerically integrate incremental constitutive equations for rate-independent inelastic materials [8]. Thus, their use for integrating the SMA thermomechanical constitutive response is typical. Rate-independence and loading history dependence allow the thermomechanical constitutive behavior to be defined by a stress-strain state relation that includes the transformation strain as an internal state variable, a differential evolution equation for the transformation strain, and a transformation function to determine the onset of phase transformation [5]. This structure implies that any proven numerical algorithm designed to integrate the rate-independent elastoplastic constitutive behavior can be used to integrate the SMA thermomechanical elastic-transformation model, where the concept of a yield criterion is replaced by the transformation criterion.

Return mapping algorithms have been studied extensively for elastoplasticity in the integration of constitutive relations. They are also called elastic predictor-plastic corrector algorithms in which a purely thermoelastic *trial state* is followed by a plastic (or transformation in the case of an SMA) corrector phase (return mapping). The corrector phase enforces consistency with the prescribed transformation flow rule. Return mapping algorithms may differ by the kind of discretization employed to numerically integrate the evolution equations and the numerical procedure adopted to solve the resultant set of nonlinear algebraic equations in the corrector phase. Some return mapping algorithms proposed in the literature are the radial return algorithm and the mean normal (mid-point) algorithm, among others [9]. Most of the return mapping algorithms employ integration rules that are particular cases of the trapezoidal and midpoint rules, suitably generalized to facilitate satisfaction of the yield (transformation) consistency condition [10, 11].

The two return mapping algorithms implemented in this work, which follows the presentation of Qidwai and Lagoudas [12] are the closest point projection algorithm [13] and convex cutting plane algorithm [14, 15]. State variables are computed for a given deformation history and both algorithms rely on the strain driven nature of the problem. They differ during the correction step. The application of the closest point projection algorithm results in a set of nonlinear algebraic equations solved using Newton's iteration method. This algorithm is unconditionally stable provided the yield surface is convex, and it is first-order accurate [13]. The convex cutting plane method is based on the explicit integration of the differential equations, and its advantage lies in its simplicity and reduction in computations.

4.2 Continuum Tangent Moduli Tensors

In this chapter, we consider incremental displacement based FEM. For such an implementation, the increment of the stress tensor is obtained through the implementation of the incremental SMA constitutive model for given increments of strain and temperature. The tensors that relate the strain and temperature increments (input) to the calculated stress increment (output) are the *tangent moduli tensors*. To formulate these tensors and begin the implementation process, the SMA constitutive model presented in Sect. 3.3.3 should be written in an incremental form as follows:

$$d\boldsymbol{\sigma} = \mathcal{L} : d\boldsymbol{\varepsilon} + \boldsymbol{\Theta}\, dT, \tag{4.2.1}$$

where $\mathcal{L}$ is the tangent stiffness tensor and $\boldsymbol{\Theta}$ is the tangent thermal moduli tensor. These two tensors are needed for the execution of the global FEM; therefore, it is important to derive the forms of these tensors.

To derive $\mathcal{L}$ and $\boldsymbol{\Theta}$, the constitutive relation in (3.3.47) is rewritten in differential form and the transformation strain evolution equation (3.3.51) is substituted to get

$$d\boldsymbol{\sigma} = \mathcal{S}^{-1} : \{d\boldsymbol{\varepsilon} - \boldsymbol{\alpha} dT - [\Delta\mathcal{S} : \boldsymbol{\sigma} + \Delta\boldsymbol{\alpha}(T - T_0) + \boldsymbol{\Lambda}]d\xi\}. \tag{4.2.2}$$

Using (3.3.57) and (3.3.58), the above equation reduces to

$$d\boldsymbol{\sigma} = \mathcal{S}^{-1} : \left(d\boldsymbol{\varepsilon} - \boldsymbol{\alpha} dT - d\xi \begin{cases} \partial_{\boldsymbol{\sigma}}\Phi; & \dot{\xi} > 0 \\ -\partial_{\boldsymbol{\sigma}}\Phi; & \dot{\xi} < 0. \end{cases} \right) \tag{4.2.3}$$

The differentiation of the transformation function (consistency condition) results in

$$d\Phi = \partial_{\boldsymbol{\sigma}}\Phi : d\boldsymbol{\sigma} + \partial_T \Phi dT + \partial_\xi \Phi d\xi = 0. \tag{4.2.4}$$

For forward transformation ($\dot{\xi} > 0$), an expression for the differential of the martensitic volume fraction, $d\xi$, is obtained by substituting $d\boldsymbol{\sigma}$ from (4.2.3a) into (4.2.4), i.e.,

$$d\xi = \frac{\partial_{\boldsymbol{\sigma}}\Phi : \mathcal{S}^{-1} : d\boldsymbol{\varepsilon} + (\partial_T\Phi - \partial_{\boldsymbol{\sigma}}\Phi : \mathcal{S}^{-1} : \boldsymbol{\alpha})dT}{\partial_{\boldsymbol{\sigma}}\Phi : \mathcal{S}^{-1} : \partial_{\boldsymbol{\sigma}}\Phi - \partial_{\xi}\Phi}. \tag{4.2.5}$$

Now (4.2.5) can be used to eliminate $d\xi$ in (4.2.3a) and obtain the relationship between the stress, strain and temperature increments as

$$\begin{aligned} d\boldsymbol{\sigma} = {} & [\mathcal{S}^{-1} - \frac{\mathcal{S}^{-1} : \partial_{\boldsymbol{\sigma}}\Phi \otimes \mathcal{S}^{-1} : \partial_{\boldsymbol{\sigma}}\Phi}{\partial_{\boldsymbol{\sigma}}\Phi : \mathcal{S}^{-1} : \partial_{\boldsymbol{\sigma}}\Phi - \partial_{\xi}\Phi}] : d\boldsymbol{\varepsilon} \\ & + \mathcal{S}^{-1} : [\partial_{\boldsymbol{\sigma}}\Phi(\frac{\partial_{\boldsymbol{\sigma}}\Phi : \mathcal{S}^{-1} : \boldsymbol{\alpha} - \partial_T\Phi}{\partial_{\boldsymbol{\sigma}}\Phi : \mathcal{S}^{-1} : \partial_{\boldsymbol{\sigma}}\Phi - \partial_{\xi}\Phi}) - \boldsymbol{\alpha}]dT. \end{aligned} \tag{4.2.6}$$

The tangent stiffness tensor, $\mathcal{L}$, and tangent thermal moduli tensor, $\boldsymbol{\Theta}$, are then given, after some simplifications of (4.2.6) by

$$\mathcal{L} = \mathcal{S}^{-1} - \mathbf{A} \otimes \mathbf{A}, \tag{4.2.7}$$

$$\boldsymbol{\Theta} = -\mathcal{L} : \boldsymbol{\alpha} - \frac{\partial_T\Phi}{a}\mathbf{A}, \tag{4.2.8}$$

where

$$\begin{aligned} a &= \sqrt{\partial_{\boldsymbol{\sigma}}\Phi : \mathcal{S}^{-1} : \partial_{\boldsymbol{\sigma}}\Phi - \partial_{\xi}\Phi}, \\ \mathbf{A} &= \frac{\mathcal{S}^{-1} : \partial_{\boldsymbol{\sigma}}\Phi}{\sqrt{\partial_{\boldsymbol{\sigma}}\Phi : \mathcal{S}^{-1} : \partial_{\boldsymbol{\sigma}}\Phi - \partial_{\xi}\Phi}} = \frac{\mathcal{S}^{-1} : \partial_{\boldsymbol{\sigma}}\Phi}{a}. \end{aligned} \tag{4.2.9}$$

Following the same methodology for the reverse phase transformation, i.e., substituting (4.2.3b) into (4.2.4) gives

$$d\xi = -\frac{\partial_{\boldsymbol{\sigma}}\Phi : \mathcal{S}^{-1} : d\boldsymbol{\varepsilon} + (\partial_T\Phi - \partial_{\boldsymbol{\sigma}}\Phi : \mathcal{S}^{-1} : \boldsymbol{\alpha})dT}{\partial_{\boldsymbol{\sigma}}\Phi : \mathcal{S}^{-1} : \partial_{\boldsymbol{\sigma}}\Phi + \partial_{\xi}\Phi}. \tag{4.2.10}$$

Using (4.2.3b) and (4.2.10), the differential of stress can be obtained in terms of $d\boldsymbol{\varepsilon}$ and dT in a similar way as described for the forward transformation, and $\mathcal{L}$ and $\boldsymbol{\Theta}$ are then given by

$$\mathcal{L} = \mathcal{S}^{-1} - \mathbf{B} \otimes \mathbf{B}, \tag{4.2.11}$$

$$\boldsymbol{\Theta} = -\mathcal{L} : \boldsymbol{\alpha} - \frac{\partial_T\Phi}{b}\mathbf{B}, \tag{4.2.12}$$

where

$$\begin{aligned} b &= \sqrt{\partial_{\boldsymbol{\sigma}}\Phi : \mathcal{S}^{-1} : \partial_{\boldsymbol{\sigma}}\Phi - \partial_{\xi}\Phi}, \\ \mathbf{B} &= \frac{\mathcal{S}^{-1} : \partial_{\boldsymbol{\sigma}}\Phi}{\sqrt{\partial_{\boldsymbol{\sigma}}\Phi : \mathcal{S}^{-1} : \partial_{\boldsymbol{\sigma}}\Phi - \partial_{\xi}\Phi}} = \frac{\mathcal{S}^{-1} : \partial_{\boldsymbol{\sigma}}\Phi}{b}. \end{aligned} \tag{4.2.13}$$

4.3 Return Mapping Algorithms

For the system of algebraic and ordinary differential equations (ODEs) given by (3.3.47) and (3.3.51), as well as constrained by (3.3.60) and (3.3.61), it is assumed that the strain and temperature histories $\boldsymbol{\varepsilon}(t)$ and $T(t)$ are known for $t \in [0, \hat{t}]$, as well as the initial conditions for $\boldsymbol{\varepsilon}^t$ and ξ at $t = 0$. Then, the evolution equations for the transformation strain are discretized using the generalized trapezoidal rule as

$$\boldsymbol{\varepsilon}^t_{n+1} = \boldsymbol{\varepsilon}^t_n + (\xi_{n+1} - \xi_n)[(1 - \lambda)\boldsymbol{\Lambda}_n + \lambda\boldsymbol{\Lambda}_{n+1}], \tag{4.3.1}$$

where λ ranges from $[0, 1]$ and subscript n indicates function evaluations at time t_n (similarly for $n + 1$), assuming time increment $\Delta t = t_{n+1} - t_n$. For different values of λ, various difference operators can be obtained. In the present formulation, two such values are chosen. In particular, for $\lambda = 1$ the implicit (backward) Euler integration rule is obtained. This implicit algorithm is implemented for the set of equations and Kuhn-Tucker conditions given by (3.3.47), (3.3.51) and (3.3.60) and (3.3.61) in Sect. 4.3.2, while in Sect. 4.3.3 the explicit (forward) Euler integration rule ($\lambda = 0$) is implemented. Both integration methods result in a set of nonlinear algebraic equations whose implicit nature requires multiple iterations to obtain a solution.

The return mapping algorithm divides the problem posed by this set of nonlinear algebraic equations in an additive split. Only the second part accounting for the transformation behavior is subject (elastic prediction and transformation correction) to iterations.

4.3.1 A General View of Thermoelastic Prediction-Transformation Correction Return Mapping

To begin, a thermoelastic predictor problem is solved that is described by assuming that the increment of the transformation strain is zero, i.e.,

$$\dot{\boldsymbol{\varepsilon}} = \dot{\boldsymbol{\varepsilon}}(t), \qquad \dot{T} = \dot{T}(t), \qquad \dot{\boldsymbol{\varepsilon}}^t = 0. \tag{4.3.2}$$

The transformation corrector problem then restores consistency if the predicted thermoelastic state is outside the transformation surface ($\Phi > 0$) given by the following evolution equation:

$$\dot{\boldsymbol{\varepsilon}} = 0, \qquad \dot{T} = 0, \qquad \dot{\boldsymbol{\varepsilon}}^t = \boldsymbol{\Lambda}\dot{\xi}. \tag{4.3.3}$$

The initial conditions are provided by the solution of the elastic predictor problem, and the increment $\dot{\xi}$ is to be found by satisfying the transformation consistency condition (4.2.4).

Before discretizing the transformation correction problem in (4.3.3) according to (4.3.1), a brief derivation is performed to address the form of this correction regardless of the integration procedure. The details of the numerical integration in the correction phase for two algorithms then follows in successive sections. Recalling that the total strain tensor, $\varepsilon(t)$, and temperature, $T(t)$, are given for each increment in FEA, the transformation function during correction will vary with only changes in $\boldsymbol{\sigma}$ and ξ. Since $d\boldsymbol{\varepsilon}$ and dT are zero during transformation correction, (4.2.3) reduces to

$$\frac{d\boldsymbol{\sigma}}{d\xi} = \begin{cases} -\mathcal{S}^{-1} : \partial_{\boldsymbol{\sigma}}\Phi; & \dot{\xi} > 0, \\ \mathcal{S}^{-1} : \partial_{\boldsymbol{\sigma}}\Phi; & \dot{\xi} < 0. \end{cases} \tag{4.3.4}$$

The evolution of stress state toward the transformation surface ($\Phi = 0$) starts at the trial thermoelastic state and follows the steepest descent path with respect to the transformation surface, Φ, defined in the metric of elasticities [16]. Transformation consistency is enforced by determining the intersection of the stress evolution curve with the boundary of the transformation surface (return mapping).

The transformation correction problem can be further understood by examining the evolution of the value of Φ from the initial thermoelastic prediction to the necessary final value of zero. Differentiating Φ with respect to ξ, recalling that T is kept constant during transformation correction, and using the expression for $\frac{d\boldsymbol{\sigma}}{d\xi}$ in (4.3.4), we get

$$\frac{d}{d\xi}\Phi(\boldsymbol{\sigma}, T|_{fixed}, \xi) = \partial_{\boldsymbol{\sigma}}\Phi : \frac{d\boldsymbol{\sigma}}{d\xi} + \partial_{\xi}\Phi, \tag{4.3.5}$$

and this reduces to

$$\frac{d}{d\xi}\Phi(\boldsymbol{\sigma}, T|_{fixed}, \xi) = \begin{cases} -\partial_{\boldsymbol{\sigma}}\Phi : \mathcal{S}^{-1} : \partial_{\boldsymbol{\sigma}}\Phi + \partial_{\xi}\Phi; & \dot{\xi} > 0, \\ \partial_{\boldsymbol{\sigma}}\Phi : \mathcal{S}^{-1} : \partial_{\boldsymbol{\sigma}}\Phi + \partial_{\xi}\Phi; & \dot{\xi} < 0 \end{cases} \tag{4.3.6}$$

which further reduces to

$$\frac{d}{d\xi}\Phi(\boldsymbol{\sigma}, T|_{fixed}, \xi) = \begin{cases} -\|\partial_{\boldsymbol{\sigma}}\Phi\|^2_{\mathcal{S}^{-1}} - \rho b^M < 0; & \dot{\xi} > 0, \\ \|\partial_{\boldsymbol{\sigma}}\Phi\|^2_{\mathcal{S}^{-1}} + \rho b^A \ > 0; & \dot{\xi} < 0. \end{cases} \tag{4.3.7}$$

Here $\partial_{\xi}\Phi$ is derived using (3.3.57) and (3.3.58), and

$$\|\partial_{\boldsymbol{\sigma}}\Phi\|^2_{\mathcal{S}^{-1}} = \sqrt{\partial_{\boldsymbol{\sigma}}\Phi : \mathcal{S}^{-1} : \partial_{\boldsymbol{\sigma}}\Phi}$$

is the norm of $\partial_{\boldsymbol{\sigma}}\Phi$ induced by the elasticity tensor, $\mathcal{S}^{-1}$. Equation (4.3.7a) implies that $\Phi(\boldsymbol{\sigma}, \xi)$ monotonically decreases with increasing ξ in the forward transformation with a negative slope. Additionally, the function Φ is convex for forward transformation as shown in Fig. 4.1[1]. These characteristics make

[1] Reprinted from "Numerical Implementation of a Shape Memory Alloy Thermomechanical Constitutive Model Using Return Mapping Algorithms", Qidwai, M.A. and Lagoudas, D.C., Copyright 2000, with permission from John Wiley & Sons Ltd.

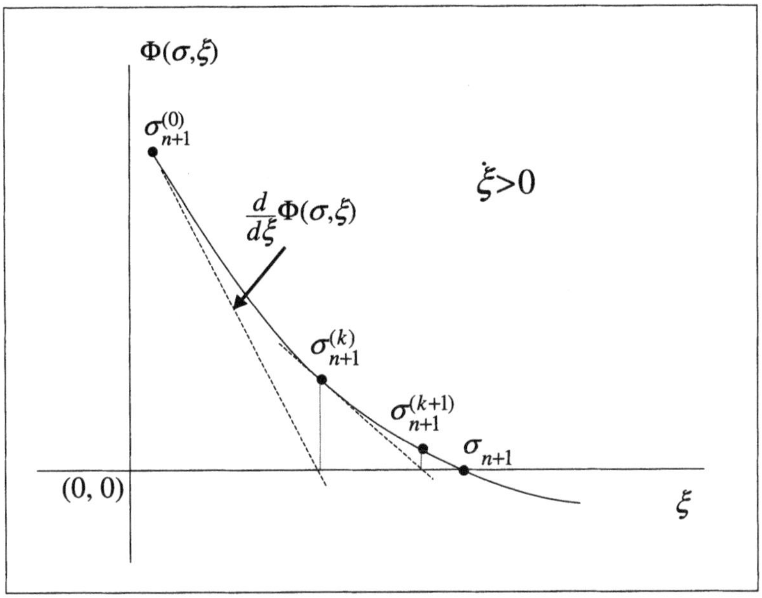

Fig. 4.1. Φ monotonically decreases with increasing ξ with a negative slope for the forward transformation which allows the use of the Newton's method to find a root as a function of ξ during transformation correction.

the transformation correction well suited for a solution by Newton's method, which is used in this work. Similarly, (4.3.7b) implies that $\Phi(\boldsymbol{\sigma}, \xi)$ vs. ξ monotonically decreases with decreasing ξ (reverse transformation) with a positive slope; however, the function Φ is concave for reverse transformation as shown in Fig. 4.2[1]. These two observations imply that the application of Newton's method for the transformation correction will be only locally convergent.

Let us now examine the application of the return mapping algorithm on the discretized SMA constitutive model based on the backward Euler and forward Euler integration methods. Again, consider the time interval $[0, \hat{t}]$ and assume that at time $t_n \in [0, \hat{t}]$, the total strain, the temperature, and the internal state variables, which are the transformation strain and the martensitic volume fraction, are given, i.e.,

$$\boldsymbol{\varepsilon}\big|_{t=t_n} = \boldsymbol{\varepsilon}_n, \quad T\big|_{t=t_n} = T_n, \quad \boldsymbol{\varepsilon}^t\big|_{t=t_n} = \boldsymbol{\varepsilon}^t_n, \quad \xi\big|_{t=t_n} = \xi_n. \tag{4.3.8}$$

Given the increments of strain and temperature, $\{\Delta\boldsymbol{\varepsilon}_{n+1} = \boldsymbol{\varepsilon}_{n+1} - \boldsymbol{\varepsilon}_n, \quad \Delta T = T_{n+1} - T_n\}$, where $\boldsymbol{\varepsilon}_{n+1} = \boldsymbol{\varepsilon}(t_{n+1})$, $T_{n+1} = T(t_{n+1})$ at time $t_{n+1} \in [0, \hat{t}]$ *and* $t_{n+1} > t_n$, the task is to update the field variables $\{\boldsymbol{\sigma}, \boldsymbol{\varepsilon}^t, \xi\}$ to t_{n+1}, in a way consistent with the constitutive relations developed in Sect. 3.3.

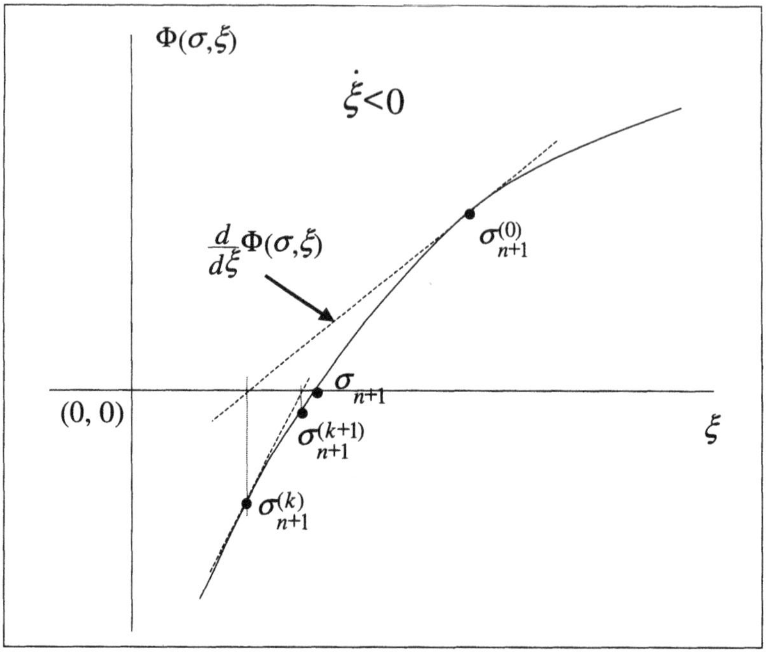

Fig. 4.2. Φ monotonically decreases with decreasing ξ with a positive slope for the reverse transformation which allows the use of the Newton's method to find a root as a function of ξ during transformation correction as shown.

4.3.2 Closest Point Projection Return Mapping Algorithm

This algorithm integrates the transformation evolution equation for the transformation correction using the backward Euler method ($\lambda = 1$ in (4.3.1)) resulting in a nonlinear algebraic set of equations that are solved using the Newton iteration method. The numerical discretization with the initial conditions given in (4.3.8) is performed on the SMA model as

$$\boldsymbol{\sigma}_{n+1} = \mathcal{S}_{n+1}^{-1} : [\boldsymbol{\varepsilon}_{n+1} - \boldsymbol{\alpha}_{n+1}(T_{n+1} - T_o) - \boldsymbol{\varepsilon}_{n+1}^t], \tag{4.3.9}$$

$$\boldsymbol{\varepsilon}_{n+1}^t = \boldsymbol{\varepsilon}_n^t + (\xi_{n+1} - \xi_n)\boldsymbol{\Lambda}_{n+1}, \tag{4.3.10}$$

constrained by the discrete Kuhn-Tucker optimality conditions

$$\begin{gathered}(\xi_{n+1} - \xi_n) > 0, \quad \Phi_{n+1}(\boldsymbol{\sigma}_{n+1}, T_{n+1}, \xi_{n+1}) \leq 0, \\ (\xi_{n+1} - \xi_n)\Phi_{n+1}(\boldsymbol{\sigma}_{n+1}, T_{n+1}, \xi_{n+1}) = 0;\end{gathered}$$

$$\begin{gathered}(\xi_{n+1} - \xi_n) < 0, \quad \Phi_{n+1}(\boldsymbol{\sigma}_{n+1}, T_{n+1}, \xi_{n+1}) \leq 0, \\ (\xi_{n+1} - \xi_n)\Phi_{n+1}(\boldsymbol{\sigma}_{n+1}, T_{n+1}, \xi_{n+1}) = 0. \end{gathered} \tag{4.3.11}$$

Thermoelastic Prediction

The thermoelastic prediction problem is given by

$$\begin{aligned}
\boldsymbol{\varepsilon}_{n+1} &= \boldsymbol{\varepsilon}_n + \Delta\boldsymbol{\varepsilon}_{n+1}, \\
T_{n+1} &= T_n + \Delta T_{n+1}, \\
\boldsymbol{\varepsilon}^{t\,(0)}_{n+1} &= \boldsymbol{\varepsilon}^t_n, \\
\xi^{(0)}_{n+1} &= \xi_n,
\end{aligned} \tag{4.3.12}$$

where $\{\Delta T_{n+1}, \Delta\boldsymbol{\varepsilon}_{n+1}\}$ are the specified strain tensor and temperature increments, respectively, over the time step $[t_n, t_{n+1}]$, and $(\cdot)^{(0)}$ denotes the values obtained in the prediction stage. The above set of equations ((4.3.9)–(4.3.11)) can be reformulated to fit computational implementation (operator split) by introducing the following trial thermoelastic state:

$$\boldsymbol{\sigma}^{(0)}_{n+1} = \mathcal{S}^{-1}_n : [\boldsymbol{\varepsilon}_{n+1} - \boldsymbol{\alpha}_n(T_{n+1} - T_o) - \boldsymbol{\varepsilon}^t_n], \tag{4.3.13}$$

$$\Phi^{(0)}_{n+1} = \Phi[\boldsymbol{\sigma}^{(0)}_{n+1}, T_{n+1}, \xi_n], \tag{4.3.14}$$

where the material properties $\mathcal{S}$ and $\boldsymbol{\alpha}$ are based on ξ_n. This problem is solved using the converged internal state variable of the previous time step and the increment of the total strain tensor. Physically, the reformulation means that this trial state is obtained by neglecting the transformation during the time step, i.e., $\Delta\xi_{n+1} = 0$. At this stage, if the transformation criterion is satisfied, that is, $\Phi(\boldsymbol{\sigma}^{(0)}_{n+1}, T_{n+1}, \xi_n) \leq 0$, then this trial thermoelastic state is taken to be the final state; however, if $\Phi(\boldsymbol{\sigma}^{(0)}_{n+1}, T_{n+1}, \xi_n) > 0$, the Kuhn-Tucker conditions are violated and the trial state lies outside the transformation surface. Transformation correction employing backward Euler integration of the evolution equation is then used to restore consistency and the solution $[\boldsymbol{\sigma}^{(0)}_{n+1}, T_{n+1}, \xi_n]$ is taken as the initial condition for the transformation corrector phase.

Transformation Correction

If the trial stress, $\boldsymbol{\sigma}^{(0)}_{n+1}$, and the converged internal state variables, $\boldsymbol{\varepsilon}^t_n$ and ξ_n, violate the transformation conditions in (4.3.11), (i.e., $\Phi^{(0)}_{n+1} > 0$), the following transformation correction procedure is adopted at the loading increment $(n+1)^{th}$ for the solution of (4.3.9)–(4.3.10) with initial guess (4.3.13) and constraints given by (4.3.11). Replacing $\boldsymbol{\sigma}_n$ by $\boldsymbol{\sigma}^{(0)}_{n+1}$ implies that the implicit integration of the transformation strain evolution equation is carried from the initial thermoelastic prediction to the final state.

The nonlinear system of these algebraic equations is solved by defining the transformation condition valid for the transformation corrector phase and transformation strain residual based on implicit backward Euler integration from (4.3.10) for the kth iteration as follows:

$$\Phi_{n+1}^{(k)} = \Phi(\boldsymbol{\sigma}_{n+1}^{(k)}, T_{n+1}, \xi_{n+1}^{(k)}), \tag{4.3.15}$$

$$\mathbf{R}_{n+1}^{t\ (k)} = -\boldsymbol{\varepsilon}_{n+1}^{t\ (k)} + \boldsymbol{\varepsilon}_{n}^{t} + \boldsymbol{\Lambda}_{n+1}^{(k)}[\xi_{n+1}^{(k)} - \xi_n], \tag{4.3.16}$$

and then linearizing them using the Newton-Raphson iteration method [11, 16], where $\Phi_{n+1}^{(k)}$ and $\mathbf{R}_{n+1}^{t\ (k)}$ should converge to zero at the end of the iteration process. Equations (4.3.15) and (4.3.16) are linearized in the following way:

$$\Phi_{n+1}^{(k)} + \partial_{\boldsymbol{\sigma}}\Phi_{n+1}^{(k)} : \Delta\boldsymbol{\sigma}_{n+1}^{(k)} + \partial_{\xi}\Phi_{n+1}^{(k)}\Delta\xi_{n+1}^{(k)} = 0, \tag{4.3.17}$$

$$\begin{aligned} \mathbf{R}_{n+1}^{t\ (k)} &- \Delta\boldsymbol{\varepsilon}_{n+1}^{t\ (k)} + \boldsymbol{\Lambda}_{n+1}^{(k)}\Delta\xi_{n+1}^{(k)} + \\ &+ \begin{cases} (\xi_{n+1}^{(k)} - \xi_n)\ \partial_{\boldsymbol{\sigma}}\boldsymbol{\Lambda}_{n+1}^{(k)} : \Delta\boldsymbol{\sigma}_{n+1}^{(k)}; & \dot{\xi} > 0 \\ \mathbf{0}; & \dot{\xi} < 0 \end{cases} = 0, \end{aligned} \tag{4.3.18}$$

where (4.3.18) is obtained from the Kuhn-Tucker condition when $(\xi_{n+1} - \xi_n) \neq 0$. Note that $\Delta\boldsymbol{\Lambda}_{n+1}^{(k)} = 0$ for reverse transformation because during reverse transformation, $\boldsymbol{\Lambda}_{n+1}^{(k)}$ remains constant, for $k = 1, 2, ...$, with its components determined at the end of the previous forward transformation. Equations (4.3.18) and (4.3.21) provide seven equations to solve for 13 unknowns, i.e., $\{\Delta\boldsymbol{\sigma}_{n+1}^{(k)}, \Delta\boldsymbol{\varepsilon}_{n+1}^{t\ (k)}, \Delta\xi_{n+1}^{(k)}\}$. The other six equations are obtained by taking the increment of the stress-elastic strain state relation, (4.3.9), and expressing it for the total increment of transformation strain

$$\begin{aligned} \Delta\boldsymbol{\sigma}_{n+1} &= \Delta\mathcal{S}_{n+1} : \boldsymbol{\sigma}_{n+1} + \mathcal{S}_{n+1} : \Delta\boldsymbol{\sigma}_{n+1} \\ &+ \Delta\boldsymbol{\alpha}_{n+1}(T_{n+1} - T_0) + \boldsymbol{\alpha}_{n+1}\Delta T_{n+1} + \Delta\boldsymbol{\varepsilon}_{n+1}^{t}. \end{aligned} \tag{4.3.19}$$

Recalling the definitions of $\mathcal{S}$ and $\boldsymbol{\alpha}$ from (3.3.42) and (3.3.43) , we get

$$\begin{aligned} \Delta\mathcal{S}_{n+1} &= \Delta\mathcal{S}\ \Delta\xi_{n+1}, \\ \Delta\boldsymbol{\alpha}_{n+1} &= \Delta\boldsymbol{\alpha}\ \Delta\xi_{n+1}, \end{aligned} \tag{4.3.20}$$

where $\Delta\mathcal{S}$ and $\Delta\boldsymbol{\alpha}$ are defined in (3.3.42) and (3.3.43). Using (4.3.20) and noting that $\boldsymbol{\varepsilon}$ and T are fixed during the return mapping (transformation corrector) stage, (4.3.19) can be written in incremental form for $\Delta\boldsymbol{\varepsilon}_{n+1}^{t}$ at the kth iteration as follows:

$$\Delta\boldsymbol{\varepsilon}_{n+1}^{t\ (k)} = -\mathcal{S}_{n+1}^{(k)} : \Delta\boldsymbol{\sigma}_{n+1}^{(k)} - [\Delta\mathcal{S} : \boldsymbol{\sigma}_{n+1}^{(k)} + \Delta\boldsymbol{\alpha}(T_{n+1} - T_0)]\Delta\xi_{n+1}^{(k)}. \tag{4.3.21}$$

(4.3.17), (4.3.18), and (4.3.21) form a system of equations to solve

$$\{\Delta\boldsymbol{\sigma}_{n+1}^{(k)}, \Delta\boldsymbol{\varepsilon}_{n+1}^{t\ (k)}, \Delta\xi_{n+1}^{(k)}\}$$

for both forward and reverse transformations.

For the forward transformation ($\dot{\xi} > 0$), (4.3.17) is substituted into (4.3.21) to obtain

$$\Delta\boldsymbol{\sigma}^{(k)}_{n+1} = \mathcal{E}^{(k)}_{n+1} : [-\mathbf{R}^{t\ (k)}_{n+1} - \Delta\xi^{(k)}_{n+1}\partial_{\boldsymbol{\sigma}}\Phi^{(k)}_{n+1}], \tag{4.3.22}$$

where

$$\mathcal{E}^{(k)}_{n+1} = [\mathcal{S}^{(k)}_{n+1} + (\xi^{(k)}_{n+1} - \xi_n)\partial_{\boldsymbol{\sigma}}\boldsymbol{\Lambda}^{(k)}_{n+1}]^{-1}, \tag{4.3.23}$$

and

$$\partial_{\boldsymbol{\sigma}}\Phi^{(k)}_{n+1} = \Delta\mathcal{S} : \boldsymbol{\sigma}^{(k)}_{n+1} + \Delta\boldsymbol{\alpha}(T_{n+1} - T_0) + \boldsymbol{\Lambda}^{(k)}_{n+1}. \tag{4.3.24}$$

After substituting (4.3.22) into (4.3.18), the increment of the martensitic volume fraction is obtained for the kth iteration as

$$\Delta\xi^{(k)}_{n+1} = \frac{\Phi^{(k)}_{n+1} - \partial_{\boldsymbol{\sigma}}\Phi^{(k)}_{n+1} : \mathcal{E}^{(k)}_{n+1} : \mathbf{R}^{t\ (k)}_{n+1}}{\partial_{\boldsymbol{\sigma}}\Phi^{(k)}_{n+1} : \mathcal{E}^{(k)}_{n+1} : \partial_{\boldsymbol{\sigma}}\Phi^{(k)}_{n+1} - \partial_{\xi}\Phi^{(k)}_{n+1}}. \tag{4.3.25}$$

A similar analysis can be performed for the reverse transformation ($\dot{\xi} < 0$) to obtain

$$\Delta\boldsymbol{\sigma}^{(k)}_{n+1} = \mathcal{S}^{(k)\ -1}_{n+1} : [-\mathbf{R}^{t\ (k)}_{n+1} + \Delta\xi^{(k)}_{n+1}\ \partial_{\boldsymbol{\sigma}}\Phi^{(k)}_{n+1}]. \tag{4.3.26}$$

Following the strategy to solve for $\Delta\xi^{(k)}_{n+1}$ in the forward transformation, we get

$$\Delta\xi^{(k)}_{n+1} = \frac{\Phi^{(k)}_{n+1} - \partial_{\boldsymbol{\sigma}}\Phi^{(k)}_{n+1} : \mathcal{S}^{(k)\ -1}_{n+1} : \mathbf{R}^{t\ (k)}_{n+1}}{-\partial_{\boldsymbol{\sigma}}\Phi^{(k)}_{n+1} : \mathcal{S}^{(k)\ -1}_{n+1} : \partial_{\boldsymbol{\sigma}}\Phi^{(k)}_{n+1} - \partial_{\xi}\Phi^{(k)}_{n+1}}. \tag{4.3.27}$$

Furthermore, the term $\partial_{\boldsymbol{\sigma}}\boldsymbol{\Lambda}^{(k)}_{n+1}$, appearing in the expression for $\mathcal{E}^{(k)}_{n+1}$, can be derived as

$$\partial_{\boldsymbol{\sigma}}\boldsymbol{\Lambda}^{(k)}_{n+1} = \sqrt{\frac{3}{2}}\frac{H}{\|\boldsymbol{\sigma}'\|}[\mathbf{I} - \frac{1}{3}\mathbf{1}\otimes\mathbf{1} - \frac{\boldsymbol{\sigma}'}{\|\boldsymbol{\sigma}'\|}\otimes\frac{\boldsymbol{\sigma}'}{\|\boldsymbol{\sigma}'\|}]. \tag{4.3.28}$$

Here $\mathbf{I}$ is the most general rank four symmetric identity tensor and in indicial notation is given by

$$\mathbf{I} = \frac{1}{2}[\delta_{ik}\delta_{jl} + \delta_{il}\delta_{jk}]\ \mathbf{e}_i\otimes\mathbf{e}_j\otimes\mathbf{e}_k\otimes\mathbf{e}_l, \tag{4.3.29}$$

where $\otimes$ denotes the tensor product and $\mathbf{1}$ is the second order identity tensor given by

$$\mathbf{1} = \delta_{ij}\mathbf{e}_i\otimes\mathbf{e}_j. \tag{4.3.30}$$

In (4.3.21), (4.3.22), and (4.3.25) for forward transformation and (4.3.21), (4.3.26), and (4.3.27) for reverse transformation, we have a complete set of equations for the 13 unknown variables $\Delta\xi^{(k)}_{n+1}$, $\Delta\boldsymbol{\sigma}^{(k)}_{n+1}$ and $\Delta\boldsymbol{\varepsilon}^{t\ (k)}_{n+1}$. The next step is to update the transformation strain, $\boldsymbol{\varepsilon}^{t\ (k)}_{n+1}$, and the martensitic volume fraction, $\xi^{(k)}_{n+1}$, by

$$\begin{aligned} \boldsymbol{\varepsilon}^{t\ (k+1)}_{n+1} &= \boldsymbol{\varepsilon}^{t\ (k)}_{n+1} + \Delta\boldsymbol{\varepsilon}^{t\ (k)}_{n+1}, \\ \xi^{(k+1)}_{n+1} &= \xi^{(k)}_{n+1} + \Delta\xi^{(k)}_{n+1}. \end{aligned} \tag{4.3.31}$$

This is an implicit algorithm involving the solution of a 6×6 system of equations. A systematic procedure for carrying out the algorithm is outlined in Table 4.1[1].

Geometric Interpretation

Before deriving the consistent tangent moduli tensors for the SMA constitutive model, let us understand the geometric meaning of the transformation correction in stress space. For the $(k+1)^{th}$ iteration at t_{n+1}, substituting (4.3.10) into (4.3.9), we get

$$\begin{aligned} \boldsymbol{\sigma}^{(k+1)}_{n+1} = \mathcal{S}^{(k+1)^{-1}}_{n+1} : \{\boldsymbol{\varepsilon}_{n+1} - \boldsymbol{\alpha}^{(k+1)}_{n+1}(T_{n+1} - T_o) - \boldsymbol{\varepsilon}^t_n \\ - [\xi^{(k+1)}_{n+1} - \xi_n]\boldsymbol{\Lambda}^{(k+1)}_{n+1}\}. \end{aligned} \tag{4.3.32}$$

Pre-multiplying (4.3.32) and (4.3.13) by $\mathcal{S}^{(k+1)}_{n+1}$ and $\mathcal{S}^{-1}_n$, respectively, and then subtracting the resulting equations, the expression for the stress at the end of the $(k+1)^{th}$ iteration in terms of the trial thermoelastic stress at t_{n+1} can be obtained after some algebraic manipulations as

$$\begin{aligned} \boldsymbol{\sigma}^{(k+1)}_{n+1} = \boldsymbol{\sigma}^{(0)}_{n+1} - \Delta\xi^{(k)}_{n+1}\, \mathcal{S}^{(k+1)^{-1}}_{n+1} : \{\partial_{\boldsymbol{\sigma}}\Phi^{(k+1)}_{n+1} \\ - \Delta\mathcal{S} : [\boldsymbol{\sigma}^{(k)}_{n+1} - \boldsymbol{\sigma}^{(0)}_{n+1}]\}. \end{aligned} \tag{4.3.33}$$

The above equation describes the return (relaxation) path of the stress tensor from the thermoelastic prediction with each iteration. The return path is not arbitrary but is defined by the projection of the thermoelastic prediction on the transformation surface, Φ^{k+1}_{n+1}, in a way defined by (4.3.33). The stress state is updated during the iterative procedure in such a manner that at the $(k+1)^{th}$ iteration, $\boldsymbol{\sigma}^{(k+1)}_{n+1}$ lies on Φ^{k+1}_{n+1} in the six-dimensional stress space defined by the metric of elasticities, $\mathcal{S}^{(k+1)^{-1}}_{n+1}$. Furthermore, the vector joining $\boldsymbol{\sigma}^{(k+1)}_{n+1}$ and $\boldsymbol{\sigma}^{(0)}_{n+1}$ is normal to Φ^{k+1}_{n+1} for the case when $\Delta\mathcal{S} = 0$. That is, if the change in material properties is not significant, the stress will be updated normal to each iterative transformation surface. Note that the implicit integration of the transformation strain evolution equation results in the above mentioned characteristic of the return path.

Consistent Tangent Moduli Tensors

As already shown, the solution to the rate-type SMA constitutive equations requires numerical integration over a number of time steps. The stress tensor,

Table 4.1. Closest point projection implicit algorithm for shape memory alloy constitutive models.

1. Let $k=0$, $\xi_{n+1}^{(0)}=\xi_n$, $\boldsymbol{\varepsilon}_{n+1}^{t\ (0)}=\boldsymbol{\varepsilon}_n^t$, $\mathcal{S}_{n+1}^{(0)}=\mathcal{S}_n$, $\boldsymbol{\alpha}_{n+1}^{(0)}=\boldsymbol{\alpha}_n$

2. Calculate the thermoelastic prediction, and evaluate the transformation condition and transformation strain residual

$$\begin{aligned}
\boldsymbol{\sigma}_{n+1}^{(k)} &= \mathcal{S}_{n+1}^{(k)\ -1} : [\boldsymbol{\varepsilon}_{n+1} - \boldsymbol{\alpha}_{n+1}^{(k)}(T_{n+1}-T_o) - \boldsymbol{\varepsilon}_{n+1}^{t\ (k)}] \\
\Phi_{n+1}^{(k)} &= \Phi[\boldsymbol{\sigma}_{n+1}^{(k)}, T_{n+1}, \xi_{n+1}^{(k)}] \\
\mathbf{R}_{n+1}^{t\ (k)} &= -\boldsymbol{\varepsilon}_{n+1}^{t\ (k)} + \boldsymbol{\varepsilon}_n^t + \boldsymbol{\Lambda}_{n+1}^{(k)}[\xi_{n+1}^{(k)} - \xi_n]
\end{aligned}$$

If $|\Phi_{n+1}^{(k)}| \le tolerance_1$ and $\|\mathbf{R}_{n+1}^{t\ (k)}\| \le tolerance_2$ then

Retain values above and return to global iterations.

Else

Continue to forward or reverse transformation.

3. Compute elastic/algorithmic tangent moduli

Forward Transformation ($\dot{\xi}>0$):

$$\mathcal{E}_{n+1}^{(k)} = [\mathcal{S}_{n+1}^{(k)} + (\xi_{n+1}^{(k)} - \xi_n)\partial_{\boldsymbol{\sigma}}\boldsymbol{\Lambda}_{n+1}^{(k)}]^{-1}$$

Reverse Transformation ($\dot{\xi}<0$):

Invert $\mathcal{S}_{n+1}^{(k)}$

4. Compute increment of volume fraction, stress and transformation strain

Forward Transformation ($\dot{\xi}>0$):

$$\Delta\xi_{n+1}^{(k)} = \frac{\Phi_{n+1}^{(k)} - \partial_{\boldsymbol{\sigma}}\Phi_{n+1}^{(k)} : \mathcal{E}_{n+1}^{(k)} : \mathbf{R}_{n+1}^{t\ (k)}}{\partial_{\boldsymbol{\sigma}}\Phi_{n+1}^{(k)} : \mathcal{E}_{n+1}^{(k)} : \partial_{\boldsymbol{\sigma}}\Phi_{n+1}^{(k)} - \partial_{\xi}\Phi_{n+1}^{(k)}}$$

$$\Delta\boldsymbol{\sigma} = \mathcal{E}_{n+1}^{(k)} : [-\mathbf{R}_{n+1}^{t\ (k)} - \Delta\xi_{n+1}^{(k)}\partial_{\boldsymbol{\sigma}}\Phi_{n+1}^{(k)}]$$

Reverse Transformation ($\dot{\xi}<0$):

$$\Delta\xi_{n+1}^{(k)} = \frac{\Phi_{n+1}^{(k)} - \partial_{\boldsymbol{\sigma}}\Phi_{n+1}^{(k)} : \mathcal{S}_{n+1}^{(k)\ -1} : \mathbf{R}_{n+1}^{t\ (k)}}{-\partial_{\boldsymbol{\sigma}}\Phi_{n+1}^{(k)} : \mathcal{S}_{n+1}^{(k)\ -1} : \partial_{\boldsymbol{\sigma}}\Phi_{n+1}^{(k)} - \partial_{\xi}\Phi_{n+1}^{(k)}}$$

$$\Delta\boldsymbol{\sigma} = \mathcal{S}_{n+1}^{(k)\ -1} : [-\mathbf{R}_{n+1}^{t\ (k)} + \Delta\xi_{n+1}^{(k)}\,\partial_{\boldsymbol{\sigma}}\Phi_{n+1}^{(k)}]$$

For either transformation:

$$\Delta\boldsymbol{\varepsilon}_{n+1}^{t\ (k)} = -\mathcal{S}_{n+1}^{(k)} : \Delta\sigma - [\Delta\mathcal{S} : \sigma + \Delta\boldsymbol{\alpha}(T_{n+1} - T_0)]\Delta\xi_{n+1}^{(k)}$$

5. Update martensitic volume fraction and transformation strain

$$\begin{aligned}
\xi_{n+1}^{(k+1)} &= \xi_{n+1}^{(k)} + \Delta\xi_{n+1}^{(k)} \\
\boldsymbol{\varepsilon}_{n+1}^{t\ (k+1)} &= \boldsymbol{\varepsilon}_{n+1}^{t\ (k)} + \Delta\boldsymbol{\varepsilon}_{n+1}^{t\ (k)}
\end{aligned}$$

6. Update elastic and thermal moduli, $\mathcal{S}_{n+1}^{(k)\ -1}$ and $\boldsymbol{\alpha}_{n+1}^{(k)}$

Let $k=k+1$ & Return to step 2

as a result of this process, is defined as a function of the deformation history up to a given instant. The tangent tensors that appear in the linearized problem (the global FEM problem employing Newton's method) must be obtained by employing the response function associated with the integration algorithm. This is important if the quadratic rate of asymptotic convergence in the global solution scheme is to be maintained. In other words, the notion of a consistent tangent stiffness tensor and tangent thermal moduli tensor arises due to the enforcement of the consistency condition on the discrete algorithmic problem. On the other hand, the continuum tangent stiffness and tangent thermal moduli tensors result from the enforcement of the classical consistency condition on the continuum problem. For the case of associative/non-associative elastoplasticity, it has been shown using numerical examples [17] that the use of the continuum tangent stiffness leads to loss of the quadratic rate of asymptotic convergence which characterizes Newton's iteration method as commonly used in the finite element method. In this section, only the final result for forward transformation is presented. The consistent tangent tensors for the reverse transformation are exactly the same as the continuum tangent tensors given in (4.2.11-4.2.12) in Sect. 4.2. This will be the focus of Problem 4.7.

For the forward phase transformation, the algorithmic tangent stiffness tensor, $\mathcal{L}$, and the algorithmic tangent thermal moduli tensor, $\boldsymbol{\Theta}$, are derived as

$$\mathcal{L} = \mathcal{E}_{n+1} - \mathbf{H}_{n+1} \otimes \mathbf{H}_{n+1}, \tag{4.3.34}$$

$$\boldsymbol{\Theta} = -\mathcal{L} : \boldsymbol{\alpha}_{n+1} - \frac{\partial_T \Phi_{n+1}}{c} \mathbf{H}_{n+1}, \tag{4.3.35}$$

where

$$\begin{aligned} c &= \sqrt{\partial_{\boldsymbol{\sigma}} \Phi_{n+1} : \mathcal{E}_{n+1} : \partial_{\boldsymbol{\sigma}} \Phi_{n+1} - \partial_{\xi} \Phi_{n+1}}, \\ \mathbf{H}_{n+1} &= \frac{\mathcal{E}_{n+1} : \partial_{\boldsymbol{\sigma}} \Phi_{n+1}}{\sqrt{\partial_{\boldsymbol{\sigma}} \Phi_{n+1} : \mathcal{E}_{n+1} : \partial_{\boldsymbol{\sigma}} \Phi_{n+1} - \partial_{\xi} \Phi_{n+1}}} = \frac{\mathcal{E}_{n+1} : \partial_{\boldsymbol{\sigma}} \Phi_{n+1}}{c}, \\ \mathcal{E}_{n+1} &= [\mathcal{S}_{n+1} + (\xi_{n+1} - \xi_n) \partial_{\boldsymbol{\sigma}} \boldsymbol{\Lambda}_{n+1}]^{-1}. \end{aligned} \tag{4.3.36}$$

If expressions for the continuum and consistent tangent tensors are compared for forward transformation, the difference between the tangent tensors ((4.2.7),(4.2.8) and (4.3.34),(4.3.35)) arises due to the algorithmic stiffness tensor defined in (4.3.36c). As the time step $\Delta t = t_{n+1} - t_n \to 0$, then $\xi_{n+1} - \xi_n \to 0^+$, i.e., as

$$\Delta t \to 0 \Rightarrow \begin{cases} \mathcal{E} \to \mathcal{S}^{-1} \\ \mathbf{H} \to \mathbf{A} \\ c \to a \end{cases}, \tag{4.3.37}$$

where ($\mathbf{A}$, a) and ($\mathbf{H}$, c) are defined in (4.2.9) and (4.3.36), respectively. Therefore, the consistent tangent stiffness tensor and consistent tangent thermal moduli tensor reduce to their continuum counterparts as $\Delta t \to 0$ for the forward transformation. This shows that the closest point projection algorithm is consistent with the continuum problem. It is also observed that for large (loading) time increments, much transformation may result in ξ_{n+1} being much larger than ξ_n. This will result in significant deviation of the algorithmic tensors from the continuum tensors.

4.3.3 Convex Cutting Plane Return Mapping Algorithm

As it is noticed in the closest point algorithm, the use of backward Euler integration leads to systems of nonlinear equations, whose solution by Newton's iteration method requires evaluation of the gradients of transformation flow direction. The addition of another internal state variable in the model (e.g., [18]) will require further gradients to account for. The task of evaluating such quantities may prove difficult and computationally expensive. The cutting plane return mapping algorithm is proposed to avoid the need for computing the above-mentioned gradients [14, 15]. The present algorithm is formulated solely on the basis of the transformation function, the consistency condition, direction of the transformation flow and the elastic stiffness tensors without evaluating any gradient.

The idea relies on integrating the transformation correction (4.3.7) in an explicit manner and linearizing the consistency condition. The transformation residual, $\mathbf{R}_{n+1}^{t\ (k)}$, and the terms containing the derivatives of the transformation tensor, $\mathbf{\Lambda}_{n+1}^{(k)}$, do not appear in the formulation and Newton's iteration method is only applied to calculate the increment of martensitic volume fraction, $\Delta\xi_{n+1}^{(k)}$. The algorithm converges towards the final value of the state variables at a quadratic rate, but convergence is not generally guaranteed. The thermoelastic part of the algorithm is the same as given in Sect. 4.3.2 so, the focus will be on the transformation correction. This is the main difference between the closest point and cutting plane algorithms.

Transformation Correction

Here, we assume transformation loading; that is, the thermoelastic prediction does not satisfy the transformation condition and $\Phi_{n+1}^{(0)} > 0$. Recalling that the total strain tensor, $\boldsymbol{\varepsilon}$, and temperature, T, are fixed, the transformation evolution equation is discretized explicitly as

$$\Delta\boldsymbol{\varepsilon}_{n+1}^{t\ (k)} = \Delta\xi_{n+1}^{(k)}\mathbf{\Lambda}_{n+1}^{(k)}. \tag{4.3.38}$$

This equation can also be obtained from (4.3.18) in Sect. 4.3.2 by neglecting the residual, $\mathbf{R}_{n+1}^{t\ (k)}$, and the last term containing the gradient of the transformation tensor, $\mathbf{\Lambda}_{n+1}^{(k)}$. Also, this explicit integration is being carried out from the initial thermoelastic state to the final unknown state. Now, using the discretized evolution equation, (4.3.21), and (4.3.24), the stress increment for the kth iteration during the $(n+1)^{th}$ increment can be written as

$$\Delta\boldsymbol{\sigma}_{n+1}^{(k)} = -\Delta\xi_{n+1}^{(k)} \mathcal{S}_{n+1}^{(k)\ -1} : \begin{cases} \partial_{\boldsymbol{\sigma}} \Phi_{n+1}^{(k)}; & \dot{\xi} > 0, \\ -\partial_{\boldsymbol{\sigma}} \Phi_{n+1}^{(k)}; & \dot{\xi} < 0. \end{cases} \tag{4.3.39}$$

Linearizing the discrete transformation function for the kth iteration and equating it to zero due to the Kuhn-Tucker condition, i.e., $\xi_{n+1} - \xi_n \neq 0$, we get

$$\Phi_{n+1}^{(k)} + \partial_{\boldsymbol{\sigma}} \Phi_{n+1}^{(k)} : \Delta\boldsymbol{\sigma}_{n+1}^{(k)} + \partial_{\xi} \Phi_{n+1}^{(k)} \Delta\xi_{n+1}^{(k)} = 0. \tag{4.3.40}$$

Substituting (4.3.39) into (4.3.40) and solving for the increment of ξ, the following expression is obtained:

$$\Delta\xi_{n+1}^{(k)} = \frac{\Phi_{n+1}^{(k)}}{\pm\partial_{\boldsymbol{\sigma}} \Phi_{n+1}^{(k)} : \mathcal{S}_{n+1}^{(k)\ -1} : \partial_{\boldsymbol{\sigma}} \Phi_{n+1}^{(k)} - \partial_{\xi} \Phi_{n+1}^{(k)}}, \tag{4.3.41}$$

where the + sign is for the forward transformation and − sign for the reverse transformation. In (4.3.41), (4.3.39), and (4.3.38), we have a complete set of equations in terms of the 13 variables $\Delta\xi_{n+1}^{(k)}$, $\Delta\boldsymbol{\sigma}_{n+1}^{(k)}$ and $\Delta\boldsymbol{\varepsilon}_{n+1}^{t\ (k)}$ for both forward and reverse transformations. Unlike the closest point formulation, the incremental expressions are the same for both the forward and reverse transformations, except the change of sign in (4.3.39) and (4.3.41), saving a lot of coding details. The next step is to update the transformation strain, $\boldsymbol{\varepsilon}_{n+1}^{t\ (k)}$, and the martensitic volume fraction, $\xi_{n+1}^{(k)}$, as given below

$$\boldsymbol{\varepsilon}_{n+1}^{t\ (k+1)} = \boldsymbol{\varepsilon}_{n+1}^{t\ (k)} + \Delta\boldsymbol{\varepsilon}_{n+1}^{t\ (k)}, \tag{4.3.42}$$

$$\xi_{n+1}^{(k+1)} = \xi_{n+1}^{(k)} + \Delta\xi_{n+1}^{(k)}. \tag{4.3.43}$$

For the cutting plane algorithm, the continuum tangent tensors given in Sect. 4.2 have to be used since consistent tangent tensors cannot be obtained in closed form. This means a loss in the quadratic convergence rate of the Newton's iteration method; therefore, quasi-Newton methods are suggested for global solution strategies [16]. A detailed strategy for implementing the algorithm is given in Table 4.2[1].

Geometric Interpretation

A geometric interpretation of the algorithm can be given by expanding (4.3.39) in the following form:

$$\boldsymbol{\sigma}_{n+1}^{(k+1)} = \boldsymbol{\sigma}_{n+1}^{(k)} - \Delta\xi_{n+1}^{(k)}\ \mathcal{S}_{n+1}^{(k)\ -1} : \begin{cases} \partial_{\boldsymbol{\sigma}} \Phi_{n+1}^{(k)}; & \dot{\xi} > 0, \\ -\partial_{\boldsymbol{\sigma}} \Phi_{n+1}^{(k)}; & \dot{\xi} < 0. \end{cases} \tag{4.3.44}$$

Table 4.2. Convex cutting plane explicit algorithm for shape memory alloy constitutive models.

1. Let $k = 0$, $\xi_{n+1}^{(0)} = \xi_n$, $\boldsymbol{\varepsilon}_{n+1}^{t\ (0)} = \boldsymbol{\varepsilon}_n^t$, $\mathcal{S}_{n+1}^{(0)} = \mathcal{S}_n$, $\boldsymbol{\alpha}_{n+1}^{(0)} = \boldsymbol{\alpha}_n$

2. Calculate thermoelastic prediction and evaluate transformation condition

$$\begin{aligned} \boldsymbol{\sigma}_{n+1}^{(k)} &= \mathcal{S}_{n+1}^{(k)\ -1} : [\boldsymbol{\varepsilon}_{n+1} - \boldsymbol{\alpha}_{n+1}^{(k)}(T_{n+1} - T_o) - \boldsymbol{\varepsilon}_{n+1}^{t\ (k)}] \\ \Phi_{n+1}^{(k)} &= \Phi[\boldsymbol{\sigma}_{n+1}^{(k)}, T_{n+1}, \xi_{n+1}^{(k)}] \end{aligned}$$

If $|\Phi_{n+1}^{(k)}| \leq$ *tolerance* then

Retain solutions above and return to global iteration.

Else

Continue to forward or reverse transformation.

3. Compute increment of martensitic volume fraction and transformation strain

$$\Delta\xi_{n+1}^{(k)} = \frac{\Phi_{n+1}^{(k)}}{\pm\partial_{\boldsymbol{\sigma}}\Phi_{n+1}^{(k)} : \mathcal{S}_{n+1}^{(k)\ -1} : \partial_{\boldsymbol{\sigma}}\Phi_{n+1}^{(k)} - \partial_\xi \Phi_{n+1}^{(k)}}$$

$$\Delta\boldsymbol{\varepsilon}_{n+1}^{t\ (k)} = \Delta\xi_{n+1}^{(k)}\boldsymbol{\Lambda}_{n+1}^{(k)}$$

where $\pm$ stands for forward and reverse transformations.

4. Update martensitic volume fraction and transformation strain

$$\begin{aligned} \xi_{n+1}^{(k+1)} &= \xi_{n+1}^{(k)} + \Delta\xi_{n+1}^{(k)} \\ \boldsymbol{\varepsilon}_{n+1}^{t\ (k+1)} &= \boldsymbol{\varepsilon}_{n+1}^{t\ (k)} + \Delta\boldsymbol{\varepsilon}_{n+1}^{t\ (k)} \end{aligned}$$

Let $k = k + 1$ & Return to step 2.

The above expression implies that during transformation correction, $\boldsymbol{\sigma}_{n+1}^{(k+1)}$ is found by projecting the plane normal to $\Phi_{n+1}^{(k)}$ and finding its intersection with level iterate $\Phi_{n+1}^{(k+1)}$. The return path is approximated by a sequence of straight segments. Note that the explicit integration of the transformation strain evolution equation is directly responsible for the mentioned characteristics of the return path.

4.3.4 Summary and Comparison of Algorithms

Having derived the algorithms, we now turn to their respective features. Regarding the closest point projection algorithm:

1. It is based on the backward Euler integration of the transformation strain evolution equation resulting in a set of nonlinear algebraic equations solved using Newton's iteration method.

2. The quadratic convergence rate from the thermoelastic prediction to the final solution of Newton's iteration method is guaranteed, as observed in the variation of the transformation function with changing martensitic volume fraction during transformation correction in Figs. 4.1 and 4.2.

3. Backward Euler integrations make the local integration procedure first-order accurate and unconditionally stable.
4. The algorithm facilitates the derivation of the consistent tangent moduli tensors in closed form, thereby preserving the global quadratic convergence rate.
5. The algorithm requires the solution of a system of linear algebraic relations ((4.3.17), (4.3.18), and (4.3.21)). This results in a relatively large number of tensorial operations beside the evaluation of the gradient of the transformation tensor (4.3.28) and inversion of the 6x6 algorithmic tangent tensor (4.3.23) during forward transformation.

Similar remarks on the features of the convex cutting plane algorithm are as follows:
1. It is based on the explicit integration of the transformation evolution equation from the thermoelastic predictor. The satisfaction of the transformation condition requires the use of Newton's iteration method.
2. The convergence of the algorithm toward the final value of the state variables is obtained at a quadratic rate as shown in Figs. 4.1 and 4.2.
3. Forward Euler integration makes the load-stepping procedure first-order accurate. The algorithm is also unconditionally stable in the forward transformation because the iteration function to find the increment of martensitic volume fraction, $\Delta\xi$, is contractive and the transformation tensor is normalized (cf. Sect. 3.3). In the reverse transformation, Newton's method is only locally convergent, which may result in instability.
4. The algorithm does not allow the derivation of consistent tangent moduli tensors in closed form, thus requiring continuum tangent moduli tensors to be used. This may require usage of quasi-Newton techniques for faster convergence in the global solution scheme.
5. Its fundamental advantage is its simplicity, since it does not require computation of the gradient of the transformation tensor or inversion of algorithmic tangent tensors, and only involves a few tensorial operations and function evaluations to determine the unknowns.

4.4 Numerical Examples

To characterize the behavior of the return mapping algorithm, we will analyze several Boundary Value Problems (BVPs). Two simple uniaxial cases are studied first to demonstrate the evolution of the transformation and stress state during a given loading history. We then present two problems that consider an SMA-actuated beam. Following this, the torsional response of an SMA tube as opposed by a biasing spring is analyzed. Finally, we describe the analysis of more complex engineering systems including a morphing aerostructure and a medical stent. The boundary value problems are numerically simulated using the commercially available Abaqus Unified Finite Element Analysis suite [19].

This software was installed locally on a PC with a 64-bit AMD FX-57 processor (2.81 GHz) and 4 GB available RAM. Total run times (wallclock times) for some of the analysis cases are provided, giving some indication of computational requirements.

To implement custom material constitutive models such as the unified model in Abaqus, we use *user material subroutines*, or "UMATs". `SMA_UM` is a FORTRAN-coded numerical implementation of the unified SMA thermomechanical constitutive model as described in the previous sections. The demonstrated implementation of the subroutine uses the convex cutting plane algorithm and follows the specifications for user-material subroutines as required by Abaqus; however, the subroutine can be integrated in any other standard finite element or computational program. The subroutine can be used in 3-D, 2-D plane strain and generalized plane strain, 2-D axisymmetric, and 1-D problems. The `SMA_UM` and associated example input files are available for use at **`http://smart.tamu.edu/SMAText/`**.

In a finite element computational program, the analysis of materials with non-linear behavior involves the application of a global iterative scheme that seeks to satisfy the conservation of linear momentum (balance of forces) at each node. In the process of determining the nodal forces and displacements during a given iteration, the global stiffness matrix must be calculated by summing the entries of the local stiffness matrices for each integration point. To calculate these entries, the main FEM code calls the material subroutine, where the local tangent moduli (previously described) are calculated according to the constitutive model for the material; thus, the material subroutine is called for each integration point of the finite element model during each global iteration. When the `SMA_UM` subroutine is called, it is provided with the material state at the start of the increment (i.e, the stress, total strain, transformation strain, temperature, martensitic volume fraction) and is also given the temperature and strain increments. The output of the subroutine is the updated value of the resulting stress and transformation strain tensors, the new martensitic volume fraction and the tangent stiffness tensor. The output provided by `SMA_UM` may not entirely be required for the successive operations of the calling program, but some of it may be demanded by the user for post-processing.

At this point it is very important to note that the analysis of smart structures containing active SMA components often requires the consideration of non-linear geometric effects (i.e., moderate to large deformations). For most FEA packages, including Abaqus, this necessitates that some "non-linear geometry" analysis option be activated. In Abaqus, this is accomplished by turning on the `NLGEOM` option. Failure to do so can result in significant error in the final results because geometrically linear analysis does not consider the rotation of force vectors as the structure deforms. The software manuals should be consulted for more information on using such features. Furthermore, the accurate analysis of structures undergoing sufficiently large deformations requires a new formulation of the constitutive model itself, as shown by Anand and Gurtin [6] and Reese and Christ [7, 20], for example. Note also that in

Chapter 3 the units for temperature were "K", while in the examples that follow we will use " °C", as was the convention in the first two chapters.

4.4.1 SMA Uniaxial Thermomechanical Loading Cases

In this example, a three-dimensional prismatic bar in the form of a unit cube and composed of SMA is subjected to two differing uniaxial loading histories. The specific material properties are given in Table 4.3 and they correspond to a standard trained NiTi material with a constant current maximum transformation strain. Since the resulting stress state is uniform, the prismatic bar is divided into only two finite elements, and these elements are eight-node (linear) hexahedral elements (Abaqus designation C3D8). Initially, the material is nickel-titanium shape memory alloy in the austenitic state. Different loading conditions are applied, as described below

- **Test 1:** The first loading case demonstrates the capabilities of the subroutine to describe the pseudoelastic response of the material. The material is at a temperature higher than the zero-stress austenitic finish temperature, A_f, and the loading is applied uniaxially until full transformation is achieved. The material is then unloaded to zero stress, recovering all of the transformation strain.
- **Test 2:** The second loading case demonstrates the capabilities of the subroutine to describe temperature-induced phase transformation with applied stress. Initially, the material is at temperature above A_f. Uniaxial stress loading is applied such that the value of the applied stress is not sufficient to induce martensitic phase transformation. Next, holding the value of the applied stress constant, the material is cooled to a temperature below the zero-stress martensitic finish, M_f. During cooling, the material undergoes phase transformation and large strains are observed. Next, the material is heated back to the initial temperature, which results in reverse phase transformation and recovery of the transformation strain.

Table 4.3. SMA material parameters for uniaxial and active beam analyses.

Material Parameter	Value
E^A	70 GPa
E^M	30 GPa
ν	.33
$\alpha^A = \alpha^M$	$22 \cdot 10^{-6}$/ °C
M_s	18 °C
M_f	−2 °C
A_s	22 °C
A_f	42 °C
$C^A = C^M$	7 MPa/ °C
$H^{cur}(\sigma) = H$	0.05

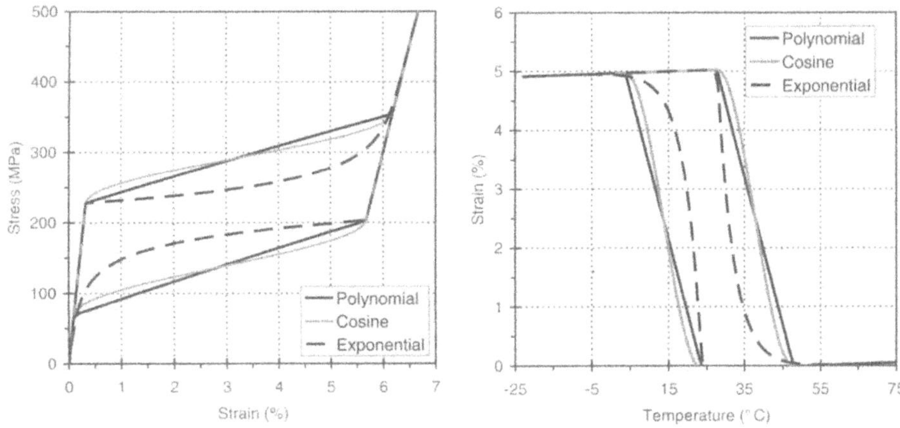

(a) (b)

Fig. 4.3. Results for uniaxial FEA test cases: (a) Test 1-constant temperature mechanical cycling, (b) Test 2-constant stress thermal cycling.

The stress-strain response of the SMA material for the first test case is shown in Fig. 4.3a. The results were obtained using all three constitutive models (see Sect. 3.4). These are: the exponential model [1], the cosine model [2], and the model with a polynomial hardening function [3]. The numerical simulations were performed at a temperature of 52 °C. The temperature-strain response of NiTi SMA for the second test is shown in Fig. 4.3b. The applied tensile load for this case was equal to 100 MPa. Note that due to the uniform stress, strain and temperature states within the bar, the results shown in Fig. 4.3 represent the 1-D constitutive model implementation as derived in the previous chapter (Sect. 3.5.1). The run time for each analysis was approximately 30 s.

4.4.2 SMA Actuated Beam

To expand the simple 1-D analysis, we investigate the operation of two active beam structures. The beam has the same geometric configuration in each case, though in one case it is composed of aluminum and actuated by an SMA wire, while in the second, it is composed of an SMA and opposed by a biasing elastic spring. The beam, which is 200 mm long and 30 mm deep, is attached to a vertical wall at the base, where it is 21 mm thick. The beam tapers at a 1:3 ratio, and the FEM mesh consists of linear 2-D continuum elements (Abaqus designation `CPE4`). The model consists of 50 elements along the thickness and five elements through the height. The beam reference configure is shown in Fig. 4.4.

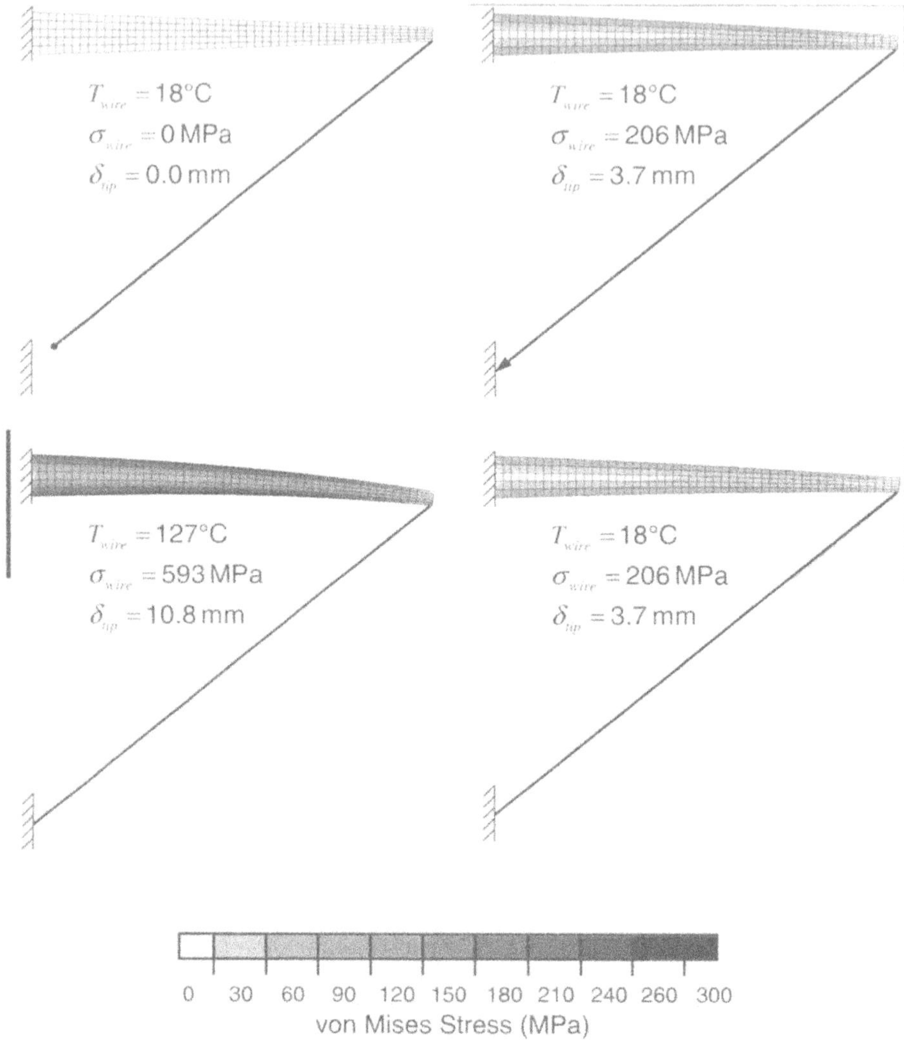

Fig. 4.4. Initial mesh and system response of SMA wire opposed to elastic beam.

Aluminum beam, SMA wire

In the first example, the SMA wire is modeled using a two-node 1-D element measuring 248.5 mm in length with a cross-sectional area of 9 mm^2. It is attached to the tip of the elastic beam and eventually fastened to a rigid point. Material properties for the SMA wire are given in Table 4.3, and the cosine form of the hardening function is assumed (for a review of the hardening function, see Sect. 3.4 in addition to the uniaxial example above, which compares possible hardening functions). The analysis steps are as follows:

1. At a temperature of 18 °C (M_s), with the wire initially in austenitic state, the free end of the wire is drawn back 17 mm and its end is fixed.
2. The temperature of the wire is raised to 127 °C.
3. The temperature of the wire is lowered to 18 °C.

The thermomechanical loading path is chosen such that initially austenitic material is stressed to form stress-induced martensite, heated into nearly full austenite, and then cooled into pure martensite. During the initial loading step, the wire experiences a total strain of 5.8%, while the stress in the wire reaches 206 MPa. This loading also forces the tip of the beam to deflect 3.7 mm downward. After the SMA wire is heated well past the zero-stress austenitic finish temperature, the stress in the wire increases to 593 MPa due to the incremental recovery of transformation strain. The beam tip deflection increases to 10.8 mm. However, on cooling into martensite, the tip deflection recovers to a total of 3.7 mm and the stress in the wire decreases to 206 MPa once again.

The von Mises stress present in the beam and the thermomechanical state of the wire at the end of each loading step can be seen in Fig. 4.4. Note the von Mises stress is identical to the effective deviatoric stress introduced in Chapter 3. The beam tip deflection is also included. The evolution of the wire thermomechanical state as plotted on the phase diagram is shown in Fig. 4.5. Note that, on heating, transformation begins at 2, and, due to the sharp increase in stress, is not completed by 3. Forward transformation does not begin until 4 (as reverse transformation did not complete), and completes at 5.

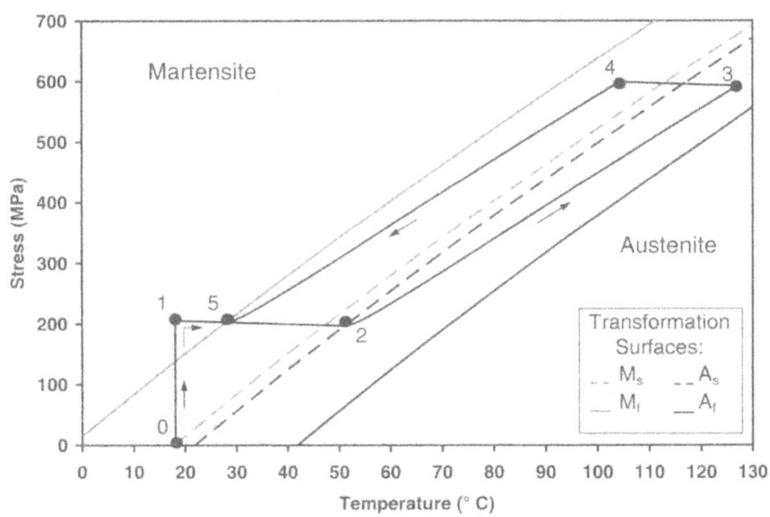

Fig. 4.5. Stress/temperature response of SMA wire opposed to elastic beam.

SMA beam, elastic bias spring

The second example considers the same beam geometric configuration, though here the beam itself is composed of SMA material. Biasing force is provided by a spring with a constant of 896 N/mm and an initial length of 240 mm. The thermomechanical loading steps are as follows:

1. At a temperature of 19 °C (M_s + 1 °C), with the beam initially purely austenitic, the free end of the spring is drawn back 17 mm and its end is fixed.
2. The temperature of the beam is raised to 127 °C.
3. The temperature of the beam is lowered to 19 °C.

The thermomechanical loading path is such that initially austenitic material (the beam) is stressed to form some stress-induced martensite, heated into full austenite, and then cooled into pure martensite. The initial downward deflection of the beam after the spring was retracted measured to 29.8 mm. After heating, the beam recovered to a total downward deflection of 18.9 mm, but returned 29.8 mm on cooling. These results can be seen in Fig. 4.6, where the contours represent the component of the transformation strain resolved along the axis of the beam (ε_{11}^t).

A comparison of the tip deflection response of the two beams (considering the specific geometry, material properties, and prescribed boundary conditions given) has been provided in Fig. 4.7. Although the results are obviously specific to the given modeling inputs, several key observations are evident which can be generalized qualitatively. First, the initial deflection of the SMA beam is much larger than the elastic beam due to the generation of transformation strain. Second, the total tip motion between the end of the heating step and the end of the cooling step is larger for the second beam as well, although the maximum transformation strain within the beam reaches only $\approx$ 0.9%, while in the wire it is $\approx$ 3%. This is due to the ability of a beam to generate significant deflection while experiencing only moderate strains.

4.4.3 SMA Torque Tube

The next numerical analysis example addresses the torsional actuation of an SMA tube fixed at one end and coupled to a biasing torsional spring at the other. The thin-walled SMA tube measures 152 mm in length, and has an outer radius of 25.4 mm with a wall thickness of 1.5 mm. It is meshed using 25 quadratic axis-symmetric elements along its length (Abaqus designation `CGAX8`). The material has the properties listed in Table 4.4 with the cosine form of transformation hardening assumed. These properties are similar to those given in Chapter 2, Table 2.4 for Ni60Ti material. The bias spring provides a moment of 192 Nm per degree of rotation. At the initiation of the analysis, the system is stress-free and the SMA is in the fully austenitic phase at a temperature of 35 °C. The thermomechanical loading steps proceed as follows:

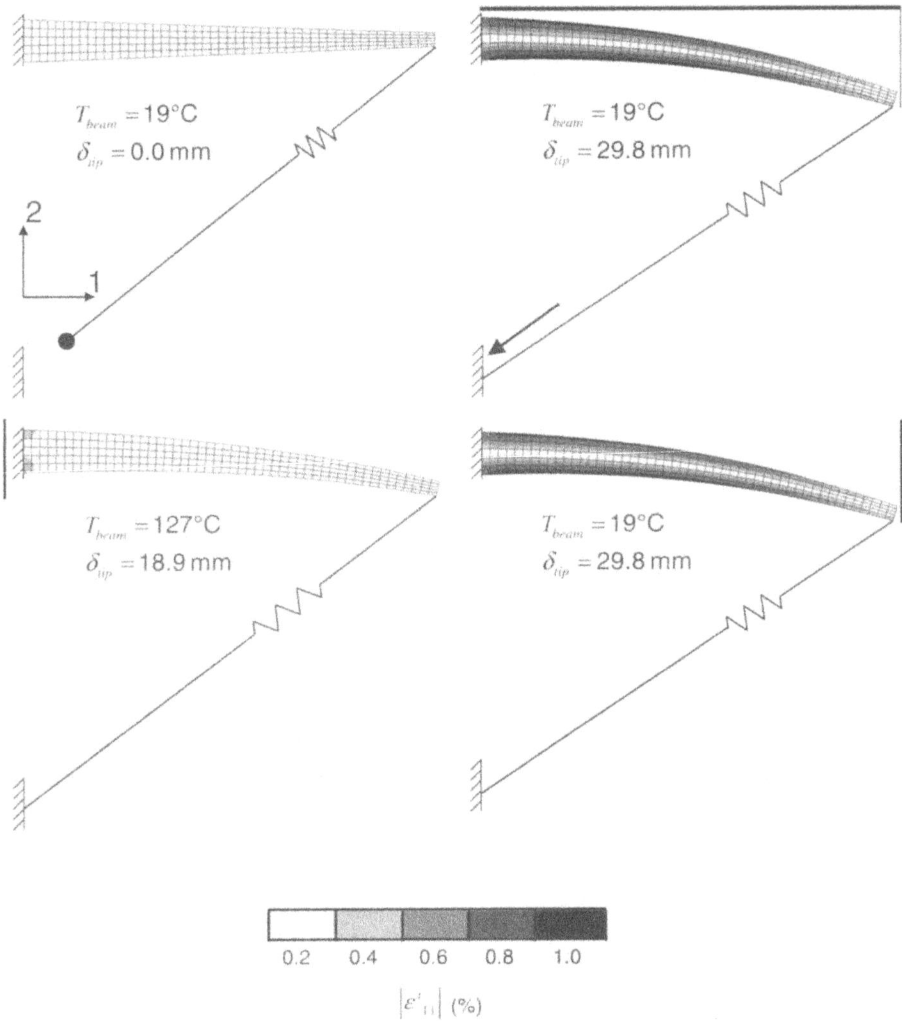

Fig. 4.6. Initial mesh and system response of SMA beam opposed to elastic spring biasing force.

1. At the initial temperature of 35 °C (M_s+1 °C), the free end of the torsional spring is rotated 11.6 °C, thus applying a load to the free end of the torque tube. The rotated end of the torsional spring is then fixed.
2. The temperature of the tube is raised to 82 °C.
3. The temperature of the tube is lowered to −18 °C.
4. The temperature of the tube is raised to 82 °C once again to assess repeatability.

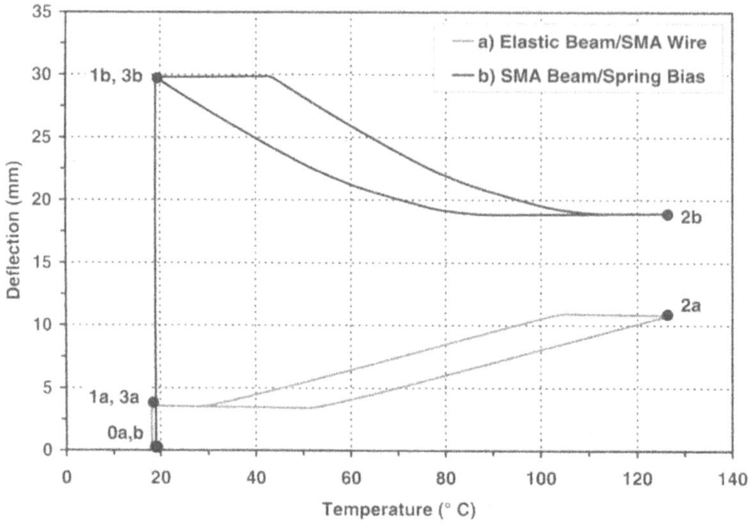

Fig. 4.7. Comparison of beam tip deflection, elastic beam/SMA wire and SMA beam/elastic bias spring.

This thermomechanical loading path causes initially austenitic material to be stressed into partial SIM, which is then heated into pure austenite. The material is finally cooled into pure martensite. When the spring is initially pre-rotated, the top of the tube rotates 5.14°. During heating into austenite, the tube recovers 43% of this initial rotation and a residual rotation of 2.92° is observed. When the temperature is lowered below the zero stress martensitic finish temperature, the tube end rotation increases to 8.67°, yielding

Table 4.4. SMA material parameters: Torque tube analysis

Material Parameter	Value
E^A	90 GPa
E^M	63 GPa
ν	0.33
$\alpha^A = \alpha^M$	$10 \cdot 10^{-6}/$ °C
M_s	34 °C
M_f	−17 °C
A_s	23 °C
A_f	57 °C
$C^A\|_{\sigma=300\text{MPa}}$	14.9 MPa/ °C
$C^M\|_{\sigma=300\text{MPa}}$	10.6 MPa/ °C
$H^{cur}(\sigma)$	$= 0.0135[1 - \exp(-720\sigma/E^A)]$

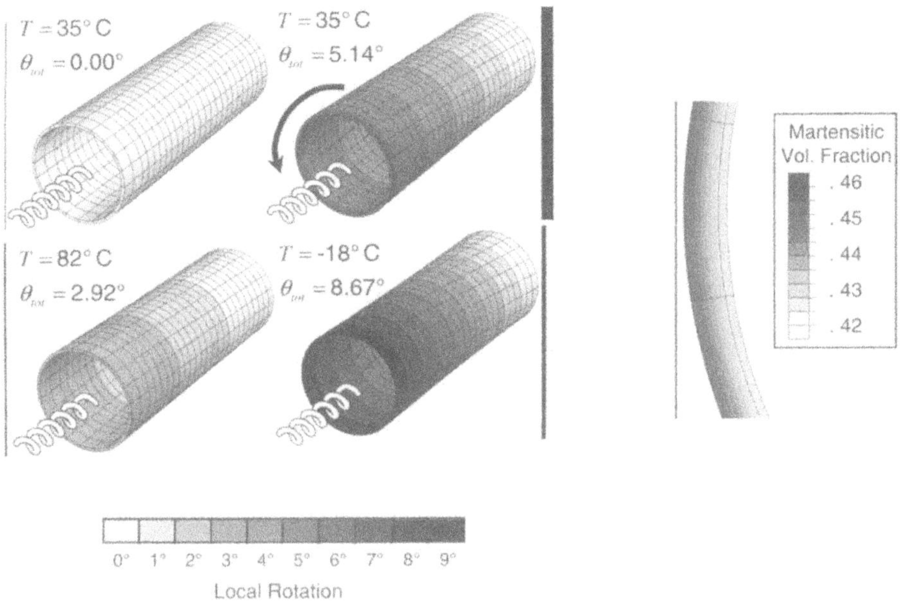

(a) (b)

Fig. 4.8. Initial mesh and response of the SMA torque tube: (a) system response at end of each loading step, (b) martensitic volume fraction distribution during heating at $T = 53\,°\mathrm{C}$.

an actuation range of motion of 5.75° as the tube material undergoes full transformation cycles.

Fig. 4.8 illustrates the predicted system response to such a series of thermomechanical loads. In Fig. 4.8a, the initial mesh is illustrated and the deformation response at the end of each step is shown. Figure 4.8b shows the nonhomogeneous distribution of martensitic volume fraction through the thickness of the tube during heating at $T = 53\,°\mathrm{C}$. The thermomechanical response of a representative torque tube element to all loading steps is shown superimposed on the phase diagram in Fig. 4.9.

4.4.4 SMA Actuated Variable Geometry Jet Engine Chevron

The Boeing Variable Geometry Chevron (VGC) was introduced in Sect. 1.10.1 and is an assemblage of three curved SMA beams installed on an elastic substrate as shown in Fig. 4.10. The three beams are attached at their centers to the upper side of the substrate and nonpenetration is enforced at the beam tips. The beams are modeled using 3-D quadratic hexahedral elements with reduced integration points (Abaqus designation `C3D20R`) and the constitutive properties given in Table 4.4 while assuming the cosine form of transformation hardening. A total of 2100 elements are arranged with 70 along the length, five

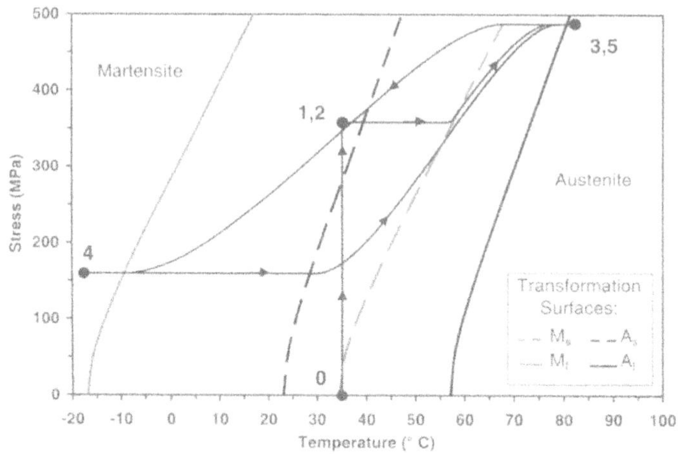

Fig. 4.9. Stress/temperature response of representative SMA torque tube element on phase diagram.

across the width, and six through the depth. The thermomechanical loading steps are as follows:

1. At an initial temperature of 18 °C and in a fully martensitic state, the beams are drawn down until flush with the chevron substrate.
2. The temperature of all three beams is homogeneously raised such that reverse transformation is completed ($T = 80\,^{\circ}\mathrm{C}$).
3. The temperature of the three beams is lowered to 20 °C.

The total computation time needed to complete this analysis, which included both non-linear material and non-linear deformations in addition to contact, was 5.8 hrs. To validate the predictions of the model, a series

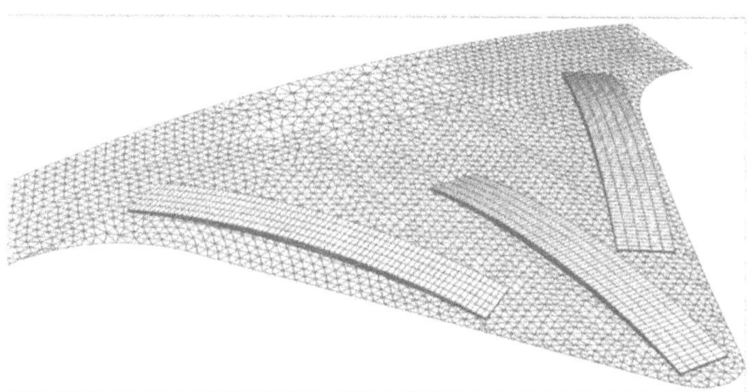

Fig. 4.10. Finite Element Analysis model of the Variable Geometry Chevron.

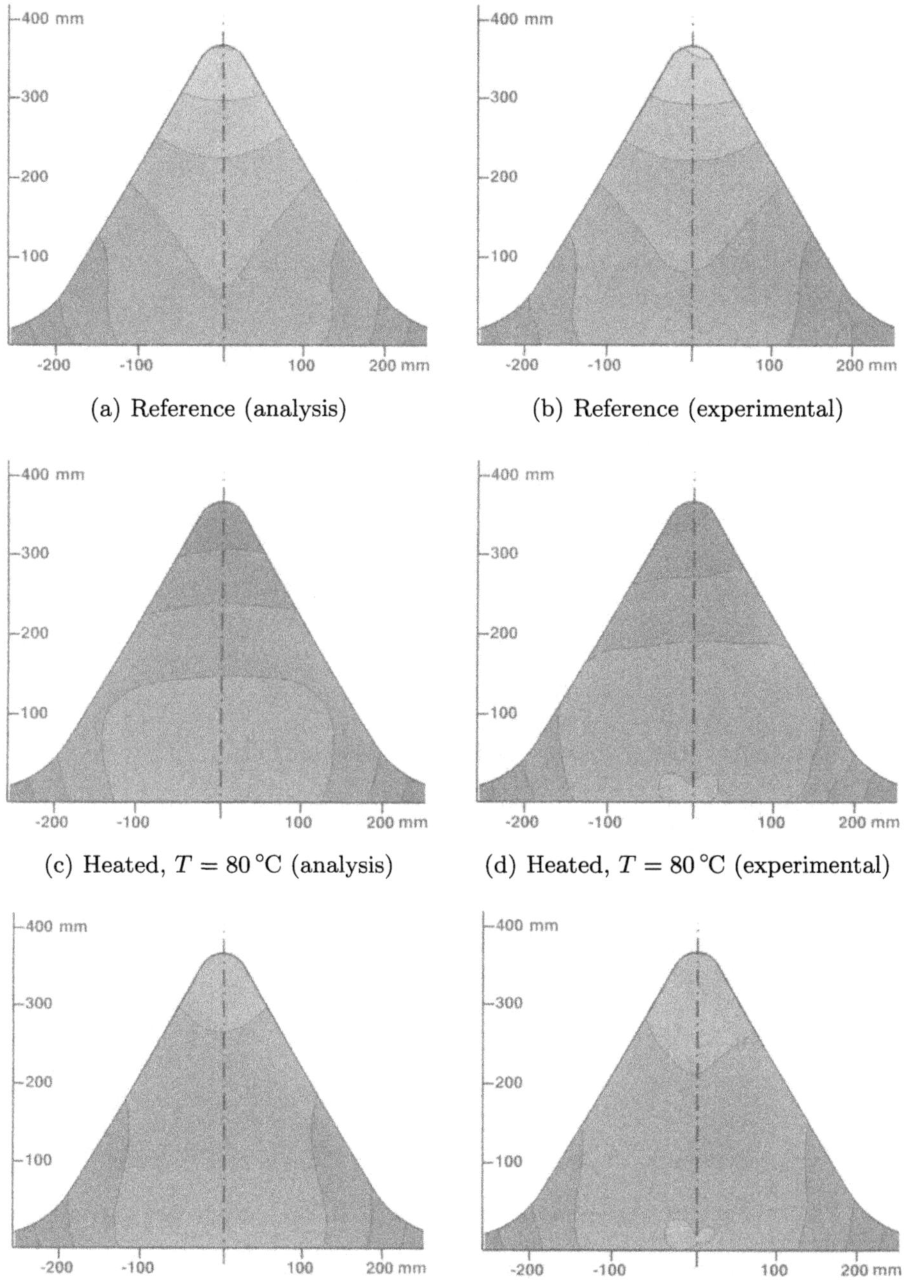

(a) Reference (analysis)

(b) Reference (experimental)

(c) Heated, $T = 80\,°\mathrm{C}$ (analysis)

(d) Heated, $T = 80\,°\mathrm{C}$ (experimental)

(e) Cooled, $T = 20\,°\mathrm{C}$ (analysis)

(f) Cooled, $T = 20\,°\mathrm{C}$ (experimental)

Fig. 4.11. Topography of the VGC: reference state, heated SMA beams, and cooled SMA beams.

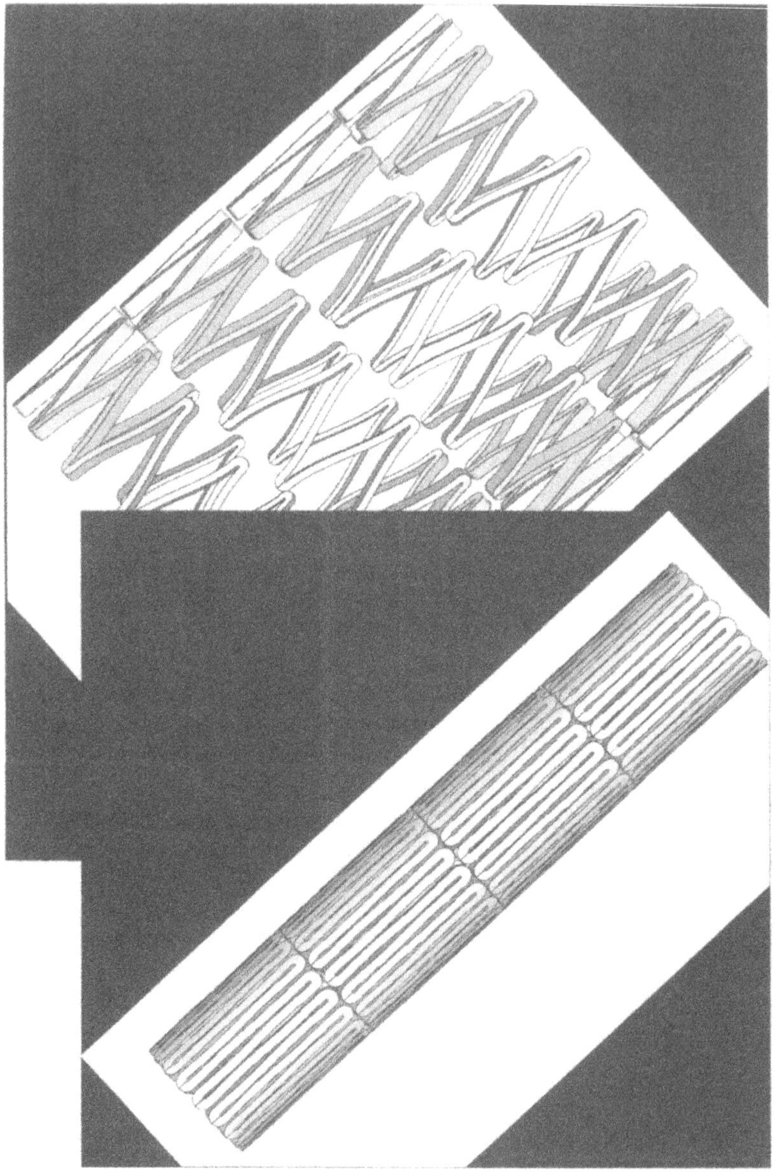

Fig. 4.12. Geometric configuration of the analyzed stent, stress-free and crimped configuration.

of experiments was performed. Thermomechanical loading steps identical to those listed above were applied to a prototype VGC. Photogrammetry was used to monitor the system response at several locations on the chevron surface. A contour plot of the surface topography (5 mm contour intervals) was generated from the measured coordinates of each discrete photogrammetry point. The analytical/experimental predictions are compared in Fig. 4.11.

4.4.5 SMA Medical Stent

The use of shape memory alloys in the fabrication of self-expanding stents was also discussed in Sect. 1.10.2. In this final numerical example, the thermomechanical behavior of such a stent is modeled considering the constraining step ("crimping") and the heating of the stent to body temperature. The cylindrical stent of interest has a stress-free outer diameter of 6 mm and a circumferentially periodic structure as shown in Fig. 4.12 where the expanded and constrained configurations can be seen.

There are a total of 18 full "Vs" which make up the stent circumference. Due to the obvious periodicity, only one "V" was meshed and analyzed. The solid model was derived from an Abaqus-provided stent model, which is publicly available. The SMA material was assumed to have the properties listed in Table 4.5 (cf. Sect. 2.5.1, Table 2.2). The modeled portion is meshed using 672 linear hexahedral continuum elements (Abaqus designation `C3D8`). The stent is considered to be initially stress-free, austenitic, and at a temperature of −27 °C. Thermomechanical loading then proceeds as follows:

1. The stent is crimped to an outer diameter of 1.5 mm.
2. The stent temperature is homogeneously raised to body temperature (37 °C).

During crimping, certain sections begin transformation into martensite resulting in relatively low stresses considering the large deformations. Upon heating, however, some of the martensitic portions of the stent begin transformation into austenite, which causes a significant increase in stress throughout the stent. The results of this analysis, as well as the reference mesh are shown in Fig. 4.13. The contours represent the level of von Mises stress throughout

Table 4.5. Material parameters: Medical stent analysis

Material Parameter	Value
E^A	55 GPa
E^M	46 GPa
ν	46 GPa
$\alpha^A = \alpha^M$	$22 \cdot 10^{-6} /\,°\mathrm{C}$
M_s	−28 °C
M_f	−43 °C
A_s	−3 °C
A_f	7 °C
$C^A = C^M$	7.4 MPa/ °C
$H^{cur}(\sigma) = H^{max}$	0.056

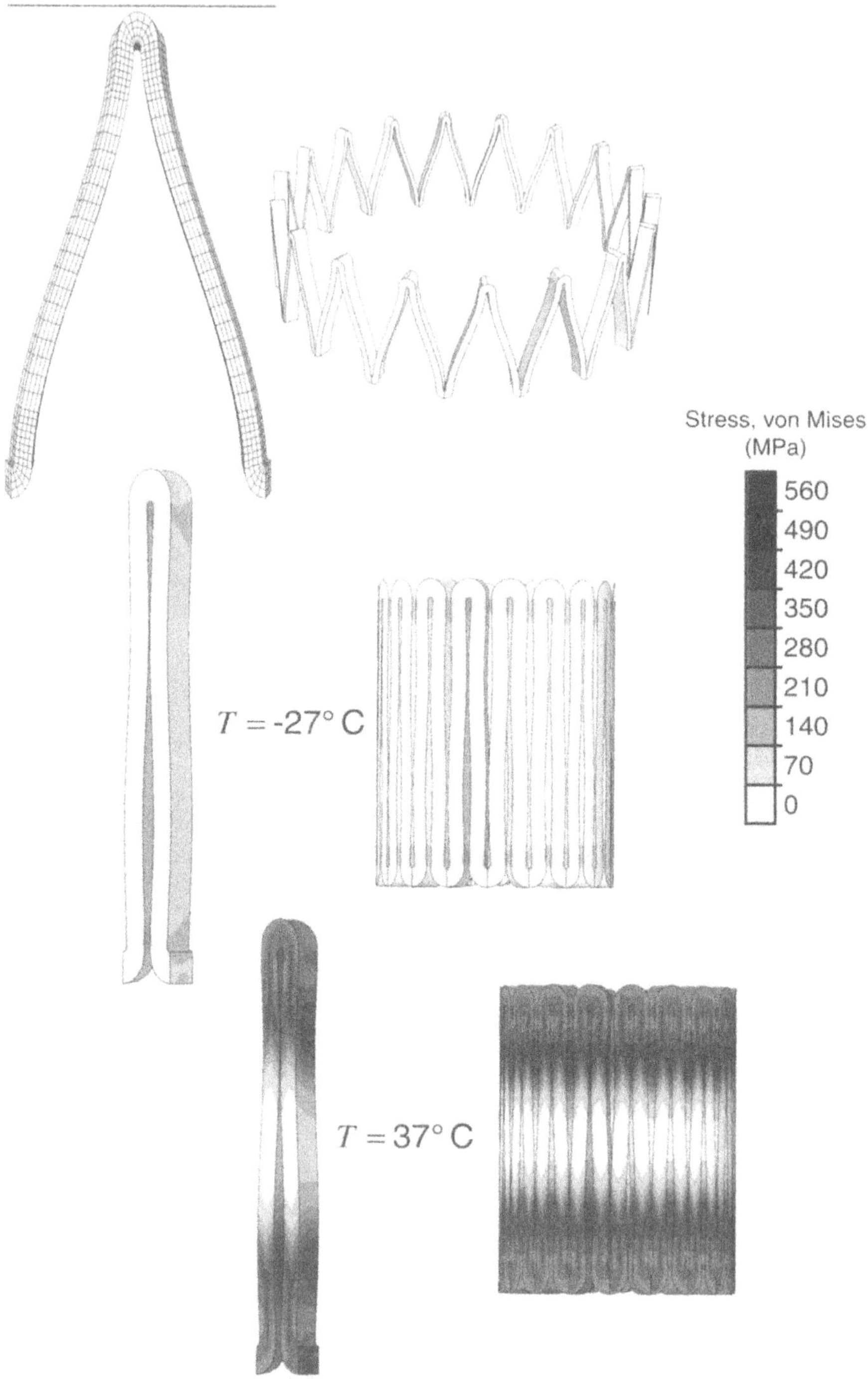

Fig. 4.13. Stent reference mesh and analysis results after crimping ($T = -27\,°C$) and heating ($T = 37\,°C$)

the stent. Note the low stress levels after the crimping step, and the much higher stress levels at body temperature. The total wallclock time needed to complete this analysis was 285 s.

4.5 Summary

This chapter presented methods for implementing thermomechanical models of shape memory alloys using return mapping algorithms. Two return mapping algorithms were presented in detail: the closest point projection algorithm and the convex cutting plane algorithm. Each of these algorithms includes an elastic predictor and a plastic (or transformation) corrector step. The general transformation correction (return mapping) structure and its geometrical interpretation were presented. The subject of tangent moduli was also addressed and it was shown that while the consistent tangent moduli can be derived for the case of closest point projection algorithm, the same is not possible for the convex cutting plane algorithm.

Various boundary value problems based on SMA applications were presented that exemplify the capabilities of both the SMA unified model and this particular implementation methodology. The analyses are of varying complexity, from simple cubes (2 SMA elements) to a multi-component aerostructure (6300 SMA elements), which included multiple components interfacing via consideration of contact. Experimental validation data is provided for the VGC case, and good agreement is observed, illustrating the usefulness of these tools in designing and modeling engineering applications of shape memory alloys. As illustrated in Fig. 4.14 for the case of the Boeing VGC, these tools will continue to streamline the traditional engineering "design-build-test" methodology by providing accurate system performance predictions while minimizing prototype construction and testing.

For additional resources related to the topics described in this chapter, visit `http://smart.tamu.edu/SMAText/`.

4.6 Problems

4.1. Starting with the rate form of the total strain (3.3.47),

$$\dot{\boldsymbol{\varepsilon}}(\boldsymbol{\sigma}, T, \xi) = \partial_{\boldsymbol{\sigma}}\boldsymbol{\varepsilon} : \dot{\boldsymbol{\sigma}} + \partial_T \boldsymbol{\varepsilon}\, \dot{T} + \partial_\xi \boldsymbol{\varepsilon}\, \dot{\xi},$$

show that the evolution equation for the inelastic strain evolution is associative, such that

$$\dot{\boldsymbol{\varepsilon}}^{in} = \begin{cases} \dot{\xi}\partial_{\boldsymbol{\sigma}}\Phi; & \dot{\xi} > 0, \\ -\dot{\xi}\partial_{\boldsymbol{\sigma}}\Phi; & \dot{\xi} < 0. \end{cases}$$

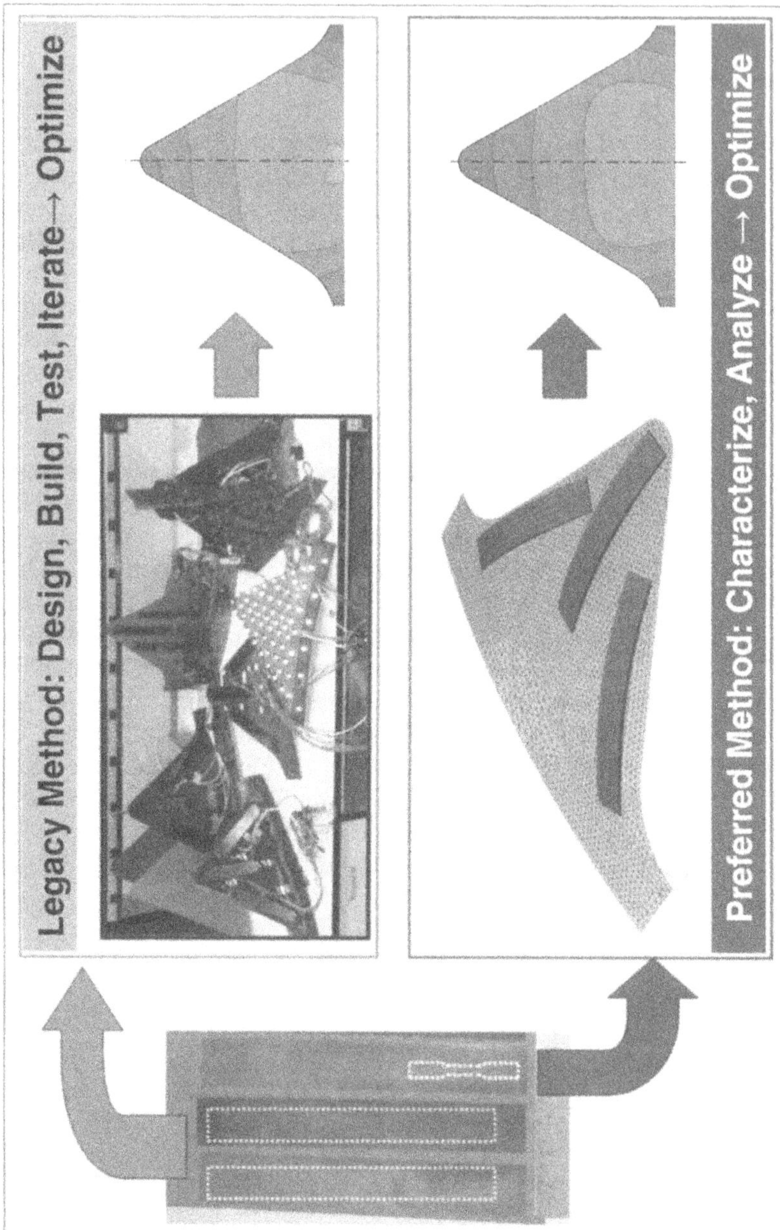

Fig. 4.14. The legacy and preferred methods of active structures engineering design.

4.2. Using the material properties shown in Table 4.4 and the unified model presented in Chapter 3, derive the analytical form of the phase diagram as is shown in Fig. 4.9. Is it possible to use the 1-D reduction of the unified model? Justify your answer.

4.3. Using the material properties shown in Table 4.3, the simple 1-D model presented in Sect. 2.6.2, and Euler-Bernoulli beam theory, estimate the response of the SMA wire-actuated aluminum beam at the end of the heating step (see Fig. 4.4). Furthermore, using the results from Prob. 2.12, predict the response after cooling.

4.4. Using the 1-D reduction of the unified model provided in Chapter 3 and Tables 4.1 and 4.2, implement both the closest point algorithm and convex cutting plane algorithm. Analyze the response of the uniaxial prismatic bar segment (cube) as shown in the numerical analysis examples. Is there a difference in the response?

4.5. Re-derive the 1-D formulation in Chapter 3 for the case of pure shear (i.e., $\sigma_{12} = \sigma_{21} \neq 0$ only). Use this model to approximate the results of the SMA torque tube loading as described in Sect. 4.4.3.

4.6. Using the unified model as given in Chapter 3, derive a 2-D reduction wherein

$$[\boldsymbol{\sigma}] = \begin{bmatrix} \sigma_{11} & 0 & 0 \\ 0 & \sigma_{22} & 0 \\ 0 & 0 & 0 \end{bmatrix}.$$

a) For this model, use Table 4.2 to implement the convex cutting plane algorithm.
b) Using this algorithm, analyze the stress-strain response of a plate with properties given in Table 4.5 undergoing full pseudoelastic 2-D loading at $T = 37\,°\mathrm{C}$ during which $\sigma_{22} = 2\sigma_{11}$. Plot von Mises stress vs. ε_{11} and ε_{22}.

4.7. Using the derivations given in Sect. 4.3.2 as a guide, derive the consistent tangent moduli tensors for the *reverse* transformation. Show that they are identical to those given in (4.2.11-4.2.12) in Sect. 4.2.

Additional Problems: Analysis of SMAs in "Active" Structures

The following problems address the use of SMA components in various active structures applications. Assumptions can be made for each problem such that a 1-D SMA model reduction can be used to approximate the solution. Furthermore, for readers with access to Abaqus Unified Finite Element Analysis software (with user subroutine capabilities), the user material subroutine discussed in Sect. 4.4 can be used to obtain an FEA solution. The UMAT and accompanying manual can be found at `http://smart.tamu.edu/SMAText/`.

4.8. Recall that one of the earliest industrial applications of SMAs was a coupling for piping [21]. Consider such a coupling installed at the joint of two steel pipes as shown in Fig. 4.15. The pipe outer diameter is 1 in (25.4 mm), and the inner diameter is 0.734 in (18.6 mm). An SMA coupling is formed from a material with the properties given in Table 4.5 and has a wall thickness of 0.266 in (6.76 mm). The coupling is pre-stretched into full martensite at $T = -10\,°C$ (loaded radially and then relaxed) such that the inner diameter matches the outer diameter of the pipe. It is then installed while remaining at a temperature of $-10\,°C$. Heating the coupling to above $-3\,°C$ begins the process of tightening. The pipe/coupling unit will be used in an environment for which the ambient temperature is 25 °C. Using the above information, perform the following tasks:

a) Calculate the stress-free inner diameter of the SMA coupling in the austenitic state.
b) Determine the pressure applied by the coupling on the pipes when the system temperature reaches the ambient temperature by using a 1-D model reduction. Solve for two cases: i) the steel pipe is rigid; ii) the steel pipe responds elastically. *Hint: Use a 1-D reduction in polar coordinates which considers only the hoop stress in the SMA coupling.*
c) Using FEA, determine the pressure applied by the SMA coupling on the pipes when the system temperature reaches the ambient temperature. Solve using an axisymmetric model.

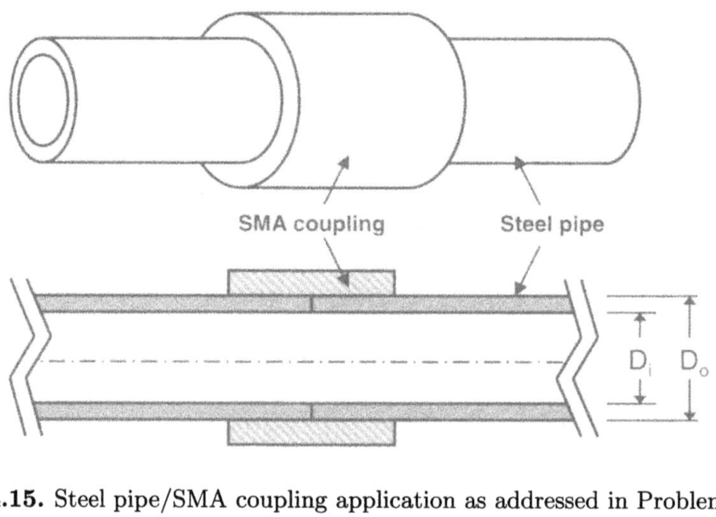

Fig. 4.15. Steel pipe/SMA coupling application as addressed in Problem 4.8.

4.9. A pressurized steel pipe ($R_o = 50\,\text{mm}$, $R_i = 45\,\text{mm}$, $L = 300\,\text{mm}$) is sealed on each end by a 12 mm thick steel plate. Four SMA rods in the austenitic state (properties given in Table 4.3) and having a diameter of 15 mm

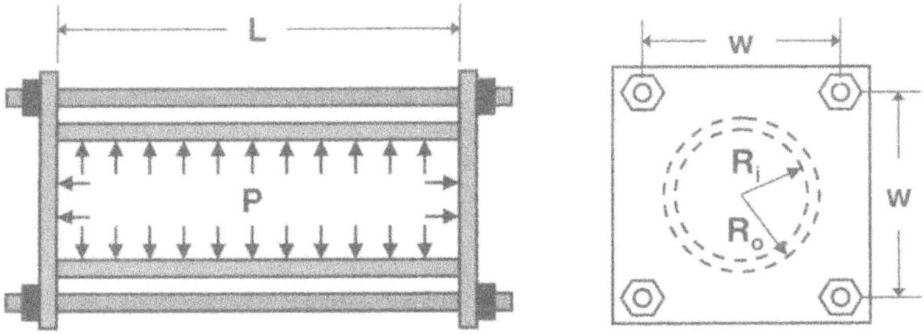

Fig. 4.16. Internally pressurized pipe constrained by four pretensioned rods as addressed in Problem 4.9.

are installed such that $w = 150\,\text{mm}$ and then equally pretensioned at 43 °C such that the martensitic volume fraction in each rod is 50%. The system configuration is shown in Fig. 4.16. Determine the internal pressure such that the system begins to leak (i.e., force no longer exists between the pipe and steel plate) for the following three cases:

a) Assuming the end plates to be rigid and the constrained pipe to be elastic, use a 1-D SMA model reduction to approximate the solution. Plot internal pressure vs. interface force between the plates and constrained pipe up to the leakage pressure.
b) Use a 3-D FEA model (with non-rigid end plates and 1-D SMA elements for the rods) to determine a solution. Plot internal pressure vs. interface force between the plates and constrained pipe.
c) Consider a similar problem wherein the four rods and end plates are steel and the pipe is SMA. Temperatures and material properties remain the same and pretensioning is applied until the pipe material is 50% martensitic. What is the internal pressure required to cause leakage in this case and how does it compare with the previous case?

4.10. This problem addresses an "antagonistic" SMA actuator, which places one SMA component in opposition to another to allow for two-way motion [22]. An aluminum beam ($L = 200\,\text{mm}$, $W = 25\,\text{mm}$, $H = 3\,\text{mm}$ thick) is cantilevered at one end. SMA strips ($l = 50\,\text{mm}$, $h = 1\,\text{mm}$ thick) with properties as given in Table 4.3 are prestrained lengthwise (loaded and relaxed) such that the martensitic volume fraction is 75% and are then installed symmetrically about the center of the beam. The installation temperature is 20 °C and the SMA strips are assumed to be perfectly bonded. The system configuration is schematically shown in Fig. 4.17. Consider the following four analysis methods:

a) Use a 1-D SMA model reduction and Euler-Bernoulli beam theory to analyze the structure (i.e., assume a constant stress-strain throughout the SMA strip cross-section).

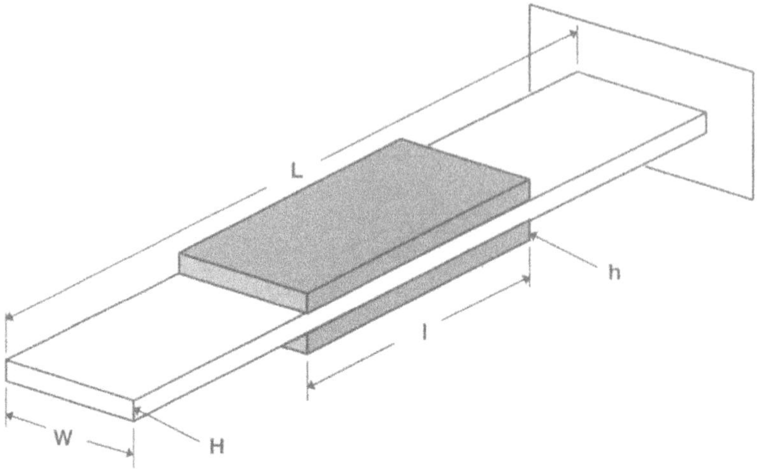

Fig. 4.17. Antagonistic SMA strips installed on aluminum beam as addressed in Problem 4.10.

b) Use a 2-D (plane strain) SMA model reduction and beam theory to analyze the structure.
c) Numerically analyze using 2-D (plane strain) FEA.
d) Repeat using 3-D FEA. Compare the solutions and computation times (2-D vs. 3-D).

For each method, plot the deflection of beam tip if the temperature of the *top* SMA strip is raised to 70 °C and then lowered back to 20 °C, after which the *bottom* SMA strip is raised to 70 °C and then lowered back to 20 °C (see [23] for a discussion of heating/cooling methods). Repeat for the case where a 30 N downward tip load is applied to the beam after installation of the SMA strips but prior to the heating/cooling steps.

4.11. An elastomeric round beam ($L = 200\,\text{mm}$, $R = 15\,\text{mm}$) is constructed such that it contains an embedded SMA wire (properties given in Table 4.3) with a diameter of 0.3 mm [24]. The SMA wire is prestrained (loaded and unloaded) to full martensite at 20 °C before installation. When installed, it is not bonded to the elastomer along its length but is assumed to apply a force to the round beam only at the ends. This configuration is schematically shown in Fig. 4.18. The elastomer/SMA system is cantilevered at one end while the other end is used to support a 1.5 N transverse tip load. Consider the following two analysis methods:

a) Use a combination of 1-D SMA model reduction and Euler-Bernoulli beam theory to analyze the structure.
b) Numerically analyze using 3-D FEA modeling (with 1-D SMA elements).

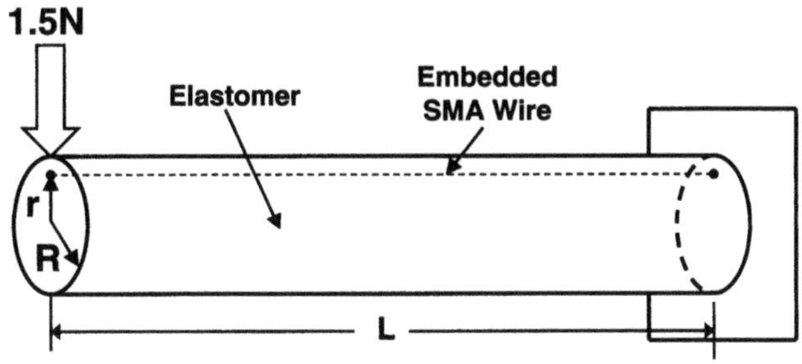

Fig. 4.18. Elastomer beam containing SMA wire element embedded off the beam centroid as addressed in Problem 4.11.

For each method, determine an optimal combination of elastomer material (e.g., solidified silicon) and wire embedding location r such that: i) the deflection of the beam at its loaded end is minimized when $T = 20\,°\text{C}$ *and*; ii) the tip displacement when the SMA wire is heated to 70 °C is maximized. Provide appropriate references for the elastomer properties chosen and plot the deflection of beam tip vs. SMA wire temperature. Compare the results to the case in which the SMA wire is assumed to be bound to the elastomer along the entire length of the wire.

4.12. A number of researchers have proposed using SMA wires as active components in a self-healing composite material (see [25], for example). This problem addresses the closing of a an elliptical "crack" in a infinitely long polymeric prismatic bar containing SMA wire components embedded transverse to the crack. The bar has a square cross-section with a side length $b = 100\,\text{mm}$. The ellipsoidal crack has a length of $2a = 0.6b$ and a width $w = 4\,\text{mm}$. SMA wires with properties as given in Table 4.3 are prestrained (loaded and unloaded) into full martensite at a temperature of $-10\,°\text{C}$. These are then installed below A_s such that they bisect the center of the crack and are evenly distributed along the length of the bar with a spacing of $s = 10\,\text{mm}$. The configuration of a segment from the bar is schematically shown in Fig. 4.19. Consider each of the following two cases:

a) The SMA wire elements only apply loads to the bar at their intersection with the outer surface (i.e., the wires are not bonded to the interior of the bar).
b) The SMA wire elements are perfectly bonded to the bar throughout its interior.

For each case, determine a suitable Representative Volume Element (RVE) to use in analyzing the structure. Find the amount by which the crack closes when the wires are heated to $A_f + 10\,°\text{C}$. Plot crack width vs. SMA wire temperature. Though a 1-D SMA model could be used to predict the wire response, an FEA model will be required to predict the response of the RVE

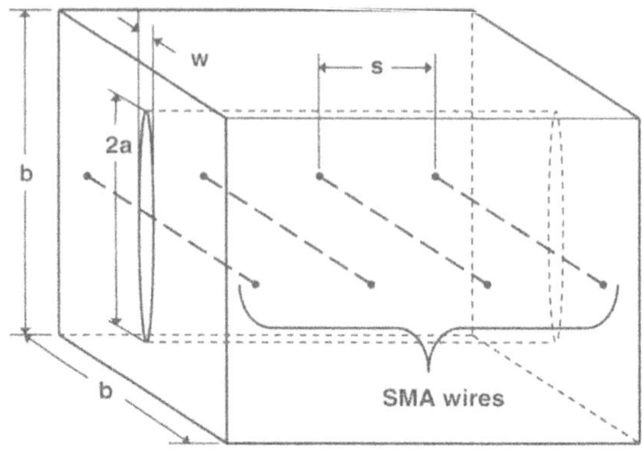

Fig. 4.19. SMA wires used in the construction of a "self-healing" structure with an elliptical "crack" as addressed in Problem 4.12.

matrix. Note that the four sides of the bar are traction-free while periodic boundary conditions should be imposed on the two periodic faces of the RVE. Assume that the RVE matrix is composed of hardened epoxy material.

4.13. Transversely loaded pseudoelastic SMA tubes have been studied for use as components in vibration isolation applications [26, 27]. Consider such loading of SMA tubes with properties as given in Table 4.5 and having an outer radius of 3 mm, a wall thickness of 0.17 mm and a length of 10 mm. Assume standard room temperature.

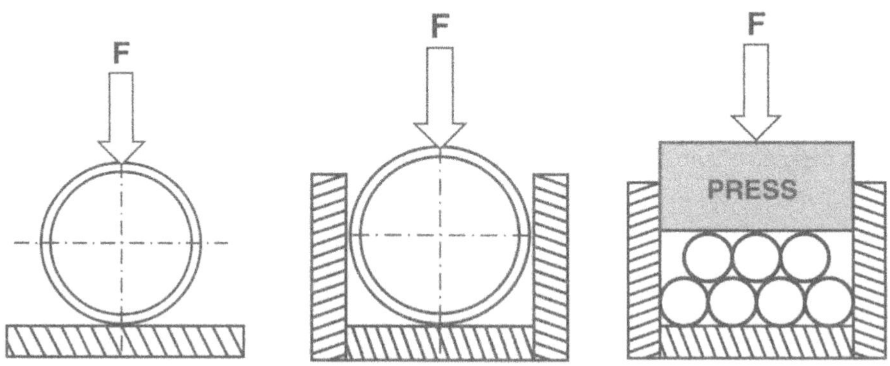

(a) Single pseudoelastic tube

(b) Single pseudoelastic tube, constrained

(c) Multiple pseudoelastic tubes under load

Fig. 4.20. Transverse loading of pseudoelastic SMA tubes as addressed in Problem 4.13.

a) Using 2-D (plane strain) modeling, determine the force/deflection results if a single such tube is transversely loaded as seen in Fig. 4.20a. Perform a mesh refinement study to ensure that the results are not mesh-dependent. Note that contact must be considered in modeling this problem.
b) Consider a similar single tube constrained on its sides as seen in Fig. 4.20b. Compare the results of a) and b) and discuss the influence of lateral confinement on the response of the SMA tubes.
c) Consider a more complex configuration in which two rows of tubes (seven total tubes) are constrained as seen in Fig. 4.20c and a force is applied from above. Using 2-D (plane strain) or 3-D modeling, determine the downward force/motion of the "press" seen in this figure. Plot the force vs. deflection curve up to the point when all SMA tubes are folded.

References

[1] K. Tanaka, S. Kobayashi, Y. Sato, Thermomechanics of transformation pseudoelasticity and shape memory effect in alloys, International Journal of Plasticity 2 (1986) 59–72.
[2] C. Liang, C. A. Rogers, A multi-dimensional constitutive model for shape memory alloys, Journal of Engineering Mathematics 26 (1992) 429–443.
[3] J. G. Boyd, D. C. Lagoudas, A thermodynamical constitutive model for shape memory materials. Part I. The monolithic shape memory alloy, International Journal of Plasticity 12 (6) (1996) 805–842.
[4] L. Brinson, R. Lammering, Finite-element analysis of behavior of shape memory alloy materials, International Journal of Solids and Structures 30 (1993) 3261–3280.
[5] D. C. Lagoudas, Z. Bo, M. A. Qidwai, A unified thermodynamic constitutive model for SMA and finite element analysis of active metal matrix composites, Mechanics of Composite Materials and Structures 3 (1996) 153–179.
[6] L. Anand, M. E. Gurtin, Thermal effects in the superelasticity of crystalline shape-memory materials, Journal of the Mechanics and Physics of Solids 51 (2003) 1015–1058.
[7] D. Christ, S. Reese, Is it necessary to model shape memory alloys within the scope of large strains?, Proceedings of SPIE, Smart Structures and Materials: Active Materials: Behavior and Mechanics, San Diego, CA, 26 February–2 March 2006 6170.
[8] W. Moss, On the computational significance of the strain space formulation of plasticity theory, International Journal for Numerical Methods in Engineering 20 (1984) 1703–1709.
[9] J. C. Simo, T. J. R. Hughes, Computational Inelasticity, Vol. 7 of Interdisciplinary Applied Mathematics, Springer-Verlag, New York, 1998.

[10] M. Ortiz, E. P. Popov, Accuracy and stability of integration algorithms for elastoplastic constitutive relations, International Journal for Numerical Methods in Engineering 21 (1985) 1561–1576.

[11] J. Lubliner, Plasticity Theory, Macmillan Publishing Company, New York, 1990.

[12] M. A. Qidwai, D. C. Lagoudas, Numerical implementation of a shape memory alloy thermomechanical constitutive model using return mapping algorithms, International Journal for Numerical Methods in Engineering 47 (2000) 1123–1168.

[13] M. Ortiz, P. Pinsky, Global analysis methods for the solution of elastoplastic and viscoplastic dynamic problems, Tech. rep., Department of Civil Engineering, Universtiy of California at Berkeley (1981).

[14] J. Simo, M. Ortiz, A unified approach to finite deformation elastoplastic analysis based on the use of hyperelastic consitituve equations, Computer Methods in Applied Mechanics and Engineering 49 (1985) 221–245.

[15] M. Ortiz, J. Simo, An analysis of a new class of integration algorithms for elastoplastic constitutive relations, International Journal for Numerical Methods in Engineering 23 (1986) 353–366.

[16] J. C. Simo, T. J. R. Hughes, General return mapping algorithms for rate-independent plasticity, in: C. S. Desai (Ed.), Constitutive Laws for Engineering Materials: Theory and Applications, Elsevier Science Publishers, 1987, pp. 221–231.

[17] J. Simo, R. Taylor, Consistent tangent operators for rate-independent elastoplacticity, Computer Methods in Applied Mechanics and Engineering 48 (1985) 101–118.

[18] Z. Bo, D. C. Lagoudas, Thermomechanical modeling of polycrystalline SMAs under cyclic loading, Part I: Theoretical Derivations, International Journal of Engineering Science 37 (1999) 1089–1140.

[19] Dassault Systmes of America Corp., Woodlands Hills, CA, ABAQUS/Standard User's Manual (2006).

[20] S. Reese, D. Christ, Finite deformation pseudo-elasticity of shape memory alloys – Constitutive modelling and finite element implementation, International Journal of Plasticity (In Press) doi:10.1016/j.ijplas.2007.05.005.

[21] K. R. Melton, General applications of shape memory alloys and smart materials, in: K. Otsuka, C. M. Wayman (Eds.), Shape Memory Materials, Cambridge University Press, Cambridge, 1999, Ch. 10, pp. 220–239.

[22] D. C. Lagoudas, D. Moorthy, M. A. Qidwai, J. N. Reddy, Modeling of the thermomechanical response of active laminates with sma strips using the layerwise finite element method, Journal of Intelligent Material Systems and Structures 8 (1997) 476–488.

[23] A. Bhattacharyya, D. C. Lagoudas, Y. Wang, V. K. Kinra, On the role of thermoelectric heat transfer in the design of SMA actuators: Theoretical modeling and experiment, Smart Materials and Structures 4 (1995) 252–263.

[24] B. de Blonk, D. Lagoudas, Actuation of elastomeric rods with embedded two-way shape memory alloy actuators, Smart Materials and Structures 7 (1998) 771–783.

[25] D. Burton, X. Gao, L. Brinson, Finite element simulation of a self-healing shape memory alloy composite, Mechanics of Materials 38 (5–6) (2006) 525–537.

[26] D. Lagoudas, M. Khan, J. Mayes, B. Henderson, Pseudoelastic SMA spring elements for passive vibration isolation, Part II: Simulations and experimental correlations, Journal of Intelligent Material Systems and Structures 15 (6) (2004) 443–470.

[27] M. Khan, D. Lagoudas, J. Mayes, B. Henderson, Pseudoelastic SMA spring elements for passive vibration isolation, Part I: Modeling, Journal of Intelligent Material Systems and Structures 15 (6) (2004) 415–441.

5

Modeling of Transformation-Induced Plasticity in SMAs

P. B. Entchev and D. C. Lagoudas

Many engineering applications of shape memory alloys require material that exhibits a repeatable thermomechanical response, and Chapters 1 and 2 discussed the stabilization of material through training. Chapters 3 and 4 addressed the derivation and implementation of models designed to simulate such stable behavior. However, the ability to account for evolving material behavior caused by induced plastic strains during transformation is also important. Here, we focus on the development of a 3-D constitutive model for SMAs that continue to exhibit substantial cyclic evolution of irrecoverable strains. A 1-D reduction of the model and comparison with experimental results is also presented.

5.1 Introduction

In reviewing the body of knowledge concerning the modeling of SMAs, one will find that the vast majority of theories and implementations address the stable, repeatable hysteretic behavior exhibited by appropriately trained SMA materials. This is obviously sufficient for most applications where repeatable behavior is required, and Chapters 1–4 addressed this important topic. However, shape memory alloys exhibit evolving hysteretic behavior (i.e., training) during initial cycling. Having presented a thermodynamic and numerical framework for modeling the stable constitutive behavior of SMAs, it is now appropriate to discuss the modification of these methods to account for other, more complex, material behaviors including the evolution of material response.

Experimental observations of SMAs undergoing cyclic loading via thermal activation under constant stress or operating in the pseudoelastic regime have shown that a significant part of the developed strain is not recovered on unloading and accumulates with every transformation cycle. This effect has been attributed to the development of irrecoverable plastic strains during the thermomechanical cycling of SMAs undergoing phase transformation. This *transformation-induced plasticity*, the result of a number of permanent microstructural changes, is often known simply as TRIP. As the number of loading cycles increases, the evolution of TRIP slows and, in some cases, ceases

D.C. Lagoudas (ed.), *Shape Memory Alloys*, DOI: 10.1007/978-0-387-47685-8_5,

after a large number of subsequent cycles until the transformation induced fatigue limit is reached (see Chapter 1, Sect. 1.7). Many models found in the literature describing the development of plastic strains in SMAs have a 1-D formulation. These models are limited in their usefulness as many important applications require modeling capabilities extending beyond 1-D stress states. In this chapter we will demonstrate the derivation and implementation of a 3-D model that captures the evolution of both transformation and plastic strains simultaneously (TRIP).

The 3-D model development [1] will follow the methodology used for the 1-D case presented by Bo and Lagoudas [2–5]. This earlier work described the behavior of SMAs undergoing thermally-induced phase transformation, and here we adapt the formulation for transformation-induced plasticity in the case of stress-induced martensitic transformation.

Although the model for TRIP derived in this chapter follows from the work of Bo and Lagoudas, the current model is capable of simulating the 3-D behavior of SMAs. The original model by Bo and Lagoudas [2] was implemented for the case of 1-D SMA components only. In addition, the calibration of the model parameters is different. While the previous publications [2–5] have been devoted exclusively to characterizing the behavior of SMA wires undergoing thermally-induced phase transformation and transformation-induced plasticity under cyclic thermal loading, the current work is aimed at characterizing the SMAs undergoing stress-induced phase transformation and simultaneous transformation and plasticity evolution with the number of mechanical cycles. Thus, the procedure for estimation of the material parameters (Sect. 5.3) utilizes data for SMAs undergoing stress-induced phase transformation. While the present model can still be used to model thermally-induced phase transformation, material parameters may need to be recalibrated to obtain accurate results. The work presented in this chapter follows the presentation of Lagoudas and Entchev [1]. It will be derived as an extension of the formulation presented in Chapter 3 (Sect. 3.3). However, a modified form of the Gibbs free energy will be used and new internal variables will be introduced. Therefore, this chapter provides a new example of phenomenological SMA constitutive model derivation considering new effects.

5.1.1 Experimental Motivation: Polycrystalline SMAs Undergoing Cyclic Loading

To motivate the model of interest, we first consider some examples of materials exhibiting TRIP behavior. Chapter 2 provides an example of plastic strain evolution during constant stress thermal transformation (Fig. 2.15) and an example of pseudoelastic TRIP behavior (Fig. 2.27). Many examples exist throughout the literature. A set of experimental results showing the SMA response undergoing cyclic stress-induced (pseudoelastic) transformation is

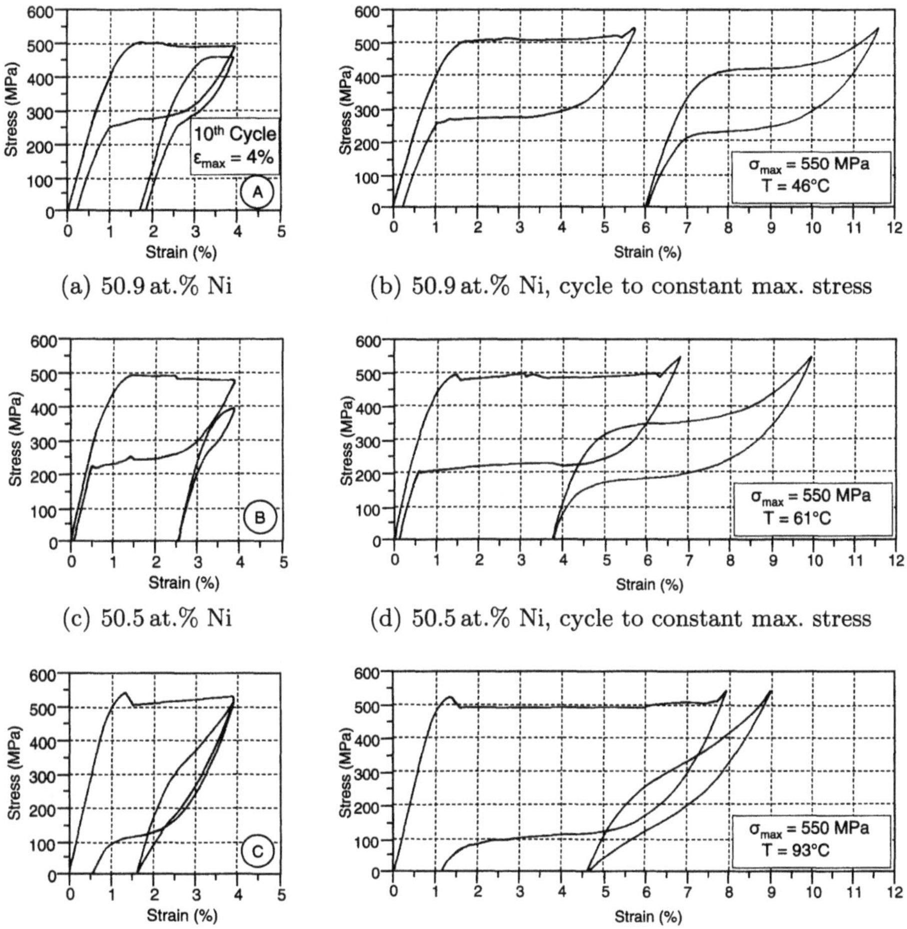

(a) 50.9 at.% Ni

(b) 50.9 at.% Ni, cycle to constant max. stress

(c) 50.5 at.% Ni

(d) 50.5 at.% Ni, cycle to constant max. stress

(e) 50.0 at.% Ni

(f) 50.0 at.% Ni, cycle to constant max. stress

Fig. 5.1. Cyclic stress-strain results for NiTi SMA strips of various compositions undergoing pseudoelastic loading (first and last cycles shown): (left) cycling up to a constant maximum value of strain; (right) cycling up to a constant maximum value of stress [6].

shown in Fig. 5.1[1]. The results are for three different NiTi alloys and the tests have been performed above the austenitic finish temperature. Two different uniaxial pseudoelastic tests were performed: cyclic loading with a constant maximum value of strain and cyclic loading with a constant maximum

[1] Reprinted from *Materials Science and Engineering A*, Vol. 203, Strnadel, B., Ohashi, S., Ohtsuka, H., Miyazaki, S. and Ishihara, T., pp. 187–196, Copyright 1995, with permission from Elsevier.

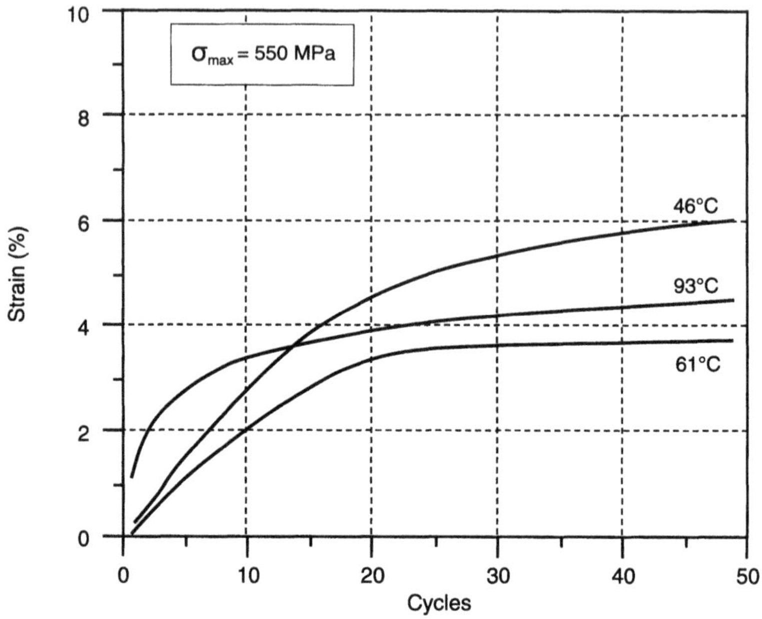

Fig. 5.2. Residual strain accumulated during cyclic loading of NiTi SMA strips. Results for cycling up to a constant maximum value of stress are shown [6].

value of stress. Both sets of results are shown in the figure. The accumulated transformation-induced plastic strain during loading is shown in Fig. 5.2[1].

Several observations can be made from these figures. First, during the cycling loading, a substantial amount of irrecoverable plastic strain accumulates. As shown in Fig. 5.2, the rate of accumulation of plastic strain is high during the initial cycles and asymptotically goes to almost zero with the increase of the number of cycles, as the plastic strain reaches a saturation value. A second observation is that the value of critical stress for onset of the transformation decreases with the number of cycles. A third observation is the substantial increase in transformation hardening. Additionally, in some cases the value of the maximum transformation strain decreases with the number of cycles. Finally, the area enclosed by the transformation hysteresis loop decreases.

The issue of TRIP observed during constant stress thermally-induced cyclic transformation, more common for actuator applications of SMAs, has also been addressed [4]. Here, various levels of constant stress were applied to NiTi wires and the evolution of the irrecoverable plastic strain was monitored over many transformation cycles. Fig. 5.3[2] shows the total strain as

[2] Reprinted from *International Journal of Engineering Science*, Vol. 37, Issue 9, Bo, Z. and Lagoudas, D.C., pp. 1175–1203, Copyright 1999, with permission from Elsevier.

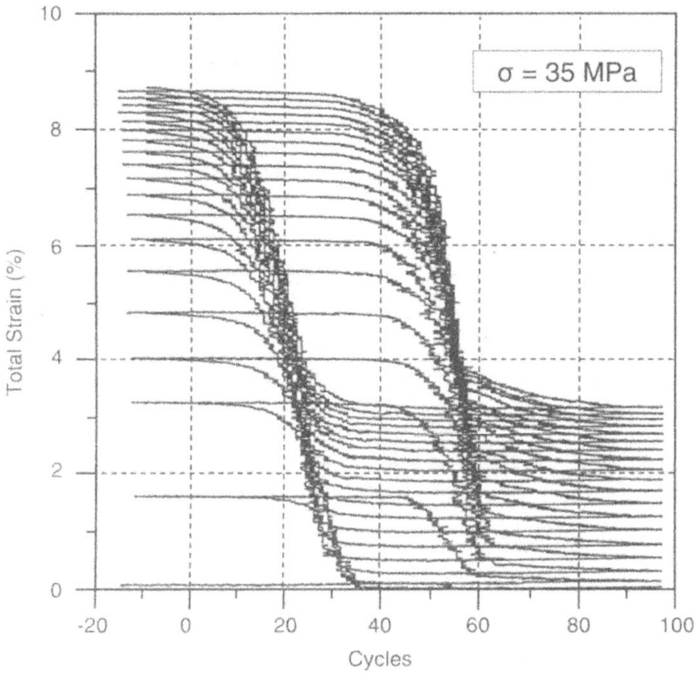

Fig. 5.3. Cyclic strain-temperature response during cyclic thermal transformation of NiTi SMA wires under a constant 35 MPa tensile load, 20 cycles shown [4].

a function of temperature for the first 20 cycles of loading at a constant force, resulting in 35 MPa uniaxial stress. The TRIP is evident in this plot by examining the residual strain present when the material is heated into austenite ($T > 350\,\mathrm{K}$). The evolution of TRIP as observed during constant stress loading at other load levels is shown in Fig. 5.4[2]. While the incremental accumulation of plastic strain observed in Fig. 5.2 for pseudoelastic testing seems to saturate, the same is not observed in the constant stress results summarized in Fig. 5.4. For these results, the response never truly stabilizes until final failure. Similar results are also observed in Fig. 1.15, which discusses the topic of thermally-induced transformation fatigue.

Similar observations have also been reported by other researchers (see, for example, the works of McCormick and Liu [7], Strnadel et al. [8], Lim and McDowell [9], Kato et al. [10] and Sehitoglu et al. [11]). Thus, the constitutive model presented in this work will address the effects described above, which are common for most polycrystalline NiTi SMAs.

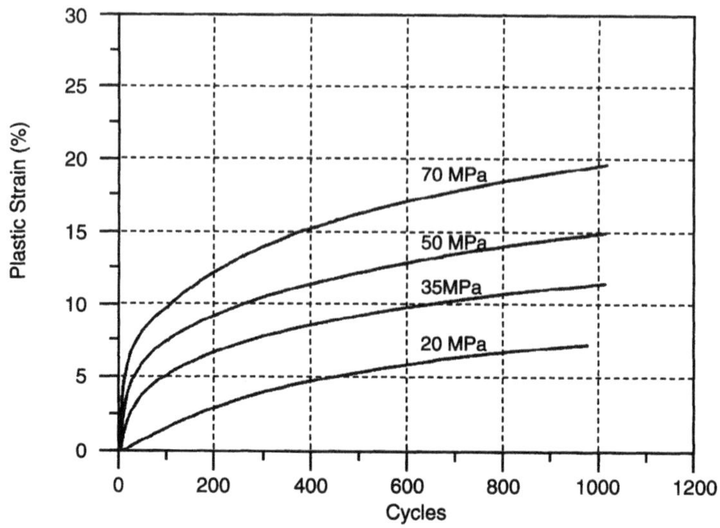

Fig. 5.4. Residual plastic strain accumulated during cyclic thermal transformation of NiTi SMA wires under various constant tensile loads [4].

5.2 Three Dimensional Constitutive Model for SMAs Experiencing TRIP

To expand on the derivation of 3-D constitutive models for SMAs as presented in Chapter 3, we must first review any additional internal variables needed to account for the evolution of transformation-induced plastic strain considering both isotropic and kinematic hardening. Discussions of appropriate internal variables for this case are in many plasticity texts (see, for example, [12]). In short, we choose the common options of plastic strain, $\boldsymbol{\varepsilon}^p$, the back stress, $\boldsymbol{\beta}$, and the drag stress, η. The back stress accounts for the kinematic portion of the plastic hardening and is a second-order tensor while the drag stress accounts for the isotropic portion and is a scalar. Using these possible internal variables and motivated by detailed developments presented elsewhere [2], we can postulate a new Gibbs free energy for the total SMA material, which includes the effects of TRIP. This is given as (cf. Sect. 3.3):

$$\begin{aligned} G(\boldsymbol{\sigma}, T, \xi, \boldsymbol{\varepsilon}^t, \boldsymbol{\varepsilon}^p, \boldsymbol{\beta}, \eta) = & -\frac{1}{2\rho}\boldsymbol{\sigma} : \mathbf{S} : \boldsymbol{\sigma} - \frac{1}{\rho}\boldsymbol{\sigma} : \left[\boldsymbol{\alpha}(T - T_0) + \boldsymbol{\varepsilon}^t + \boldsymbol{\varepsilon}^p\right] \\ & - \frac{1}{\rho}\int_0^{\xi} \left(\boldsymbol{\beta} : \frac{\partial \boldsymbol{\varepsilon}^t}{\partial \tau} + \eta\right) \mathrm{d}\tau + c\left[T - T_0 - T\ln\left(\frac{T}{T_0}\right)\right] \\ & - s_0(T - T_0) + u_0 + f(\xi). \end{aligned} \tag{5.2.1}$$

In comparing this to the form given in Chapter 3, specifically in (3.3.41), one notes that there are only two new terms which account for the energetic consequences of plasticity. These are the $\boldsymbol{\sigma} : \boldsymbol{\varepsilon}^p$ and martensitic volume fraction integral terms. In determining the material constants (i.e., $\mathbf{S}$, $\boldsymbol{\alpha}$, etc.), the rule of mixtures as discussed in (3.3.42)–(3.3.46) is used. Assumptions related to the new plasticity-based quantities are described in the following subsection.

5.2.1 Modifications Needed to Account for TRIP

The evolution equations for the new plasticity state variables $\boldsymbol{\varepsilon}^p$, $\boldsymbol{\beta}$ and η are presented here. Because this model addresses plastic deformation distinctly caused by *transformation*, it is a key assumption that the back stress and drag stress each evolve with the martensitic volume fraction ξ. Additional TRIP model assumptions also necessitate modification of the *transformation tensor* $\boldsymbol{\Lambda}$ as shown below.

Plastic Strain

The plastic strain considered here is different from conventional plasticity in metals. The observable macroscopic plastic strain is developed simultaneously with the transformation strain during martensitic phase transformation. This is a result of the accommodation of different martensitic variants during the phase transformation. Due to the misfit between the austenite-martensite interfaces, significant distortion is created. In addition, in a polycrystalline SMA, different grains transform in different manners, which causes additional distortion at the grain boundaries. These two phenomena act in concert and the final result is an observable macroscopic plastic strain, which occurs at stress levels much lower than the plastic yield limit of the material without phase transformation [13]. This model does not address the plastic strain evolution initiated when pure austenite or martensite is subjected to stresses that exceed the critical stress for slip, but is focused on plasticity caused by cyclic transformation only.

Similar to the evolution of transformation strain (see (3.3.51)), the direction of plastic strain is determined by the direction of the applied stress; although, another factor must be taken into account. Lim and McDowell [9] have suggested that the plastic strain rate depends on the magnitude of $\dot{\xi}$. However, in this work, a different approach as outlined by Bo and Lagoudas [4], where the self-accommodating martensitic phase transformation is assumed not to result in plastic strain development will be used. Therefore, the development of the plastic strain is connected to the detwinned martensitic volume fraction via the relation given in (2.6.9) as

$$\xi^d \equiv \frac{H^{cur}\left(\bar{\sigma}'\right)}{H^{max}}\xi$$

recall that $H^{cur}(\bar{\sigma}')$ is the maximum current transformation strain which will be generated during forward phase transformation at the current applied stress level. This topic was discussed in Sect. 2.2.3 and functional forms are suggested throughout Chapter 2. Recall also that the constant H^{max} denotes the maximum value of $H^{cur}(\bar{\sigma}')$ obtained during forward transformation at large values of the applied stress. Next, the *accumulated detwinned martensitic volume fraction* ζ^d is introduced as:

$$\zeta^d = \int_0^t |\dot{\xi}^d(\tau)| \, d\tau \, . \tag{5.2.2}$$

From its definition, one finds that $\dot{\zeta}^d = |\dot{\xi}^d|$, and the following evolution equation for $\boldsymbol{\varepsilon}^p$ is proposed:

$$\dot{\boldsymbol{\varepsilon}}^p = \widetilde{\boldsymbol{\Lambda}}^p \dot{\zeta}^d = \widetilde{\boldsymbol{\Lambda}}^p \frac{H^{cur}(\bar{\sigma}')}{H^{max}} |\dot{\xi}|. \tag{5.2.3}$$

The quantity $\widetilde{\boldsymbol{\Lambda}}^p$ is the *transformation-induced plasticity tensor* (or TRIP tensor) whose functional form is discussed next.

One form of $\widetilde{\boldsymbol{\Lambda}}^p$ is suggested by Bo and Lagoudas [4] for the 1-D case. This form depends on the value of the applied stress, the accumulated detwinned martensitic volume fraction, and the value of the plastic strain itself. As explained by Bo and Lagoudas [4], the plastic strain predicted by their model does not reach a saturation value. While this is observed during thermally-induced phase transformation (see Fig. 5.3), for the case of stress-induced transformation (pseudoelasticity), the experimental results reported by Strnadel et al. [6] indicate that the plastic strain reaches a saturation value (Fig. 5.2). Thus, to model stress-induced phase transformation in addition to thermally induced transformation, the following form of $\widetilde{\boldsymbol{\Lambda}}^p$ is proposed [1]:

$$\widetilde{\boldsymbol{\Lambda}}^p = \begin{cases} \frac{3}{2} C_1^p \frac{\boldsymbol{\sigma}^{\text{eff}\prime}}{\bar{\sigma}^{\text{eff}}} e^{-\frac{\zeta^d}{C_2^p}}; \; \dot{\xi} > 0, \\ C_1^p \frac{\boldsymbol{\varepsilon}^{t-r}}{\bar{\varepsilon}^{t-r}} e^{-\frac{\zeta^d}{C_2^p}}; \; \dot{\xi} < 0. \end{cases} \tag{5.2.4}$$

where $\boldsymbol{\sigma}^{\text{eff}}$ denotes the *effective stress tensor*, which is defined as

$$\boldsymbol{\sigma}^{\text{eff}} = \boldsymbol{\sigma} + \boldsymbol{\beta}. \tag{5.2.5}$$

The back stress $\boldsymbol{\beta}$ will be introduced in more detail shortly. The *deviatoric part of the effective stress* $\boldsymbol{\sigma}^{\text{eff}\prime}$ and the *von Mises equivalent effective stress* $\bar{\sigma}^{\text{eff}}$ are defined as

$$\boldsymbol{\sigma}^{\text{eff}\prime} = \boldsymbol{\sigma}^{\text{eff}} - \frac{1}{3}(\text{tr}(\boldsymbol{\sigma}^{\text{eff}}))\mathbf{1}, \quad \bar{\sigma}^{\text{eff}} = \sqrt{\frac{3}{2}\|\boldsymbol{\sigma}^{\text{eff}\prime}\|^2}, \tag{5.2.6}$$

where $\|\cdot\|^2$ is the inner product of the enclosed quantity and $\mathbf{1}$ is the second order identity tensor (c.f. (3.3.53), (3.3.54)).

It can be seen from comparing (5.2.3) and (3.3.51) that the rates of the transformation strain and the plastic strain during phase transformation are proportional. An evolution equation for the plastic strain during stress-induced phase transformation, similar to (5.2.3), has also been proposed by Lim and McDowell [9]. However, in their equation the governing parameter is the accumulated martensitic volume fraction. For the equation proposed in this work, the governing parameter is the accumulated detwinned martensitic volume fraction. In addition, the equation proposed here is for the 3-D case. The above form of the TRIP tensor enables a saturation of the plastic strain after a certain number of cycles. The material parameters C_1^p and C_2^p govern the saturation value as well as the number of cycles necessary for the plastic strain to saturate.

Note that, given the form of the TRIP tensor in (5.2.4) above, we can eliminate the absolute value condition from the evolution equation for $\boldsymbol{\varepsilon}^p$ as seen in (5.2.3). To do so, we reformulate the TRIP tensor as follows:

$$\boldsymbol{\Lambda}^p = \begin{cases} \frac{3}{2} C_1^p \frac{H^{cur}(\bar{\sigma}')}{H^{max}} \frac{\boldsymbol{\sigma}^{\text{eff}\prime}}{\bar{\sigma}^{\text{eff}}} e^{-\frac{\zeta^d}{C_2^p}}; \ \dot{\xi} > 0, \\ -C_1^p \frac{H^{cur}(\bar{\sigma}')}{H^{max}} \frac{\boldsymbol{\varepsilon}^{t-r}}{\bar{\varepsilon}^{t-r}} e^{-\frac{\zeta^d}{C_2^p}}; \ \dot{\xi} < 0, \end{cases} \tag{5.2.7}$$

which now also includes the ratio $H^{cur}(\bar{\sigma}')/H^{max}$ term out of convenience. The introduction of the modified TRIP tensor allows us to write:

$$\dot{\boldsymbol{\varepsilon}}^p = \boldsymbol{\Lambda}^p \dot{\xi}. \tag{5.2.8}$$

Back, Drag Stresses and Effective Stress

The *back* stress, $\boldsymbol{\beta}$, and the *drag* stress, η, control the transformation hardening during the martensitic phase transformation. They are physically related to the local residual stresses that are developed in the material due to material heterogeneity. As explained by Bo and Lagoudas [4], $\boldsymbol{\beta}$ and η take into account the effects of both the initial material imperfections and heterogeneities, (grain boundaries, crystal lattice imperfections, precipitates) as well as the transformation-induced heterogeneities (transformation eigenstrains) and misfit between martensite-austenite interfaces. In addition, the back stress $\boldsymbol{\beta}$ accounts for the evolution of the maximum transformation strain during thermomechanical cycling.

In the current formulation, the evolution of the back and drag stresses is described using algebraic equations in terms of the martensitic volume fraction. Since the back stress is a tensorial quantity, its functional form also contains the direction of the accumulated transformation strain. Alternatively, an evolution equation can be specified that will have a form similar to the evolution equation for the transformation strain (see (3.3.51)).

In this work, the back stress is assumed to have the following polynomial functional representation:

$$\boldsymbol{\beta} = -\frac{\boldsymbol{\varepsilon}^t}{\bar{\varepsilon}^t} \sum_{i=1}^{N^b} D_i^b (H^{cur}(\bar{\sigma}')\,\xi)^{(i)}, \tag{5.2.9}$$

where N^b is the degree of the polynomial and D_i^b are the coefficients associated with the back stress. Note that the form of the back stress used here differs from the expression used by Bo and Lagoudas [4], where a logarithmic function has been used. The use of a polynomial expression significantly simplifies the estimation of the material parameters and the calibration of the model. Using (5.2.9), the back stress parameters can be calibrated by implementing a least square fit of the experimental data, while the logarithmic expression used by Bo and Lagoudas [4] will result in a non-linear optimization problem.

The expression for η used in this work is similar to the one used by Bo and Lagoudas [4]:

$$\eta = -D_1^d\,[-\ln(1-\xi)]^{\frac{1}{m_1}} + D_2^d \xi, \tag{5.2.10}$$

where D_1^d, D_2^d and m_1 are parameters governing the evolution of the drag stress.

Evolution of Transformation Strain

The evolution of the transformation strain in the current TRIP model for SMAs is assumed to follow the same functional form as (3.3.51) where $\boldsymbol{\Lambda}$ is the transformation tensor. During forward transformation of SMA material in which plastic strains are not considered (i.e., the model of Chapter 3), this tensor is assumed to depend on the applied stress level. In the current TRIP model, however, we must consider the contribution of the back stress. Therefore, it is the effective stress tensor $\boldsymbol{\sigma}^{\text{eff}}$ which is assumed to drive the direction of recoverable strain generation during forward transformation. This yields the following for $\boldsymbol{\Lambda}$ (c.f. 3.3.52):

$$\boldsymbol{\Lambda} = \begin{cases} \frac{3}{2} H^{max} \dfrac{\boldsymbol{\sigma}^{\text{eff}\prime}}{\bar{\sigma}^{\text{eff}}}\,; & \dot{\xi} > 0 \\ H^{max} \dfrac{\boldsymbol{\varepsilon}^{t-r}}{\bar{\varepsilon}^{t-r}}\,; & \dot{\xi} < 0 \end{cases}. \tag{5.2.11}$$

In the current model it is also generally assumed that the maximum transformation strain varies with stress ($H^{cur}(\bar{\sigma}')$) and this must also be considered in the formulation of $\boldsymbol{\Lambda}$. This is straightforward during forward transformation. During reverse transformation, however, $\boldsymbol{\Lambda}$ is assumed to depend only on the microstructural state of the material at the reversal of forward transformation. The current final form of the transformation tensor is then given by

$$\boldsymbol{\Lambda} = \begin{cases} \frac{3}{2} H^{cur}(\bar{\sigma}') \dfrac{\boldsymbol{\sigma}^{\text{eff}\prime}}{\bar{\sigma}^{\text{eff}}}\,; \dot{\xi} > 0, \\ H^{cur-r} \dfrac{\boldsymbol{\varepsilon}^{t-r}}{\bar{\varepsilon}^{t-r}}\,; \dot{\xi} < 0. \end{cases}. \tag{5.2.12}$$

Here H^{cur-r} denotes the *maximum current transformation strain at the reversal of forward transformation.* Recall from Chapter 3 that ε^{t-r} denotes the transformation strain at the reversal of forward phase transformation. The condition for evaluation of $H^{cur}(\bar{\sigma}')$ used here is the one suggested by Lagoudas and Bo [4], i. e., where $H^{cur}(\bar{\sigma}')$ is evaluated from the condition that the effective applied stress $\bar{\sigma}'$ (see (3.3.54)) is equal to the effective back stress $\bar{\beta}$ at the value of the martensitic volume fraction $\xi = 1$:

$$\bar{\sigma}' = \bar{\beta}\big|_{\xi=1} . \tag{5.2.13}$$

As shown in Sect. 5.2.1, the back stress, $\boldsymbol{\beta}$, is a function of both ξ and $H^{cur}(\bar{\sigma}')$. Thus, for $\xi = 1$, (5.2.13) becomes an equation of one variable, $H^{cur}(\bar{\sigma}')$, which can be solved to determine its value. This also implies that when the value of stress changes, a new value of the current maximum transformation strain should be calculated. The latter case arises not only during stress-induced phase transformation, but also during thermally-induced phase transformation when the material is constrained, e.g., in applications where the SMA acts as an actuator.

5.2.2 Complete Constitutive Model for TRIP

Given the discussion in Sect. 5.2.1, we see that the back stress and drag stress are dependent on the value of the martensitic volume fraction (i.e., $\boldsymbol{\beta}(\xi)$ and $\eta(\xi)$). The Gibbs free energy as stated in (5.2.1) should then be rewritten as a functional wherein the TRIP variables ($\boldsymbol{\beta}$ and η) are no longer independent but rather depend on ξ. Recalling the rule of mixtures as given in Sect. 3.3, we see that the material parameters for elastic moduli, thermal expansion, etc. are also functions of ξ. The careful reader will note that no explicit hardening function $f(\xi)$ is included in this new form of the Gibbs free energy, though one can be seen in (5.2.1). This is because the energetic consequences of phase mixing have been captured in the formulations of $\boldsymbol{\beta}$ and η. This reformulation yields

$$\begin{aligned} G[\boldsymbol{\sigma}, T, \boldsymbol{\varepsilon}^t, \boldsymbol{\varepsilon}^p, \boldsymbol{\beta}(\xi), \eta(\xi); \xi] = &-\frac{1}{2\rho}\boldsymbol{\sigma} : \mathbf{S}(\xi) : \boldsymbol{\sigma} - \frac{1}{\rho}\boldsymbol{\sigma} : \left[\boldsymbol{\alpha}(\xi)(T - T_0) + \boldsymbol{\varepsilon}^t + \boldsymbol{\varepsilon}^p\right] \\ &- \frac{1}{\rho}\int_0^{\xi} \left(\boldsymbol{\beta}(\tau) : \frac{\partial \boldsymbol{\varepsilon}^t}{\partial \tau} + \eta(\tau)\right) \mathrm{d}\tau + c(\xi)\left[T - T_0 - T\ln\left(\frac{T}{T_0}\right)\right] \\ &- s_0(\xi)(T - T_0) + u_0(\xi). \end{aligned} \tag{5.2.14}$$

Applying the second law of thermodynamics to the independent variables ($\boldsymbol{\sigma}$, T, ξ, $\boldsymbol{\varepsilon}^t$, and $\boldsymbol{\varepsilon}^p$) in the same manner described in Sect. 3.3, we arrive at the following constitutive equations:

$$\boldsymbol{\varepsilon} = -\rho\frac{\partial G}{\partial \boldsymbol{\sigma}} = \mathbf{S} : \boldsymbol{\sigma} + \boldsymbol{\alpha}(T - T_0) + \boldsymbol{\varepsilon}^t + \boldsymbol{\varepsilon}^p, \tag{5.2.15}$$

$$s = -\frac{\partial G}{\partial T} = \frac{1}{\rho}\boldsymbol{\sigma} : \boldsymbol{\alpha} + c \ln\left(\frac{T}{T_0}\right) + s_0. \tag{5.2.16}$$

Having thus defined relations for the entropy and total strain, the remaining local strong form of the second law of thermodynamics yields:

$$-\rho\frac{\partial G}{\partial \xi}\dot{\xi} - \rho\frac{\partial G}{\partial \boldsymbol{\varepsilon}^t} : \dot{\boldsymbol{\varepsilon}}^t - \rho\frac{\partial G}{\partial \boldsymbol{\varepsilon}^p} : \dot{\boldsymbol{\varepsilon}}^p \geq 0. \tag{5.2.17}$$

We then invoke the evolution equations for $\boldsymbol{\varepsilon}^t$ and $\boldsymbol{\varepsilon}^p$ stated in (3.3.51) and (5.2.8), respectively. Equation (5.2.17) is then rewritten as (cf. (3.3.56)):

$$-\rho\frac{\partial G}{\partial \xi}\dot{\xi} - \rho\frac{\partial G}{\partial \boldsymbol{\varepsilon}^t} : \boldsymbol{\Lambda}\dot{\xi} - \rho\frac{\partial G}{\partial \boldsymbol{\varepsilon}^p} : \boldsymbol{\Lambda}^p\dot{\xi} = \pi\dot{\xi} \geq 0. \tag{5.2.18}$$

In the above equation, π is the thermodynamic force conjugate to ξ and is given by

$$\begin{aligned}\pi = {} & \frac{1}{2}\boldsymbol{\sigma} : \Delta\mathbf{S} : \boldsymbol{\sigma} + \boldsymbol{\sigma} : \Delta\boldsymbol{\alpha}(T - T_0) + \boldsymbol{\sigma} : \boldsymbol{\Lambda} + \boldsymbol{\sigma} : \boldsymbol{\Lambda}^p \\ & + \boldsymbol{\beta}(\xi) : \boldsymbol{\Lambda} + \eta(\xi) - \rho\Delta c\left[T - T_0 - T\ln\left(\frac{T}{T_0}\right)\right] \\ & + \rho\Delta s_0(T - T_0) + \Delta u_0.\end{aligned} \tag{5.2.19}$$

The material parameter M^{0s} is now introduced as a combination of other parameters such that

$$M^{0s} = T_0 + \frac{1}{\rho\Delta s_0}(Y + \rho\Delta u_0), \tag{5.2.20}$$

with Y as a material constant representing a measure of the internal dissipation during phase transformation [2]. Recall that the effective stress $\boldsymbol{\sigma}^{\text{eff}}$ is defined as a sum of the applied stress, $\boldsymbol{\sigma}$, and the back stress, $\boldsymbol{\beta}$. These two considerations lead to a final form for the thermodynamic force:

$$\begin{aligned}\pi = {} & \frac{1}{2}\boldsymbol{\sigma} : \Delta\mathbf{S} : \boldsymbol{\sigma} + \boldsymbol{\sigma} : \Delta\boldsymbol{\alpha}(T - T_0) + \boldsymbol{\sigma}^{\text{eff}} : \boldsymbol{\Lambda} + \boldsymbol{\sigma} : \boldsymbol{\Lambda}^p + \eta(\xi) \\ & - \rho\Delta c\left[T - T_0 - T\ln\left(\frac{T}{T_0}\right)\right] + \rho\Delta s_0(T - M^{0s}) + Y.\end{aligned} \tag{5.2.21}$$

The evolution of the martensitic volume fraction and conditions on transformation can be found from a standard formalism of thermodynamic dissipation potentials (see, for example, [14]). Following the approach of Chapter 3, this gives:

$$\Phi = \begin{cases} \pi - Y; & \dot{\xi} > 0, \\ -\pi - Y; & \dot{\xi} < 0. \end{cases}$$

which was first shown in (3.3.58). Constraints on the evolution of ξ are expressed in terms of the Kuhn-Tucker conditions originally given in (3.3.60) and (3.3.61) for the forward and reverse phase transformations as

$$\begin{aligned} &\dot{\xi} \geq 0, \qquad \Phi \leq 0, \qquad \Phi\dot{\xi} = 0, \\ &\dot{\xi} \leq 0, \qquad \Phi \leq 0, \qquad \Phi\dot{\xi} = 0. \end{aligned}$$

The current 3-D formulation of the model will not properly take into account the development of two-way shape memory effect in its full generality, but only in special cases. The 1-D reduced model, however, will not be able to account for the TWSME. This limitation of the model is caused by the fact that the current choice for calculating $H^{cur}(\bar{\sigma}')$ cannot take into account the direction of the developed TWSME. To properly model the training and development of TWSME, a tensorial quantity must be introduced, which will replace $H^{cur}(\bar{\sigma}')$. Note, however, that for the case of stress-induced martensitic phase transformation the current formulation is still suitable. This is because the transformation strain, during stress-induced phase transformation, will develop in the direction of the applied stress.

5.2.3 Evolution of the Hysteretic Response of an SMA Undergoing Cyclic Loading

During mechanical cyclic loading, several characteristic changes of the thermomechanical response of SMAs exist. Along with the accumulation of non-recoverable plastic strain, changes of the hysteresis loop have also been experimentally observed [6, 9]. The hysteresis loop progressively evolves with the number of cycles, until a stabilization point is reached. Some of the characteristic changes of the hysteresis loop are:

1. decrease of the stress level necessary for the onset of the transformation;
2. increase of the transformation hardening;
3. decrease the width of the hysteresis loop;
4. decrease of the maximum transformation strain.

In addition, as noted in the literature [4, 7], the martensitic start temperature at zero applied stress M_s can also change during the transformation cycling.

The accumulation of the plastic strain has been addressed in the previous section, where an evolution equation for $\boldsymbol{\varepsilon}^p$ has been proposed. This section addresses the evolution of the hysteresis loop. The approach taken here is to identify two sets of parameters; the first set for the material that has not undergone any thermodynamic loading and the second set for the material that has undergone transformation cycling and for which the hysteresis loop has stabilized. Then, having identified these two sets of material parameters, evolution equations are proposed such that during the cycling the material parameters evolve from the first set to the second set. The procedure is described in detail in a sequel.

First, the evolution of the back stress parameter D_i^b is prescribed. The initial and final values of the back stress parameters are denoted by $(D_i^b)^{\text{init}}$ and $(D_i^b)^{\text{fin}}$. Following the work of Bo and Lagoudas [4], the evolution of the parameter D_i^b is assumed to be governed by the same equation, which is selected as

$$D_i^b = (D_i^b)^{\text{fin}} + \left((D_i^b)^{\text{init}} - (D_i^b)^{\text{fin}}\right) e^{-\lambda_1 \zeta^d}. \tag{5.2.22}$$

As seen from (5.2.22), the back stress parameters are assumed to change with the evolution of the accumulated detwinned martensitic volume fraction ζ^d. In their work, Bo and Lagoudas [4] have assumed that the evolution of D_i^b is governed by the change in plastic strain. However, as indicated by (5.2.8) and (5.2.7), the plastic strain ε^p and ζ^d are connected. Choosing ζ^d as the governing parameter for the change of D_i^b simplifies the model calibration, because for stress-induced transformation where full detwinning takes place, ζ^d is proportional to the number of cycles, i. e., $\zeta^d = 2N$, where N is the number of cycles. The parameter λ_1 in equation (5.2.22) is a positive material constant that governs the increasing rate of D_i^b.

As explained in Sect. 5.2.2, the current maximum transformation strain $H^{cur}(\bar{\sigma}')$ is calculated using the effective back stress $\bar{\alpha}$. Since the maximum transformation strain H^{max} is a limit value of $H^{cur}(\bar{\sigma}')$ the change of H^{max} is assumed to obey the same governing equation as the change of D_i^b. Therefore, H^{max} is given by

$$H^{max} = H^{\text{fin}} + \left(H^{\text{init}} - H^{\text{fin}}\right) \mathrm{e}^{-\lambda_1 \zeta^d}. \tag{5.2.23}$$

Similar evolution equations are proposed for the drag stress parameter D_i^d. The evolution of D_i^d is described by

$$D_i^d = (D_i^d)^{\text{fin}} + \left((D_i^d)^{\text{init}} - (D_i^d)^{\text{fin}}\right) e^{-\lambda_2 \zeta} \tag{5.2.24}$$

where λ_2 is a material parameter governing the evolution of D_i^d. As seen from (5.2.24), the evolution of the drag stress parameters is governed by the total accumulated martensitic volume fraction, ζ, and not by the detwinned portion, ζ^d. This is related to the fact that microstructural changes can be induced by cyclic self-accommodating phase transformation [4]. For stress-induced phase transformation with large values of the applied stress, (5.2.22) and (5.2.24) are identical (if, of course, $\lambda_1 = \lambda_2$), since in this case $\zeta = \zeta^d$.

Finally, the evolution of the material parameters Y, M^{0s} and $\rho\Delta s_0$ is considered. The equations governing the change of these parameters are similar to (5.2.24):

$$Y = Y^{\text{fin}} + \left(Y^{\text{init}} - Y^{\text{fin}}\right) e^{-\lambda_2 \zeta}, \tag{5.2.25}$$

$$M^{0s} = (M^{0s})^{\text{fin}} + \left((M^{0s})^{\text{init}} - (M^{0s})^{\text{fin}}\right) e^{-\lambda_2 \zeta}, \tag{5.2.26}$$

$$\rho\Delta s_0 = (\rho\Delta s_0)^{\text{fin}} + \left((\rho\Delta s_0)^{\text{init}} - (\rho\Delta s_0)^{\text{fin}}\right) e^{-\lambda_2 \zeta}. \tag{5.2.27}$$

The value for the parameter λ_1 can be obtained by performing cyclic loading during which the material undergoes stress-induced transformation, while the value of λ_2 can be determined by performing thermal cycling with no applied stress. The evolution equation for the back stress parameters (see (5.2.22) and (5.2.23)) includes the accumulated detwinned martensitic volume fraction, ζ^d, and the other evolution equations (5.2.24)–(5.2.27) include the accumulated total martensitic volume fraction, ζ. In self-accommodated thermally induced transformation cycling, the value of ζ^d is zero. Any change of the material parameters will be caused by the change of ζ, which will allow the determination of the value of λ_2 from any of (5.2.24)–(5.2.27), assuming that λ_2 remains the same for all evolution (5.2.24)–(5.2.27).

As discussed above, two sets of the material parameters need to be identified: the initial set, characterizing the initial response of the annealed material, and the final set, characterizing the stable material response. Having identified these two sets, the material parameters continuously change according to the evolution equations. However, this situation poses a problem in identifying the initial and final values of the parameters. Indeed, if the material parameters change continuously during the identification of the first set, it would be extremely difficult to take into account the change during the first cycle due to the nonlinearity introduced by that change.

This problem is addressed by keeping the value of the material parameters constant during forward or reverse phase transformation. The parameters will be recalculated according to the evolution equations when a reversal of the phase transformation occurs. Thus, the change in sign of the martensitic volume fraction rate $\dot{\xi}$ triggers the change of the material parameters. However, the above procedure is applied for only the material parameters. The plastic strain during cyclic loading is continuously calculated during both forward and reverse phase transformation, according to the evolution equation (5.2.8).

5.2.4 Modeling of Minor Hysteresis Loops

An important part of the thermomechanical constitutive modeling of SMAs is accounting for the minor hysteresis loops. In the presented model, a *major loop* is characterized by a full transformation cycle with the martensitic volume fraction, ξ, monotonically increasing from 0 to 1 and then monotonically decreasing from 1 to 0. Conversely, during a *minor loop* the martensitic volume ξ has an initial value strictly greater than 0 and less than 1.

The approach to modeling the minor hysteresis loops used in this work is the one presented by Bo and Lagoudas [5]. The main idea behind the modeling of minor loops is the modification of the transformation function, depending on whether the loading path follows a major or a minor loop. The details of the approach can be found in the work of Bo and Lagoudas [5]. Here it will only be mentioned that an additional parameter γ controlling the shape of the minor loops is introduced.

5.3 Estimation of Material Parameters

In this section, the determination of the material parameters entering the model are discussed as a final step to characterizing the thermomechanical behavior of SMAs undergoing cyclic loading. Three groups of material parameters are identified. First, the parameters that are necessary to describe a stable transformation cycle are determined. In the current model, the stable transformation cycle is defined as a thermomechanical loading cycle during which no plastic strains are developed and the material parameters remain constant[3]. Some of these material parameters, such as the elastic moduli, thermal expansion coefficients, etc, are material constants and do not change during transformation cycling, while others, such as the back stress parameters, drag stress parameters, etc, are material functions and evolve with transformation cycling.

The second group includes the material parameters that describe the behavior of SMAs under cyclic loading. These material parameters govern the evolution of plastic strains as well as the evolution of the material functions from the first group.

Finally, the third group encompasses the parameters governing the SMA behavior during minor hysteresis loops. The material parameters will be determined using uniaxial tests. Therefore, the 1-D reduction of the model is given next.

5.3.1 1-D Reduction of the Model

During uniaxial loading in x_1 direction, the stress tensor has one non-zero component. Thus, the components of the stress tensor are given by

$$\sigma_{11} = \sigma \neq 0, \quad \sigma_{ij} = 0 \text{ for all other } i, j. \tag{5.3.28}$$

In the equation above, σ is the applied uniaxial stress. The transformation and plastic strain components are given by

$$\varepsilon^t_{11} = \varepsilon^t, \quad \varepsilon^t_{22} = \varepsilon^t_{33} = -\frac{1}{2}\varepsilon^t, \quad \varepsilon^t_{ij} = 0 \text{ for all other } i, j, \tag{5.3.29}$$

$$\varepsilon^p_{11} = \varepsilon^p, \quad \varepsilon^p_{22} = \varepsilon^p_{33} = -\frac{1}{2}\varepsilon^p, \quad \varepsilon^p_{ij} = 0 \text{ for all other } i, j, \tag{5.3.30}$$

where ε^t and ε^p are the uniaxial transformation and plastic strains, respectively, assuming that both result in isochoric deformations. Assuming isotropic elastic properties, the constitutive equations (5.2.15) in the 1-D form become:

$$\varepsilon_{11} = \varepsilon = S\sigma + \alpha(T - T_0) + \varepsilon^t + \varepsilon^p, \tag{5.3.31}$$

[3] Such transformation cycle is obtained when the material parameters have reached their asymptotic limit, e. g., within 1%.

where ε is the uniaxial total strain, α is the thermal expansion coefficient, and S is the elastic compliance given by $S = 1/E$, where E is Young's elastic modulus.

The evolution equations for the transformation and plastic strain become:

$$\dot{\varepsilon}^t = \Lambda\dot{\xi}, \tag{5.3.32}$$

$$\dot{\varepsilon}^p = \Lambda^p\dot{\zeta}^d. \tag{5.3.33}$$

The uniaxial components of the transformation and TRIP tensors are given by:

$$\Lambda = \Lambda_{11} = \begin{cases} H^{cur}(\sigma)\dfrac{\sigma^{\text{eff}}}{|\sigma^{\text{eff}}|} & ;\dot{\xi} > 0, \\ H^{cur-r}\dfrac{\varepsilon^t_{\text{max}}}{|\varepsilon^t_{\text{max}}|} & ;\dot{\xi} < 0, \end{cases} \tag{5.3.34}$$

$$\Lambda^p = \Lambda^p_{11} = \begin{cases} C^p_1\dfrac{\sigma^{\text{eff}}}{|\sigma^{\text{eff}}|}e^{-\frac{\zeta^d}{C^p_2}} & ;\dot{\xi} > 0, \\ C^p_1\dfrac{\varepsilon^t_{\text{max}}}{|\varepsilon^t_{\text{max}}|}e^{-\frac{\zeta^d}{C^p_2}} & ;\dot{\xi} < 0. \end{cases} \tag{5.3.35}$$

The uniaxial effective stress σ^{eff} is defined in terms of the applied stress, σ, and the back stress, β, as:

$$\sigma^{\text{eff}} = \sigma + \beta, \quad \beta = -\frac{3}{2}\frac{\varepsilon^t}{|\varepsilon^t|}\sum_{i=1}^{N^b} D^b_i(H^{cur}(\sigma)\,\xi)^{(i)}. \tag{5.3.36}$$

The expression for the drag stress, η, is the same as given by (5.2.10)

$$\eta = -D^d_1\left[-\ln(1-\xi)\right]^{\frac{1}{m_1}} + D^d_2\xi. \tag{5.3.37}$$

The transformation function is given by:

$$\Phi = \begin{cases} \pi - Y = 0;\ \dot{\xi} > 0, \\ -\pi - Y = 0;\ \dot{\xi} < 0, \end{cases} \tag{5.3.38}$$

where the thermodynamic driving force, π, is given by:

$$\begin{aligned} \pi = {} & \frac{1}{2}\sigma^2\Delta S + \sigma\Delta\alpha(T - T_0) + \sigma^{\text{eff}}\Lambda + \eta \\ & - \rho\Delta c\left[T - T_0 - T\ln\left(\frac{t}{T_0}\right)\right] + \rho\Delta s_0(T - M^{0s}) + Y. \end{aligned} \tag{5.3.39}$$

The quantity ΔS is the difference between the elastic compliances of the austenitic and martensitic phases and is given by:

$$\Delta S = \frac{1}{E^M} - \frac{1}{E^A}, \tag{5.3.40}$$

and $\Delta\alpha$ is the difference between the thermal expansion coefficients of the austenite and martensite. The uniaxial tangent stiffness is given by:

$$\mathcal{L} = E - \frac{\chi E^2 \partial_\sigma \Phi}{\chi E \partial_\sigma \Phi - \partial_\xi \Phi}, \tag{5.3.41}$$

where $\partial_\sigma \Phi$ and $\partial_\xi \Phi$ are the derivatives of the transformation function with respect to stress and the martensitic volume fraction, respectively, and the quantity χ is given by:

$$\chi = \sigma \Delta S + \Delta\alpha(T - T_0) + \Lambda + \text{sign}(\dot{\xi}) \frac{H^{cur}(\sigma)}{H^{max}} \Lambda^p. \tag{5.3.42}$$

The determination of the material parameters of all three groups is described in the following subsections. The necessary tests will be discussed and a parametric study for selected material parameters will be performed. The material parameters for NiTi reported by Bo et al. [15] and shown in Table 5.1[4] will be used during the parametric study. The parameters shown in Table 5.1 are obtained for NiTi wires undergoing thermally-induced phase transformation.

5.3.2 Material Parameters for a Stable Transformation Cycle

This group of material parameters includes the elastic compliance tensors of both austenitic and martensitic phases, $\mathbf{S}^A$ and $\mathbf{S}^M$, respectively, their

Table 5.1. Material parameters for NiTi SMA characterizing a stable transformation cycle [15]

Physical constants	
$E^A = 70.0$ GPa	$\alpha^A = 11.0 \times 10^{-6}$ K^{-1}
$E^M = 30.0$ GPa	$\alpha^M = 6.6 \times 10^{-6}$ K^{-1}
$\rho c^A = 2.12$ MJ/(m^{3}K)	$\nu = 0.33$
$\rho c^M = 2.12$ MJ/(m^{3}K)	
Parameters characterizing the phase transformation	
$M^{0s} = 311.0$ K	$\rho\Delta s_0 = -0.422$ MJ/(m^{3}K)
$H^{max} = 0.069$	$Y = 6.0$ MJ/m^3
$D_1^b = 3.40 \times 10^3$ MPa	$D_1^d = 8.0$ MPa
$D_2^b = -2.23 \times 10^5$ MPa	$D_2^d = 1.7$ MPa
$D_3^b = 8.32 \times 10^6$ MPa	$m_1 = 3.5$
$D_4^b = -1.50 \times 10^8$ MPa	
$D_5^b = 1.03 \times 10^9$ MPa	
Minor loop parameter	
$\gamma = 3.0$	

[4] Reprinted from *Mechanics of Materials*, Vol. 36, Issue 9, Lagoudas, D.C., and Entchev, P., pp. 865–892, Copyright 2004, with permission from Elsevier.

thermal expansion coefficient tensors, $\boldsymbol{\beta}^A$ and $\boldsymbol{\beta}^M$, and the specific heat, c^A and c^M. Additional parameters include martensitic start temperature, M_s, the maximum transformation strain, H^{max}, the difference between the specific entropy per unit volume, $\rho\Delta s_0$, the material parameter, Y, which provides a measure of the internal dissipation during phase transformation, and the parameters associated with the back and drag stresses.

To determine the elastic compliance tensors $\mathbf{S}^A$ and $\mathbf{S}^M$, one must assume that the SMA material behaves isotropically at macroscale. This assumption is reasonable because of the random orientation of the grains in a polycrystalline SMA. Therefore, one only needs to find the Young's elastic moduli, E^A and E^M, and Poisson's ratios, ν^A and ν^M, of both phases. To determine the elastic constants, a uniaxial pseudoelastic test must be performed. The elastic stiffness E^A is determined by calculating the initial slope of the stress-strain curve for a uniaxial pseudoelastic test, as shown in Fig. 5.5. The elastic stiffness of the martensite phase E^M is given by the slope of the stress-strain curve at the point of initial unloading (see Fig. 5.5)[4]. A general assumption in the literature is that the Poisson's ratios of austenite and martensite are equal, with a typical reported value of $\nu^A = \nu^M = 0.33$ [3, 16].

The thermal expansion coefficient tensors, $\boldsymbol{\beta}^A$ and $\boldsymbol{\beta}^M$, are fully represented by two scalar constants, α^A and α^M. These constants can be estimated by measuring the slope of the strain-temperature curve under constant stress at high temperature for the austenitic phase and at low temperature for the martensitic phase. The specific heat constants, c^A and c^M, can be obtained from a calorimetric test. These constants (α^A, α^M, c^A, c^M) are not needed to model stress-induced phase transformation under constant temperature, but they become important when temperature changes.

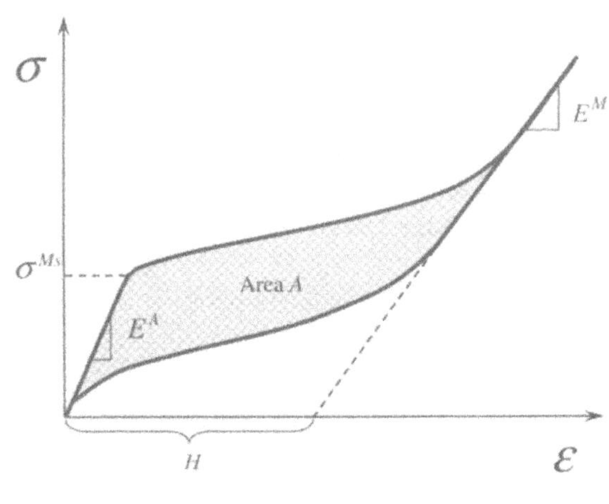

Fig. 5.5. Schematic of a uniaxial pseudoelastic test.

The martensitic start temperature, M_s, can be obtained from a DSC test. The maximum transformation strain, H^{max}, is obtained by performing a pseudoelastic uniaxial test. The value is estimated by extending the unloading part of the stress-strain curve using the elastic stiffness of the martensitic phase, E^M, as shown in Fig. 5.5 or given in (2.5.7).

During the initial loading cycles, observable plastic strain can develop. One way to separate the plastic strain from the transformation strain is to subtract a portion of the residual plastic strain from the measured value of the transformation strain. Since the plastic strain develops during both forward and reverse transformation, the subtracted value must be adjusted to take into account the plastic strain developed during the reverse phase transformation. As experimentally observed and reflected in the evolution equation for the plastic strain, initially the rate of accumulation of the plastic strain can be approximated by a linear function. Therefore, a reasonable assumption is that the amount of the plastic strain developed during the forward phase transformation is equal to one-half of the total residual plastic strain after one cycle. Thus, in the presence of plastic strains, the value of the maximum transformation strain, H^{max}, is obtained by extending the unloading part of the stress-strain curve and subtracting one-half of the total residual strain from the obtained number.

The material parameter, $\rho\Delta s_0$, can also be obtained from the pseudoelastic stress-strain curve, schematically shown in Fig. 5.5. Equations (5.3.38) and (5.3.39) for the transformation function, Φ, at the onset of phase transformation ($\xi = 0$) lead to:

$$\rho\Delta s_0 = -\frac{\frac{1}{2}(\sigma^{Ms})^2\Delta S + \sigma^{Ms}H^{cur}(\sigma)}{T - M^{0s}}. \tag{5.3.43}$$

The value of $H^{cur}(\sigma)$ used in (5.3.43) should be calculated for the corresponding value of stress, which, in this case, is equal to σ^{Ms}. A slightly different procedure for determination of M_s and $\rho\Delta s_0$ is reported by Bo et al. [15]. In their work, the quantity $\rho\Delta s_0$ is calculated from the DSC test and is related to the latent heat released during forward phase transformation and absorbed during reverse phase transformation. The martensitic start temperature M_s is calculated in their work using the strain-temperature curve obtained during an isobaric experiment.

To demonstrate the effect of the value of $\rho\Delta s_0$ on the stress-strain response, loading cases with different numerical values of $\rho\Delta s_0$ are simulated. The results of the parametric study are shown in Fig. 5.6[4] where only the loading part is shown. With the increasing magnitude of $\rho\Delta s_0$, the value of stress for the onset of the transformation also increases. Therefore, the parameter $\rho\Delta s_0$ is connected to the slope of the transformation line on the stress-temperature phase diagram.

The material parameter, Y, can also be calculated using a pseudoelastic stress-strain curve. The value of Y is related to the total area A enclosed by

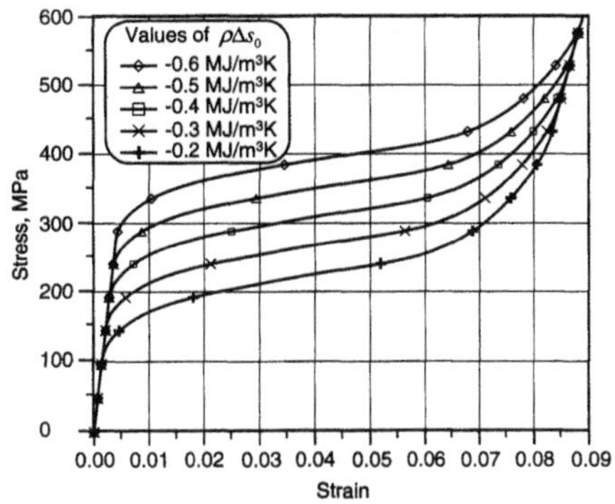

Fig. 5.6. Effect of the parameter $\rho\Delta s_0$ on the stress-strain response during forward phase transformation.

the hysteresis curve during a complete phase transformation as:

$$A = 2Y. \tag{5.3.44}$$

The effect of the parameter Y on the stress-strain response is shown in Fig. 5.7[4]. The unloading parts of the stress-strain curves for different values of Y are shown in the figure while the loading parts are the same for all values of Y. Furthermore, with the increase of the value of Y the reverse phase transformation starts at lower values of the applied stress. This, in effect, causes the increase of the area of the hysteresis loop, as seen in Fig. 5.7.

As mentioned earlier in Sect. 5.2.2 [cf. (5.2.13)], the effective back stress is used to calculate the value of the current transformation strain $H^{cur}(\sigma)$. Thus, an experimental result showing the dependance of $H^{cur}(\sigma)$ on the value of the applied stress, σ, is utilized to obtain the back stress parameter, D_i^b. To obtain this dependance, several isobaric tests with thermally-induced phase transformation must be performed for different values of the applied stress. Each test gives one value of the current maximum transformation strain $H^{cur}(\sigma)$

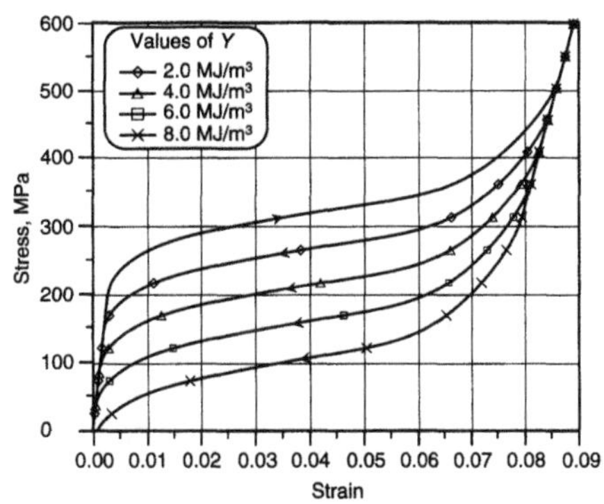

Fig. 5.7. Effect of the parameter Y on the size of the hysteresis loop.

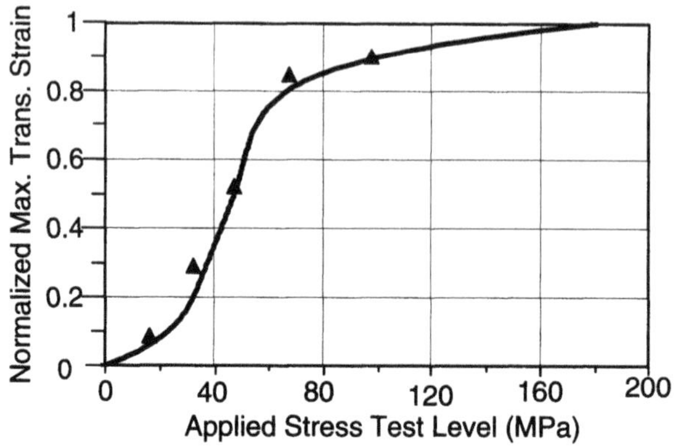

Fig. 5.8. Normalized maximum transformation strain $(H^{cur}(\sigma)/H^{max})$ for thermally-induced phase transformation versus the applied stress: experimental data for NiTi SMA and a polynomial least-square fit. A polynomial of degree 5 is used.

corresponding to the value of the applied stress (see Fig. 5.8)[4]. Then, the back stress parameter, D_i^b, is obtained using the condition that the effective stress σ^{eff} at $\xi = 1.0$ vanishes and a least-square fit of the following equation is performed:

$$\sigma = \frac{3}{2}\frac{\varepsilon^t}{|\varepsilon^t|}\sum_{i=1}^{N^b} D_i^b [H^{cur}(\sigma)]^{(i)}. \tag{5.3.45}$$

After the value of $H^{cur}(\sigma)$ reaches the value of the maximum transformation strain, H^{max}, one can assume that a further increase in stress does not yield further increase of $H^{cur}(\sigma)$. After that point, the value of $H^{cur}(\sigma)$ is considered constant and equal to H^{max}. Note that this result can be compared to the experimental results shown in Chapter 2, Fig. 2.17 and Fig. 2.35.

To estimate the material parameters entering the expression for the drag stress, η, the tangent stiffness during an isothermal uniaxial pseudoelastic test is used. Assuming that all of the material parameters, except the drag stress parameters, are known, the tangent stiffness, $\mathcal{L}$, given by (5.3.41) becomes a function of the drag stress parameters D_1^d, D_2^d and m_1 and the martensitic volume fraction ξ:

$$\mathcal{L}(\xi; D_1^d, D_2^d, m_1) = E - \frac{E^2[\sigma \Delta S + H^{cur}(\sigma)]^2}{\chi}, \tag{5.3.46}$$

where the denominator is given by:

$$\chi = E(\sigma \Delta S + H^{cur}(\sigma))^2 + \frac{3}{2}(H^{cur}(\sigma))^2 \sum_{i=1}^{N^b} i D_i^b [H^{cur}(\sigma)\,\xi]^{(i-1)} + \frac{1}{m_1(1-\xi)} D_1^d [-\ln(1-\xi)]^{\frac{1-m_1}{m_1}} - D_2^d. \tag{5.3.47}$$

The experimental values of the uniaxial tangent stiffness are obtained by performing a pseudoelastic test. Thus, having obtained the experimental values of the tangent stiffness, a least-square fit of (5.3.46) is performed to obtain the drag stress parameters D_1^d, D_2^d and m_1. The internal variable ξ cannot be directly measured. Based on numerous numerical experiments performed using the current model, one can reasonably assume a linear relationship between the rate of total strain and the rate of the martensitic volume fraction during transformation.

To illustrate the effect of the drag stress parameters on the stress-strain results, a parametric study for different values of D_1^d and D_2^d has been performed. The loading part of the pseudoelastic stress-strain curve for different values of D_1^d is shown in Fig. 5.9[4]. From the figure, the transformation hardening increases with the increase of the value of D_1^d. In addition, the value of D_1^d has a very strong effect on the initial transformation hardening. The effect of the parameter D_2^d on the stress-strain results is shown in Fig. 5.10[4].

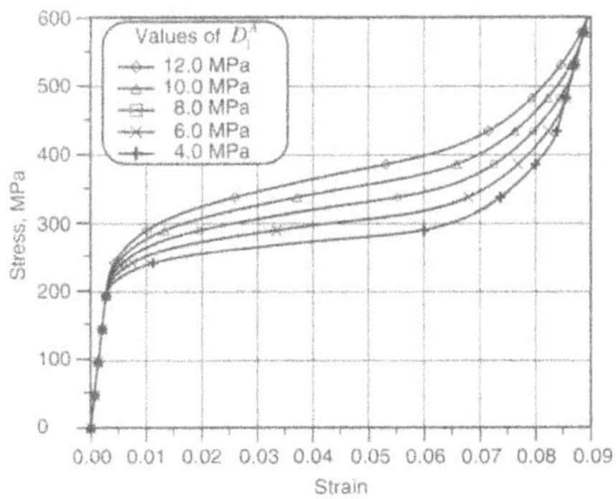

Fig. 5.9. Effect of the parameter D_1^d on the stress-strain response during forward phase transformation.

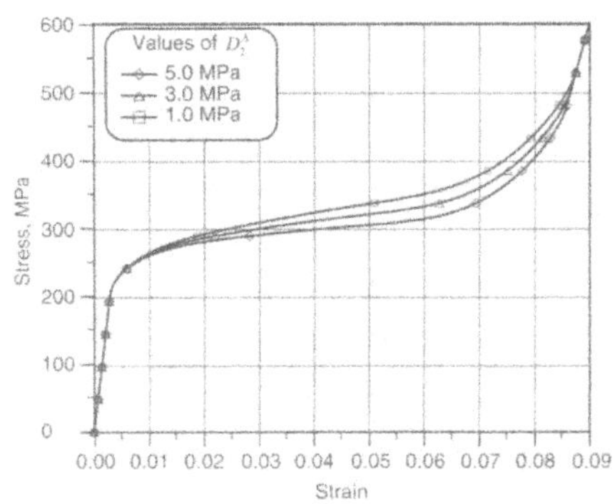

Fig. 5.10. Effect of the parameter D_2^d on the stress-strain response during forward phase transformation.

Furthermore, higher values of the parameter D_2^d leads to a smaller slope of the stress-strain curve. This parameter, however, has very little effect on the initial transformation hardening. The effect of D_2^d can be explained by comparing the expression for the drag stress in the current model to the analogous expression given by Lagoudas et al. [16] for a model with a polynomial hardening function. If the parameter D_1^d is taken to be zero, then the parameter D_2^d corresponds to the linear hardening parameter in the model by Lagoudas et al. [16]. Therefore, it has the same effect for all values of ξ and it equally affects the slope of the stress-strain curve during the phase transformation.

5.3.3 Material Parameters for Cyclic Loading

This group of parameters includes the terms C_1^p and C_2^p, entering the expression for Λ^p (5.3.35) and constant λ_1, which determines the evolution of the material parameters during cyclic loading. To understand the physical meaning of the constants, C_1^p and C_2^p (5.3.33) can be integrated to find the following expression for ε^p as a function of ζ^d:

$$\varepsilon^p = C_1^p C_2^p \left(1 - e^{-\frac{\zeta^d}{C_2^p}}\right) = \varepsilon_{\text{sat}}^p \left(1 - e^{-\frac{\zeta^d}{C_2^p}}\right). \tag{5.3.48}$$

As seen from (5.3.48) the product of C_1^p and C_2^p gives the maximum (saturation) value of the plastic strain:

$$C_1^p C_2^p = \varepsilon_{\text{sat}}^p. \tag{5.3.49}$$

Further, recall that for stress-induced martensite $\zeta^d = 2N$. Therefore, C_2^p can be estimated by enforcing the condition that after a given number of cycles the plastic strain approaches its maximum value, or more precisely, C_2^p can be found from the condition:

$$e^{-\frac{2N_{\text{sat}}^p}{C_2^p}} = \delta, \tag{5.3.50}$$

where N_{sat}^p is the number of cycles to reach the saturation value of plastic strain and δ is a small number. A value of 0.01 has been used in the current work; however, values of up to 0.1 are also reasonable. Next, parameter λ_1 is considered. From (5.2.22), λ_1 determines the number of cycles N_{sat}^b until the material parameters reach their final values. Thus, λ_1 can be found using the following equation, similar to (5.3.50):

$$e^{-2N_{\text{sat}}^b \lambda_1} = \delta. \tag{5.3.51}$$

The value of the parameter λ_2 is determined in a similar way as the value of λ_1. For the case of stress-induced transformation, however, it is difficult to distinguish between the parameters λ_1 and λ_2. Therefore, in all further calculations in this chapter, these parameters are taken to be equal.

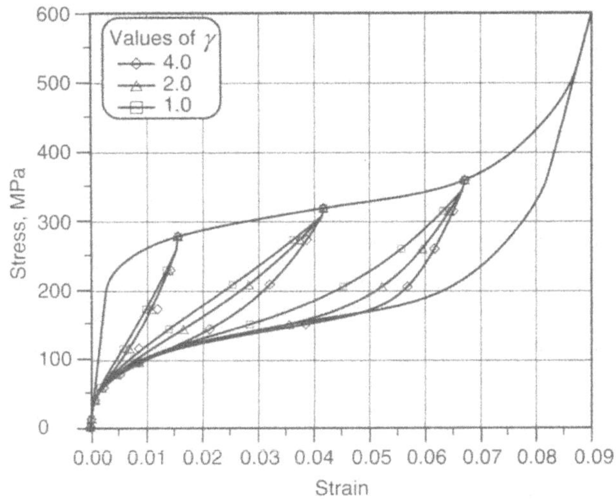

Fig. 5.11. Effect of the parameter γ on the minor loop curvature. The minor loops occur due to incomplete forward phase transformation.

5.3.4 Material Parameters for Minor Loop Modeling

Only one material parameter belongs to this last group. This is the constant γ, dictating the size of the minor hysteresis loop. A smaller value of γ leads to a smaller area occupied by the minor loop, while a larger value increases the area. An experimental stress-strain curve with one minor loop is sufficient to estimate the value of γ, which lies in the range of 1 to 4. To illustrate the effect of this parameter on the shape of the minor loop branches, the results for different values of γ are shown in Fig. 5.11[4].

Further details on the numerical implementation is provided in Appendix C.

5.4 Sample Loading Cases

To demonstrate the capabilities of the model, different boundary-value problems (BVPs) are presented in this section. First, a uniaxial isothermal pseudoelastic case is presented. The second BVP represents a uniaxial thermally-induced transformation under constant stress. The third BVP models the response of a 3-D bar under combined torsion-compression loading. Finally, the response of a torque tube is simulated in the fourth BVP. The material parameters for the first three cases are the ones given by Bo et al. [15] and are presented in Table 5.1[4]. The material parameters for the fourth case are presented in Table 5.2[4].

The results in this and the following sections are obtained using Abaqus [17]. To perform the calculations, the model is numerically implemented in a user material subroutine UMAT. The details of the implementation are not discussed here, but the numerical implementation follows the procedure described in Chapter 4.

5.4.1 Uniaxial Isothermal Pseudoelastic Loading

The schematic of the BVP for this loading case is shown in Fig. 5.12[4]. The domain is chosen to be a unit cube (1 × 1 × 1 m). Since the stress is constant in the whole domain during the loading, a finite element mesh of only one linear element is chosen. Initially, the material is in a stress-free state. During the first step of the loading, the stress is increased to 600 MPa, which results in stress-induced phase transformation. During the unloading step, a reverse phase transformation occurs. The stress-strain response of the material for this case is shown in Fig. 5.13[4] for a temperature of 70°C.

In the case of loading at $T = 70$°C, the forward phase transformation starts at the value of applied stress approximately equal to 210 MPa and is completed at 510 MPa. At the point of completion of the forward transformation, the value of the transformation strain is equal to the value of the maximum transformation strain: $H^{max} = 0.069$. The onset of the reverse phase transformation is at 350 MPa with the transformation completed at 40 MPa. During the reverse phase transformation, the transformation strain changes from its maximum value to zero.

5.4.2 Uniaxial Constant Stress Thermally-Induced Transformation

The BVP describes a thermal cycle of a uniaxial SMA specimen under constant applied stress. The schematic of the BVP and the loading history are shown in Fig. 5.14[4]. Initially, the material is isothermally loaded to a given

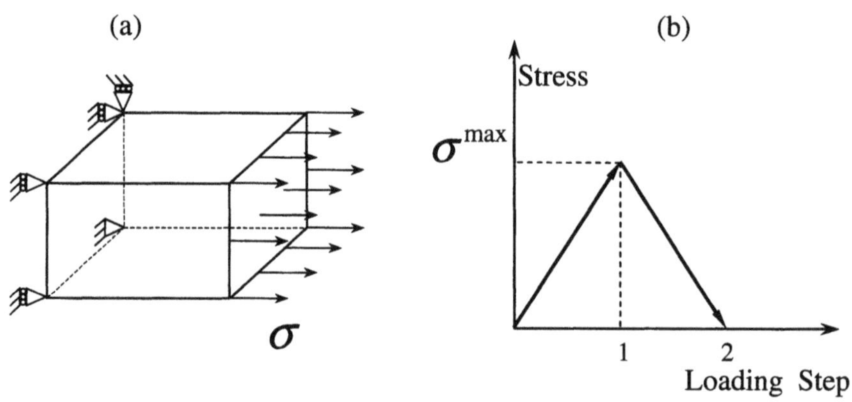

Fig. 5.12. Schematic of the BVP for uniaxial pseudoelastic SMA response: (a) geometry and boundary conditions; (b) loading history.

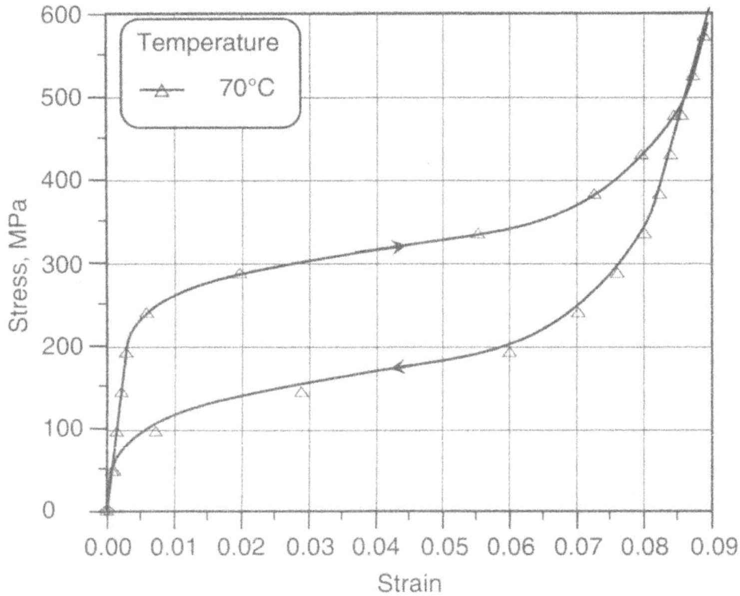

Fig. 5.13. Pseudoelastic uniaxial SMA response.

stress level, which is then kept constant. The first loading step involves heating of the SMA from the initial temperature of 0°C to the final temperature of 100°C. During this loading step, a reverse phase transformation occurs and the transformation strain is recovered. The final value of the recovered transformation strain, after the completion of the phase transformation, is dictated

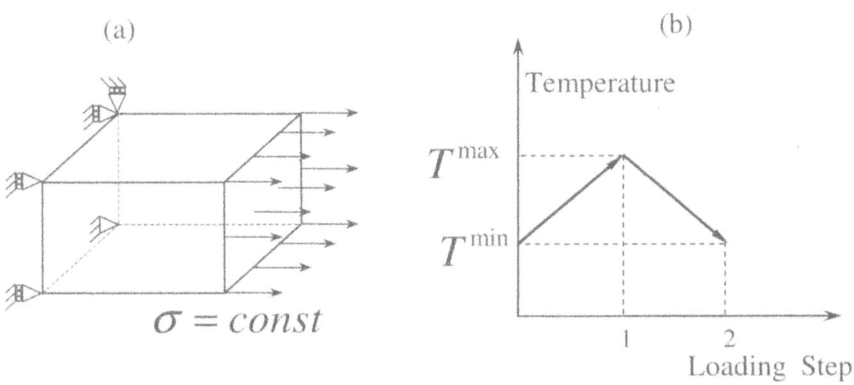

Fig. 5.14. Schematic of the BVP for uniaxial isobaric thermally-induced transformation: (a) geometry and boundary conditions; (b) loading history.

by the level of the applied stress. To illustrate the dependance of $H^{cur}(\sigma)$ on the applied stress, cases with different values of the applied stress are performed. During the second loading step, the material is cooled back to 0°C, and the transformation strain is recovered.

The results for the loading cases are shown in Fig. 5.15[4]. Two different effects due to the level of the applied stress are observed. First, as mentioned above, the maximum transformation strain during the phase transformation increases with the increase of the value of the applied stress. The second effect is the shift of the initial transformation temperature; the transformation starts at higher temperature for higher value of the applied stress. This shift of the transformation temperatures can be explained by recalling the expression for the transformation function, given by (5.3.38) and (5.3.39). As seen from (5.3.39), with higher value of the applied stress, the transformation criterion will be satisfied for higher temperatures.

5.4.3 Torsion-Compression Loading

To demonstrate the model's capability to handle 3-D loading cases, the response of a cylindrical SMA bar subjected to combined torsion-compression

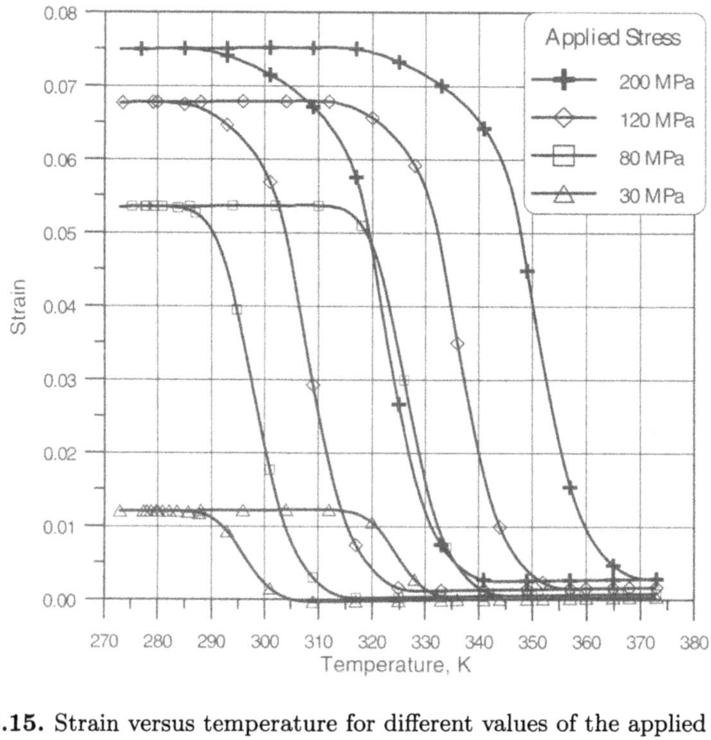

Fig. 5.15. Strain versus temperature for different values of the applied stress.

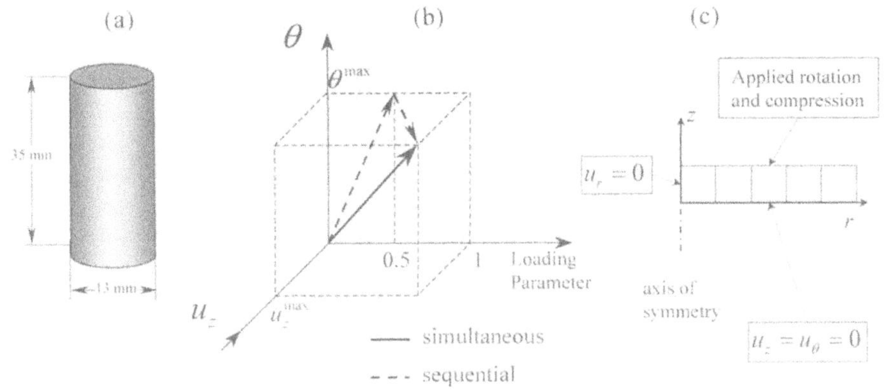

Fig. 5.16. Schematic of the BVP for multiaxial loading of an SMA bar; (a) geometry; (b) loading history; (c) finite element mesh and boundary conditions.

and torsion followed by compressive loading is modeled. The schematic of the BVPs is shown in Fig. 5.16[4]. The specimen is chosen to have a diameter of 13 mm and a length of 35 mm. The domain is discretized using eight-node quadratic axisymmetric elements with a twist degree of freedom (elements `CGAX8`, for details see [17]). Since there is no stress variation in the axial direction, only one layer of elements is used in that direction with a length of 1.75 mm, while five elements are used in the radial direction, as shown in Fig. 5.16c. The maximum value of the applied rotation (see Fig. 5.16) is taken to be $\theta^{\max} = 0.02$ rad ($\approx 1.15°$), which corresponds to 11.4 rad/m rotation ($\approx$ 1.8 full revolutions per meter length), while the maximum value of the displacement in the axial direction is taken to be $u_z^{\max} = -0.125$ mm, corresponding to axial compressive strain of $\varepsilon_{zz} = 0.071$. Both compression and rotation boundary conditions are applied on the top surface of the specimen, while the bottom surface is held fixed in the $z-$ and $\theta-$ directions. Traction-free boundary condition in the radial direction is applied. The numerical simulations are performed at 60°C.

First, the results for the sequential torsion-compression loading are presented. During the first loading step when the bar is subjected to torsion, a nonhomogeneous stress state is created. As the stress increases, the critical stress for the onset of the phase transformation is first reached at the outer surface of the bar. The phase transformation front then propagates toward the center of the bar. To illustrate this, the contour plot of the martensitic volume fraction for the cross-section of the bar at the end of the torsional loading (when the value of the loading parameter is equal to 0.5) is shown in Fig. 5.17[4]. Notice that the lines of constant value of ξ form concentric circles.

The history of the axial and shear stress components during both loading steps is shown in Fig. 5.18[4]. The *average* values of these quantities are shown

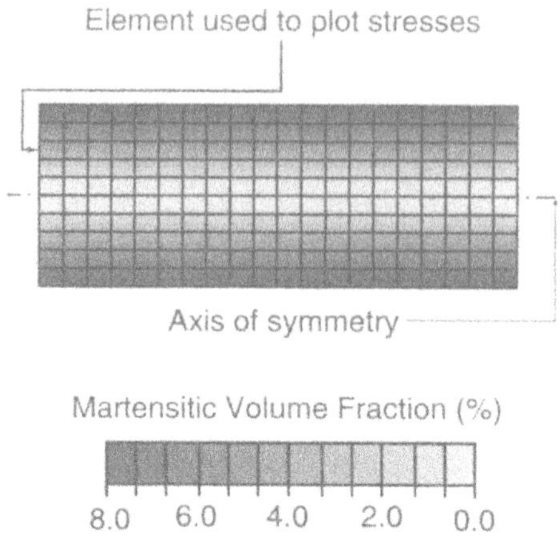

Fig. 5.17. Contour plot of the martensitic volume fraction in the SMA bar at the end of the torsional loading.

for the element indicated in Fig. 5.18. During the initial torsional loading, the shear stress linearly increases until a critical value is reached. At this point (at the value of the loading parameter of ≈ 0.1), the phase transformation initiates. During the first loading step, the axial stress component remains

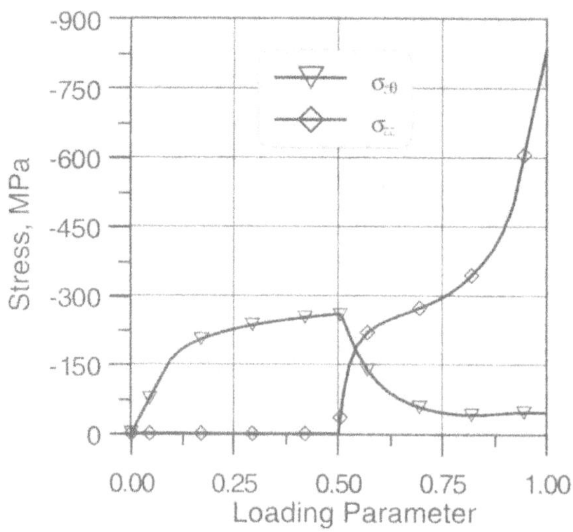

Fig. 5.18. History of the axial and shear stress components during sequential torsion-compression loading.

zero. During the second loading step, the phase transformation continues until it is fully completed (at the value of the loading parameter of ≈ 0.9). The axial stress increases while the shear stress is partially relaxed. After the completion of the phase transformation, the material is fully in the martensitic phase and behaves linearly.

To explain the partial relaxation of the shear stress during the compressive axial loading, consider the shear and axial components of the transformation strain, shown in Fig. 5.19[4]. During the second loading step, when a compressive loading is applied, the shear component of the transformation strain continues to increase. This is caused by the choice of the functional form of the transformation direction tensor, $\mathbf{\Lambda}$, given by (5.2.12). Since the shear component of the stress is non-zero, the shear component of the transformation strain keeps developing during the second loading step at a smaller rate. Conversely, since the applied rotation and, therefore, the total strain, is kept constant during the compressive loading, the shear component of the stress decreases according to the constitutive equation (5.2.15).

The history of the axial and shear stress components for the simultaneous torsion-compression loading case are presented in Fig. 5.20[4]. The results for this loading case resemble superposition of the results under compression and torsion applied independently. The stresses increase linearly until a critical value is reached and transformation initiates (the value of the loading parameter is approximately equal to 0.05). Then, the transformation proceeds

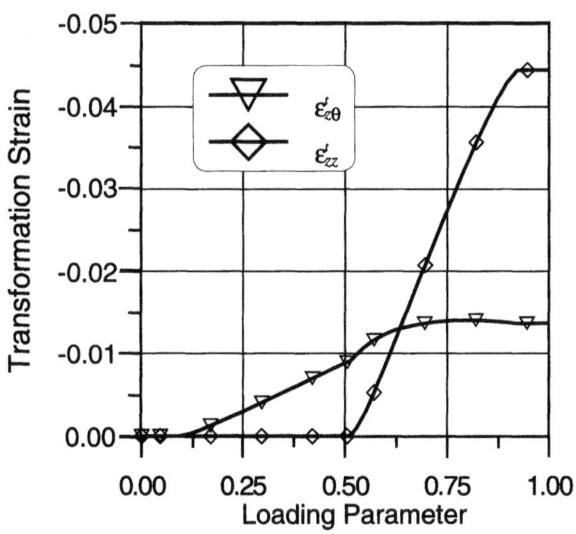

Fig. 5.19. History of the axial and shear transformation strain components during sequential torsion-compression loading.

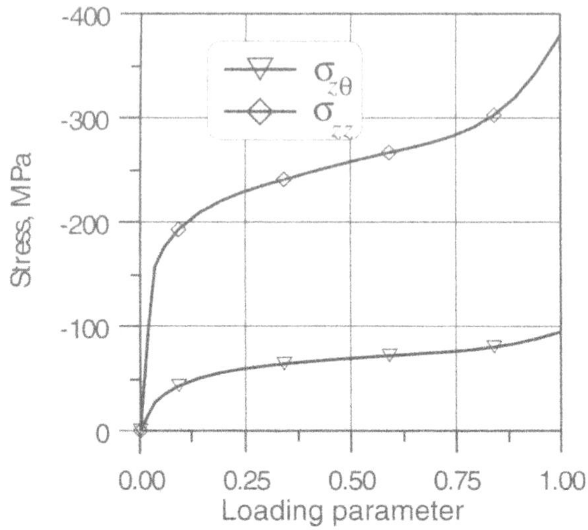

Fig. 5.20. History of the axial and shear stress components during simultaneous torsion-compression loading.

until the material is fully transformed at the value of the loading parameter approximately equal to 0.9. Once complete, the response of the material is again linear with the elastic properties of the martensite.

5.4.4 Response of an SMA Torque Tube

In this section, the capabilities of the model to handle loading cases beyond uniaxial loading are tested by simulating an SMA torque tube. The material parameters for this particular NiTi alloy are presented in Table 5.2[4]. The dimensions of the tube used are: outer diameter $d_o = 6.34$ mm and inner diameter $d_{in} = 5.0$ mm. The reason for selecting these dimensions is to model a tube that geometrically resembles commercially available tubes. The diameters used here have also been used by Qidwai [18] and are based on the specifications of torque tubes manufactured by Memry Corp.

Based on the small thickness of the tube wall, only one quadratic element in radial direction is used. In addition, since the stress is constant in the axial direction, one element in the axial direction is sufficient to obtain accurate results. To obtain an appropriate aspect ratio, the length in the axial direction has been chosen as 0.67 mm, equal to the wall thickness. An axisymmetric finite element with a rotational degree of freedom (element `CGAX8` from the Abaqus element library, see [17]) was used. The schematic of the mesh and the boundary conditions is shown in Fig. 5.21[4]. The bottom part of the tube is fixed and rotation is applied to the top part. The maximum value of the applied rotation is 2.2×10^{-2} rad. The rotation is applied cyclically in both

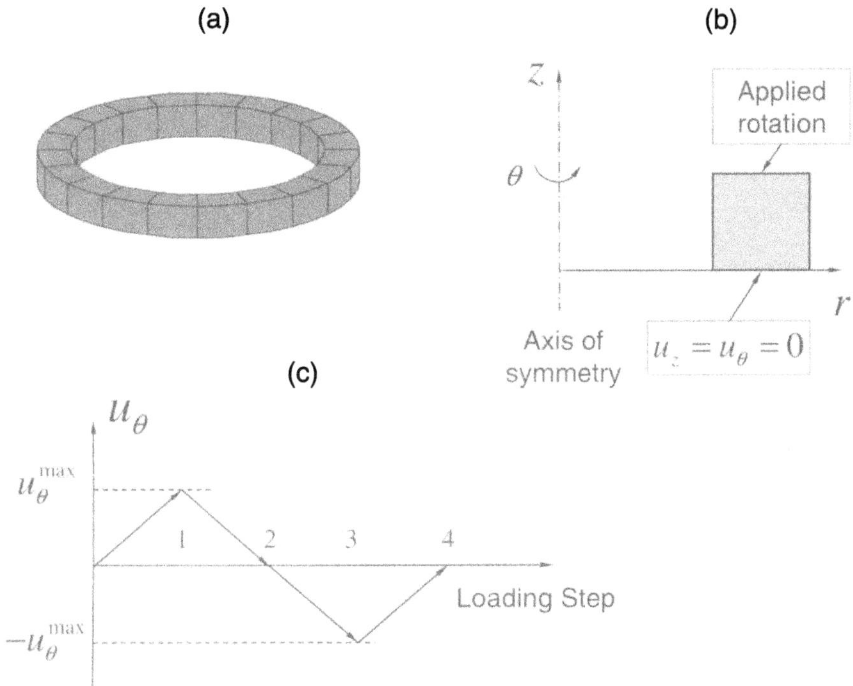

Fig. 5.21. NiTi SMA torque tube: (a) geometry; (b) boundary conditions; (c) loading history for the first loading cycle.

directions. The loading history for one full cycle is also shown in Fig. 5.21. Ten full rotational loading/unloading cycles in both directions have been applied.

The stress-strain response of the tube is shown in Fig. 5.22[4] where the *average* shear stress in the finite element is plotted versus the *average* shear strain. The results obtained are in qualitative agreement with those reported by Lim and McDowell [19]. Since a full set of material parameters is not reported in the original work of Lim and McDowell [19], the results cannot be compared quantitatively. From these results, the hysteresis loop is seen to evolve with the number of loading cycles. One significant difference observed between these results and the uniaxial results presented in Sections 5.5.1 and 5.5.1 is the value of the plastic strain at the end of the cycling test. While the final value of the plastic strain in the uniaxial test is equal to the value of the *accumulated* plastic strain, the final value in the case of torsional loading is significantly smaller. This result can be explained as follows: In the case of torsional cycling loading with the loading history shown in Fig. 5.21(c), the direction of the loading is reversed during each cycle. Therefore, during each half of the loading cycle, the direction of the plastic strain accumulation is also reversed. The result of this effect, as shown in Fig. 5.23[4], is small total plastic strain. Note, however, that even in the presence of small observable plastic

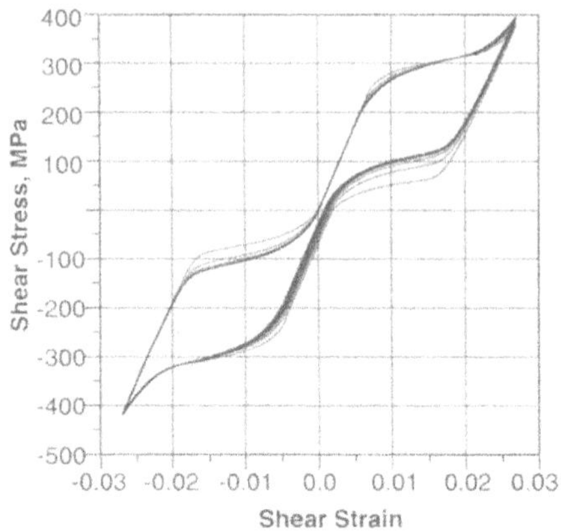

Fig. 5.22. Stress-strain response of NiTi SMA tube subjected to cycling torsional loading: *average* shear stress over *average* shear strain.

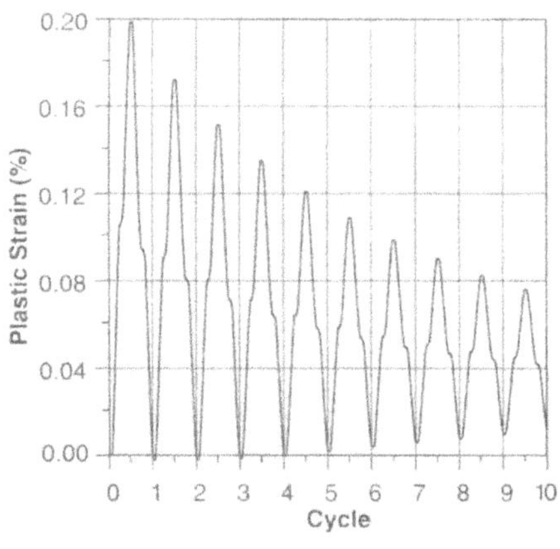

Fig. 5.23. Plastic strain evolution in NiTi SMA tube during cycling torsional loading.

strain, the material still changes during the cycling. As seen from Fig. 5.22, the hysteresis loop evolves during the cycling loading. Thus, the microstructural changes caused by the cyclic loading are taken into account by evolving the material parameters and updating the internal state variables.

5.5 Correlation with Experimental Data

The experimental results for NiTi undergoing cyclic loading will be simulated in this section. Two sets of experimental data will be used. First, in Section 5.5.1 the model will be calibrated using the results reported by Strnadel et al. [6] and presented in Section 5.1.1. Next, in Section 5.5.2 a set of experimental results obtained at Active Materials and Structures Laboratory at Texas A&M University will be used to calibrate the model and the model simulations will be presented.

5.5.1 Cyclic Behavior up to a Constant Stress or Strain

In this section, the experimental data for one of the NiTi alloys reported by Strnadel et al. [6] is used. Two types of tests were performed. The first type involved cycling, during which the material is loaded to the same value of strain during each cycle. The second type involved cycling performed to a given constant value of stress. Both types of tests have been performed at a temperature, $T = 46\,°\mathrm{C}$, higher than the austenitic finish temperature A_f. The experiments are performed on flat tensile specimens (strips) with cross section dimensions $1 \times 4\ mm^2$. Prior to the testing, the specimens are polished and vacuum annealed at 400 °C for one hour. The experimental results for NiTi alloy with composition Ti-50.9 at.%Ni for both loading cases are shown in the top portion of Fig. 5.1[4]. Based on the dimensions of the specimen and the reported elongation, the maximum value of strain during the first cycling test is estimated to be equal to $\varepsilon_{\max} = 0.04$. The maximum value of stress during the second test is reported to be $\sigma_{\max} = 550\mathrm{MPa}$. Twenty loading-unloading cycles with a constant maximum value of strain have been performed, while the number of cycles with a constant maximum value of stress is 50.

First, the material parameters for the alloy are determined based on the experimental graphs and reported data, using the procedure described in Section 5.2.4. The obtained parameters are given in Table 5.2[4]. No data exists on the dependance of the current maximum transformation strain $H^{cur}(\bar{\sigma}')$ on the value of the applied stress. Therefore, the data shown in Fig. 5.8 is used to obtain the back stress coefficients, D_i^b. The coefficients are obtained for the value of the maximum transformation strain for high values of stress $H^{max} = 0.038$ as reported by Strnadel et al. [6]. A polynomial of degree 5 is used in the expression for the back stress. Based on the experimental results, the back stress parameters are assumed to be the same for both the initial and the final state of the material and are given by $D_1^b = 6.18 \times 10^3$ MPa, $D_2^b = -7.37 \times 10^5$ MPa, $D_3^b = 4.98 \times 10^7$ MPa, $D_4^b = -1.63 \times 10^9$ MPa and $D_5^b = 2.03 \times 10^{10}$ MPa.

The initial set of material parameters has been determined using the initial transformation cycle for the case of loading up to a constant value of stress, while the final set has been determined from the stress-strain response after 50 cycles (see Fig. 5.1). In addition, as seen from Table 5.2, the parameters λ_1 and λ_2 have been selected to be the same. Their value has been determined such that the material parameters reach their saturation values after 50 cycles. As seen from Table 5.2, not all of the parameters change their values during the transformation cycling. Based on the experimental results, the maximum transformation strain is assumed to be constant. Only the initial value of the martensitic start temperature is reported in the work by Strnadel et al. [6]; thus, it is assumed to remain constant. Four of the parameters

Table 5.2. Material parameters for Ti-50.9at.%Ni alloy

Parameters remaining constant during cyclic loading	
$E^A = 46.0$ GPa	$\alpha^A = 11.0 \times 10^{-6}$ K^{-1}
$E^M = 25.0$ GPa	$\alpha^M = 6.0 \times 10^{-6}$ K^{-1}
$\nu = 0.33$	$\rho c^A = 2.12$ MJ/(m^{3}K)
$m_1 = 3.5$	$\rho c^M = 2.12$ MJ/(m^{3}K)
$\gamma = 3.5$	
Parameters changing during transformation cycling	
Initial set	*Final set*
$M^{0s} = 242.3$ K	$M^{0s} = 242.3$ K
$H^{max} = 0.038$	$H^{max} = 0.038$
$\rho\Delta s_0 = -0.24$ MJ/(m^{3}K)	$\rho\Delta s_0 = -0.17$ MJ/(m^{3}K)
$Y = 5.0$ MJ/m^3	$Y = 3.0$ MJ/m^3
$D_1^d = 5.0$ MPa	$D_1^d = 6.0$ MPa
$D_2^d = 5.22$ MPa	$D_2^d = 4.0$ MPa
Parameters characterizing the response during cyclic loading	
$C_1^p = 3.6 \times 10^{-3}$	$\lambda_1 = 0.25$
$C_2^p = 18.0$	

(α^A, α^M, c^A, c^M) are taken from the work of Lagoudas and Bo [3]. Since both works [3, 6] deal with NiTi alloys, one expects that the physical constants will be close for alloys with only slight change in the composition. As explained in Section 5.3.2, the values of these constants have no effect on the results when the temperature during the loading-unloading cycle remains constant.

Response to Cycling up to a Constant Value of Stress

The response of the material for the cyclic loading up to a constant value of stress is presented in Fig. 5.24[4]. Only the stress-strain curves for the first and 50th (last) loading cycle are shown. It can be seen that the transformation response after 50 cycles has stabilized and the plastic strain developed during the last cycle is negligible. The transformation loop has significantly evolved during the cyclic loading. First, the value of stress for the onset of phase transformation is much lower for the last loading cycle compared to the first cycle. Furthermore, the transformation hardening during the last cycle is significantly higher than the hardening during the first cycle. Also, the area enclosed by the transformation hysteresis loop is smaller for the last cycle than the area enclosed by the initial loop.

The plastic strain for this loading case has saturated after 50 cycles. Fig. 5.25[4] shows its evolution during the cycling. Further transformation cycling will result in a negligible change of the plastic strain and the shape of the transformation loop. As seen from Fig. 5.24, the modeling results are in good agreement with the experimental observations. An important point is

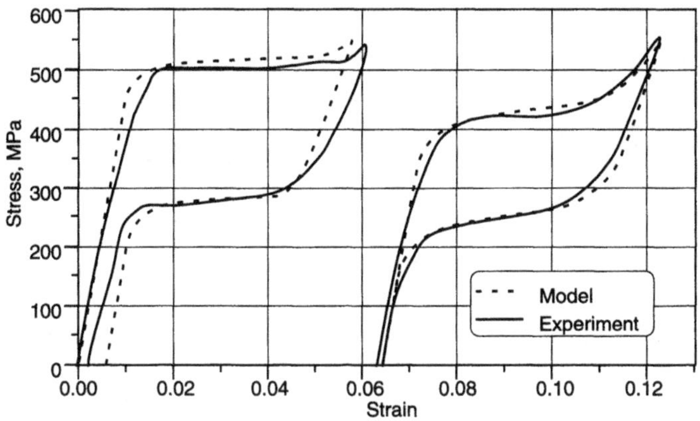

Fig. 5.24. Stress-strain response of NiTi SMA to cyclic loading up to a constant value of stress: curves for the first and 50th cycles.

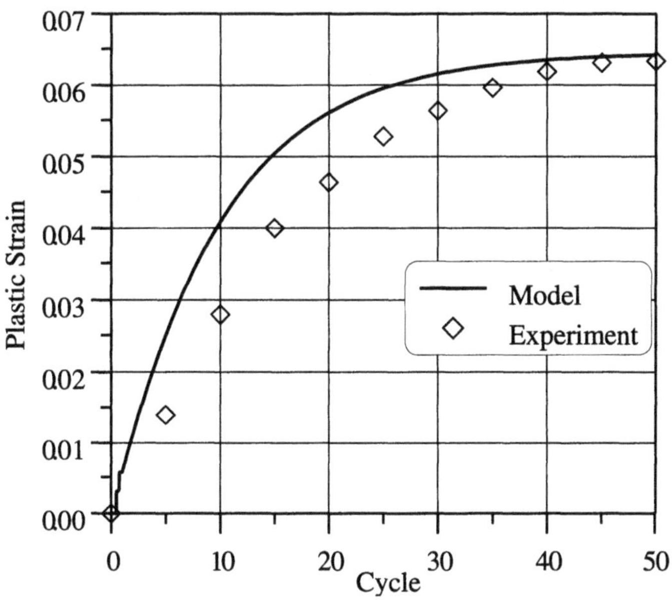

Fig. 5.25. Plastic strain evolution during loading up to a constant value of stress.

that these modeling results are not predictions but simulations of the experiment, since the experimental data has been used to estimate the material parameters.

Response to Cycling up to a Constant Value of Strain

The response of the material during cyclic loading up to a constant value of strain is shown in Fig. 5.26[4]. Two stress-strain curves are presented: the initial stress-strain curve (first cycle) as well as the stress-strain curves for the 10th cycle. From the figure, as the number of cycles increases, the value of stress at the maximum value of strain ε_{max} decreases. In addition, the area enclosed by the hysteresis loop also decreases.

Two factors acting in concert are responsible for these effects. First, the accumulation of the residual strain contributes for the smaller hysteresis area as well as for the lower value of stress. The second factor is the evolution of the material parameters with the number of cycles. Since the value of stress for the onset of the phase transformation decreases, the same value of strain for a later cycle will correspond to a lower value of stress than that for an earlier cycle.

The evolution of the residual plastic strain for this loading case is shown in Fig. 5.27[4]. The plastic strain has not reached a saturation value and continues to increase because this type of cycling results in incomplete phase transformation, therefore taking many more cycles for the plastic strain to saturate than complete transformation cycles.

In contrast to the previous case, the modeling results presented for cyclic loading up to a constant value of strain are predictions, since the experimental data for this loading case has not been used to calibrate the model. The comparison of the modeling results with the experimental curves, presented in Figs. 5.26 and 5.27, shows that the results are in a good agreement. Both the stress-strain responses as well as the plastic strain evolution are predicted with good accuracy. Also, the shape of the minor hysteresis loops predicted by the model is very close to the shape of the experimental loops.

5.5.2 Experiments on Large Diameter NiTi SMA Actuators

In addition to the experimental results found in the literature, a set of experimental data were obtained in the Active Materials and Structures Laboratory

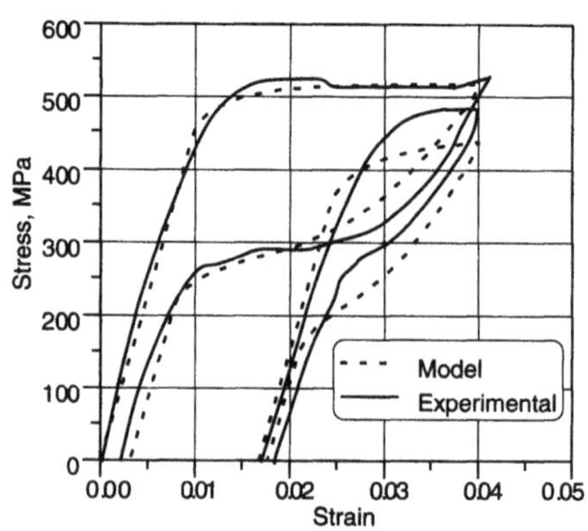

Fig. 5.26. Stress-strain response of NiTi SMA to cyclic loading up to a constant value of strain: curves for the first and 10th cycles.

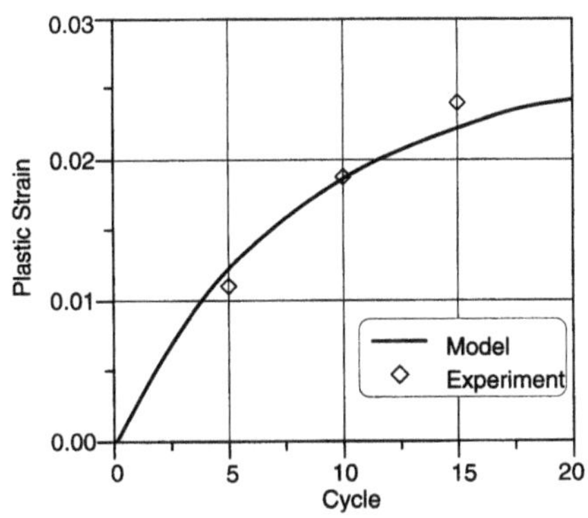

Fig. 5.27. Plastic strain evolution during loading up to a constant value of strain.

at Texas A&M University. Tensile tests were performed on large diameter (2.16 mm) Ti-49.7 at.%Ni wires. Prior to the testing, the material was heat-treated to enhance its pseudoelastic response with different times and temperatures considered during the heat treatment. Based on the results, the material was heat treated in atmospheric conditions at 400 °C for five minutes. The wires received were mechanically polished, and no additional surface treatment was performed prior to testing.

As in the previous section, two types of results were obtained. First, cyclic loading up to a constant value of stress, and second, cyclic loading up to a constant value of strain, were performed. The results of the first loading case were used to calibrate the model, while the results of the second loading case were used to verify the predictions of the model.

Cyclic Loading up to a Constant Value of Stress

A set of experiments with cycling up to a constant stress level was performed on NiTi wires. Twenty loading/unloading cycles were performed at a temperature of 70 °C, above the austenitic finish temperature. The maximum stress value achieved during loading was selected to be equal to 670 MPa. A representative stress-strain result for this loading case is shown in Fig. 5.28[4]. An important result is the evolution of the stress-strain response follows the same general trend as the stress-strain curve in the results presented in Section 5.5.1. The shape and the characteristics of the pseudoelastic loops are similar for both cases. However, it should be pointed out that the amount of plastic strain observed from the results presented in the current section is significantly lower than the one observed in Section 5.5.1. The small plastic strain, which is desirable in SMA actuator applications, is due to the different composition, cold work and heat treatment of the material.

The material parameters for the model are obtained following the procedure described in Section 5.2.4 and the experimental results shown in Fig. 5.28. The set of parameters is summarized in Table 5.3[4]. The comparison of the

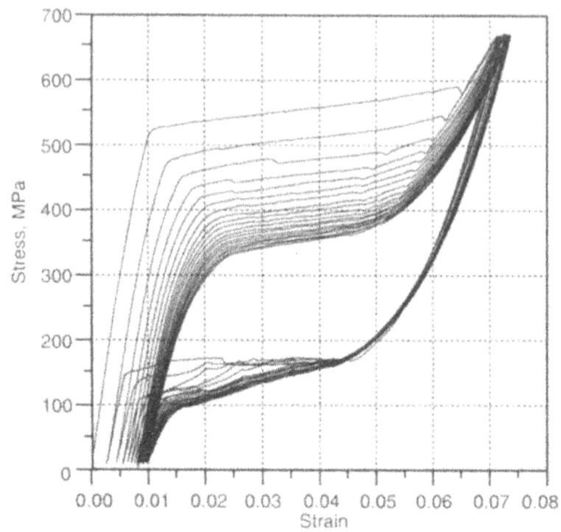

Fig. 5.28. Stress-strain response of Ti-49.7at.%Ni wire specimen to cycling up to a constant value of stress at 70 °C.

Table 5.3. Material parameters for Ti-49.7at.%Ni alloy.

Parameters remaining constant during cyclic loading	
$E^A = 52.0$ GPa	$\alpha^A = 11.0 \times 10^{-6}$ K^{-1}
$E^M = 30.0$ GPa	$\alpha^M = 6.0 \times 10^{-6}$ K^{-1}
$\nu = 0.33$	$\rho c^A = 2.12$ MJ/(m^{3}K)
$m_1 = 3.5$	$\rho c^M = 2.12$ MJ/(m^{3}K)
$\gamma = 3.5$	
Parameters changing during transformation cycling	
Initial set	*Final set*
$M^{0s} = 309.0$ K	$M^{0s} = 309.0$ K
$H^{max} = 0.048$	$H^{max} = 0.042$
$\rho\Delta s_0 = -0.77$ MJ/(m^{3}K)	$\rho\Delta s_0 = -0.23$ MJ/(m^{3}K)
$Y = 11.0$ MJ/m^3	$Y = 4.7$ MJ/m^3
$D_1^d = 1.0$ MPa	$D_1^d = 10.0$ MPa
$D_2^d = 0.7$ MPa	$D_2^d = 0.5$ MPa
$D_1^b = 4.89 \times 10^3$ MPa	$D_1^b = 5.59 \times 10^3$ MPa
$D_2^b = -4.62 \times 10^5$ MPa	$D_2^b = -6.03 \times 10^5$ MPa
$D_3^b = 2.47 \times 10^7$ MPa	$D_3^b = 3.69 \times 10^7$ MPa
$D_4^b = -6.39 \times 10^8$ MPa	$D_4^b = -1.09 \times 10^9$ MPa
$D_5^b = 6.30 \times 10^9$ MPa	$D_5^b = 1.23 \times 10^{1}0$ MPa
Parameters characterizing the response during cyclic loading	
$C_1^p = 1.0 \times 10^{-3}$	$\lambda_1 = 0.1$
$C_2^p = 10.0$	

model simulations with the experimental stress-strain data for the first and the last cycles is shown in Fig. 5.29[4]. The evolution of plastic strain with the number of cycles is also shown in Fig. 5.30. The model is observed to capture the characteristics of the material behavior well.

Cyclic Loading up to a Constant Value of Strain

To test the predictive capabilities of the current model, a set of experiments with cyclic loading up to a constant value of strain were performed on Ti-49.7at.%Ni wires, which were from the same batch of wires tested up to a constant value of stress (see Section 5.5.2). Twenty loading/unloading cycles were performed at a temperature of 80 °C. The maximum strain level was

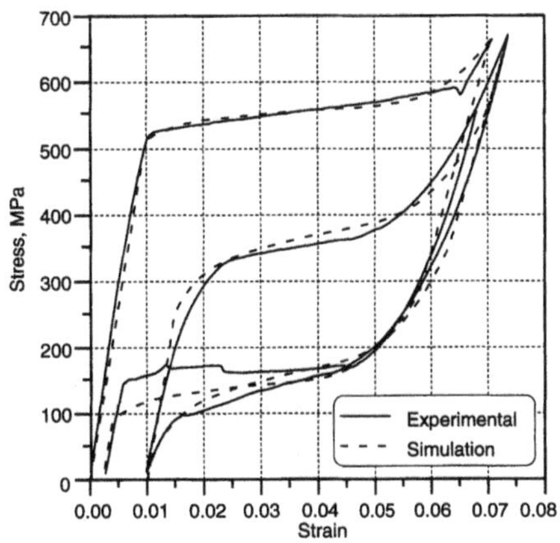

Fig. 5.29. Comparison of the model simulations with the experimental data: stress-strain response at 70 °C during the first and 20th cycles.

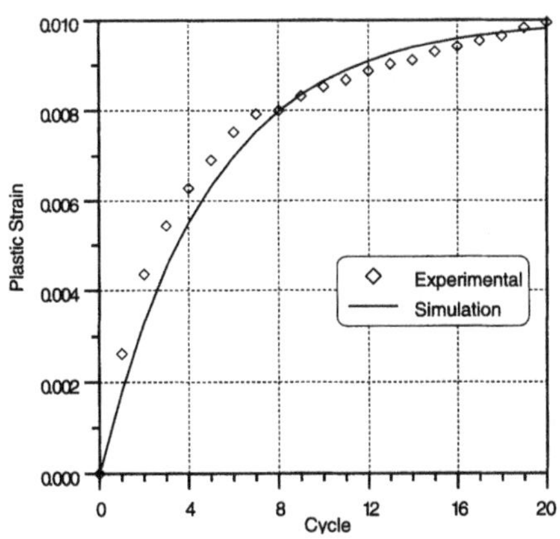

Fig. 5.30. Evolution of plastic strain during constant maximum stress cycling at 70 °C.

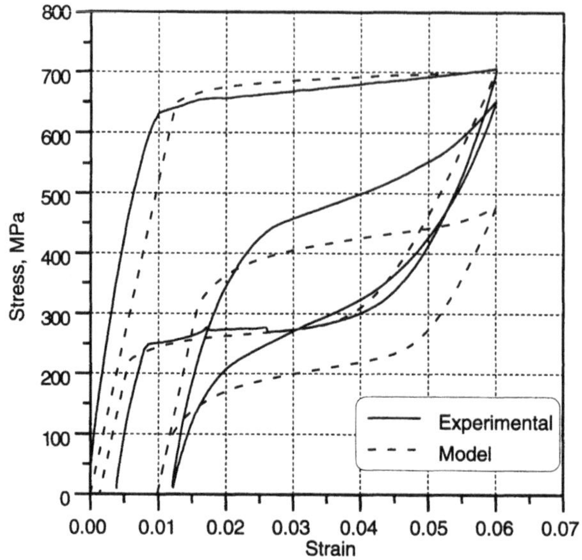

Fig. 5.31. Cyclic loading up to a constant value of strain at 80 °C: comparison of the model predictions with the experimental data for the first and 20th cycles.

chosen to be $\varepsilon_{\max} = 0.06$. The stress-strain response was predicted using the model and the material parameters obtained earlier and shown in Table 5.3. The comparison of the model predictions and the experimental results for the first and the last loading cycles is shown in Fig. 5.31[4]. The modeling results are in good qualitative agreement with the experimental observations.

5.6 Summary

This chapter presented a model for transformation-induced plasticity (TRIP) in shape memory alloys that describes the thermomechanical response of an SMA under cyclic loading. The description of such behavior is important for SMA actuators undergoing repeated actuation cycles and also for addressing the "training" of SMAs. The development of the 3-D model followed the presentation of earlier chapters, with the addition of internal state variables representing the inelastic strain induced by the transformation but not recoverable upon heating to above A_f. Several examples were presented in this chapter demonstrating TRIP cases in SMAs, including uniaxial pseudoelastic loading, constant stress thermally induced phase transformation and combined cyclic loading cases. Finally, the correlation with experimental data was discussed.

5.7 Problems

5.1. Express the transformation function, i.e., the thermodynamic force given by (5.2.6), in terms of the effective stress defined by (5.2.20).

5.2. Plot the transformation function in the in the σ_{11}–σ_{22} stress plane for a temperature $T = A_f + 10\,°\text{C}$ and values for material constants located in Table 4.3. Assume all other stress components are zero and that the drag stress, η, is zero. The back stress components are all assumed to be zero except for $\beta_{11} = 50$ MPa.

5.3. Following a similar approach with the 1-D reduction of the model presented in Sect. 5.3.1, derive the explicit form of the constitutive model for the case of:

a) pure shear
b) biaxial loading in plane stress
c) plane strain

5.4. Derive an analytical expression for the equations appropriate for the example given in Sect. 5.4.1 of uniaxial isothermal pseudoelastic loading and plot the stress–strain pseudoelastic response similar to Fig. 5.13.

5.5. Derive the analytical expressions for the uniaxial adiabatic pseudoelastic loading and compare them with the isothermal case by plotting the stress–strain pseudoelastic response under the same conditions and material parameters as with Problem 5.4 and Fig. 5.13.

5.6. Derive the response of an SMA torque tube similar to Sect. 5.4.4 and Fig. 5.21 with the same loading history. In order to derive analytical expressions, assume the torque tube is thin and the stress is constant through the thickness. Generate similar plots to Fig. 5.22 and Fig. 5.23.

5.7. Repeat Problem 2.14 and Problem 2.15 from Chapter 2 which address differential SMA actuators by incorporating some evolution of TRIP. Use the material properties from Table 5.2.

5.8. Starting with the 1-D reduction of the unified model given in Chapter 3, add the appropriate modifications to account for the generation of permanent plastic strain which occurs when the stress exceeds some critical value (i.e., the yield stress). What additional internal variable(s) are required? Using the properties for NiTi from Table 2.5, plot the stress-strain response of a material loaded at $T = A_f$ to beyond the yield stress. Assume a form of linear plastic hardening.

5.9. The characteristic temperature, M_d, is commonly used to describe the maximum temperature at which full pseudoelastic loading can be applied to a particular SMA material without reaching the critical stress for slip (plastic yielding). How can M_d be experimentally measured and what microstructural parameters influence its values for SMAs?

References

[1] D. Lagoudas, P. Entchev, Modeling of transformation-induced plasticity and its effect on the behavior of porous shape memory alloys: Part I: Constitutive model for fully dense SMAs, Mechanics of Materials 36 (2004) 865–892.

[2] Z. Bo, D. C. Lagoudas, Thermomechanical modeling of polycrystalline SMAs under cyclic loading, Part I: Theoretical Derivations, International Journal of Engineering Science 37 (1999) 1089–1140.

[3] D. C. Lagoudas, Z. Bo, Thermomechanical modeling of polycrystalline SMAs under cyclic loading, Part II: Material characterization and experimental results for a stable transformation cycle, International Journal of Engineering Science 37 (1999) 1141–1173.

[4] Z. Bo, D. C. Lagoudas, Thermomechanical modeling of polycrystalline SMAs under cyclic loading, Part III: Evolution of plastic strains and two-way shape memory effect, International Journal of Engineering Science 37 (1999) 1175–1203.

[5] Z. Bo, D. C. Lagoudas, Thermomechanical modeling of polycrystalline SMAs under cyclic loading, Part IV: Modeling of minor hysteresis loops, International Journal of Engineering Science 37 (1999) 1205–1249.

[6] B. Strnadel, S. Ohashi, H. Ohtsuka, S. Miyazaki, T. Ishihara, Effect of mechanical cycling on the pseudoelasticity characteristics of Ti-Ni and Ti-Ni-Cu alloys, Material Science & Engineering A 203 (1995) 187–196.

[7] P. G. McCormick, Y. Liu, Thermodynamic analysis of the martensitic transformation in TiNi—II. Effect of transformation cycling, Acta Metallurgica et Materialia 42 (7) (1994) 2407–2413.

[8] B. Strnadel, S. Ohashi, H. Ohtsuka, T. Ishihara, S. Miyazaki, Cyclic stress-strain characteristics of Ti-Ni and Ti-Ni-Cu shape memory alloys, Material Science & Engineering A 202 (1995) 148–156.

[9] T. J. Lim, D. L. McDowell, Degradation of an NiTi alloy during cyclic loading, in: Proceedings of the 1994 North American Conference on Smart Structures and Materials, SPIE, 1994, pp. 153–165.

[10] H. Kato, T. Ozu, S. Hashimoto, S. Miura, Cyclic stress-strain response of superelastic Cu-Al-Mn alloy single crystals, Material Science and Engineering A 264 (1999) 245–253.

[11] H. Sehitoglu, R. Anderson, I. Karaman, K. Gall, Y. Chumlyakov, Cyclic deformation behavior of single crystal NiTi, Material Science & Engineering A 314 (1–2) (2001) 67–74.

[12] A. S. Khan, S. Huang, Continuum Theory of Plasticity, John Wiley & Sons, New York, 1995.

[13] T. Saburi, Ti-Ni shape memory alloys, in: K. Otsuka, C. M. Wayman (Eds.), Shape Memory Materials, Cambridge University Press, Cambridge, 1999, Ch. 3, pp. 49–96.

[14] D. G. B. Edelen, On the characterization of fluxes in nonlinear irreversible thermodynamics, International Journal of Engineering Science 12 (1974) 397–411.

[15] Z. Bo, D. C. Lagoudas, D. Miller, Material characterization of SMA actuators under nonproportional thermomechanical loading, Journal of Engineering Materials and Technology 121 (1999) 75–85.

[16] D. C. Lagoudas, Z. Bo, M. A. Qidwai, A unified thermodynamic constitutive model for SMA and finite element analysis of active metal matrix composites, Mechanics of Composite Materials and Structures 3 (1996) 153–179.

[17] Dassault Systmes of America Corp., Woodlands Hills, CA, ABAQUS/Standard User's Manual (2006).

[18] M. A. Qidwai, Thermomechanical constitutive modeling and numerical implementation of polycrystalline shape memory alloy materials, Ph.D. thesis, Texas A&M University, Department of Aerospace Engineering (1999).

[19] J. T. Lim, D. L. McDowell, Mechanical behavior of a Ni-Ti shape memory alloy under axial-torsional proportional and nonproportional loading, Journal of Engineering Materials and Technology 121 (1999) 9–18.

6

Extended SMA Modeling

P. Popov and D. C. Lagoudas

This chapter is dedicated to an extended SMA model that addresses not only the martensitic transformation between austenite and detwinned martensite but simultaneously considers the detwinning of self-accommodated martensite in polycrystalline SMAs. As is often the case, modeling a more complex system requires additional experimental measurements specific to the SMA response when a mixture of the three phases is present. In this chapter we use both new experimental results and carefully revisit results in the literature on SMA characterization to aid the development of the extended model. The analysis of the relevant experimental work is reflected in an additional layer of complexity added to the SMA phase diagram.

6.1 Introduction

As we have seen in the previous chapters, several models exist for capturing various aspects of the martensitic phase transformation in SMAs. Two key SMA characteristics that require only the austenite-detwinned martensite ($A \leftrightarrow M^d$) transformation are pseudoelasticity and the shape memory effect under stress. In Chapters 1-4, it was shown that these two effects underly many applications of shape memory alloys. Chapter 5 further addressed the modeling of $A \leftrightarrow M^d$ transformation under cyclic loading by including the effects of transformation-induced plasticity. However, the reality of the material behaviors behind the shape memory response throughout the entirety of the phase diagram is much more complex than transformation between austenite and detwinned martensite alone. For example, modeling of the conventional stress-free shape memory effect (see Fig. 1.5) requires one to consider the reorientation (detwinning) of twinned martensite before shape recovery takes place by heating. One should then account for the existence of *twinned martensite* or self-accommodated martensite (see Chapter 3, Sect. 3.6), especially because some loading paths involve co-existence of all the three microstructural configurations (austenite, twinned martensite and detwinned martensite). The $A \leftrightarrow M^d$ SMA models discussed in the previous chapters, while useful in the proper contexts, do not have the ability to handle this additional level of complexity. This is also true for the majority of the very extensive body of SMA models available in the literature (see the table which concludes Chapter 3).

D.C. Lagoudas (ed.), *Shape Memory Alloys*, DOI: 10.1007/978-0-387-47685-8_6,

The early attempts to combine the austenite-detwinned martensite material response with detwinning of martensite were formulated in the context of one-dimensional response [1–3]. These models used two internal state variables to model $A \leftrightarrow M^d$ transformation and detwinning. In addition, Brinson [1] proposed a uniaxial phase diagram in stress-temperature space, which conveniently defines the thermodynamically stable domains for the three "phases" and the possible transformations between them. The work was further refined by Bekker and Brinson [4, 5] who incorporated minor loops for the pseudoelastic transformation. However, this basic phase diagram does not account for certain loading paths, especially those that traverse the regions where the three phases can co-exist.

Three dimensional thermodynamics-based models of combined detwinning and $A \leftrightarrow M^d$ transformation have been proposed in references [6–8]. The models of Leclercq and Lexcellent [6] and Lagoudas and Shu [7] used two scalar volume fractions for twinned and detwinned martensite. While formulated in 3-D, they were implemented and tested only on 1-D examples. Furthermore, complex loading paths that involve a mixture of the three phases were also not tested. The model of Juhasz and co-workers [8] used the entire transformation strain as a tensorial internal variable instead of the volume fraction of detwinned martensite. All three models used phase diagrams that were based on the work of Brinson [1]. While attempts were made to overcome some of its basic limitations, current work attempts to present an extended phase diagram that refines existing concepts and also incorporates new experimental results.

In this chapter, a three-dimensional, thermodynamics based model with three internal variables is formulated for the simultaneous modeling of $A \leftrightarrow M^d$ transformation and detwinning of self-accommodated martensite in polycrystalline SMAs. The model is consistent with an extended uniaxial phase diagram. The novel characteristics of this model are: *(i)* integration into the phase diagram, new experimental results that demonstrate that twinned and detwinned martensite transform to austenite at different temperatures; *(ii)* refinement of the phase diagram with respect to loading paths that involve a mixture of the three phases; *(iii)* the use of three independent internal variables (in contrast to the usual two, typically used in this class of models) which provides a new approach to modeling the training of SMA materials and the associated evolution of the phase diagram; *(iv)* numerical implementation that tests complex loading paths, including ones that involve a mixture of the three phases.

This chapter, which follows the presentation of Popov and Lagoudas [9][1], begins with experimental results demonstrating that, at zero stress, twinned and detwinned martensite transform to austenite at different temperatures

[1] Reprinted from *International Journal of Plasticity*, Vol. 23, Issue 10–11, Popov, P. and Lagoudas, D.C., pp. 1679–1720, Copyright 2007, with permission from Elsevier.

(Sect. 6.2). The phase diagram is constructed in Sect. 6.3 based on these observations, as well as a careful re-examination of published experimental data on detwinning of twinned martensite and the conversion of twinned martensite to austenite. The 3-D constitutive model is presented in Sect. 6.4. A discussion of how to identify the material parameters used in the model from experimentally observable quantities is given in Sect. 6.5.2. The numerical implementation of the model is quite involved and is presented separately in Appendix B. Finally, several numerical examples are given in Sect. 6.6.

To simplify the presentation, throughout this chapter the three phases are denoted by A, M^t and M^d for austenite, twinned martensite and detwinned martensite, respectively. The five possible phase transformations are denoted by $A \to M^t$, $A \to M^d$, $M^t \to A$, $M^d \to A$ and $M^t \to M^d$ for austenite to twinned martensite, austenite to detwinned martensite, twinned martensite to austenite, detwinned martensite to austenite and twinned to detwinned martensite, respectively. The detwinning of twinned martensite $M^t \to M^d$ does not involve phase transformation and is, in fact, an inelastic deformation process of reorientation of martensitic variants (cf. [10]). For the sake of simplicity, the collective term transformations is applied to it whenever the distinction is not important. Note also, that the transformation $M^d \to M^t$ from detwinned to twinned martensite (the rubber-like effect, see Sect. 1.5) is not thermodynamically stable and it is not considered. Finally, the critical start and finish transformation temperatures at zero stress level are denoted as follows: M_s and M_f for the $A \to M^t$ transformation, A_s^t and A_f^t for the $M^t \to A$ transformation and A_s^d and A_f^d for the $M^d \to A$ transformation. The clarification that these temperatures are at zero stress level will be omitted, and only the term transformation temperatures will be used.

6.2 Experimental Results on the Transformation Temperatures of Twinned and Detwinned Martensite to Austenite.

In a recent paper, Sakamoto [11] questioned the assumption of many researchers which states that, at zero stress, the transformation temperatures for $M^t \to A$ and $M^d \to A$ coincide. He introduced the concept of shape change stress, which is a local stress field generated at the interface between twinned martensitic variants and the surrounding matrix. In stress-induced martensite, this elastic stress field is absent, and a detailed analysis of the magnitude of this shape change stress with respect to specimen and martensitic plate sizes leads to the conclusion of different transformation temperatures for twinned and detwinned martensite. In this section, mechanical testing combined with calorimetric measurements are used to confirm this idea.

6.2.1 Setup and Experimental Procedure

A 2.16 mm diameter $Ni_{50.3}Ti_{49.7}$ wire was used in the experiment. Two specimens were annealed at 800 °C for 30 min, slowly cooled to 0 °C, and then brought to room temperature (22 °C). DSC measurements in a Perkin-Elmer Pyrus 1 apparatus were performed in order to establish the transformation temperatures for the $A \rightarrow M^t$ and $M^t \rightarrow A$ transformations and characterize the material state after the annealing. It was found that the transformation temperatures were $M_s = 45$ °C, $M_f = 3$ °C, $A_s^t = 40$ °C and $A_f^t = 76$ °C. Since the austenitic start temperature was well above room temperature, it was concluded that, after the heat treatment the wire was entirely in the M^t state. Note, that the transformation temperatures A_s^d and A_f^d (assumed different from A_s^t and A_f^t) cannot be determined from a DSC sweep that involves only the $A \rightarrow M^t$ transition. The remainder of this section details the measurement of A_s^d and A_f^d for this SMA material. It will be shown that they are substantially different from the $M^d \rightarrow A$ temperatures.

After establishing the transformation temperatures for the $A \leftrightarrow M^t$ transition, the two specimens were mechanically loaded at room temperature in an MTS 810 loading frame (Figure 6.1(b)). Due to the initial state of the specimens (M^t), the self-accommodated martensite underwent detwinning ($M^t \rightarrow M^d$) and the specimen was loaded until the entire length of the specimen was detwinned. Upon elastic unloading, large inelastic strain of about 7.2% was observed. Note that there was no strain recovery during unloading, indicating the A_s^d (to be determined by subsequent DSC testing) is higher than room temperature. In order to quantify the amount of inelastic strain due to detwinning of M^t and the amount due to plastic deformations, the first specimen was heated to about 150 °C. During the process, about 5.2% of the inelastic strain was recovered, indicating that it was due to detwinning and the remaining 2% was due to plastic deformations.

The second specimen, immediately after unloading and hence entirely in the M^d state, was subjected to further DSC testing (Fig. 6.2), described below. Care was taken to prepare the DSC sample so that the material state (M^d) achieved at the end of the mechanical unloading step was not altered in the sample preparation process, that is, the specimen was always kept at room temperature, which is below A_s^d.

A total of five thermal loading steps were executed in the DSC apparatus. The actual heat flow observed in the specimen during the first three DSC steps is shown in Fig. 6.2(a). The corresponding latent heat is shown in Fig. 6.2(b). The specimen was first heated from room temperature (the temperature at which the mechanical test was performed) to 200 °C. The first signs of the forward $M^d \rightarrow A$ transformation were observed at $A_s^d = 82$ °C, the peak of the transformation was at approximately 96 °C and the transformation ended at approximately $A_f^d = 108$ °C. At this point, the sample was in the austenitic phase. The sample was then cooled from 200 °C to −60 °C. During the cooling, a single peak was observed at approximately 28 °C, corresponding

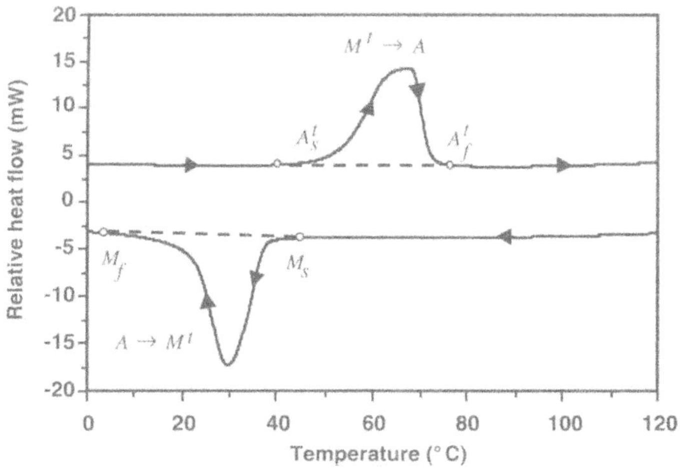

(a) Initial DSC test

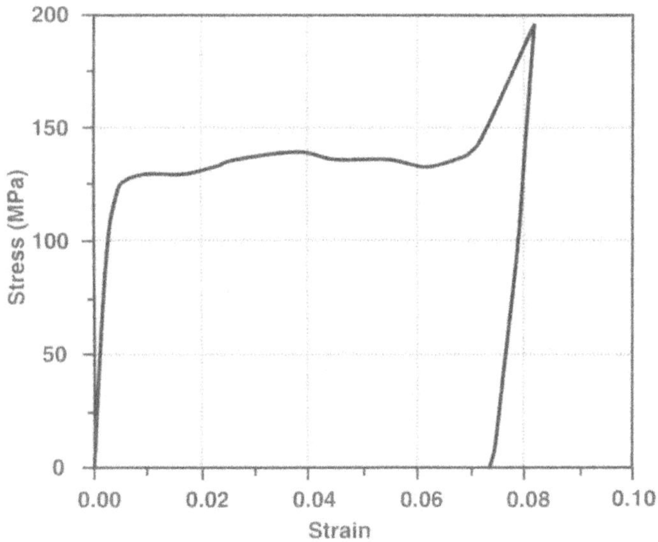

(b) Mechanical loading path

Fig. 6.1. Results from quasistatic mechanical testing of annealed 2.16 mm NiTi wire followed by DSC test. The transformation temperatures for $A \leftrightarrow M^d$, as annealed, are obtained from an initial DSC test (a). A mechanical loading path is then performed (b).

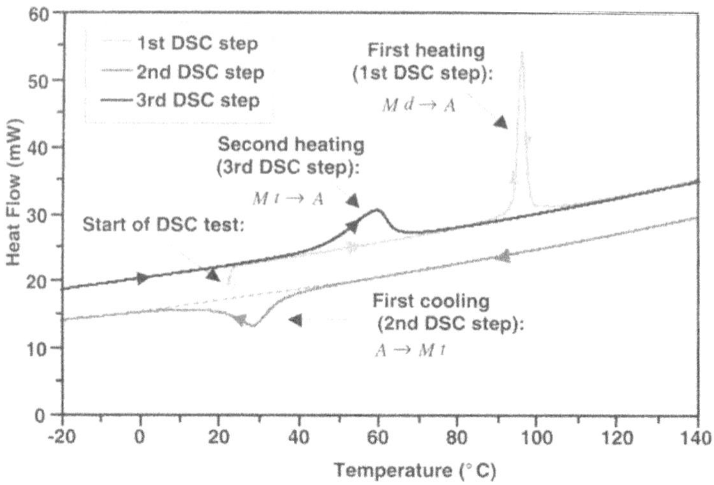

(a) DSC test sequence results

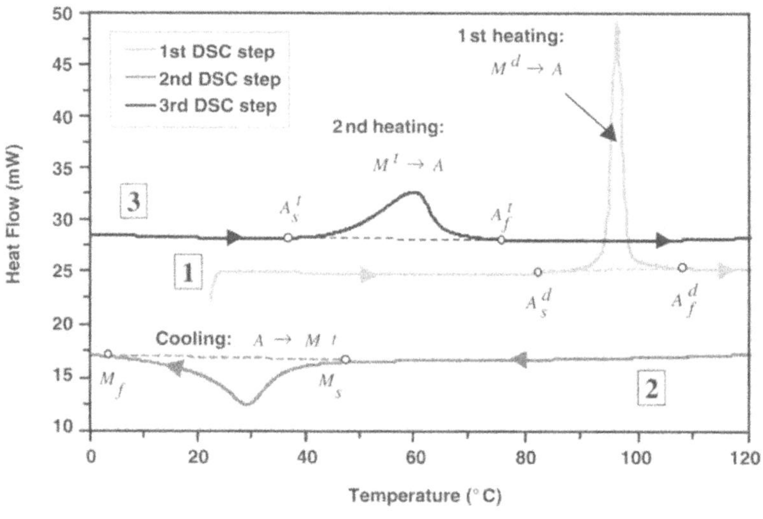

(b) Latent heat of transformation

Fig. 6.2. Results from DSC testing sequence of an untrained NiTi wire performed immediately after a single mechanical loading. The initial state of the wire is M^d. The wire is first heated, revealing the transformation temperatures for the $M^d \to A$, followed by cooling, during which the wire undergoes $A \to M^t$ transformation, followed by a second heating that shows the transformation temperatures for $M^t \to A$.

to the $A \to M^t$ transformation. Note that, due to the nature of a DSC test, the sample always remains stress-free. The beginning of the reverse transformation indicated $M_s = 47\,°\text{C}$ and $M_f = 3\,°\text{C}$, which is consistent with the first

DSC test performed before the wire was subjected to mechanical loading. A repeatability in the $A \rightarrow M^t$ temperatures was therefore observed.

The third thermal loading step was again heating from −60 °C to 200 °C. The transformation temperatures were markedly different from the first heating step: $A_s^t = 35\,°\text{C}$, $A_f^t = 76\,°\text{C}$ with the peak at 59 °C. At the beginning of this step, the sample was entirely in the M^t state (the natural state after the DSC cooling step), therefore the transformation temperatures correspond to the $M^t \rightarrow A$ transformation. Two more loading steps, not shown in Fig. 6.2, were performed. These included an additional cooling and a heating cycle. Due to the stress-free state of the SMA, the transformations involved were $A \rightarrow M^t$ and $M^t \rightarrow A$, respectively. Results were close to those from the second (cooling) and third (heating) cycles, respectively, indicating repeatability of the $A \rightarrow M^t$ transformation temperatures. The latter are substantially different from the $M^d \rightarrow A$ temperatures.

The same type of mechanical loading followed by the above sequence of DSC tests was performed for SMA materials with different annealing history and pseudoelastic training. In all cases, similar results of markedly different critical temperatures for the $M^t \rightarrow A$ and $M^d \rightarrow A$ were observed [12]. The simplest conclusion from these experiments is that the $M^t \rightarrow A$ and $M^d \rightarrow A$ transformation temperatures at zero stress are, generally, different. A qualitative explanation for these results can be done as follows: the twinned martensite requires some energy input to transform back to austenite; the detwinned martensite also requires this energy input, but in addition it also needs more energy in order to reverse the inelastic strains that are present (note this always happens in the presence of a local stress field, even when its macroscopic average is zero). Thus, the reverse phase transformation occurs at higher temperatures, compared to twinned martensite. The theoretical study [11] arrives at the same conclusion with the help of microstructural arguments and by analyzing the local stresses around the martensite/austenite interfaces which are different for twinned and detwinned martensite. These experimental results motivate a re-examination of the commonly used SMA phase diagram (next section) and the proposed constitutive model takes into account the different temperatures A_s^t, A_s^d, A_f^t and A_f^d.

6.3 Modified SMA Phase Diagram

The phase transformations from austenite to martensite, as well as the detwinning of self-accommodated martensite, occur due to thermomechanical loading. A convenient way of describing general thermomechanical loading paths leading to the different transformations is to use a phase diagram in stress-temperature space (Fig. 6.3 and 6.4). Such phase diagrams include the stable domains of A, M^t and M^d in stress-temperature space, as well as transformation strips in which the various transformations take place. The proposed SMA model is based on the 1-D phase diagram shown in Fig. 6.4. This phase diagram incorporates both the new data presented in Sect. 6.2, as well as certain modifications in comparison with other works. The aim is two-fold: first, to take into account the different critical transformation temperatures for the $M^t \to A$ and $M^d \to A$ transformation, which has not been considered previously; second, to define the transformation strips, in agreement with available experimental data, so that non-physical behavior is eliminated for all possible loading paths. In this section, the proposed phase diagram is presented and compared with other common choices in the literature (an example is the diagram in Fig. 6.3).

Several SMA models that attempt to take into account both the development of M^t and M^d (cf. e.g [1, 4, 6–8]) use phase diagrams. The phase diagram shown in Fig. 6.3 was used by Brinson [1] and works well for pure pseudoelastic paths (path 1 on the figure, no M^t is ever produced) and pure SME paths (path 2, complete $M^t \to M^d$ transformation, stress is zero during heating). However, as more complicated loading paths are considered (3a,b,c, for example) certain non-physical behavior becomes possible, mostly in the intermediate regions where a mixture of the three phases can exist. In particular, there is no agreement how the $M^t \to M^d$ strip looks in that intermediate region and what is the shape of the $M^t \to A$ strip. Depending on the thermomechanical loading paths of interest, different assumptions and modifications

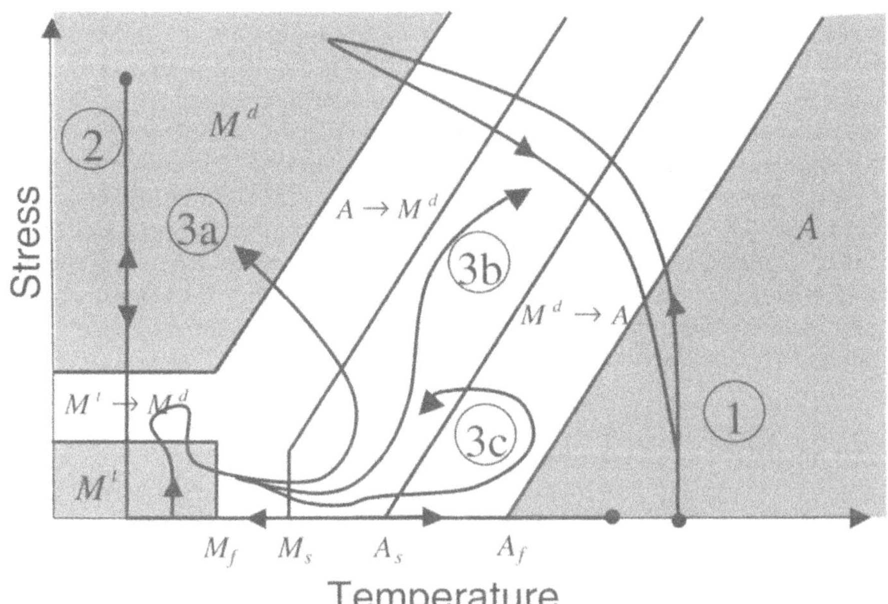

Fig. 6.3. Complete stress-temperature phase diagram for an SMA showing different thermomechanical loading paths.

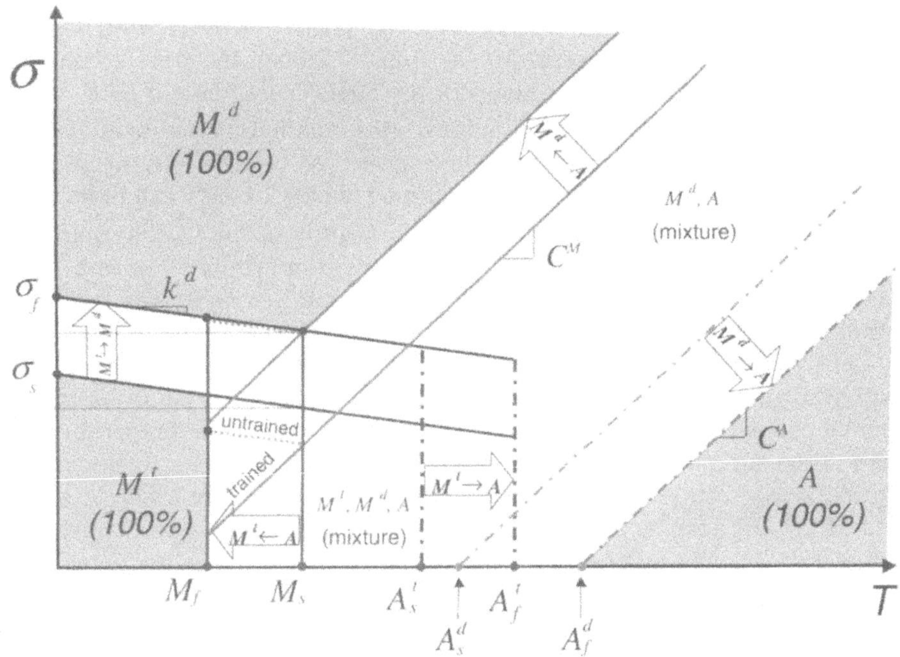

Fig. 6.4. The SMA phase diagram used for the extended SMA model. All the three pure phase regions (A, M^d and M^t) are enclosed by transformation strips. The diagram is completely defined by the respective transformation temperatures M_s, M_f, A_s^t, A_f^t, A_s^d, A_f^d, the critical stresses for detwinning σ_s and σ_f and the slopes $C^M = C^A$ and k^d.

are used by subsequent studies [6–8]. As a result, there is no unambiguous understanding of how the phase diagram should look like.

The phase diagram proposed in this work (Fig. 6.4) follows the established literature in assuming three regions where only the pure phases A, M^t and M^d can exist (cf. Fig. 6.3). These regions are shaded and labeled A, M^t and M^d, respectively. The three regions are separated by transformation strips, labeled according to the transformations ($A \to M^t$, $A \to M^d$, $M^t \to A$, $M^d \to A$, $M^t \to M^d$) which take place. Note that some of these strips overlap, and in an overlap region multiple transformations are possible. In the non-shaded region of the phase diagram, various mixtures can exist.

The critical temperatures for the start and finish of the $A \to M^t$ transformation are denoted by M_s and M_f. Based on the experimental results of Sect. 6.2, the critical start and finish temperatures at zero stress for the $M^t \to A$ transformation are denoted by A_s^t and A_f^t. They are assumed different from the corresponding critical temperatures at zero stress for the $M^d \to A$ transformation, which are denoted by A_s^d and A_f^d. The start and finish lines for the forward and reverse transformations $A \leftrightarrow M^t$ are vertical and pass through the critical temperatures M_s, M_f, A_s^t, A_f^t, respectively. The start and finish lines for the reverse strip $M^d \rightarrow A$ pass through the critical temperatures A_s^d and A_f^d and exhibit a temperature dependence, defined by the positive slope $C^M = C^A$. The critical uniaxial start and finish stresses at $T = M_s$ required for detwinning of twinned martensite ($M^t \rightarrow M^d$) are denoted by σ_s and σ_f, respectively. The transformation strip $M^t \rightarrow M^d$ exhibits a mild temperature dependence characterized by a negative slope k^d. The start and finish lines for the forward $A \rightarrow M^d$ transformation exhibits the same temperature dependence as the reverse transformation $M^d \rightarrow A$. The finish line for $A \rightarrow M^d$ passes through or below the point (M_s, σ_f).

There are several modifications to this phase diagram as compared to the one by Brinson [1]. First, and most importantly, based on the experimental results of the previous Sect. 6.2, the critical start and finish temperatures at zero stress for the $M^t \rightarrow A$ are assumed different from the corresponding critical temperatures at zero stress for the $M^d \rightarrow A$ transformation. Secondly, the $M^t \rightarrow M^d$ strip is a single, well-defined strip for the entire temperature range $T < A_f^t$. The original work of Brinson [1] assumes that the detwinning strip $M^t \rightarrow M^d$ has the same form as the one assumed here for temperatures $T < M_s$ but coincides with the strip for stress-induced martensite $A \rightarrow M^d$ at temperatures $T > M_s$. This can lead to the existence of twinned martensite at high stress levels (above σ_f, cf. path 3b in Fig. 6.3), which is not physically realistic. This problem is critically examined in Sect. 6.3.2. With the help of the experimental study [13], it is shown that a single transformation strip extending to temperatures as high as A_s^t and possibly to A_f^t, as done in this work (Fig. 6.4), is a more natural assumption.

Secondly, there is a disagreement in the literature on the shape of the reverse $M^t \rightarrow A$ strip. In the work of Brinson [1] and later papers, it is assumed to coincide with the $M^d \rightarrow A$ strip while other authors [6–8] have used a vertical $M^t \rightarrow A$ strip, which is independent of stress. An argument can be made (Sect. 6.3.1) that the latter is a more natural choice. Furthermore, there is an ambiguity in the definition of the $A \rightarrow M^d$ strip at low stresses and temperatures ($T < M_s$ and $\sigma < \sigma_s$). Some authors have extended it to zero-stress level [4], while others [7] suggest, that in the region $T < M_s$ the dependence on temperature disappears and there is a critical stress below which $A \rightarrow M^d$ does not occur. There are two possible ways of completing it, depending mainly on the training history of the material. In this work, for trained materials, it will be assumed that the $A \rightarrow M^d$ transformation strip extends all the way to zero stress level (the solid lines in Fig. 6.4). For untrained SMA materials, it is assumed that the lines change slope at M_s and at temperature below M_s they have slope k^d (dashed lines in Fig. 6.4).

To fully define the phase diagram, one also has to consider the relationship between the $A \rightarrow M^d$, $A \rightarrow M^t$ and $M^t \rightarrow M^d$ strips in the vicinity of M_s. The general assumption by most authors is that there exists a triple point (M_s, σ_s) where the three onset lines intersect and another point (M_s, σ_f)

where the three finish lines intersect [1, 4, 6–8]. The experimental evidence cited is usually inconclusive at drawing such a strong relation between the three transformation strips. However, if the finish line for $A \rightarrow M^d$ passes above the intersection point of the other two finish lines, then one can find a particular isobaric cooling path leading to jump discontinuities in the strain as the temperature is lowered. This is demonstrated in Sect. 6.5.4. Therefore, one has to assume that the finish line for the $A \rightarrow M^d$ transformation passes through or is below the intersection of the other two lines. In the absence of sufficiently clear experimental data, we assume that the $A \rightarrow M^d$ finish line is below the intersection of the other two. (In Fig. 6.4 the extreme case of a triple point is shown.) This, along with other restrictions on the relative locations of the transformation strips that arise in the development of the thermodynamically consistent model, are discussed in Sect. 6.5.4. The remainder of this section presents a detailed description of the proposed extensions and modifications of the phase diagram of Fig. 6.4.

6.3.1 Austenite to Martensite ($A \leftrightarrow M^t$, $A \leftrightarrow M^d$)

An early observation in quasi-static isothermal loading tests was that the transformation surfaces for $A \leftrightarrow M^d$ exhibit a strong temperature dependence [10, 13, 14]. These, and many other experimental results, show that the critical transformation stress required for initiation and completion of both the $A \rightarrow M^d$ and $M^d \rightarrow A$ forward and reverse transformations increase relatively linearly, with an increase in temperature. The reason for this dependence on temperature is the development of transformation strain during the transformation and the associated work expended by the SMA. The theoretical derivation of the precise functional dependence of the critical transformation stress for detwinning is based on a Clausius-Clapeyron relation [15]. After some simplifying assumptions, such as equal stiffness and thermal expansion coefficient of austenite and martensite, a linear dependence on temperature is obtained [10, 15]. This has been observed consistently by many experimentalists ever since the work of Cross and co-workers [13]. Virtually any constitutive model for pseudoelastic SMA response, including the current work, takes this into account.

Unlike the $A \leftrightarrow M^d$ transitions, the phase transformation from A to M^t does not involve generation of macroscopic strains. At zero stress level, the $A \rightarrow M^t$ phase transformation begins when a critical temperature M_s is reached and is completed when a second, and lower, critical temperature M_f is reached. Due to the lack of transformation strain, a Clausius-Clapeyron argument suggests that there is no dependence of the critical temperatures M_s and M_f on stress. As a consequence, one can expect that the transformation strip $A \rightarrow M^t$ is nearly vertical when plotted in the stress-temperature space (Fig. 6.4). This fact has been used in most models that take into account the separate development of twinned and detwinned martensite (cf. e.g. [1, 4, 6–8]). There is, however, disagreement on what the shape of the reverse transformation strip $M^t \rightarrow A$ should be. Some [1, 4] assume the same stress-temperature dependence as for the $M^d \rightarrow A$ transformation. Others [6–8] take the $M^t \rightarrow A$ strip to be stress-independent.

There are not many experiments reported in the literature, which aim at determining the shape of the $A \leftrightarrow M^t$ strips. It should be noted that, due to lack of macroscopically observable mechanical quantities, such as inelastic strains, it is very difficult to experimentally detect the formation of twinned martensite under applied stress. Differential scanning calorimetry measurements, which are usually employed for revealing the transformation temperatures at zero stress level, cannot be directly used under applied stress. The two direct methods of measuring the progress of martensitic transformation under applied load that have been used by researchers are electrical resistivity measurements [16, 17] and in-situ neutron diffraction measurements [18]. In both cases, sophisticated testing procedures in a precisely controlled thermal environment in a MTS-type testing frame are required. The focus of these and other direct measurement studies however was not the stress dependence of the critical temperatures for the $A \leftrightarrow M^t$ transformation.

An alternative indirect method, used specifically for determining the $M^t, M^d \rightarrow A$ transformation temperatures at nonzero stress levels during heating and cooling cycles has recently been performed by Tsoi and co-workers [19]. The experiment is done by first loading an SMA wire and embedding it in a epoxy matrix, as loaded. After the epoxy has cured, the SMA is kept deformed without the need for external apparatus. The composite can further be cut into a small enough specimen, suitable for DSC measurements. The tests included pre-strain levels low enough that only $M^t \rightarrow A$ transformation can be expected during heating. While the DSC results are difficult to interpret conclusively, it can be inferred that the $M^t \rightarrow A$ temperatures do not depend on applied stress. Thus, due to the implications of lack of inelastic strains associated with the $M^t \rightarrow A$ transformation, and based on some recent experimental indications [19], in this chapter, it will be assumed that both $M^t \rightarrow A$ and $A \rightarrow M^t$ are stress-independent. In Sect. 6.6.1 a different indirect experimental method, based on the different stiffness of the pure martensitic and austenitic phases, will be proposed.

6.3.2 Detwinning of Self-Accommodated Martensite ($M^t \rightarrow M^d$)

The three pure phases regions (A, M^t and M^d) are separated by transformation strips that indicate which transformation occurs ($A \rightarrow M^d$, $A \rightarrow M^t$, etc). In the original phase diagram of Brinson [1], the transformation strip $M^t \rightarrow M^d$ is not defined at temperatures above $T > M_s$. If the initial conditions are such that M^t is not present and once it is produced the temperature is never increased beyond M_s, this will not cause any problems. This is the case with a major class of SME paths where all the M^t is depleted via the $M^t \rightarrow M^d$ deformation before the temperature is increased above M_s (cf. e.g.

path 2 in Fig. 6.3). Since these types of SME loading paths are quite important in characterization and testing of SMAs, the possibility that M^t may be present at temperatures in the range $M_s < T < A_f^t$ (for example by detwinning only part of the M^t) has generally been overlooked. As a result, in [1] it is assumed for simplicity that the transformation strip for $M^t \to M^d$ coincides with the $A \to M^d$ strip in this temperature range. This assumption creates the inconvenience of having a concave transformation surface in stress-temperature space. Furthermore, at $T > M_s$ and high stress it is not clear how a single transformation surface can be used to determine the evolution of a two-phase mixture that involves two transformations: $M^t \to A$ and $M^d \to A$. It can also be argued that the detwinning of martensite is an inelastic deformation process does not involve change in the crystal lattice. Therefore, the temperature dependence of the detwinning surface should not change drastically, as suggested, from slightly decreasing yield stress as the temperature is raised in the range $T < M_s$ to rapidly increasing with increase of temperature for $M_s < T < A_f^t$.

More importantly this does not seem to be supported by experimental results. A careful review of the pioneering work of Cross and co-workers [13] suggests that it extends to temperatures higher than M_s. The reader is referred specifically to Fig. 16 on page 26 of [13], which reports two sets of experiments. In both cases, the material is loaded mechanically under isothermal conditions, at several different temperatures. The difference is that, prior to the mechanical loading in the first set, the material is cooled from a high temperature and once the prescribed temperature is reached, it is fixed and the SMA is mechanically loaded. In the second set, the material is heated from a low temperature and then loaded. The initial yield stress is recorded in both cases. A look at the transformation temperatures reported by the authors shows that, for the first set of experiments, the initial material state is A, while for the second it is M^t. The latter implies that the initial yield stress measured in the second set corresponds to the beginning of the $M^t \to M^d$ deformation over the entire range $T \leq A_s^t$. The results in the range $A_s^t \leq T \leq A_f^t$ cannot be easily interpreted since, in this range, the material before loading is a mixture of A and M^t. The observed values for the critical stress for detwinning exhibit only very slight dependence on temperature, decreasing slowly as temperature is increased. For the first experimental set, the initial material state is A, therefore a transformation surface for $M^t \to M^d$ can be inferred from the yield stress results only in the range $T \leq M_f$. Observe that the measured yield stress in the range $T \leq M_f$ for both sets of experiments is the same. This is a consistent experimental result since, at these temperatures, the material for both experimental sets is pure M^t when the loading begins.

Based upon this analysis, it is assumed in this work that the shape of the $M^t \to M^d$ has the same dependence on temperature, both for temperatures below and above M_s (Fig. 6.4). Note that the region of the phase diagram covered both by the $M^t \to M^d$ and $M^t \to A$ (to be discussed next) completely surrounds the region where pure M^t can exist. Therefore, there is no possibility that a loading path may lead to the existence of M^t at high temperature or high-stress regions of the phase diagram.

6.3.3 Combined Austenite to Detwinned Martensite at Low Stresses

As was explained in the previous section, it is difficult to determine experimentally when the transformation to twinned martensite is occurring. Therefore another outstanding question, for which there is little experimental information, is what is the shape of the $A \rightarrow M^d$ surface at low stress $\sigma < \sigma_s$ and temperatures. In this region of the phase diagram, it can be expected that both $A \rightarrow M^d$ and $A \rightarrow M^t$ occur. Note that the $A \rightarrow M^d$ is measured experimentally by observing the critical transformation stress required for the $A \rightarrow M^d$ transformation.

Lagoudas and co-workers [20, 21] have measured the development of transformation strain during isobaric heating and cooling of annealed NiTiCu wires at different, constant, stress levels. Such a test can be represented by a horizontal line on the phase diagram and assist in determining the location of the $A \rightarrow M^d$ (during cooling) and $M^d \rightarrow A$ (during heating) transformation surfaces. The results for *untrained* specimen suggest that $A \rightarrow M^d$ does not take place at stress levels below 40 MPa. They argue that there is a critical stress level, below which detwinned martensite cannot form. This has usually been incorporated into SMA models [1, 7] by assuming the $M^d \rightarrow A$ surface is independent of temperature below $T < M_s$.

This SMA behavior at low temperatures and stresses, however, is heavily influenced by the material composition, manufacturing process (e.g. cold work), heat treatments, etc. If a wire is trained for pseudoelastic regime, then development of transformation strain is observed even at zero stress level, which implies that the $M^d \rightarrow A$ surface should extend to zero stress. In order to take into account both types of behavior, the model developed here will include both the capability to proceed with the $A \rightarrow M^d$ transformation at an arbitrary stress level and the possibility of a critical stress below which production of M^d does not happen. In the first case, the $A \rightarrow M^d$ transformation strip would reach zero stress (dotted line in Fig. 6.4), while in the second, it becomes horizontal at $T < M_s$

6.4 Description of the SMA Constitutive Model

In order to simplify the presentation, the term "transformation" will be used to denote both the phase transformation from austenite to twinned and detwinned martensite, as well as the detwinning deformation of self-accommodated martensite. We start with the volume fractions c_i, $i = 1, 2, 3$ of the self-accommodated martensite M^t, stress-induced martensite M^d and austenite A, respectively. The volume fractions are subject to the constraints:

$$c_1 + c_2 + c_3 = 1, \tag{6.4.1}$$

$$0 \le c_i \le 1, \quad \text{for} \quad i = 1,2,3. \tag{6.4.2}$$

While the state of the material is represented completely by the three volume fractions c_i, it is also useful to know how this state was achieved. To do this, the total amount ξ_1 of M^t produced from A, the amount ξ_2 of M^d produced from A, and the amount ξ_3 of M^d produced from M^t, are introduced. They are connected to the three volume fractions c_i by:

$$c_1 = c_{10} + \xi_1 - \xi_3, \tag{6.4.3}$$
$$c_2 = c_{20} + \xi_2 + \xi_3, \tag{6.4.4}$$
$$c_3 = c_{30} - \xi_1 - \xi_2, \tag{6.4.5}$$

where c_{i0}, $i = 1,2,3$ are the initial volume fractions of the three phases, subject to the constraint

$$c_{10} + c_{20} + c_{30} = 1.$$

These two representations of the phase state of the material are schematically portrayed in Fig. 6.5. The two *phase* transformation $A \leftrightarrow M^t$ and $A \leftrightarrow M^d$ can proceed both ways, hence, $\dot{\xi}_1$, $\dot{\xi}_2$ can take arbitrary real values. The detwinning deformation $M^t \rightarrow M^d$, however, is assumed irreversible, e.g. the rubberlike effect (see [10] for definition) is not considered. Therefore, $\dot{\xi}_3 \ge 0$. Observe that equations (6.4.3)-(6.4.5) automatically satisfy the constraint (6.4.1).

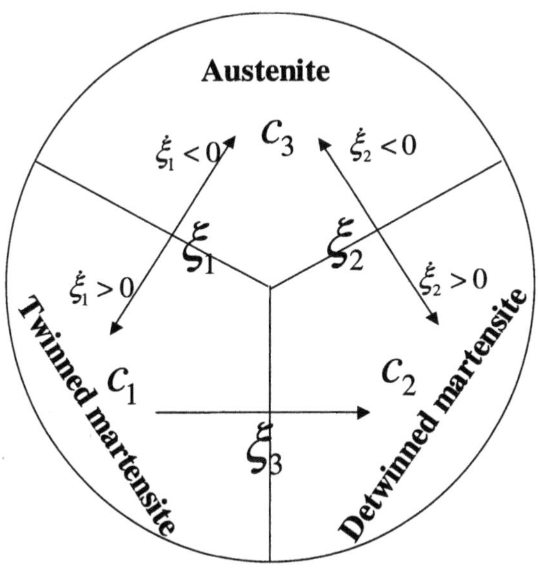

Fig. 6.5. Schematic of the three phases and the possible transitions between them.

With these preliminary definitions, we now move to the question of selecting the inelastic internal variables. In this work, the internal variables that describe the phase state of the material are selected to be:

$$\xi_i, \quad i = 1, 2, 3. \tag{6.4.6}$$

This selection of three independent internal variables requires further discussion. A common choice in the literature is to select two of the three volume fractions c_i, say c_1 and c_2. Such a choice is appropriate when the phase diagram does not change during cyclic loading and when three simultaneous transformations cannot occur (for example $A \to M^d$, $A \to M^t$, $M^t \to M^d$). Observe that, if the three rates $\dot{c}_i$, $i = 1, 2, 3$ are known, equations (6.4.3)-(6.4.5) cannot be used to determine how exactly the three species transformed, that is, the rates $\dot{\xi}_i$, $i = 1, 2, 3$. Therefore, if three simultaneous transformations occur, one has to use the three rates $\dot{\xi}_i$ to have complete knowledge of how the species transformed. This may be necessary, for example, during a simultaneous transformation $A \to M^d$, $A \to M^t$ and $M^t \to M^d$. In such a case, the inelastic strains associated with the $A \to M^d$ and $M^t \to M^d$ processes may evolve differently from each other and it is needed to keep track of the individual transformations. While some authors [8] argue that such situations should be forbidden, this in itself is an additional assumption that should not be excluded *a priori*.

Another reason to use ξ_i is that they provide a complete loading history of the material and thus can be used to account for various cyclic loading effects. For example, at any instance of time, ξ_3 denotes the total amount of detwinned martensite produced from self-accommodated martensite. So, if one has a cyclic SME path and wants to account for plastic strains accumulated over all cycles, ξ_3 would be a suitable variable. Similarly, $\int_0^T |\dot{\xi}_2| dt$ is a suitable indicator of the total amount of $A \leftrightarrow M^d$ transformation. The latter variable has been used [20, 22–24] to account for transformation-induced plasticity. This issue of modeling cyclic effects is related to the choice of hardening functions and will be revisited in Sect. 6.4.5.

6.4.1 Kinematic Assumptions

A large class of applications involving SMAs can easily be accommodated within the framework of small deformations. Thus, for the sake of simplicity, the constitutive theory is formulated for linearized strains and the total strain tensor $\boldsymbol{\varepsilon}$ is given by:

$$\boldsymbol{\varepsilon} = \frac{1}{2} \left(\nabla \mathbf{u} + \nabla \mathbf{u}^T \right),$$

as was shown in (3.2.4) where $\mathbf{u}$ is the displacement. Examples of SMA models formulated in terms of finite strains can be found, for example, in [25–27]. The theory is presented from a macroscopic point of view, therefore, all quantities involved are macroscopic ones, considered over a suitable representative volume element. Further, it is assumed that the strain can be decomposed additively into thermoelastic strain $\boldsymbol{\varepsilon}^{th}$ and inelastic strain $\boldsymbol{\varepsilon}^{in}$ components (c.f. (3.3.39)):

$$\boldsymbol{\varepsilon} = \boldsymbol{\varepsilon}^{th} + \boldsymbol{\varepsilon}^{in}.$$

The inelastic strain $\boldsymbol{\varepsilon}^{in}$ is produced during the forward and reverse stress-induced phase transformations ($\dot{\xi}_2 \neq 0$) and during the detwinning deformation ($\dot{\xi}_3 > 0$). Consequently, it can be further decomposed into:

$$\boldsymbol{\varepsilon}^{in} = \boldsymbol{\varepsilon}^{t} + \boldsymbol{\varepsilon}^{d}, \tag{6.4.7}$$

where $\boldsymbol{\varepsilon}^t$ is the stress-induced transformation strain (produced during $A \rightarrow M^d$ transformation) and $\boldsymbol{\varepsilon}^d$ is the inelastic strain generated during detwinning ($M^t \rightarrow M^d$). Note that while the formation of martensitic twins in the $A \rightarrow M^t$ transformation does involve local strain fields, the macroscopic strain (averaged over a large representative volume element) is zero. Since this chapter deals only with macroscopic description of the SMA, the transformation $A \leftrightarrow M^t$ is associated with zero (macroscopic) strain.

Finally, it is assumed that the transformation and detwinning strains obey the following two transformation/detwinning evolution equations:

$$\dot{\boldsymbol{\varepsilon}}^{t} = \boldsymbol{\Lambda}^{t}\dot{\xi}_2, \tag{6.4.8}$$

$$\dot{\boldsymbol{\varepsilon}}^{d} = \boldsymbol{\Lambda}^{d}\dot{\xi}_3, \tag{6.4.9}$$

where $\boldsymbol{\Lambda}^t$ is the transformation tensor for the stress-induced martensitic transformation ($A \leftrightarrow M^d$) and $\boldsymbol{\Lambda}^d$ is the detwinning tensor for the detwinning of twinned martensite ($M^t \rightarrow M^d$). In general, the transformation tensors $\boldsymbol{\Lambda}^t$ and $\boldsymbol{\Lambda}^d$ are different [7]. Their specific form is discussed in Sect. 6.4.6 in conjunction with the definition of transformation surfaces.

The last assumptions (6.4.8) and (6.4.9) allow the formulation of the constitutive theory in terms of ξ_i, $i = 1, 2, 3$ as the only internal variables. It is convenient to use vector notation $\boldsymbol{\xi} = (\xi_1, \xi_2, \xi_3)^T$ for the internal variables. The internal variables can be thought of as a time-like parameter because of the following relations:

$$\boldsymbol{\varepsilon}^{in} = \int_0^t \left(\boldsymbol{\Lambda}^t\dot{\xi}_2 + \boldsymbol{\Lambda}^d\dot{\xi}_3\right) d\tau = \int_0^{\xi_2} \boldsymbol{\Lambda}^t d\eta + \int_0^{\xi_3} \boldsymbol{\Lambda}^d d\eta, \tag{6.4.10}$$

that is, the inelastic strain $\boldsymbol{\varepsilon}^{in}$ can be considered as a path-dependent functional of $\boldsymbol{\xi}$.

6.4.2 Free Energy for Polycrystalline SMAs

The following form of the Gibbs energy, based on the work of Lagoudas and co-workers [7, 20] is assumed:

$$G\left(\boldsymbol{\sigma}, T, \boldsymbol{\xi}, \operatorname{sgn}(\dot{\xi}_1), \operatorname{sgn}(\dot{\xi}_2)\right) = (c_1 + c_2)G^M(\boldsymbol{\sigma}, T, \boldsymbol{\xi}) + c_3 G^A(\boldsymbol{\sigma}, T, \boldsymbol{\xi}) + G^{\text{mix}}\left(\boldsymbol{\sigma}, T, \boldsymbol{\xi}, \operatorname{sgn}(\dot{\xi}_1), \operatorname{sgn}(\dot{\xi}_2)\right), \quad (6.4.11)$$

where G^M is the thermoelastic free energy of both martensitic phases (twinned and detwinned), G^A is the thermoelastic component of the free energy of austenite, and G^{mix} is the free energy of mixing, which is responsible for the transformation behavior of the SMA. The mixing energy, and thus the entire free energy, depends also on the direction of the $A \leftrightarrow M^t$ and $A \leftrightarrow M^d$ transformations, which is expressed through the sign function as defined in (3.5.98). This kind of dependence allows the model to take into account the different material behavior during forward ($\dot{\xi}_i > 0$) and reverse ($\dot{\xi}_i < 0$) phase transformation, $i = 1, 2$.

In the pseudoelastic SMA literature, this is often done implicitly, without including $\operatorname{sgn}(\dot{\xi}_2)$ in the list of parameters of G [cf. e.g. 28] in order to provide different hardening during loading and unloading. An argument is then made that such a free energy is allowed to take values from two distinct, possibly discontinuous branches (one for loading, one for unloading) and the laws of thermodynamics should be verified for each branch alone. This argument is made mathematically rigorous in the current work by including $\operatorname{sgn}(\dot{\xi}_2)$ explicitly in the parameter list (cf. the discussion after equation (6.4.17)). This is also generalized to the $A \leftrightarrow M^t$ transformation by including dependence on $\operatorname{sgn}(\dot{\xi}_1)$.[2]

Next, the two thermoelastic components are given by:

$$G^A(\boldsymbol{\sigma}, T, \boldsymbol{\xi}) = -\frac{1}{2\rho}\boldsymbol{\sigma} : \boldsymbol{\mathcal{S}}^A : \boldsymbol{\sigma} - \frac{1}{\rho}\boldsymbol{\alpha}^A : \boldsymbol{\sigma}(T - T_0) - \frac{1}{\rho}\boldsymbol{\sigma} : \boldsymbol{\varepsilon}^{in} + c^A\left[(T - T_0) - T\ln\left(\frac{T}{T_0}\right)\right] - s_0^A T + u_0^A, \quad (6.4.12)$$

$$G^M(\boldsymbol{\sigma}, T, \boldsymbol{\xi}) = -\frac{1}{2\rho}\boldsymbol{\sigma} : \boldsymbol{\mathcal{S}}^M : \boldsymbol{\sigma} - \frac{1}{\rho}\boldsymbol{\alpha}^M : \boldsymbol{\sigma}(T - T_0) - \frac{1}{\rho}\boldsymbol{\sigma} : \boldsymbol{\varepsilon}^{in} + c^M\left[(T - T_0) - T\ln\left(\frac{T}{T_0}\right)\right] - s_0^M T + u_0^M, \quad (6.4.13)$$

where $\boldsymbol{\mathcal{S}}^i$, $\boldsymbol{\alpha}^i$, c^i, s_0^i and u_0^i are the compliance tensor, thermal expansion coefficient tensor, specific heat, specific entropy and the specific internal energy at the reference state of the individual phases with the superscript $i = A$ for austenitic and $i = M$ for martensite, respectively. It is assumed that the

[2] Regardless of whether the dependence is implicit or explicit, continuum thermodynamics of irreversible processes dictates that all fields, including G, should be independent of the rates of the internal variables. However, this classical result [cf. e.g. 29] is applicable only to smooth functions G, that is, functions that are continuous and all their partial derivatives are also continuous. In the current case, G is not smooth since $\operatorname{sgn}(\cdot)$ is a discontinuous, non-differentiable function.

material properties of the two martensitic phases are the same. Note that this assumption, and correspondingly, the selection of the same energy for M^t and M^d, is guided by the fact that from a metallurgical point of view, the two phases are indistinguishable [6]. However, as was discussed in the beginning of this section, it is the macroscopic mechanical behavior of twinned and detwinned martensite that is different, which is reflected in the kinematic considerations of Sect. 6.4.1.

Note that in light of equation (6.4.10), the two free energies G^A and G^M and consequently, G, are path-dependent functionals of $\boldsymbol{\xi}$. This approach of making G a path-dependent functional is chosen for convenience [as is done by many other authors cf. e.g. 20, 30], and does not lead to any mathematical inconsistencies in the thermodynamic treatment of the constitutive model.

Further, upon substituting equations (6.4.12) and (6.4.13) into (6.4.11), and by using equations (6.4.3)-(6.4.5) and the constraint (6.4.1), the following expression is obtained for the free energy:

$$G = -\frac{1}{2\rho}\boldsymbol{\sigma} : \boldsymbol{\mathcal{S}}(c_1 + c_2) : \boldsymbol{\sigma} - \frac{1}{\rho}\boldsymbol{\sigma} : \left[\boldsymbol{\alpha}(c_1 + c_2)(T - T_0) + \boldsymbol{\varepsilon}^{in}\right] + c(c_1 + c_2) \left[(T - T_0) - T\ln\left(\frac{T}{T_0}\right)\right] - s_0(c_1 + c_2)T + u_0(c_1 + c_2) + G^{\text{mix}}, \quad (6.4.14)$$

where $\boldsymbol{\mathcal{S}}(c_1 + c_2)$, $\boldsymbol{\alpha}(c_1 + c_2)$, $c(c_1 + c_2)$, $s_0(c_1 + c_2)$ and $u_0(c_1 + c_2)$ are the effective compliance tensor, thermal expansion coefficient tensor, specific heat, specific entropy and the specific internal energy at the reference state, respectively. These effective material properties are calculated in terms of the total martensitic volume fraction $c_1 + c_2$ using the rule of mixtures:

$$\boldsymbol{\mathcal{S}}(\boldsymbol{\xi}) = \boldsymbol{\mathcal{S}}(c_1 + c_2) = \boldsymbol{\mathcal{S}}^A + (c_1 + c_2)(\boldsymbol{\mathcal{S}}^M - \boldsymbol{\mathcal{S}}^A) = \boldsymbol{\mathcal{S}}^A + (c_1 + c_2)\Delta\boldsymbol{\mathcal{S}}, \quad (6.4.15a)$$

$$\boldsymbol{\alpha}(\boldsymbol{\xi}) = \boldsymbol{\alpha}(c_1 + c_2) = \boldsymbol{\alpha}^A + (c_1 + c_2)(\boldsymbol{\alpha}^M - \boldsymbol{\alpha}^A) = \boldsymbol{\alpha}^A + (c_1 + c_2)\Delta\boldsymbol{\alpha}, \quad (6.4.15b)$$

$$c(\boldsymbol{\xi}) = c(c_1 + c_2) = c^A + (c_1 + c_2)(c^M - c^A) = c^A + (c_1 + c_2)\Delta c, \quad (6.4.15c)$$

$$s_0(\boldsymbol{\xi}) = s_0(c_1 + c_2) = s_0^A + (c_1 + c_2)(s_0^M - s_0^A) = s_0^A + (c_1 + c_2)\Delta s_0, \quad (6.4.15d)$$

$$u_0(\boldsymbol{\xi}) = u_0(c_1 + c_2) = u_0^A + (c_1 + c_2)(u_0^M - u_0^A) = u_0^A + (c_1 + c_2)\Delta u_0. \quad (6.4.15e)$$

A detailed discussion of the functional form (6.4.14) for the free energy and the resulting rule of mixtures (6.4.15), based on micromechanical averaging over a representative volume element of the polycrystalline SMA, can be found in [20]. Note that, in view of relations (6.4.3)-(6.4.5), the effective parameters can be viewed either as functions of the total volume fraction of martensite $(c_1 + c_2)$ or as functions of the internal variables $\boldsymbol{\xi}$. The latter notation is more convenient when performing the algebraic manipulations of this section, while the former gives a better physical understanding of the quantities involved.

The mixing term G^{mix} in the free energy (cf. equations (6.4.11), (6.4.14)) is defined as follows:

$$G^{\text{mix}}\left(\boldsymbol{\xi}, \operatorname{sgn}(\dot{\xi}_1), \operatorname{sgn}(\dot{\xi}_2)\right) = \frac{1}{\rho}\int_0^t [\Big(f_1(\boldsymbol{\xi}; \operatorname{sgn}(\dot{\xi}_1))\dot{\xi}_1(\tau) + f_2(\boldsymbol{\xi}; \operatorname{sgn}(\dot{\xi}_2))\dot{\xi}_2(\tau) + f_3(\boldsymbol{\xi})\dot{\xi}_3(\tau)\Big)]d\tau, \tag{6.4.16}$$

where the yet to be defined functions f_i , $i = 1, 2, 3$ are responsible for the hardening during the $A \leftrightarrow M^t$, $A \leftrightarrow M^d$ and $M^t \rightarrow M^d$ transformations, respectively (see next section). Since many SMAs exhibit different hardening behavior during loading and unloading [28] ,it is necessary to allow f_1 and f_2 to take different values depending on whether one has forward or reverse transformation, hence the dependence on $\operatorname{sgn}(\dot{\xi}_1)$, $\operatorname{sgn}(\dot{\xi}_2)$. It should be noted that, in the case of pseudoelasticity ($\dot{\xi}_1 = \dot{\xi}_3 = 0$), the above mixing energy leads to free energy, which is equivalent to the one used by Lagoudas and co-workers [28].

In order to apply the second law of thermodynamics to the constitutive theory (as was done in Sect. 3.2.3 for the SMA model of Chapter 3), it is necessary to first derive the rate of change of the free energy (6.4.11). It is given by

$$\dot{G} = \dot{\boldsymbol{\sigma}}\frac{\partial G}{\partial \boldsymbol{\sigma}} + \dot{T}\frac{\partial G}{\partial T} + \dot{\boldsymbol{\xi}} \cdot \frac{\partial G}{\partial \boldsymbol{\xi}}. \tag{6.4.17}$$

The derivation of this relation is not straightforward. The usual way to derive such identities is to consider G as a function of time and apply the chain rule to the definition of G, in our case, equations (6.4.11) and (6.4.16). However, in the current case, G is not a smooth function of all its internal variables. Indeed, consider a point in the state space, where the $A \leftrightarrow M^t$ or $A \leftrightarrow M^d$ transformation changes sign. At such a point of transformation reversal, G is not differentiable with respect to the two rates $\dot{\xi}_1$ and $\dot{\xi}_2$, (cf. equation (3.5.98)). Moreover, the derivatives $\partial_{\dot{\xi}_1} G$ and $\partial_{\dot{\xi}_2} G$ are discontinuous when $\dot{\xi}_1$ or $\dot{\xi}_2$ change sign, (see equation (6.4.20) in Sect. 6.4.3). As a result, the chain rule cannot be applied directly to equation (6.4.11).

To obtain (6.4.17), first consider a point in state space where both $\dot{\xi}_1$ and $\dot{\xi}_2$ do not change sign. In the neighborhood of this point G, does not depend on $\dot{\boldsymbol{\xi}}$ and it is clearly smooth. Then, (6.4.17) is obtained using the chain rule. At points of transformation reversal, where G is not smooth, one has to differentiate G directly with respect to time, the details for which are given in the following section. 6.4.3.

6.4.3 Evolution of the Rate of the Gibbs Free Energy Function

We consider the set of thermodynamical processes for which the functions f_1, f_2 and f_3 are smooth in their first argument ($\boldsymbol{\xi}$). Furthermore we assume that either of the phase transformation from A to M^t and M^d can be reversed a countable number of times. That is, there exist (possibly infinitely many) occasions in time $T_0 < T_1 \leq T_3 \cdots \leq T_{2n} < T_{2n+1} \cdots$ where $\dot{\xi}_i = 0$, for $i = 1$ and/or $i = 2$ and is strictly positive or strictly negative on each interval (T_{2n}, T_{2n+1}). Observe that only the term G^{mix} in the definition of the free energy (6.4.14) is not a smooth function of its variables. Therefore, to show the identity (6.4.17), it is sufficient to show that

$$\dot{G}^{\text{mix}} = \sum_{i=1}^{3} \frac{\partial G^{\text{mix}}}{\partial \xi_i} \dot{\xi}_i(t). \tag{6.4.18}$$

First, using the definition of G^{mix} given by equation (6.4.16), its rate is directly computed as:

$$\dot{G}^{\text{mix}} \left(\boldsymbol{\xi}, \operatorname{sgn}(\dot{\xi}_1), \operatorname{sgn}(\dot{\xi}_2)\right) = \frac{1}{\rho} \sum_{i=1}^{3} \left[f_i \left(\boldsymbol{\xi}; \operatorname{sgn}(\dot{\xi}_i)\right) \dot{\xi}_i(t) \right]. \tag{6.4.19}$$

Assume now that $t \in [T_{2n}, T_{2n+1}]$ and let the values of ξ_i at the beginning of this interval be denoted by Ξ_i, $i = 1, 2$, that is, $\Xi_1 = \xi_1(T_{2n})$ and $\Xi_2 = \xi_2(T_{2n})$. Without loss of generality, suppose that $\dot{\xi}_i > 0$ in the interval $[T_{2n}, T_{2n+1}]$. Then one has:

$$\int_{T_{2n}}^{t} f_i \left(\boldsymbol{\xi}, \operatorname{sgn}(\dot{\xi}_i)\right) \dot{\xi}_i(\tau) d\tau = \int_{\Xi_i}^{\xi_i} f_i \left(..., \eta, ..., 1\right) d\eta.$$

On the other hand,

$$\frac{\partial}{\partial \xi_i} \int_{\Xi_i}^{\xi_i} f_i \left(..., \eta, ..., 1\right) d\eta = f_i \left(\xi_1, \xi_2, \xi_3, 1\right), \text{ for all } \xi_i \in [\Xi_1, \xi_i(T_{2n+1})].$$

Now, G^{mix} is a path-dependent functional of $\boldsymbol{\xi}$. Therefore, the derivative $\partial_{\xi_i} G^{\text{mix}}$ with respect to the current state $\boldsymbol{\xi}$ (and the path that led to it) can only be meaningfully defined by continuing the current path through $\boldsymbol{\xi}$ and allowing only ξ_i to change. That is, $\dot{\xi}_i \neq 0$ and $\dot{\xi}_j = 0$, $j \neq i$ for times greater then t. For such paths, only the term containing $\dot{\xi}_i$ in the integral expression in equation (6.4.16) will change past the point $\boldsymbol{\xi}$ and therefore:

$$\begin{aligned} \rho \frac{\partial G^{\text{mix}}}{\partial \xi_i} &= \frac{\partial}{\partial \xi_i} \left[\int_0^t f_i \left(\boldsymbol{\xi}; \operatorname{sgn}(\dot{\xi}_i)\right) \dot{\xi}_i(\tau) d\tau \right] = \frac{\partial}{\partial \xi_i} \left[\int_{\Xi_i}^t f_i \left(\boldsymbol{\xi}; \operatorname{sgn}(\dot{\xi}_i)\right) \dot{\xi}_i(\tau) d\tau \right] \\ &= f_i \left(\boldsymbol{\xi}; \operatorname{sgn}(\dot{\xi}_i)\right). \end{aligned} \tag{6.4.20}$$

The last equation, combined with equation (6.4.19), leads to the identity (6.4.18). This in turn implies that equation (6.4.17) holds for arbitrary loading path. Thus, the particular type of path dependence of G demonstrates that the explicit inclusion of $\operatorname{sgn}\dot{\xi}_1$ and $\operatorname{sgn}\dot{\xi}_2$ in the list of parameters of G does not introduce derivatives with respect to $\dot{\xi}_i$ in the expression for $\dot{G}$.

Observe that the definition of the free energy (3.3.41) together with the assumption (3.3.63) introduces an implicit dependence on $\operatorname{sgn}(\dot{\xi})$ for the basic model introduced in Chapter 3. However, the above general derivation for a vector $\boldsymbol{\xi}$ shows that the similar application of the chain rule in Sect. 3.2.3 (equation (3.2.32)) also holds for arbitrary loading paths. It is left as an exercise to the reader to reformulate the free energy in Sect. 3.2.3, incorporating directly the evolution of the transformation strain, and using a similar argument as in this section, derive equation (3.2.32) at point of transformation reversal.

Note that one may want to avoid defining the derivatives $\partial_{\xi_i} G^{\text{mix}}$ along specific paths altogether and instead split the free energy G into a potential part (in our case, $(c_1+c_2)G^M(\boldsymbol{\sigma},T,\boldsymbol{\xi})+c_3G^A(\boldsymbol{\sigma},T,\boldsymbol{\xi})$) and a path-dependent functional (G^{mix}). The path-dependent part is then defined as a function of time only, i.e. $G^{\text{mix}}(t) = G^{\text{mix}}(\boldsymbol{\xi}(t))$, whose rate can be explicitly calculated. Thus, the rate $\dot{G}$ is known and can be used in the entropy inequality (6.4.21).

6.4.4 Thermodynamics and Constitutive Relations

Every thermomechanical process must satisfy the second law of thermodynamics, which, in its strong form reads (see Chapter 3, equation (3.2.31)):

$$-\rho\dot{G} - \dot{\boldsymbol{\sigma}} : \boldsymbol{\varepsilon} - \rho s\dot{T} \geq 0. \tag{6.4.21}$$

Now, by substituting equation (6.4.17) into (6.4.21), the inequality becomes:

$$-\left(\boldsymbol{\varepsilon} + \rho\frac{\partial G}{\partial \boldsymbol{\sigma}}\right) : \dot{\boldsymbol{\sigma}} - \rho\left(s + \frac{\partial G}{\partial T}\right)\dot{T} - \rho\frac{\partial G}{\partial \boldsymbol{\xi}} \cdot \dot{\boldsymbol{\xi}} \geq 0.$$

Following the standard procedure outlined in Sect. 3.2.3, the following two constitutive relations are established for the strain and entropy:

$$\boldsymbol{\varepsilon} = -\rho\frac{\partial G}{\partial \boldsymbol{\sigma}}, \tag{6.4.22}$$

$$s = -\frac{\partial G}{\partial T}, \tag{6.4.23}$$

With the help of equations (6.4.14) and (6.4.16), the above relations are explicitly written as:

$$\boldsymbol{\sigma} = \boldsymbol{\mathcal{S}}(\boldsymbol{\xi})^{-1} : \left(\boldsymbol{\varepsilon} - \boldsymbol{\alpha}(\boldsymbol{\xi})(T - T_0) - \boldsymbol{\varepsilon}^{in}\right), \tag{6.4.24}$$

$$s = \frac{1}{\rho}\boldsymbol{\alpha}(\boldsymbol{\xi})\boldsymbol{\sigma} + c(\boldsymbol{\xi})\ln(T/T_0) + s_0(\boldsymbol{\xi}). \tag{6.4.25}$$

Further, the entropy inequality (6.4.21) also implies:

$$-\rho \frac{\partial G}{\partial \boldsymbol{\xi}} \cdot \dot{\boldsymbol{\xi}} \geq 0. \tag{6.4.26}$$

Note that the above equation is similar to equation (3.3.49). The difference is that in Sect. 3.2.3 the transformation strain $\boldsymbol{\varepsilon}^t$ was assumed to be an independent internal variable, later being connected to the single scalar volume fraction ξ by equation (3.3.51). In contrast, we have a vector of three independent internal variables $\boldsymbol{\xi} = (\xi_1, \xi_2, \xi_3)$ and we have already used equation (6.4.10) to obtain an expression for the inelastic strain $\boldsymbol{\varepsilon}^{in}$ in terms of $\boldsymbol{\xi}$.

Let the thermodynamic forces, conjugate to $\boldsymbol{\xi}$, be denoted by $\boldsymbol{\pi} = (\pi_1, \pi_2, \pi_3)^T$. With the help of (6.4.14), (6.4.15) and (6.4.20) they are given by:

$$\pi_1 = -\rho \frac{\partial G}{\partial \xi_1} = \tilde{\pi}(\boldsymbol{\sigma}, T) - f_1\left(\boldsymbol{\xi}, \operatorname{sgn}(\dot{\xi}_1)\right), \qquad \text{whenever } \dot{\xi}_1 \neq 0, \tag{6.4.27}$$

$$\pi_2 = -\rho \frac{\partial G}{\partial \xi_2} = \boldsymbol{\sigma} : \boldsymbol{\Lambda}^t + \tilde{\pi}(\boldsymbol{\sigma}, T) - f_2\left(\boldsymbol{\xi}, \operatorname{sgn}(\dot{\xi}_2)\right), \qquad \text{whenever } \dot{\xi}_2 \neq 0, \tag{6.4.28}$$

$$\pi_3 = -\rho \frac{\partial G}{\partial \xi_3} = \boldsymbol{\sigma} : \boldsymbol{\Lambda}^d - f_3(\boldsymbol{\xi}), \qquad \text{whenever } \dot{\xi}_3 > 0. \tag{6.4.29}$$

where $\tilde{\pi}$ is:

$$\begin{aligned}\tilde{\pi}(\boldsymbol{\sigma}, T) =& \frac{1}{2}\boldsymbol{\sigma} : \Delta \boldsymbol{S} : \boldsymbol{\sigma} + \Delta \boldsymbol{\alpha} : \boldsymbol{\sigma}(T - T_0) \\ &- \rho \Delta c\left[(T - T_0) - T \ln\left(\frac{T}{T_0}\right)\right] + \rho \Delta s_0 T - \rho \Delta u_0.\end{aligned} \tag{6.4.30}$$

6.4.5 Transformation Hardening Functions

The hardening function f_1 for the $A \leftrightarrow M^t$ transformation is assumed to depend on c_1, and may be different for the forward and reverse transformation:

$$f_1 = \begin{cases} \Delta_1^+ f_1^+(c_1) \text{ for } \dot{\xi}_1 > 0 \\ \Delta_1^- f_1^-(c_1) \text{ for } \dot{\xi}_1 < 0 \end{cases}. \tag{6.4.31}$$

Here $f_1^{\pm}(c_1)$ are two arbitrary monotonously increasing functions in the interval $[0, 1]$ for the forward and reverse transformations $A \to M^t$ and $M^t \to A$, respectively, which can be determined from experimental measurements. The two material constants $\Delta_1^{\pm}$ serve as a scaling factors for $f_1^{\pm}(c_1)$ so that

$$f_1^{\pm}(0) = 0, \quad f_1^{\pm}(1) = 1. \tag{6.4.32}$$

The hardening function $f_2^{\pm}$, f_3 for the stress-induced martensitic transformation $A \leftrightarrow M^d$ and the reorientation of twinned martensite $M^t \to M^d$, respectively, are assumed to depend on the volume fraction of twinned martensite c_2:

$$f_2 = \begin{cases} \Delta_2^+ f_2^+(c_2) \text{ for } \dot{\xi}_2 > 0 \\ \Delta_2^- f_2^-(c_2) \text{ for } \dot{\xi}_2 < 0 \end{cases}, \quad f_3 = \Delta_3 f_3(c_2) \quad \text{for } \dot{\xi}_3 > 0. \qquad (6.4.33)$$

Similarly to equation (6.4.31), the material constants $\Delta_2^{\pm}$ and Δ_3 are scaling factors for the monotonous functions $f_2^{\pm}$ and f_3, respectively, and:

$$f_2^{\pm}(0) = 0, \quad f_2^{\pm}(1) = 1, \qquad (6.4.34)$$

$$f_3(0) = 0, \quad f_3(1) = 1. \qquad (6.4.35)$$

Several things should be noted about this selection of hardening functions. The choice of c_2 as the independent variable for $f_2^{\pm}$ and f_3 has generally been accepted in the literature. The choice of c_1 as the unknown variable for $f_1^{\pm}$, while often used [1, 6, 8], is not the only possible option. The total amount of austenite c_3 may be an equally suitable choice for certain classes of SMA materials.

The specific form of the functions f_i (e.g. polynomials, trigonometric functions, exponents, etc.) is material-dependent and should be treated as part of the material specifications. Through the rest of this chapter, for the sake of simplicity, it is assumed that the hardening functions are linear:

$$f_1^{\pm}(c_1) = c_1, \quad f_2^{\pm}(c_2) = c_2, \quad f_3(c_2) = c_2. \qquad (6.4.36)$$

This selection[3] is typical for the pseudoelastic and detwinning response of polycrystalline NiTi SMAs [28]. In principle, however, the model allows for arbitrary monotonous functions that can be curve-fitted from experiments (Sect. 6.5.2).

Finally, the hardening functions depend indirectly on $\boldsymbol{\xi}$ through the volume fractions c_i (equations (6.4.3)-(6.4.5)). The volume fractions c_i have fixed bounds (cf. equation (6.4.2)). Hence, a hardening function that depends explicitly on c_i will have the property that the transformation strips (see next section) will not change with cyclic thermomechanical loading. It should be kept in mind that the position of the transformation strips in the phase diagram do evolve with cyclic repetition of thermomechanical loading paths. Such effects can be accounted for by specifying an explicit dependency of $f_i^{\pm}$ on ξ_1, ξ_2 and ξ_3. For example, if $f_1^+ = \xi_1 - (1 + \lambda)\xi_3$ is selected, with $\lambda > 0$ as a small positive parameter, every full SME cycle will increase the M_s and M_f temperatures by a fixed amount. For this type of modeling, the evolution of SMA material response, was outside the scope of this work.

[3] This choice of f_2 is consistent with the hardening function of [31, 32], that used the derivative of a quadratic polynomial, which is a linear function.

6.4.6 Transformation Surfaces and Evolution Equations

It is assumed that a closed elastic domain is associated with each possible transformation, bounded by a transformation surface. The five surfaces are:

$$\Phi_1^+(\boldsymbol{\sigma},T,\boldsymbol{\xi}) = 0, \quad \text{whenever } A \to M^t \text{ takes place,} \tag{6.4.37a}$$

$$\Phi_1^-(\boldsymbol{\sigma},T,\boldsymbol{\xi}) = 0, \quad \text{whenever } M^t \to A \text{ takes place,} \tag{6.4.37b}$$

$$\Phi_2^+(\boldsymbol{\sigma},T,\boldsymbol{\xi}) = 0, \quad \text{whenever } A \to M^d \text{ takes place,} \tag{6.4.37c}$$

$$\Phi_2^-(\boldsymbol{\sigma},T,\boldsymbol{\xi}) = 0, \quad \text{whenever } M^d \to A \text{ takes place,} \tag{6.4.37d}$$

$$\Phi_3(\boldsymbol{\sigma},T,\boldsymbol{\xi}) = 0, \quad \text{whenever } M^t \to M^d \text{ takes place,} \tag{6.4.37e}$$

and the elastic domains in stress-temperature space, for given $\boldsymbol{\xi}$, with respect to $\dot{\xi}_i$ are defined implicitly by the inequalities:

$$\left\{(\boldsymbol{\sigma},T)|\Phi_i^+(\boldsymbol{\sigma},T,\boldsymbol{\xi}) \leq 0, \quad \dot{\xi}_i > 0\right\}, \quad \text{for } i = 1,2, \tag{6.4.38}$$

$$\left\{(\boldsymbol{\sigma},T)|\Phi_i^-(\boldsymbol{\sigma},T,\boldsymbol{\xi}) \leq 0, \quad \dot{\xi}_i < 0\right\}, \quad \text{for } i = 1,2, \tag{6.4.39}$$

$$\left\{(\boldsymbol{\sigma},T)|\Phi_3(\boldsymbol{\sigma},T,\boldsymbol{\xi}) \leq 0, \quad \dot{\xi}_3 > 0\right\}. \tag{6.4.40}$$

The first two inequalities (6.4.38) describe the elastic domains of the two *forward* transformations, while the second two inequalities (6.4.39) the elastic domains of the two *reverse* transformations. The last inequality (6.4.40) describes the elastic domain for the $M^t \to M^d$ transformation. In contrast to conventional plasticity, the phase transformation terminates whenever the constraints (6.4.2) are violated. Therefore, the elastic domain associated with given phase transformation is assumed to be the entire space, when the transformation is complete or there is no more material to transform.

Following Lagoudas and Shu [7, 26], the following form of the transformation surfaces is used:

$$\Phi_1^+(\boldsymbol{\sigma},T,\boldsymbol{\xi}) = \pi_1 - Y_1^+, \qquad \Phi_1^-(\boldsymbol{\sigma},T,\boldsymbol{\xi}) = -\pi_1 - Y_1^-, \tag{6.4.41}$$

$$\Phi_2^+(\boldsymbol{\sigma},T,\boldsymbol{\xi}) = \pi_2 - Y_2^+, \qquad \Phi_2^-(\boldsymbol{\sigma},T,\boldsymbol{\xi}) = -\pi_2 - Y_2^-, \tag{6.4.42}$$

$$\Phi_3(\boldsymbol{\sigma},T,\boldsymbol{\xi}) = \pi_3 - Y_3, \tag{6.4.43}$$

where and $Y_1^{\pm}$, $Y_2^{\pm}$, Y_3 are measures of internal dissipation of the respective transformations. We further assumed that $Y_i^{\pm}$, $i = 1,2,3$ are constants, independent of $\boldsymbol{\sigma}$, T and $\boldsymbol{\xi}$. This, due to the inequalities (6.4.38)-(6.4.40), implies that the appropriate conjugate forces π_i remain constant during the transformation. It also implies that the entropy production due to a phase transformation is proportional to $\dot{\xi}_i$, with $Y_i^{\pm}$ being the proportionality constant (cf. equations (6.4.26) and (6.4.27)-(6.4.29)).

The functions f_i defined by (6.4.31) and (6.4.33) appear in the definition of the transformation function (6.4.41)-(6.4.43) through the constitutive relations (6.4.27)-(6.4.29). They are the only terms in the transformation functions which are dependent on the internal variables $\boldsymbol{\xi}$, hence, they are responsible for the transformation hardening.

In order to complete the model, the transformation tensors in the evolution equations (6.4.8) and (6.4.9) should be specified. The detwinning tensor is taken to be of the form

$$\boldsymbol{\Lambda}^d = \frac{3}{2} H^d(\bar{\sigma}') \frac{\boldsymbol{\sigma}'}{\bar{\boldsymbol{\sigma}}'}, \tag{6.4.44}$$

where H^d is the maximal uniaxial inelastic strain, assumed to be a material constant, and $\bar{\boldsymbol{\sigma}}'$ and $\boldsymbol{\sigma}'$ have been defined in Chapter 3.

For the sake of simplicity, the transformation tensor used for the $A \leftrightarrow M^d$ transformation is, following [28, 32], taken to be

$$\boldsymbol{\Lambda}^t = \begin{cases} \frac{3}{2} H^t(\bar{\boldsymbol{\sigma}}') \frac{\boldsymbol{\sigma}'}{\bar{\boldsymbol{\sigma}}'} \ ; & \dot{\xi}_2 > 0 \\ H^{t-r} \frac{\boldsymbol{\varepsilon}^{in-r}}{\bar{\boldsymbol{\varepsilon}}^{in-r}} \ ; & \dot{\xi}_2 < 0 \end{cases} \tag{6.4.45}$$

where H^t is a material constant having the meaning of maximum current uniaxial transformation-induced strain (equivalent to $H^{cur}(\bar{\sigma}')$). The variable $\bar{\boldsymbol{\varepsilon}}^{in-r}$ is analogous to $\bar{\boldsymbol{\varepsilon}}^{t-r}$ as defined in Chapter 3 and H^{t-r} is the maximum transformation at the reversal of phase transformation. It should be noted that the forward evolution equation is a simple J_2 based one, which has been used in many of the early works on modeling pseudoelasticity [cf. e.g. 28]. A number of alternative transformation surfaces have been proposed in the literature [cf. e.g. 26, 33–36], which account for the observed tension-compression asymmetry of SMA materials, as well as the development of a small volumetric strain during the $A \to M^d$ phase transformation. Due to the large number of different SMA alloys, the selection of an appropriate transformation surface can be a difficult task and is specific for each alloy. The simple choice of transformation surface also helps in the Sect. 6.5, where the necessary relations are found for the material parameters so that the model is consistent with the selected phase diagram. Since the main goal of the current chapter is the formulation of a consistent model capable of accounting for phase transformation and detwinning of self-accommodated martensite over a wide range of stresses and temperatures, the choice of more accurate transformation detwinning surfaces was not addressed in detail.

The reverse transformation tensor of the last equation (6.4.45) also deserves some attention. The reason why two different transformation tensors are used for loading and unloading is the need to account for reorientation in multiaxial loading paths. In general, if the direction of the stress state is changed, some martensitic variants will reorient in the new direction, thus changing the direction of the inelastic strain. A constitutive model with a single volume fraction for all detwinned variants of martensite cannot account for this process. If the same transformation tensor is used for forward and reverse transformations, it may happen that residual inelastic strain is present after unloading to austenite (e.g. $c_3 = 1$ and the stress becomes zero) from a non-proportional loading path. The unloading criterion used above ensures that when $c_3 = 1$, the inelastic strain becomes zero. It reduces to the same transformation tensor used by Qidwai and Lagoudas [32] when $\dot{\xi}_3 = 0$.

6.5 One-Dimensional Reduction and Material Parameter Determination

Having defined the model in the three dimensional case, we are presented with two interrelated tasks: determine the material parameters from physically observable quantities and verify that the model conforms to the phase diagram of Sect. 6.3. To achieve both goals we need to examine how the model behaves for uniaxial stress states.

6.5.1 Reduction of the Model to the Uniaxial Stress State

In this section, we will explicitly write the transformation surfaces in the uniaxial stress case. Furthermore, we will also show that the material parameters

$$Y_1^{\pm}, Y_2^{\pm}, Y_3, \Delta_1^{\pm}, \Delta_2^{\pm}, \Delta_3, \tag{6.5.46}$$

are directly related to the shape of the phase diagram and can be completely determined if it is known.

To begin, assume that we have an SMA rod, subjected to a uniaxial tension/compression. Assume further that $H^d(\bar{\sigma}') = H^t(\bar{\sigma}') = H$ is a constant. Then, at any material point of this rod, the stress state is:

$$\sigma_{11} = \sigma, \quad \sigma_{12} = ... = \sigma_{33} = 0. \tag{6.5.47}$$

Since uniaxial loading is always proportional, any combination of detwinning $M^t \to M^d$, forward $A \to M^d$ or reverse $M^d \to A$ by virtue of (6.4.44) and/or (6.4.45) will result in a transformation direction tensor:

$$\mathbf{\Lambda}_{11}^{t,d} = H, \quad \mathbf{\Lambda}_{22}^{t,d} = \mathbf{\Lambda}_{33}^{t,d} = -\frac{1}{2}H, \quad \mathbf{\Lambda}_{12}^{t,d} = \mathbf{\Lambda}_{13}^{t,d} = \mathbf{\Lambda}_{23}^{t,d} = 0. \tag{6.5.48}$$

With this in mind, and in light of equations (6.4.27)-(6.4.29) and (6.4.41)-(6.4.43), the inequalities (6.4.38)-(6.4.40) take the form:

$$\hat{\pi}(\sigma, T) - \Delta_1^+ f_1^+(c_1) \leq Y_1^+ \tag{6.5.49}$$

$$-\hat{\pi}(\sigma, T) + \Delta_1^- f_1^-(c_1) \leq Y_1^- \tag{6.5.50}$$

$$\sigma H + \hat{\pi}(\sigma, T) - \Delta_2^+ f_2^+(c_2) \leq Y_2^+ \tag{6.5.51}$$

$$-\sigma H - \hat{\pi}(\sigma, T) + \Delta_2^- f_2^-(c_2) \leq Y_2^- \tag{6.5.52}$$

$$\sigma H - \Delta_3 f_3(c_2) \leq Y_3 \tag{6.5.53}$$

where $\hat{\pi}$ denotes $\tilde{\pi}$ for the uniaxial stress case and it has the evaluation

$$\begin{aligned}\hat{\pi}(\sigma,T) =& \Delta S\sigma^2 + \Delta\alpha\sigma(T-T_0)\\ &- \rho\Delta c\left[(T-T_0) - T\ln\left(\frac{T}{T_0}\right)\right] + \rho\Delta s_0 T - \rho\Delta u_0.\end{aligned} \tag{6.5.54}$$

We now want to show that the parameters (6.5.46) are responsible for the relative position and width of the transformation strips in the phase diagram. The transformation strips, on the other hand, are completely determined by the start and finish detwinning stresses σ_s and σ_f , the transformation temperatures M_s, M_f, A_s^t, A_f^t, A_s^d, A_f^d, as well as the start and finish temperatures $T_s(\sigma_f)$ and $T_f(\sigma_f)$ for the $A \to M^d$ transformation at constant (uniaxial) stress $\sigma = \sigma_f$[4].

To show that (6.5.46) can be determined based on σ_s, σ_f, M_s, M_f, A_s^t, A_f^t, A_s^d, A_f^d, $T_s(\sigma_f)$ and $T_f(\sigma_f)$ we assume that the material constants ΔS, $\Delta\alpha$, Δs_0 and Δu_0, entering the expression for $\hat{\pi}$, are known from either mechanical or calorimetric measurements (See Sect. 6.5.2). Therefore, $\hat{\pi}$ is a well defined function of σ and T. Whenever one or more transformations are taking place (that is, $\dot{\xi}_i \neq 0$, $i = 1, 2, 3$), the respective inequalities (6.5.49)-(6.5.53) turn into equalities (cf. conditions (6.4.37)).

The parameters (6.5.46) are then established as follows: Consider a loading path in which a purely twinned SMA ($c_1 = 1$, $c_2 = c_3 = 0$ is loaded at temperature below M_f. As the detwinning deformation progresses, $\dot{\xi}_3 > 0$, and the inequality (6.5.53) becomes an equality:

$$\sigma H - \Delta_3 f_3(c_2) = Y_3, \tag{6.5.55}$$

Therefore, at the beginning of the detwinning deformation one has $\sigma = \sigma_s$, $c_2 = 0$, and the last equation, together with (6.4.35), implies:

$$Y_3 = \sigma_s H.$$

Similarly, upon completion of the deformation, one has $\sigma = \sigma_f$, $f_3(1) = 1$ and $\Phi_3 = 0$, hence:

$$\Delta_3 = H(\sigma_f - \sigma_s).$$

The material parameters $Y_1^{\pm}$, $\Delta_1^{\pm}$ for the $A \leftrightarrow M^t$ are determined with the help of a zero stress cooling/heating cycle. During cooling, the forward transformation surface (6.5.49) turns into equality:

$$\hat{\pi}(\sigma,T) - \Delta_1^+ f_1^+(c_1) = Y_1^+, \tag{6.5.56}$$

which, in conjunction with (6.4.32), yields:

[4] In this model, there is no assumption of a triple point, so the $A \to M^d$ strip can be located at or to the right of the intersection point of the finish lines for the $A \to M^t$ and $M^t \to M^d$ transformations.

$$
\begin{aligned}
Y_1^+ &= \hat{\pi}(0, M_s) = \\
&- \rho \Delta u_0 + \rho \left(M_s \Delta s_0 - \Delta c \left[(M_s - T_0) - M_s \ln\left(\frac{M_s}{T_0}\right) \right] \right), \qquad (6.5.57) \\
\Delta_1^+ &= \hat{\pi}(0, M_f) - Y_1^+ = \hat{\pi}(0, M_f) - \hat{\pi}(0, M_s) \\
&= \rho \left((M_f - M_s) \Delta s_0 - \Delta c \left[M_f - M_s + M_s \ln\left(\frac{M_s}{T_0}\right) - M_f \ln\left(\frac{M_f}{T_0}\right) \right] \right). \qquad (6.5.58)
\end{aligned}
$$

Similarly, during the heating (6.5.50) becomes:

$$
-\hat{\pi}(\sigma, T) + \Delta_1^- f_1^-(c_1) = Y_1^-, \qquad (6.5.59)
$$

hence Y_1^- and Δ_1^- can be determined:

$$
\begin{aligned}
Y_1^- &= -\hat{\pi}(0, A_f^t) \\
&= \rho \Delta u_0 - \rho \left(A_f^t \Delta s_0 - \Delta c \left[(A_f^t - T_0) - A_f^t \ln\left(\frac{A_f^t}{T_0}\right) \right] \right), \qquad (6.5.60) \\
\Delta_1^- &= \hat{\pi}(0, A_s^t) + Y_1^- = \hat{\pi}(0, A_s^t) - \hat{\pi}(0, A_f^t) \\
&= \rho \left((A_s^t - A_f^t) \Delta s_0 - \Delta c \left[A_s^t - A_f^t + A_f^t \ln\left(\frac{A_f^t}{T_0}\right) - A_s^t \ln\left(\frac{A_s^t}{T_0}\right) \right] \right). \qquad (6.5.61)
\end{aligned}
$$

Determining the parameters for the stress-induced martensitic transformation is done by considering two loading paths. First, assume a fully detwinned state at some temperature below A_s^d and at zero stress (this can be obtained by loading isothermally at $T \le M_f$ until all the material has detwinned and then unloading until zero stress is reached) and heat, while maintaining the material stress-free. Then $\dot{\xi}_2 < 0$ and (6.5.52) becomes an equality:

$$
-\sigma H - \hat{\pi}(\sigma, T) + \Delta_2^- f_2^-(c_2) = Y_2^-. \qquad (6.5.62)
$$

Noting that $\sigma = 0$ throughout the loading path, and with the help of (6.4.34), Y_2^- and Δ_2^- are found to be:

$$
\begin{aligned}
Y_2^- &= -\hat{\pi}(0, A_f^d) \\
&= \rho \Delta u_0 - \rho \left(A_f^d \Delta s_0 - \Delta c \left[(A_f^d - T_0) - A_f^d \ln\left(\frac{A_f^d}{T_0}\right) \right] \right), \qquad (6.5.63) \\
\Delta_2^- &= \hat{\pi}(0, A_s^d) + Y_2^- = \hat{\pi}(0, A_s^d) - \hat{\pi}(0, A_f^d) \\
&= \rho \left((A_s^d - A_f^d) \Delta s_0 - \Delta c \left[A_s^d - A_f^d + A_f^d \ln\left(\frac{A_f^d}{T_0}\right) - A_s^d \ln\left(\frac{A_s^d}{T_0}\right) \right] \right). \qquad (6.5.64)
\end{aligned}
$$

Finally, in order to determine Y_2^+ and Δ_2^+, load the material in austenite to some stress level, for example, σ_f and then cool the material. Let the critical temperatures for the $A \to M^d$ transformation at this stress level be $T_s(\sigma_f)$ for the start and $T_f(\sigma_f)$ for the finish. Then the constraint (6.5.51) becomes:

$$\sigma H + \hat{\pi}(\sigma, T) - \Delta_2^+ f_2^+(c_2) = Y_2^+, \tag{6.5.65}$$

which results in

$$Y_2^+ = \sigma_f H + \hat{\pi}(\sigma_f, T_s(\sigma_f)), \tag{6.5.66}$$
$$\Delta_2^+ = \sigma_f H + \hat{\pi}(\sigma_f, T_f(\sigma_f)) - Y_2^+. \tag{6.5.67}$$

Note that it is necessary to load to a stress equal or higher than σ_f, in order to avoid development of twinned martensite.

6.5.2 Determination of Material Parameters

A successful implementation of a material model depends on the ability to express the material parameters from physically observable quantities. The material parameters entering the present model can be divided into two groups. The first group is parameters with direct physical interpretation:

$$\boldsymbol{\mathcal{S}}^i, \boldsymbol{\alpha}^i, c^i, s_0^i, u_0^i, H^t, H^d, \tag{6.5.68}$$

and parameters related to the structure of the phase diagram. The transformation hardening of the material form the second group:

$$Y_1^{\pm}, Y_2^{\pm}, Y_3, \Delta_1^{\pm}, \Delta_2^{\pm}, \Delta_3, f_1^{\pm}, f_2^{\pm}, f_3. \tag{6.5.69}$$

In both groups, the index i takes the values A, M for austenite and martensite, respectively.

The first group of parameters can be measured directly. A polycrystalline SMA, unlike the single crystal SMAs, is an isotropic material. Therefore, the compliances $\boldsymbol{\mathcal{S}}^A$, $\boldsymbol{\mathcal{S}}^M$ are determined if the Young's modulus E^A, E^M and Poisson's ratio ν^A, ν^M of the two phases are available. These can be determined from a standard uniaxial pseudoelastic test. The thermal expansion coefficient $\boldsymbol{\alpha}^A$, $\boldsymbol{\alpha}^M$ for an isotropic material are scalars and are determined from an isobaric test and the specific heats c^A, c^M, the change in specific entropy $\rho\Delta s_0$ between the two phases and the change of specific internal energy Δu_0 can be determined from calorimetric measurements [20, 22]. The maximum uniaxial transformation strain H can be obtained from either an isothermal test or from an isobaric test [20].

Next, in Sect. 6.5.1 the first six parameters in (6.5.69) were expressed using the critical temperatures M_s, M_f, A_s^t, A_f^t, A_s^d, A_f^d, as well as the critical stresses σ_s and σ_f, all of them defining the phase diagram. Therefore, the problem of determining them is reduced to finding the material phase diagram. The critical temperatures can be measured in the following way: the transformation temperatures M_s, M_f, A_s^t, A_f^t are easily determined from a DSC test, such as the one shown in Fig. 6.2. Knowing the critical temperatures at zero stress is sufficient to determine the $A \rightarrow M^t$ and $M^t \rightarrow A$ transformation strips. On the other, the A_s^d, A_f^d temperatures hand can be found by first detwinning the specimen completely. It is then mechanically unloaded in a way that preserves the material state and then a DSC test is performed as described in Sect. 6.2. To do the DSC test, it is necessary to perform the mechanical loading and the subsequent preparation of a DSC sample from the loaded specimen at temperatures below A_s^d, which may not always be possible. A more direct approach relies on several isothermal tests above A_f^t, which allow one to construct both the $A \rightarrow M^d$ and $M^d \rightarrow A$ strips and therefore, also the A_s^d, A_f^d temperatures. Isothermal tests at temperatures below M_f can be used to determine σ_s and σ_f and hence the $M^t \rightarrow M^d$ strip.

Finally, we have to determine the five hardening functions $f_1^{\pm}, f_2^{\pm}, f_3$. The function f_3 can be curve-fitted from a stress-strain relationship obtained in a standard isothermal loading test at some fixed temperature below M_f. If one assumes that the internal variable ξ_3 is proportional to the inelastic strain developed during detwinning, then equation (6.5.55) can be used to curve-fit $f_3(c_2)$. Similarly, equations (6.5.62) and (6.5.65), can be used to curve-fit the functional form of f_2^- and f_2^+, respectively. A curve-fit for $f_1^-(c_1)$ and $f_1^+(c_1)$ can be obtained by using a DSC measurement in conjunction with equations (6.5.59) and (6.5.56).

The presentation of the current model is concluded in the next two sections by demonstrating that it reproduces the phase diagram of Fig. 6.4 and by discussing certain restrictions on the relative position of the phase transformation strips.

6.5.3 The Uniaxial Transformation Strips and the Phase Diagram

The one dimensional reduction of the model (Sect. 6.5.1) resulted in the inequalities (6.5.49)-(6.5.53) for the elastic domains of the respective transformations. It is clear from equation (6.5.55) that the transformation strip in stress-temperature space for the $M^t \rightarrow M^d$ deformation is the horizontal strip

$$\sigma_s \leq \sigma \leq \sigma_f,$$

which is consistent with the assumptions of Sect. 6.3.

Next, assume for a moment that the elastic moduli of the two phases, the thermal moduli and the specific heats of the two phases are equal:

$$\boldsymbol{\mathcal{S}}^A = \boldsymbol{\mathcal{S}}^M, \quad \boldsymbol{\alpha}^A = \boldsymbol{\alpha}^M, \quad c^a = c^M.$$

In this case, equation (6.5.54) reduces to

$$\hat{\pi} = \rho\Delta s_0 T - \rho\Delta u_0.$$

Then, equation (6.5.56) implies that the transformation strip for the $A \rightarrow M^t$ is defined by

$$M_f \leq T \leq M_s$$

and from equation (6.5.59), the transformation strip for the $M^t \rightarrow A$ is the vertical region

$$A_s^t \leq T \leq A_f^t.$$

It can also be seen from equations (6.5.65) and (6.5.62) that, for any given c_2, the transformation line for both $A \rightarrow M^d$ and $M^d \rightarrow A$ transformation is linear and has slope

$$C^A = C^M = -\frac{\rho\Delta s_0}{H}.$$

Therefore, the $A \rightarrow M^d$ and $M^d \rightarrow A$ strips have the shape shown in Fig. 6.4, and the slope $C^M = C^A$ is given by the above formula.[5]

When the moduli for the two phases are different, the transformation lines for the $A \leftrightarrow M^t$ and $A \leftrightarrow M^d$ depart from the above linear relationships. However, the terms $\Delta S\sigma^2$, $\Delta\alpha\sigma$ and $\rho\Delta c\left[(T - T_0) - T\ln\left(\frac{T}{T_0}\right)\right]$, which will now appear in (6.5.54), are all an order of magnitude smaller than the leading term $\rho\Delta s_0 T$. The departure from a linear shape is therefore visible for high stress (several hundred MPa) for $A \leftrightarrow M^t$ transformation and for both higher stresses and away from the equilibrium temperature T_0 for the $A \leftrightarrow M^d$ transformation. It is easy to show that in the general case of different elastic and thermal moduli, the meaning of the slope $C^M = C^A$ now becomes the tangent to the transformation surface at zero stress.

6.5.4 Relative Position of the Transformation Surfaces

As mentioned in Section 6.3, the current model does not assume any triple point as is often done in the literature [1]. As a result, the $A \rightarrow M^d$ strip can be translated according to experimental measurements. However, certain restrictions, which result from the assumed functional dependence of f_1, f_2 and f_3 are still valid. To the best of the authors knowledge, two of these exist and will be mentioned briefly here. Both of them occur for certain specific material parameters.

First, the transformation strips $M^t \rightarrow A$, $M^d \rightarrow A$ and $M^t \rightarrow M^d$ must have a zero intersection. It is easy to show that if they do, the three inequalities (6.5.50), (6.5.52) and (6.5.53) cannot be satisfied simultaneously. In other words, a simultaneous transformation $M^t \rightarrow A$, $M^d \rightarrow A$ and $M^t \rightarrow M^d$ is

[5] This last formula is frequently used (for example, see [32]) as an alternative method to determine the difference in specific entropies Δs_0. See also Problem 3.3.

not possible. In light of the experimental results of Sect. 6.2, it seems unlikely that such a situation can occur. Also, it is physically difficult to explain why some twinned martensite will transform to austenite through an intermediate detwinned phase, while the rest of the twinned martensite will transform directly to austenite. This limitation can be removed by assuming a different functional dependence of f_1, for example on c_3. Note that if such a transformation is allowed, it is necessary to use all the three internal variables $\boldsymbol{\xi}$, because just knowing the rate of change of c_i is not sufficient to determine exactly how the species transformed (cf. equations (6.4.3)-(6.4.5)).

The second limitation of the theory is associated with a bifurcation in some loading paths (in $\boldsymbol{\sigma} - \boldsymbol{\varepsilon} - T$ space) and certain material parameters. Suppose that a material is cooled from pure austenite at constant stress, which is higher by a finite amount than the critical stress for the $M^t \to M^d$ deformation. As there is no available M^t, no M^d can be produced so that the inequality (6.5.53) is turned to equality. Suppose further, that, as the cooling proceeds, the $A \to M^t$ surface is reached and there is still some available A. That is, $\Phi_1^+ = 0$, $\Phi_2^+ = 0$ and $\Phi_3 > 0$. As the Φ_1 surface is first activated, some (small) amount of M^t is produced so that (6.5.49) becomes equality. This M^d must all be detwinned via $M^t \to M^d$ ($\dot{\xi}_3 > 0$) in order to relax the violation of (6.5.53). As this happens, c_1 again becomes 0, thus (6.5.49) is again violated, hence more M^t must be produced, and so forth, until all the austenite is exhausted via this transformation $A \to M^t \to M^d$. Thus an infinitesimal drop in the temperature, which activates Φ_1^+ (cf. equations (6.4.41) and (6.5.49)) and produces the first M^t, will result in a finite amount of $A \to M^d$. This implies a finite production of transformation strain and this results in strain discontinuity, which, to the best of the our knowledge, has not been observed in practice. Clearly, this situation is possible if the finish line for the $A \to M^d$ transformation passes above the point (M_s, σ_f) in stress temperature space. Then, as long as f_1 is a function of c_1 and f_3 a function of c_2, and regardless of the functional form of f_i (cf. equation (6.4.36)), a simple isobaric path at stress equal to $\boldsymbol{\sigma}_f$ will result in the above situation.

Hence, to prevent such behavior, it is necessary (but not sufficient) to require that the finish $A \to M^d$ line passes at or below the point (M_s, σ_f). It is easy to see that for the selected linear form of the hardening function (6.4.36) the above loading path discontinuity is not possible if the finish $A \to M^d$ line passes at or below the point (M_s, σ_f) and $\Delta_3 > \Delta_2^+$. For nonlinear functions, f_i, it is more difficult to derive sufficient conditions for which there is no discontinuity. Whether such materials, for which the finish $A \to M^d$ line passes above the point (M_s, σ_f), exist is an open question. Note that, for such class of materials a different functional dependence of f_1 may provide a solution to the discontinuity problem.

6.6 Numerical Examples

The numerical examples in this section were selected so that complex loading paths in stress-temperature phase space could be tested. First, a uniaxial example (Sect. 6.6.1) of a constrained SMA rod is considered. A one-dimensional setting allows is to easily determine the loading path in stress-temperature space, and solve the relevant equations by symbolic software. In the example, an SMA rod is cooled from the austenitic phase to low temperature while the strain is kept constrained. This problem allows us to demonstrate the cut-off of the $A \rightarrow M^d$ transformation in untrained SMA materials and the predominant development of M^t from A at low stress levels. Second, the constitutive model was then numerically implemented in the full 3-D setting and was tested on a plane strain problem. An SMA thick plate with a square hole was loaded at low temperature and then heated while constrained so that multiple transformations could take place. The results are described in Sect. 6.6.2. These examples allow us to compute aggregate quantities, such as recovery stress, which are important in the engineering analysis of SMA devices.

The basic material parameters used in all examples are given in Table 6.1 and represent generic SMA properties [32]. The selection of the same critical temperatures for $M^d \rightarrow A$ and $M^t \rightarrow A$ was done to maximize the intersection region of the two transformation strips (cf. the phase diagram of Fig. 6.4) and to test the model for multiple transformations in the most severe cases from a numerical point of view.

6.6.1 Constrained Cooling of an SMA Rod

To get a feeling of the thermomechanical response predicted by this model, a simple example is presented. Consider a rod in uniaxial stress state (6.5.47), (6.5.48). The rod is first loaded mechanically to 200 MPa from a stress-free, fully austenitic configuration, at a constant temperature of 47 °C. Then the two ends of the rod are fixed and it is cooled to −13 °C, well below the M_f temperature. The loading process is plotted in the stress-temperature space (Fig. 6.6).

Due to the uniform stress state, this problem is simple enough and can be solved semi-analytically. The stress in the rod is related to the strain by (cf.

Table 6.1. Material parameters used in the SMA model

Material constant	Value	Material constant	Value	Material constant	Value
E^A	70×10^9 Pa	H	0.05	A_f^t, A_f^d	42 °C
E^M	30×10^9 Pa	$C^M = C^A$	4.5×10^6 Pa/(m^{3}K)	A_s^t, A_s^d	22 °C
α^A	22×10^{-6} °C	M_f	2 °C	σ_s	100 MPa
α^M	10×10^{-6} °C	M_s	18 °C	σ_f	200 MPa

equations (6.4.15), (6.4.24) and (6.5.47)):

$$\sigma = E(c_1 + c_2)\left[\epsilon - \alpha(c_1 + c_2)(T - T_0) - \epsilon^{in}\right]. \qquad (6.6.70)$$

The maximal detwinning and transformation strains are the same, e.g. $H^t = H^d = H$ (see Table 6.1). In the uniaxial case, the inelastic strain is proportional to the volume fraction of detwinned martensite c_2 (cf. equations (6.4.7)-(6.4.9) and (6.5.48)):

$$\epsilon^{in} = Hc_2. \qquad (6.6.71)$$

For the particular example under consideration, the rod is initially in the austenitic phase up to a stress $\sigma_0 = 200\,\mathrm{MPa}$, which is below the critical stress required to initiate the forward, $A \to M^d$ phase transformation. Without loss of generality, let this be a tensile stress. Then, the inelastic strain is identically zero, i.e.,$\epsilon^{in} = 0$, and from equation (6.6.70), the rod has developed uniform elastic strain given by $\epsilon_0 = \sigma_0/E^A$.

At this point of the loading path, the strain is fixed and the SMA is gradually cooled. At first, a thermoelastic contraction of the rod slightly increases the stress. When the $A \to M^d$ transformation surface is reached, transformation strains begin to develop. At this point, one has to solve equations (6.6.70) and (6.6.71), along with the rule of mixtures (6.4.15) and the relevant transformation surfaces (B.2.23). This is done using symbolic software (Mathematica). The material state during the entire loading path is plotted in stress-temperature space in Fig. 6.6 and the relevant transformation strips are also shown. Observe that the rod is loaded in tension at the austenitic phase to a stress lower than required for phase transformation. The strain is then fixed and the rod is cooled. The rapid drop of the stress during the phase transformation is caused by the development of inelastic strains. Since the total achievable inelastic strain is an order of magnitude larger than the initial elastic strain, very little $A \to M^d$ transformation occurs. For clarity, only the $A \to M^d$, $A \to M^t$, $M^t \to M^d$ and transformation strips are shown.

Now, as the transformation strain becomes nonnegative, it will relax the stress state. Observe that the maximum possible value of the transformation strain H is an order of magnitude larger than the elastic strain ϵ_0, (which is kept fixed during the cooling). Therefore, very little phase transformation is required to produce transformation strains comparable with ϵ_0, and thus to drastically reduce the stress. In this example, the $A \to M^d$ surface terminates at some finite value of stress σ_s (which, as discussed before, is material-dependent). Slightly before this point, the $A \to M^t$ transformation surface is also reached and the material undergoes combined transformation.

As the stress decreases below the critical stress σ_s, only the $A \to M^t$ transformation proceeds. In the process, no further transformation strain is produced, but the stiffness changes. The stiffness of the martensite E^M is less than the stiffness of austenite E^A, so the effective stiffness decreases (cf. equation (6.4.15)). On the other hand, the total strain is fixed. Therefore, neglecting the thermal strains, and noting that a very small amount of

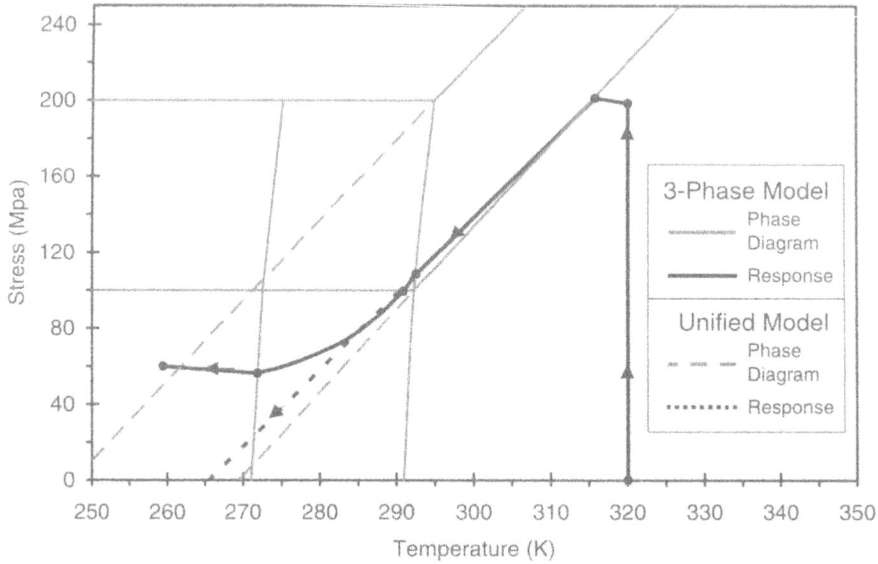

Fig. 6.6. A constrained cooling path in stress-temperature space.

$A \rightarrow M^d$ has occurred, from equation (6.6.70) it follows that the stress in the rod will decrease by a factor of E^A/E^M. This is clearly visible in Fig. 6.6. Upon completion of the transformation, the material again exhibits thermoelastic contraction, which causes small increase in stress. Additional uniaxial examples on multiple transformations while heating a constrained rod can be found in [12].

6.6.2 Thermomechanical Loading of an SMA Thick Plate with a Cylindrical Hole

The numerical example in this section involves a complex thermal and mechanical loading path applied to an SMA thick plate with a cylindrical hole. A 2-D cross-section of the geometry is shown in Fig. 6.7(a). It is assumed that in the off-plane direction the body is constrained, so that plane strain conditions are achieved in the cross-section. As is usual for plane strain, the stress is still three-dimensional, and in light of equations (6.4.45), (6.4.44) and (6.4.10), so are the inelastic strains (6.4.7). This, combined with both thermal and mechanical loading applied to the body, allows us to test the SMA model during a complex loading path. In addition, this section will show that the thermal loading, in conjunction with the plane strain conditions, also leads to evolution of a non-proportional stress state in the body.

The numerical example is solved using a displacement-based finite element method. The constitutive model was implemented numerically using return mapping algorithms (Appendix B). All discretizations use triangular

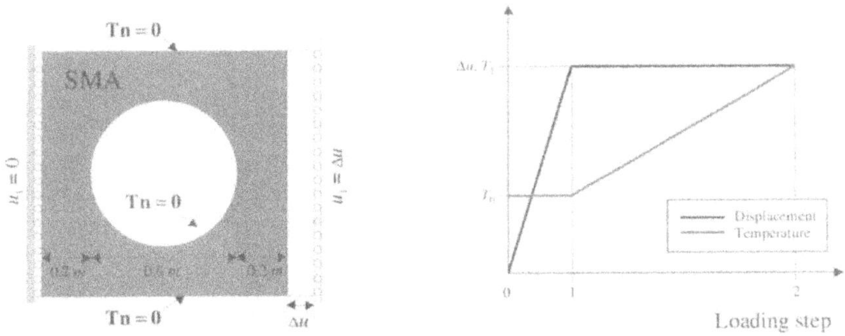

(a) Problem geometry and boundary conditions

(b) Loading path followed for the displacement boundary conditions and temperature

Fig. 6.7. A plane strain perforated SMA thick plate model problem

meshes and standard linear Lagrangian finite element spaces. The SMA material properties used represent generic NiTi SMA material used previously in the literature [32] and are given in Table 6.1.

Initially, the SMA thick plate is stress-free and in the self-accommodated phase, i.e. $c_1 = 1$ everywhere in the domain. The thermomechanical loading that the square is subjected to is shown in Fig. 6.7(b). The thick plate is first loaded mechanically at constant temperature $T = -13\,°\mathrm{C}$ as follows: the left side of the square is fixed against horizontal displacement, but is allowed to roll in the vertical direction; the right side is pulled uniformly by the amount of $0.002m$ in the horizontal direction and the side is again allowed to move in the vertical direction; the remaining part of the boundary (including the hole) is stress-free. The second loading step consists of keeping the horizontal component of the displacement fixed and uniformly raising the temperature to $T = 77\,°\mathrm{C}$.

The first loading step was used to determine a suitable mesh size for the entire simulation. This was done by starting with a very coarse mesh (Fig. 6.8) and then consecutively refining it (Fig. 6.9–6.12). Due to obvious symmetry consideration, only one quarter of the domain was used in the calculations. The stresses are shown as is, i.e. piecewise-constant over each element. The solution was judged to be accurate enough when the non-smoothed, piecewise-constant stress components in each element showed little variation over element boundaries. Based on this, the mesh shown in Fig. 6.10 (8964 elements, 9274 DOF) was selected for the rest of the computation. While the solution shown in Fig. 6.12 (110793 elements, 111994 DOF) is clearly the best, in complicated nonlinear problems, such as the ones involving SMAs, the cost of assembly and the memory requirements needed to save the material state at each integration point make it desirable to keep the number of elements at a

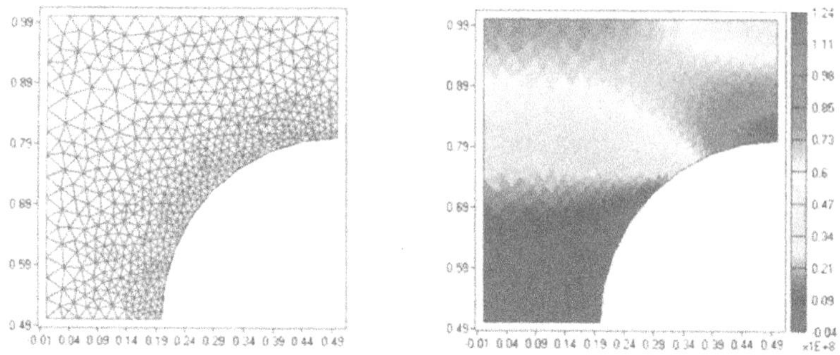

(a) Mesh, $h \sim 0.04$, 1286 DOF (b) σ_{11}

Fig. 6.8. Initial mesh, $T = 260$K, end of first loading step.

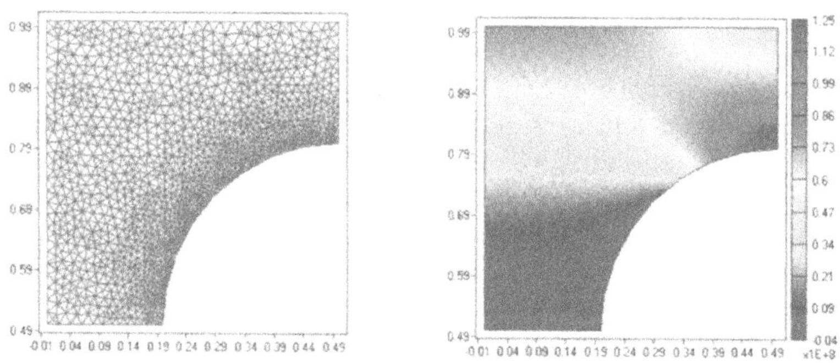

(a) Mesh, $h \sim 0.02$, 3118 DOF (b) σ_{11}

Fig. 6.9. First refinement, $T = 260$K, end of first loading step.

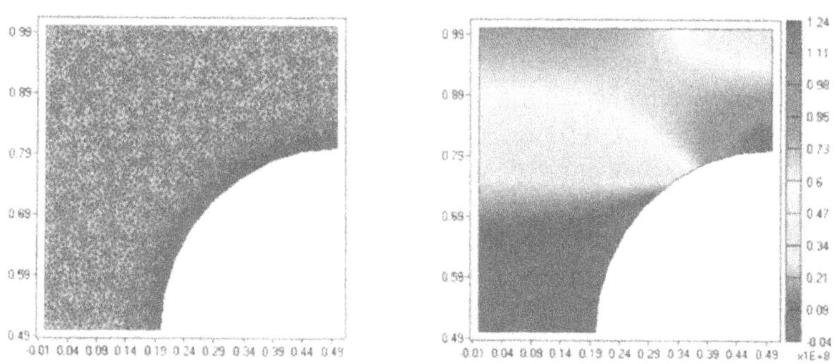

(a) Mesh, $h \sim 0.01$, 9274 DOF (b) σ_{11}

Fig. 6.10. The triangular FEM mesh selected for the entire computation and associated stresses and internal variable at the end of the first loading step ($T = 260$K).

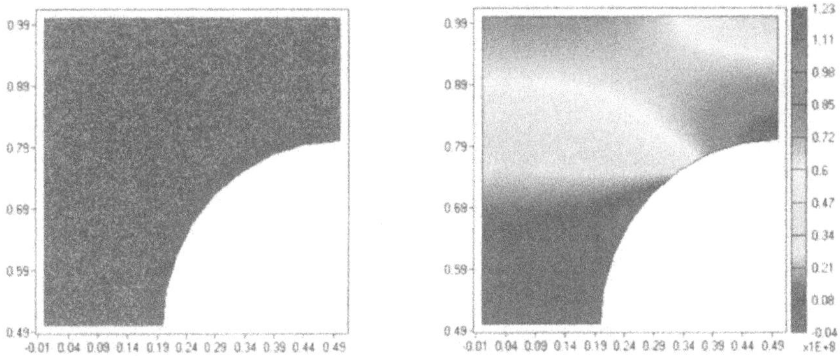

(a) Mesh, $h \sim 0.005$, 31054 DOF (b) σ_{11}

Fig. 6.11. Third refinement, $T = 260\text{K}$, end of the first loading step.

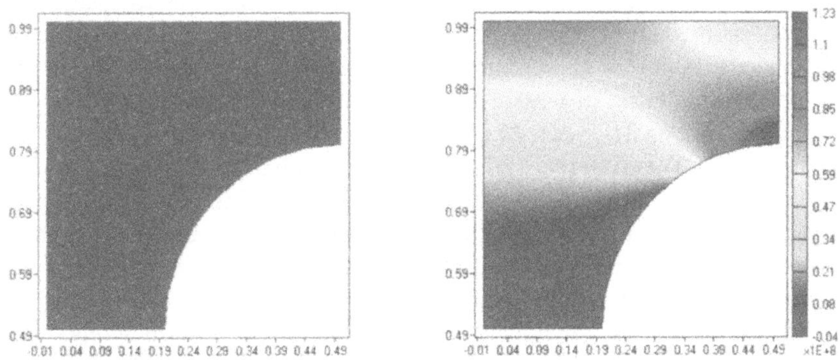

(a) Mesh, $h \sim 0.0025$, 111994 DOF (b) σ_{11}

Fig. 6.12. Fourth refinement, $T = 260\text{K}$, end of first loading step.

minimum. In this respect, the second refinement (8964 elements, 9274 DOF) is acceptable enough and was used in the rest of the simulations.

It can be seen from this first part of the simulation that stress concentrations develop near the top and bottom edge of the hole. The effective stress in these locations become sufficiently high for the detwinning of small amounts of self-accommodated martensite, as shown on Fig. 6.13. If the stress concentrations are compared to a pure elastic solution, the development of inelastic strains in the detwinning process tends to reduce this stress concentration. Note that as the heating begins, the material undergoes only thermal expansion and no transformation occurs. As a result, the von Mises stress decreases slightly by $T = 280\text{K}$ and the internal variables do not change. The heating step is shown in Fig. 6.14. As the material is slowly heated, the reverse

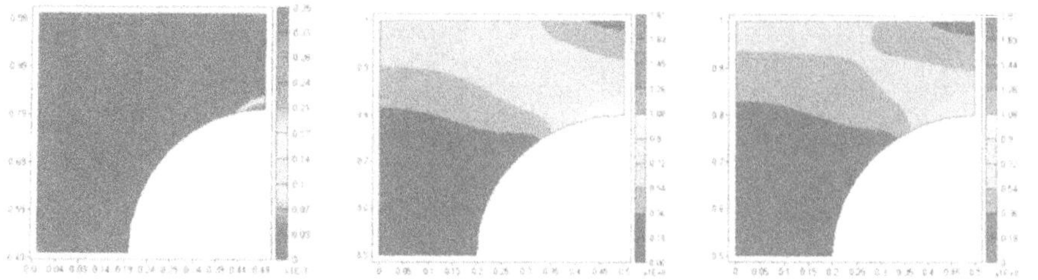

(a) c_2 at $T = 260, 280$K (b) von Mises Stress, T = 260K (c) von Mises Stress, T = 280K

Fig. 6.13. The martensitic volume fraction (a) and the von Mises stress (b) at the completion of the initial loading.

transformation $M^t \to A$ occurs (c). Since the austenite is considerably stiffer than martensite, and the material is constrained, the stresses also increase. This results in a simultaneous $M^t \to A, M^d$ transformation, which is manifested in an increase in the volume fraction of M^d

During the heating process, the material undergoes two distinct processes – first simultaneous $M^t \to A, M^d$ transformations then followed by a $M^d \to A$ transformation. When the second loading step begins, the material first experiences initial linear thermoelastic expansion. The conditions are of plane strain: the horizontal component of the displacements is fixed and the vertical displacements are not constrained during the heating. Hence, any thermal expansion will result in a nonhomogeneous change in the stress state. During the linear thermoelastic expansion ($T < A_s^t$), this results in small relaxation of the effective stress (Fig. 6.13(c)). As the critical temperature for the $M^t \to A$ transformation is reached, the self-accommodated martensite begins to transform to austenite. The stiffness of austenite is approximately 2.3 times that of martensite (see Table 6.1) and due to the fixed horizontal displacement, the stresses increase throughout the thick plate. The effective stress increases correspondingly and this causes further detwinning of martensite in some areas of the thick plate, resulting in a simultaneous $M^t \to A, M^d$ transforma-

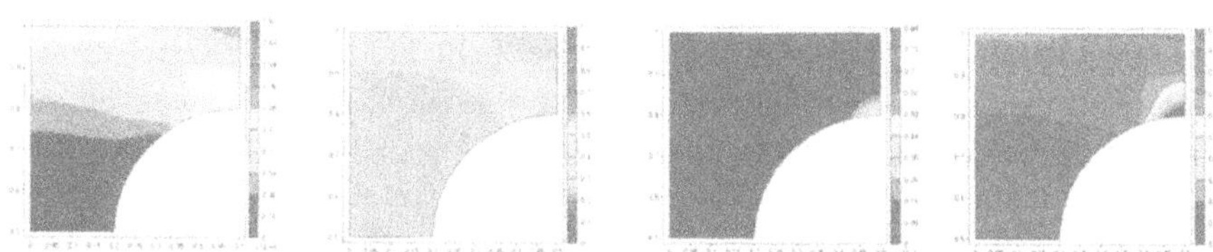

(a) von Mises stress (b) c_1 (M^t) (c) c_2 (M^d) (d) c_3 (A)

Fig. 6.14. Solution for square hole problem at $T = 310$K.

tions, mostly near the top and bottom parts of the hole (Fig. 6.14). This is manifested further as a increase in c_2 in comparison to the amount that was produced during the first loading step. The effective stress and c_2 after the completion of the $M^t \rightarrow A$ transformation are shown in Fig. 6.15.

The areas where detwinned martensite is present have generally higher effective stress (above σ_s), compared to the rest of the domain. Hence, the reverse transformation of detwinned martensite ($M^d \rightarrow A$) does not happen until a much higher temperature, because the corresponding transformation surfaces exhibit stress dependence (see Fig. 6.4). Around $T = 335$K, the $M^d \rightarrow A$ transformation begins in areas with lowest effective stress. During this phase, the inelastic strains decrease according to the transformation rule (6.4.8),(6.4.45). Again, due to the constraint on the displacements imposed by the boundary conditions, the elastic portion of the stress generally increases, which leads to a corresponding (non-uniform) increase in the stress during the reverse transformation (Fig. 6.16–6.18).

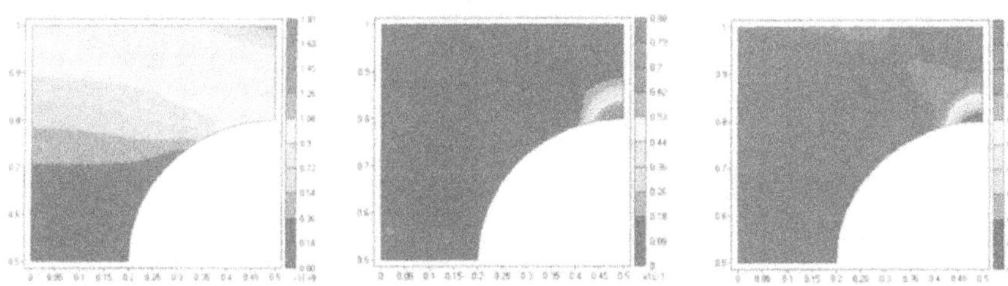

(a) von Mises Stress (b) c_2 (M^d) (c) c_3 (A)

Fig. 6.15. By $T = 320$K the twinned martensite is depleted. The volume fraction of M^d (b) has reached approximately 0.11 and the rest is in the austenitic phase (c).

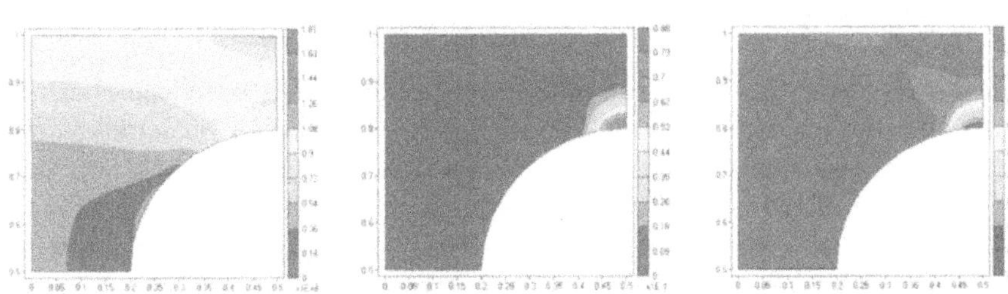

(a) von Mises Stress (b) c_2 (M^d) (c) c_3 (A)

Fig. 6.16. Around $T = 335$K the temperature is sufficient for the reverse $M^d \rightarrow A$ transformation to begin.

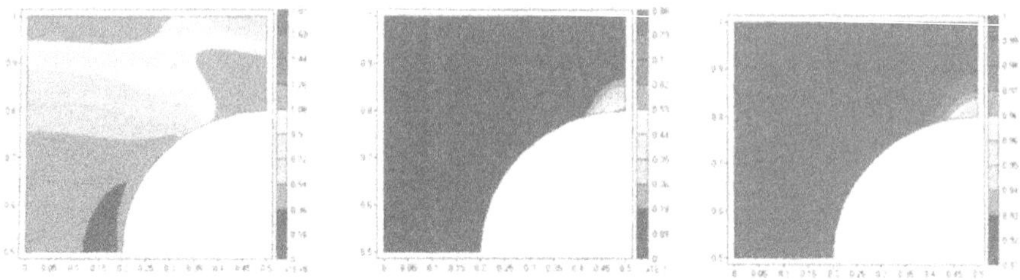

(a) von Mises Stress (b) c_2 (M^d) (c) c_3 (A)

Fig. 6.17. Further heating ($T = 340$K) causes further $M^d \to A$ transformation and subsequent decrease in c_2 (b), and the inelastic strain ε^{in}, and increase of c_3 (c). Due to the constrained displacement and the decrease of the inelastic strain, the stresses begin to increase (a).

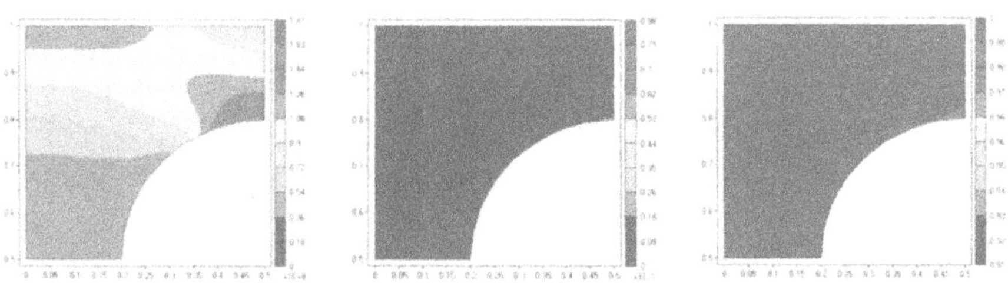

(a) von Mises Stress (b) c_2 (M^d) (c) c_3 (A)

Fig. 6.18. The $M^d \to A$ is complete almost everywhere in the domain by the time the temperature reaches $T = 350$K (cf. (c)). Note that the maximum value for the effective stress has reached increased to approximately 166 MPa (cf. (a)).

6.7 Summary

A new 3-D constitutive model for polycrystalline SMAs based on thermodynamic potentials was presented. The model can account for both development of stress-induced martensite directly from austenite (pseudoelasticity), as well as detwinning of twinned martensite. This is accomplished by describing the material state as a mixture of three phases – twinned martensite, detwinned martensite and austenite, and by using the three possible "reactions" between these phases as internal variables. The model is made consistent with a modified stress-temperature phase diagram. A key new experimental finding is the existence of separate reverse transformation temperatures for detwinned and twinned martensite. This is obtained through a series of calorimetric measurements and is incorporated in the model. The constitutive model was numerically implemented using a return mapping algorithm. The implementation was integrated into a numerical environment and tested for several model problems.

6.8 Problems

6.1. With the 1-D model derived in Sect. 6.5.1, model the thermomechanical response of the SMA material described in Table 6.1 subjected to three different loading conditions:

1. Isothermal loading to 600 MPa at a temperature above A_f.
2. Isothermal loading to 600 MPa at a temperature between A_s and A_f.
3. Cooling from A_f to a temperature below M_f followed by isothermal loading to 600 MPa.

Compare the results to those obtained using the 1-D form of the polynomial hardening model presented in Chapter 3.

6.2. Using the model derived in 6.5.1 and the material properties from Table 6.1, evolute the thermomechanical response of an SMA bar specimen under the following conditions:

1. The SMA bar is cooled from A_f to below M_f at a constant stress greater than σ^f.
2. The SMA bar is cooled from A_f to below M_f at a constant stress less than σ^f.

6.3. Using the constitutive model presented in Chapter 3 and in this chapter, and the material properties from Table 6.1, reproduce the results shown in Fig. 6.6.

6.4. An SMA wire is wrapped around a cylinder after having been prestrained while in martensite at temperature $T = M_f - 10\,°\mathrm{C}$ by a detwinning deformation resulting in $\varepsilon^d = 0.04$, and has zero stress in this initial configuration. Assuming a rigid cylinder and no friction between the SMA and the cylinder, find the axial stress in the wire as a function of temperature when the SMA is heated to $T = A_f + 50\,°\mathrm{C}$ and then cooled to its initial temperature T_0.

1. Plot the thermomechanical loading path on the stress-temperature space, similar to Fig. 6.6 and indicate the phase transformation lines on the same plot.
2. Does the final stress return to zero again on cooling to T_0 or does the SMA return to a different stress state?
3. How does the extended SMA model of this chapter differ from the one presented in Chapter 3 for the prediction of the final state upon returning to T_0?

6.5. Rework problems 3.8, 3.9 and 3.11 in Chapter 3 using the extended SMA model presented in this chapter.

6.6. The selection made for the independent variables of the hardening functions in this chapter are not the only possible ones. The current selection imposes certain constraints on the arrangement of the transformation regions in the phase diagram. Other choices, for example a dependence of the $A \leftrightarrow M^t$ hardening function on the total amount of austenite rather than twinned martensite, are possible and could be explored. Discuss the modifications necessary in the extended SMA model presented in this chapter to account for such a choice of independent variables in the hardening functions.

6.7. The phase diagram is expected to evolve as the material is cycled through a certain thermomechanical loading path. The evolution could be better understood in the case of cyclic pseudoelastic loading paths. Propose a reformulation of the SMA model presented in this chapter to account for such an evolution in the regions of the phase diagram.

References

[1] L. C. Brinson, One-dimensional constitutive behavior of shape memory alloys: Thermomechanical derivation with non-constant material functions and redefined martensite internal variable, J. of Intell. Mater. Syst. and Struct. 4 (1993) 229–242.

[2] L. C. Brinson, R. Lammering, Finite element analysis of the behavior of shape memory alloys and their applications, International Journal of Solids and Structures 30 (23) (1993) 3261–3280.

[3] J. G. Boyd, D. C. Lagoudas, Thermomechanical response of shape memory composites, in: N. W. Hagood, G. J. Knowles (Eds.), Proc. 1993 Smart Structures and Materials Conf., Smart Structures and Intelligent systems, Vol. 1917, SPIE, 1993, pp. 774–790.

[4] A. Bekker, L. C. Brinson, Temperature-induced phase transformation in a shape memory alloy: Phase diagram based kinetics approach, Journal of the Mechanics and Physics of Solids 45 (6) (1997) 949–988.

[5] A. Bekker, L. C. Brinson, Phase diagram based description of the hysteresis behavior of shape memory alloys, Acta Materialia 46 (10) (1998) 3649–3665.

[6] S. Leclercq, C. Lexcellent, A general macroscopic description of the thermomechanical behavior of shape memory alloys, Journal of the Mechanics and Physics of Solids 44 (6) (1996) 953–980.

[7] D. Lagoudas, S. Shu, Residual deformations of active structures with SMA actuators, International Journal of Mechanical Sciences 41 (1999) 595619.

[8] L. Juhasz, E. Schnack, O. Hesebeck, H. Andra, Macroscopic modeling of shape memory alloys under non-proportional thermo-mechanical loadings, Journal of Intelligent Material Systems and Structures 13 (2002) 825–836.

[9] P. Popov, D. C. Lagoudas, A 3-D constitutive model for shape memory alloys incorporating pseudoelasticity and detwinning of self-accommodated martensite, International Journal of Plasticity 23 (10–11) (2007) 1679–1720.

[10] K. Otsuka, C. M. Wayman, Shape Memory Materials, Cambridge University Press, Cambridge, 1999, Ch. Introduction, pp. 1–26.

[11] H. Sakamoto, Distinction between thermal and stress-induced martensitic transformations and inhomogeneity in internal stress, Materials Transactions 43 (9) (2002) 2249–2255.

[12] P. Popov, Constitutive modelling of shape memory alloys andupscaling of deformable porous media, Ph.D. thesis, Texas A&M University (2005).

[13] W. B. Cross, A. H. Kariotis, F. J. Stimler, Nitinol characterization study, Tech. Rep. CR-1433, NASA (1969).

[14] C. M. Jackson, H. J. Wagner, R. J. Wasilewski, 55-nitinol—The alloy with a memory: Its physical metallurgy, properties and applications, Tech. Rep. NASA SP-5110, NASA, Technology Utilization Office, Washington, D.C. (1972).

[15] P. Wollants, M. De Bonte, J. Roos, A thermodynamic analysis of the stress-induced martensitic transformation in a single crystal, Z. Metallkd. 70 (1979) 113–117.

[16] P. Šittner, D. Vokoun, G. Dayananda, R. Stalmans, Recovery stress generation in shape memory Ti50Ni45Cu5 thin wires, Material Science and Engineering A 286 (2000) 298–311.

[17] T. Kotil, H. Sehitoglu, H. Maier, Y. Chumlyakov, Transformation and detwinning induced electrical resistance variations in NiTiCu, Materials and Engineering A359 (2003) 280–289.

[18] P. Šittner, P. Lukáš, D. Neov, D. Lugovyy, Martensitic tranformations in NiTi polycryctals investigated by in-situ neutron diffraction, Materials Science Forum 426-432 (2003) 2315–2320.

[19] K. Tsoi, R. Stlmans, J. Schrooten, Transformational behavior of constrained shape memory alloys, Acta Materialia 50 (2003) 3535–3544.

[20] Z. Bo, D. C. Lagoudas, Thermomechanical modeling of polycrystalline SMAs under cyclic loading, Part I: Theoretical derivations, International Journal of Engineering Science 37 (1999) 1089–1140.

[21] D. Miller, Thermomechanical characterization of plastic deformation and transformation fatigue in shape memory alloys, Ph.D. thesis, Texas A&M University (2000).

[22] Z. Bo, D. C. Lagoudas, Thermomechanical modeling of polycrystalline SMAs under cyclic loading, Part III: Evolution of plastic strains and two-way memory effect, International Journal of Engineering Science 37 (1999) 1175–1204.

[23] Z. Bo, D. C. Lagoudas, Thermomechanical modeling of polycrystalline SMAs under cyclic loading, Part IV: Modeling of minor hysteresis loops, International Journal of Engineering Science 37 (1999) 1205–1249.

[24] D. C. Lagoudas, Z. Bo, Thermomechanical modeling of polycrystalline SMAs under cyclic loading, Part II: Material characterization and experimental results for a stable transformation cycle, International Journal of Engineering Science 37 (1999) 1205–1249.

[25] F. Auricchio, R. L. Taylor, J. Lubliner, Shape-memory alloys: Modelling and numerical simulations of the finite-strain superelastic behavior, Computer Methods in Applied Mechanics and Engineering 143 (1997) 175–194.

[26] M. A. Qidwai, D. C. Lagoudas, On thermomechanics and transformation surfaces of polycrystalline NiTi shape memory alloy material, International Journal of Plasticity 16 (2000) 1309–1343.

[27] L. Anand, M. Gurtin, Thermal effects in the superelasticity of crystalline shape-memory materials, Journal of the Mechanics and Physics of Solids 51 (6) (2003) 1015–1058.

[28] D. C. Lagoudas, Z. Bo, M. A. Qidwai, A unified thermodynamic constitutive model for SMA and finite element analysis of active metal matrix composites, Mechanics of Composite Materials and Structures 3 (1996) 153–179.

[29] B. Coleman, M. Gurtin, Thermodynamics with internal state variables, The Journal of Chemical Physics 47 (1967) 597–613.

[30] D. C. Lagoudas, P. B. Entchev, Modeling of transformation-induced plasticity and its effect on the behavior of porous shape memory alloys. Part I: Constitutive model for fully dense SMAs, Mech. Mater. 36 (9) (2004) 865–892.

[31] J. G. Boyd, D. C. Lagoudas, Thermomechanical response of shape memory composites, Journal of Intelligent Material Systems and Structures 5 (1994) 333–346.

[32] M. A. Qidwai, D. C. Lagoudas, Numerical implementation of a shape memory alloy thermomechanical constitutive model using return mapping algorithms, International Journal for Numerical Methods in Engineering 47 (2000) 1123–1168.

[33] F. Auricchio, R. L. Taylor, J. Lubliner, Shape-memory alloys: Macromodelling and numerical simulations of the superelastic behavior, Computer Methods in Applied Mechanics and Engineering 146 (1997) 281–312.

[34] Y. Gillet, E. Patoor, M. Berveiller, Calculation of pseudoelastic elements using a non symmetrical thermomechanical transformation criterion and associated rule, Journal of Intelligent Materials and Technology 9 (1998) 366–378.

[35] W. Huang, Yield surfaces of shape memory alloys and their applications, Acta Materialia 47 (9) (1999) 2769–2776.

[36] C. Lexcellent, A. Vivet, C. Bouvet, S. Calloch, P. Blanc, Experimental and numerical determinations of the initial surface of phase transformations under biaxial loading in some polycrystalline shape-memory alloys, Journal of the Mechanics and Physics of Solids (to appear).

7

Modeling of Magnetic SMAs

B. Kiefer and D. C. Lagoudas

This chapter discusses a new category of shape memory alloys, namely magnetic shape memory alloys (MSMAs). The unique magnetomechanical constitutive behavior that these materials exhibit is described in detail and is explained by connecting the macroscopic response to micro-scale mechanisms. A constitutive model for MSMAs is introduced, which follows the same phenomenological approach used in Chapters 3–6 for the modeling of conventional shape memory alloys. The modeling of MSMAs poses new challenges, since magnetomechanical coupling effects have to be addressed and additional independent and internal state variables have to be incorporated.

7.1 Introduction

Magnetic shape memory alloys (MSMAs), often referred to as ferromagnetic shape memory alloys (FSMAs) [1–3], have recently emerged as a new type of active and multifunctional material. In these alloys, recoverable strains of up to 10% can be induced by magnetic fields, which is one or even two orders of magnitude larger than the *magnetic-field-induced strains* observed in ordinary magnetostrictive materials, such as Terfenol-D [4] and Galfenol [5]. These strains are also much larger than the electric field-induced strains in piezoelectrics [6]. The macroscopically observable field-induced strains in MSMAs are caused by the microstructural *reorientation of martensitic variants*. Since the variants have different preferred directions of magnetization, applied magnetic fields can be used to select certain variants over others, which results in the macroscopic shape change. MSMAs also exhibit conventional temperature- or stress-activated shape memory behavior [7–10]. At comparable recoverable strains, however, they have an advantage over conventional shape memory alloys due to the much higher frequency range over which they can be operated, up to 1 kHz [11]. This is because their actuation is driven by the magnetic-field-induced reorientation of martensitic variants and is not limited by heat transfer [12].

D.C. Lagoudas (ed.), *Shape Memory Alloys*, DOI: 10.1007/978-0-387-47685-8_7,

The main limitation of MSMAs is the relatively low blocking stress level of typically 6–10 MPa, above which magnetic-field-induced strains are completely suppressed [1, 13]. The field-induced strain response of MSMAs is nonlinear, hysteretic, stress-dependent, and intrinsically coupled to the magnetization response of the material. The coupled macroscopic response is driven by three mechanisms: the motion of magnetic domain walls, the local rotation of magnetization vectors (both of which also occur in regular ferromagnetic materials [14–16]), and field-induced variant reorientation.

This unique coupling of mechanical and magnetic properties (and thermal properties if one considers the conventional shape memory behavior) makes MSMAs interesting materials for smart structures, as well as actuator [11, 17] and sensor applications [18, 19]. A different class of applications take advantage of the unique and adjustable magnetic properties of MSMAs in solenoid transducers [2] and voltage generators [20].

The most widely investigated magnetic shape memory materials have been Ni-Mn-Ga alloys [21]. Martensitic transformations in Ni_2MnGa alloys were first conclusively reported by Webster et al. [22]. Zasimchuk et al. [23] and Martynov and Kokorin [9] performed detailed studies on the crystal structure of martensite in the Ni_2MnGa alloy. Ullakko et al. [4] are credited with first suggesting the possibility of a magnetic field-controlled shape memory effect in these materials. They observed magnetic-field-induced strains of nearly 0.2% in stress-free experiments on martensitic Ni_2MnGa single crystals. Further work on off-stoichiometric intermetallic compounds near the composition Ni_2MnGa, in combination with thermomechanical treatments and the utilization of a better understanding of the crystallographic structure of these alloys, has yielded larger field-induced strains of 6% [1] and up to 10% [24] in single crystals. Other magnetic shape memory alloys have been studied including Fe-Pd [25–28], Fe-Ni-Co-Ti, Fe-Pt, Co-Ni-Ga, Ni-Mn-Al [29–34] and Co-Ni-Al [10, 35]. These alloys exhibit lower field-induced strains, but they can have other advantages. The largest field-induced strains that have been observed in Fe-Pd, for example, are 3.1% [28, 36], but this material is much more ductile than Ni-Mn-Ga [25].

The magnetic-field-induced strains that can be generated in polycrystalline magnetic shape memory alloys are smaller than those observed for single crystals [37–41]. One effort aimed toward increasing the strain output of polycrystals is based on creating favorable texture in these materials. Marioni et al. [42] calculated the upper bound for the achievable field-induced strain in untextured NiMnGa polycrystals to be 21% of the single-crystal value and at most 50% for textured crystals.

The phenomenon of magnetic-field-induced austenite-martensite phase transformations has also been investigated. Such transformations have been observed in Fe-Pt [43], Ni-Mn-Ga [44, 45] and Ni-Mn-Fe-Ga [37] alloys. Magnetic fields have also been shown to influence the temperature- or stress-induced austenite-martensite phase transformation in MSMAs [37]. Furthermore, Ni-Mn-Ga alloys have exhibited several different martensite morphologies and thus intermartensitic phase transformations [8, 46, 47]. The work presented here, however, is focused on the well-established MSMA behavior as caused by the magnetic-field-induced martensitic variant reorientation.

7.2 Properties of Magnetic SMAs

A general introduction to the constitutive behavior of MSMAs was given in the previous section. In this section a more detailed description of the connection between the evolving crystallographic and magnetic microstructure of MSMAs and the observed macroscopic response is provided. This knowledge will then be used to motivate the formulation of the constitutive model.

7.2.1 Magnetic-Field-Induced Strain Response of MSMAs

Since the ternary intermetallic compound Ni-Mn-Ga is the most widely investigated magnetic shape memory alloy, it shall be the focus of the following discussion, which does not imply that the basic concepts or the modeling approach presented in this work are restricted in any way to this particular alloy.

The high temperature austenite phase of Ni-Mn-Ga alloys near the composition Ni_2MnGa exhibits a $L2_1$ Heusler-type structure, in which all of the atoms are located in the sites of a body-centered cubic lattice [22]. The austenite phase is paramagnetic above the Curie temperature, which for the stoichiometric composition of Ni_2MnGa is 376 K [44], and ferromagnetic below the Curie temperature. The Curie temperature shows only a slight variation with changes in the composition [44, 48]. A strong compositional dependence, however, is observed for the austenite-martensite phase transformation start temperature [48–50], which is 202 K in stoichiometric Ni_2MnGa [44]. The martensite in these alloys can be of five-layered tetragonal (5M), seven-layered orthorhombic (7M), and non-modulated tetragonal martensite (NM) morphology [8, 46, 47]. Here only the most commonly observed tetragonal martensite of Ni_2MnGa is considered.

A simplified representation of the crystal structure, which is usually adopted for convenience [51, 52], is shown in Fig. 7.1[1]. The undeformed austenite has cube edges of length a_0, whereas the undeformed tetragonal martensite unit cell has short and long edges of lengths a and c, respectively. Typical lattice parameters for Ni_2MnGa have been reported [22, 23, 53–55].

Since this transition temperature is well below the Curie temperature, the martensitic phase is ferromagnetic such that, even in the absence of an external magnetic field, the martensitic variants are *spontaneously magnetized* [14, 16].

[1] Reprinted from *Philosophical Magazine*, Special Issue, Vol. 85, Issues 33–35, Kiefer, B. and Lagoudas, D.C., pp. 4289–4329, Copyright 2005, with permission from Taylor & Francis, `http://www.informaworld.com`.

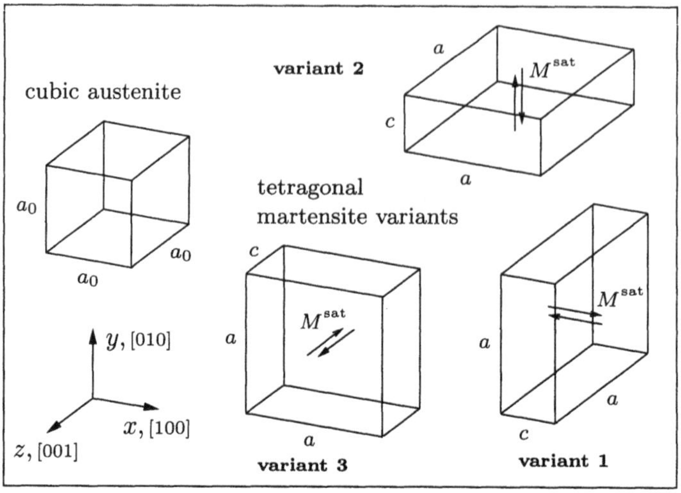

Fig. 7.1. Crystal structure of the austenitic and the tetragonal martensite phases in Ni_2MnGa. Arrows indicate possible magnetization vector orientations along the magnetic easy axis of each variant.

The local magnetization vector in each ferromagnetic variant is oriented along one preferred crystallographic direction named the *magnetic easy axis*, which in this case is aligned with the short edge c of the tetragonal unit cell. The magnetization vectors can be oriented in either the positive or negative easy axis direction.

If an external field is applied, it is energetically favorable for the magnetization vectors to align with the applied field. In MSMAs, three competing mechanisms are available to achieve this alignment. The first two, the *magnetic domain wall motion* and the *magnetization vector rotation*, are common to all ferromagnetic materials and shall be discussed shortly. The third mechanism, which is unique for magnetic shape memory alloys, is the magnetic field-driven *reorientation of martensitic variants*. This is possible since the preferred axes of the tetragonal variants are mutually perpendicular, such that an external magnetic field can be used to favor certain variants over others. The induced redistribution of variants leads to the observed macroscopic shape change.

Fig. 7.2 on the facing page schematically illustrates a typical thermo-magneto-mechanical loading sequence of an experiment in which the magnetic-field-induced strain response of MSMAs is measured[1]. Initially, the single crystalline MSMA specimen is cooled to induce the austenite to martensite phase transformation. Then a sufficiently high compressive stress is applied to produce a single variant configuration. In the depicted case, the sample has been cut such that the [100]-direction of the austenitic crystal aligns with the compression axis, which is also denoted the x-axis. The compression along this axis in the martensitic state favors variant 1 (for nomenclature see Fig. 7.1), since its short axis is along the x-direction, and the two other tetragonal

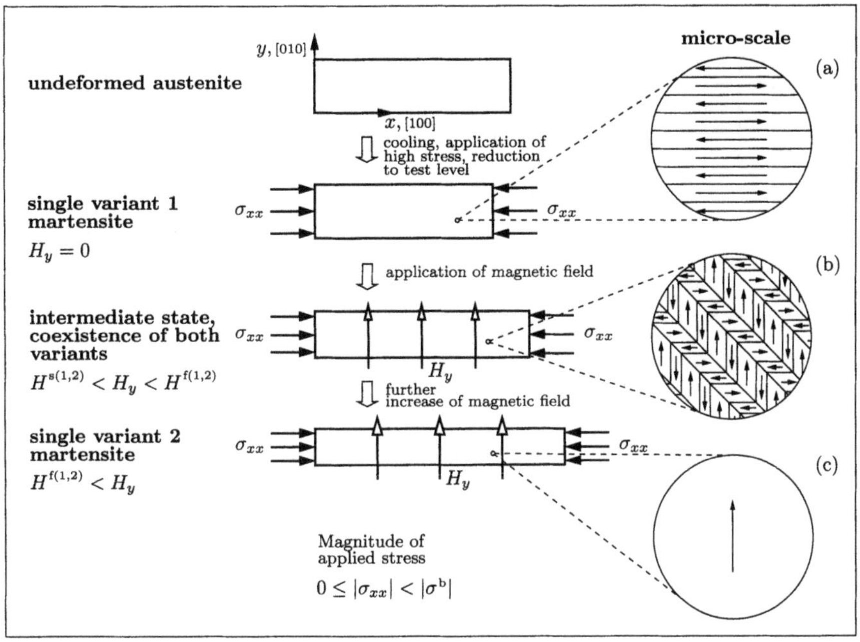

Fig. 7.2. Schematic of a typical loading sequence for magnetic-field-induced strain measurements in MSMA [56–59]. Inserts (a)–(c) depict the arrangement of martensitic variants and magnetic domains at different loading stages.

variants are eliminated. The stress is then lowered to the desired test level, during which the single variant state is preserved, and kept constant for the remainder of the experiment. The upper stress bound above which the variant reorientation is completely suppressed is called the *blocking stress* σ^b [56].

The micro-scale schematic (a) of Fig. 7.2 shows the described stress-induced single martensitic variant configuration, before a magnetic field is applied. Also depicted are magnetic domains, i. e., regions of uniform magnetization, which are separated by 180° magnetic domain walls, indicated by horizontal lines. These magnetic domains form to minimize the magnetostatic energy of the configuration [14, 15, 60, 61]. In each domain, the magnetization vectors are aligned with the magnetic easy axis of variant 1, which according to Fig. 7.1 coincides with the c-edge of the tetragonal variant and the [100]-direction. These magnetization vectors point in either the positive or negative coordinate direction. Since no external magnetic field is applied, the two domains are of equal volume fraction such that the macroscopic magnetization vanishes.

The application of a magnetic field perpendicular to the direction of mechanical loading induces the nucleation of variant 2 once a critical threshold is reached. This variant is favored by the magnetic field because its easy axis is aligned with the y-direction. The mechanism of twin-boundary motion through variant reorientation is illustrated in more detail in the schematic

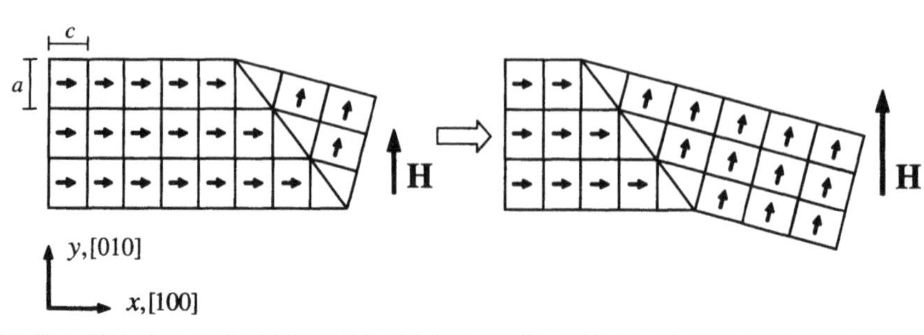

Fig. 7.3. Magnetic-field-induced twin boundary motion.

of Fig. 7.3. With increasing H_y, variant 2 grows at the expense of stress-favored variant 1. This results in an elongation of the sample along the x-axis, because the long edge a of variant 2 replaces the short edge c of variant 1 along this direction. The magnetic-field-induced strain is usually measured with respect to the initial single variant state. Typical magnetic-field-induced strain response curves are schematically plotted in Fig. 7.4[1]. Note that the vertical axis marks the total strain, not just the reorientation strain. The initial strains for the hysteresis loops at each stress level are non-zero, since the undeformed austenite was assumed as the reference configuration [57], and are comprised of elastic strain as well as austenite to partially-twinned martensite transformation strain [62, 63].

The reorientation from the stress-favored into the magnetic-field-favored variant shows a strong stress level dependence. Fig. 7.4 also explains the notation for the critical magnetic fields for the start of the forward reorientation process (variant 1 $\rightarrow$ variant 2), which is denoted $H^{s(1,2)}$ and its finish $H^{f(1,2)}$. Equivalently, $H^{s(2,1)}$ and $H^{f(2,1)}$ are defined as the critical magnetic fields to start and finish the reverse (variant 2 $\rightarrow$ variant 1) reorientation process.

The arrangement of twinned martensitic variants and magnetic domains after the activation of the reorientation process is sketched Fig. 7.2b for a generic intermediate applied magnetic field level $H^{s(1,2)} < H_y < H^{f(1,2)}$. Again, horizontal and vertical lines indicate 180° magnetic domain walls. Slanted lines indicate twin boundaries [64], which coincide with 90° domain walls. The variant reorientation process can also be visualized as occurring through the motion of these twin boundaries. Two types of magnetic domains are present in each twin band.

The volume fraction of the domain in which the magnetization vector opposes the applied field is less than that of the favorable domain, which yields a non-zero macroscopic magnetization. In large fields, the unfavorable domains are eliminated [65]. The magnetization of the material also changes due to the simultaneously occurring rotation of the magnetization vectors in variant 1 away from the magnetic easy axis. These rotations are not explicitly shown in the schematic, but they will be addressed in detail in Sect. 7.2.2.

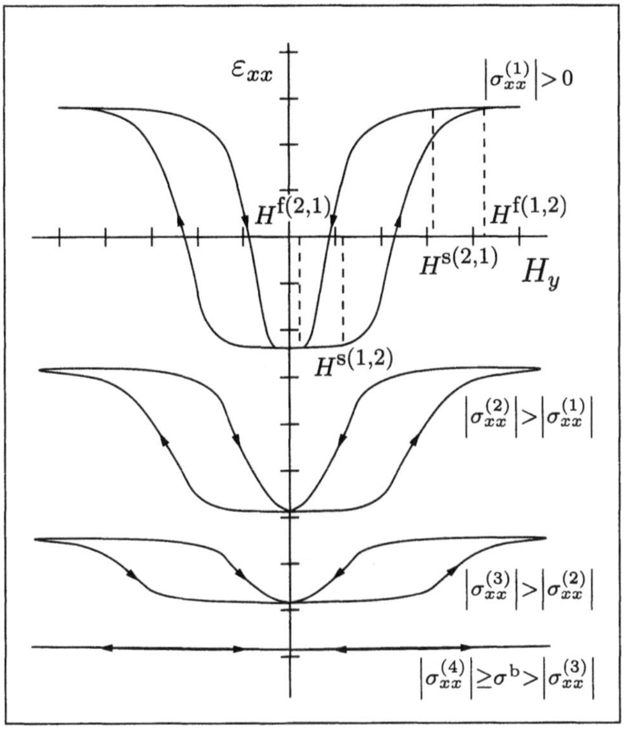

Fig. 7.4. Typical total strain-magnetic field response curves for MSMAs at different stress levels.

At low stress levels, further increase of the magnetic field completely eliminates variant 1 at the threshold value of $H^{f(1,2)}$ such that the maximum reorientation strain is achieved. If the field is increased above $H^{f(1,2)}$, the strain remains constant. The resulting single variant, single domain configuration is depicted in Fig. 7.2b. At most stress levels, however, variant 1 is not completely eliminated, such that the achievable magnetic-field-induced strain is reduced. At the blocking stress, the field-induced strain is completely suppressed. The mechanism that causes the stress level dependence of the achievable reorientation strain will be discussed in Sect. 7.4.

A typical experiment designed to apply the compressive stress and perpendicular magnetic field loading conditions to MSMA single crystals, as previously discussed and schematically shown in Fig. 7.2, and to measure the resulting magnetic-field-induced strain response, was previously reported [58, 59]. The setup described therein is shown in Figs. 7.5[2] and 7.6[2] and consists of a 2 T electromagnet, which is adjustably mounted on a mechanical

[2] Reprinted from *Journal of Magnetism and Magnetic Materials*, Vol. 312, Issue 1, Kiefer, B., Karaca H. E., Lagoudas, D.C. and Karaman, I., pp. 164–175, Copyright 2007, with permission from Elsevier.

Fig. 7.5. Magneto-thermo-mechanical setup used for MFIS measurements [58].

load frame such that the directions of applied force and magnetic field are perpendicular. The specimen is held in place by non-magnetic grips. A polymer chamber, which encloses the grips and specimen, is filled with nitrogen gas for cooling. Temperature, deformation and magnetic field measurements are taken by a thermocouple, a capacitive displacement sensor, and a Hall probe, respectively. Similar experiments have been reported by Tickle [56, 57], Heczko [46] and Shield [27], among others.

Magnetic-field-induced strain data obtained from this test frame are plotted in Figs. 7.7[2] and 7.8[2] for first and second magnetic field cycles, respectively. For reasons of comparison, all of the second cycle magnetic-field-induced strain curves plotted in Fig. 7.2 have been shifted to start at the origin. Unlike the qualitative strain curves of Fig. 7.4, these figures show the magnetic-field-induced strain as a function of the magnetic field, not the total strain, such that all curves start at the origin. These data exhibit all the characteristic features of MSMA behavior previously discussed. The observed response is nonlinear and hysteretic, which indicates that there is considerable dissipation associated with the variant reorientation. The achievable field-induced reorientation strain and the shape of the hysteresis loops show the strong dependence on stress level.

After the first magnetic cycle at low stress levels, the initial single variant configuration is not restored and residual strains are observed. Since the stress level is not raised in between cycles, the second cycle tests at low stress levels

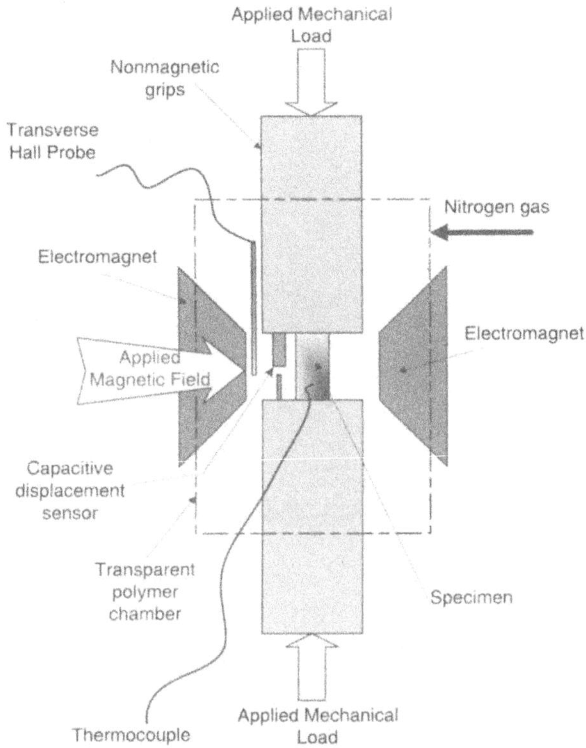

Fig. 7.6. Schematic illustration of the setup.

start from mixed variant configurations corresponding to the only partially recovered magnetic-field-induced strain of the first cycle. At low stress levels, the reorientation strain that is produced in the second cycle is then significantly reduced as compared to the corresponding test at the same stress level during the first cycle. This phenomenon has been termed the *first cycle effect* [58, 59] and will further be explained in Sect. 7.4. Hysteresis loops of subsequent cycles are experimentally observed to be nearly identical to those of the second cycle.

7.2.2 Magnetization Response of MSMAs

The previous section explained that the process of magnetizing an MSMA specimen involves three mechanisms which achieve alignment of its magnetization with the external magnetic field. These mechanisms are the redistribution of martensitic variants, the magnetic domain wall motion and the rotation of the magnetization vectors away from their preferred magnetic axes. To understand the individual mechanisms, consider a thought experiment in

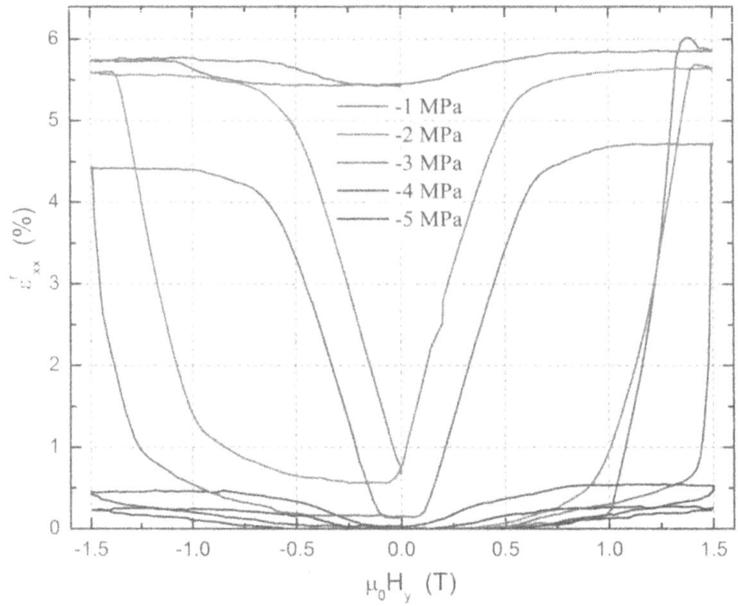

Fig. 7.7. Evolution of the MFIS in a Ni_2MnGa single crystal at different stress levels during the first magnetic cycle. (Data taken from [58].)

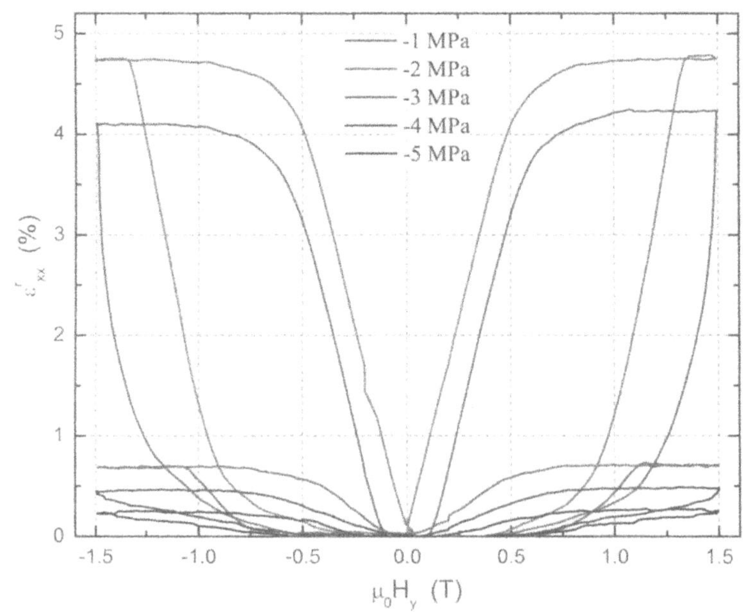

Fig. 7.8. Evolution of the MFIS in a Ni_2MnGa single crystal at different stress levels during the second magnetic cycle. (Data taken from [58].)

which these effects can be separated: If the reorientation of martensitic variants in an MSMA single crystal is completely suppressed by the application of a stress above the blocking stress, then the magnetization of the crystal can change by means of only the domain wall motion, the magnetization rotation, or combinations thereof. The magnetization process of the MSMA in this case is the same as that of a regular ferromagnetic material.

Fig. 7.9[1] shows a sketch of the initial single variant 1 configuration, which corresponds to that of Fig. 7.2, except here a stress level above the blocking stress is considered to analyze the magnetization process without variant reorientation. Next to the macroscopic view of the specimen, Fig. 7.9 also depicts schematics of magnetic domains on the micro-scale. The crystallographic scale is shown simply to indicate that magnetic domains generally span many unit cells. As discussed in the previous section, magnetic domains form to reduce the macroscopic magnetization of the material and thereby the magnetostatic energy [14–16, 66]. They are separated by magnetic domain walls. In these walls, the magnetization vectors (magnetic dipole moments) are rotated over short distances to accommodate the magnetization directions of neighboring domains. The formation of many small domains leads to an increase in the amount of domain walls, whose formation also costs energy. This competition of energy terms determines the size of the domains and also the thickness of the domain walls. Depending on the material, the domain wall thickness can range from 10 nm to 1μm [15].

If the constrained single crystal is magnetized along different crystallographic directions, one observes an anisotropy of the magnetization response. The direction along which the least amount of energy is required to magnetize the crystal is termed the *magnetic easy axis*, and, correspondingly, the *magnetic hard axis* is the direction for which the most energy needs to be expended. This anisotropic behavior can be explained by the mechanism of

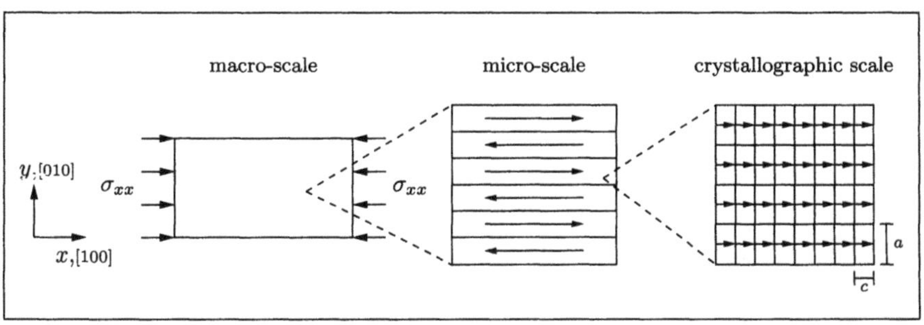

Fig. 7.9. A schematic of the initial single variant 1 martensite state. The variant reorientation is suppressed by an axial compressive stress higher than the blocking stress. Also shown, schematics of the corresponding microscopic scale and the crystallographic scale.

magnetic domain wall motion and magnetization rotation as shown in the following sections.

Magnetization by Magnetic Domain Wall Motion

Fig. 7.10[1] schematically shows the evolution of the magnetic domain distribution at different applied field levels for the magnetization of the MSMA specimen along the [100]-direction. The starting configuration (left box of Fig. 7.10) is the same microstructural view of the compressed single variant specimen that was presented in middle box of Fig. 7.9.

The applied field promotes the growth of these domains with favorably oriented magnetization vectors at the expense of the other domains. Since the external field is applied in the [100]-direction, which coincides with the magnetic easy axis of the compressive stress-favored variant 1, the magnetization to saturation can completely be achieved by 180° domain wall motion.

Magnetization by Rotation of Magnetization Vectors

Fig. 7.11[1] schematically illustrates the magnetization of the same single variant 1 sample perpendicular to the compression axis.

Since the magnetization vectors in both domains are equally unfavorable with respect to the applied field, no domain wall motion mechanism is available to accommodate the magnetization along the [010]-direction. The magnetization in both domains must be rotated away from the common easy axis. The rotation of the magnetization within a martensitic variant requires work against the magnetocrystalline anisotropy energy. The amount of energy expended in activating this mechanism is higher than that associated with domain wall motion. The [010]-direction is therefore the hard axis for this material. The magnetization of the MSMA specimen along directions in between [100] and [010] requires an intermediate amount of energy and involves the activation of both mechanisms. Unlike the motion of 180° domain walls, the rotation of the magnetization is associated with ordinary

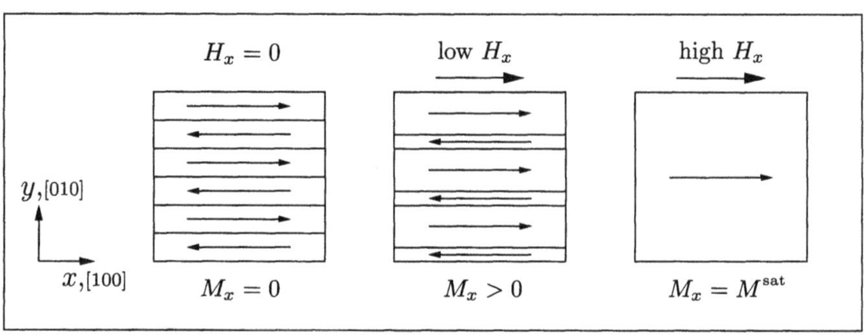

Fig. 7.10. Magnetization of the single variant specimen along the easy axis.

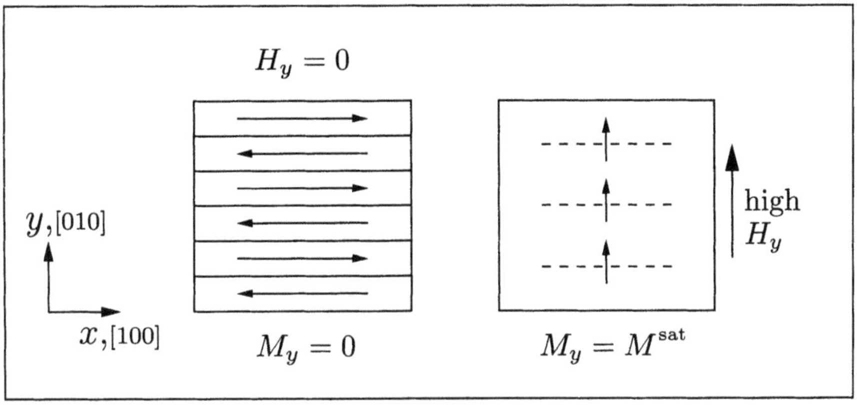

Fig. 7.11. Magnetization of the single variant specimen along the hard axis.

magnetostriction, i. e., the crystal elongates in the direction of the rotating magnetization vector [14, 16].

Fig. 7.12[1] qualitatively shows the resulting magnetization curves for the easy [100] and the hard [010]-directions. The coordinate axes are normalized by the saturation magnetization M^{sat} and an arbitrary maximum applied field value H^{max}, respectively. Data for the magnetization of constrained MSMA single crystals have been reported by Tickle and James [56], Cui et al. [25], Shield [27], Lickhachev and Ullakko [67] and Hezcko [68].

The magnetization curves in Fig. 7.12 are explained by the mechanisms discussed in the context of Figs. 7.10 and 7.11. Recall that the mechanism for alignment with the applied field is the domain wall motion, in the easy axis case, and rotation of the magnetization vectors, in the hard axis case. The hysteresis for both magnetization curves is observed to be almost negligible. This is expected for the hard axis magnetization curve since the magnetization rotation is in a reversible process. Magnetic domain wall motion, on the other hand, can be associated with dissipation. Permanent magnets, for example, are made from materials that exhibit a strong internal resistance to magnetic domain wall motion [2, 14, 15], which leads to large hysteresis effects. In MSMAs, however, the magnetic domain wall motion appears to be associated with only a very small amount of dissipation.

Magnetization by Variant Reorientation

In MSMAs, the previously described variant reorientation process, which was schematically illustrated in Figs. 7.2 and 7.3, provides an additional mechanism to change the magnetization of the material. This is because the magnetic easy axes in the martensitic variants have different directions with respect to a global coordinate system. In the presence of an external field, the structural rearrangement is therefore always coupled to a magnetization change. If the

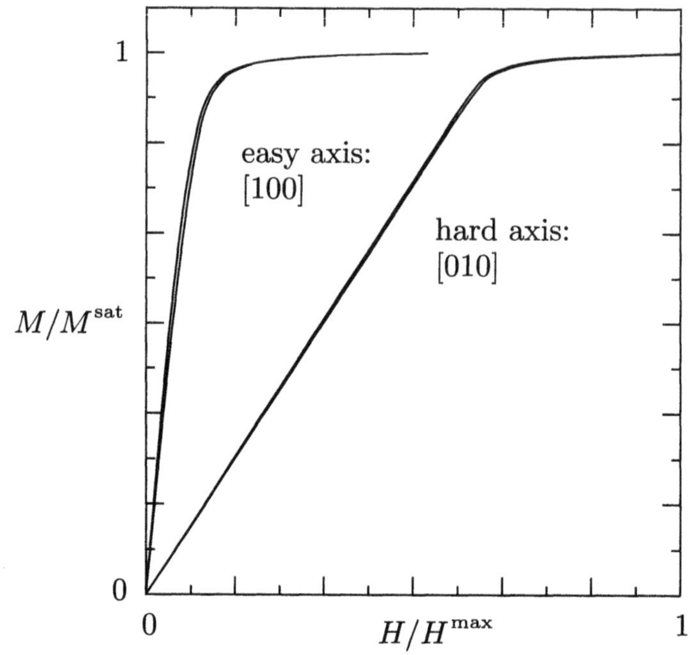

Fig. 7.12. Qualitative magnetization curves of the single variant MSMA specimen magnetized along the compression and perpendicular axes. For quantitative experimental results see [56].

reorientation process is initiated by mechanical loading instead of applying a magnetic field, and the applied field is constant, the variant reorientation is in fact the only mechanism that changes the magnetization. This aspect will be explained in more detail in connection with the model predictions presented in Sect. 7.4.

Coupling of Magnetization Mechanisms in MSMAs.

For stress levels below the blocking stress, the magnetic-field-induced change of the magnetization observed in MSMA is driven by all three mechanisms: the magnetic domain wall motion, the magnetization rotation, and the variant reorientation. Typical magnetization response curves, measured under the loading conditions depicted in Fig. 7.2, have been reported by Heczko et al. [13, 46] and are shown in Fig. 7.13. Similar data have been obtained by Likhachev and Ullakko [67].

The sequence of activation of the different mechanisms is the following: The initial linear slope of the magnetization curves in Fig. 7.13 corresponds to the magnetization of the constrained crystal along the magnetic hard axis, as shown in Fig. 7.12. The mechanism that drives the magnetization change is thus the rotation of the magnetization vectors, as was schematically illustrated in Fig. 7.11. Since both magnetic domains in the stress-favored variant are

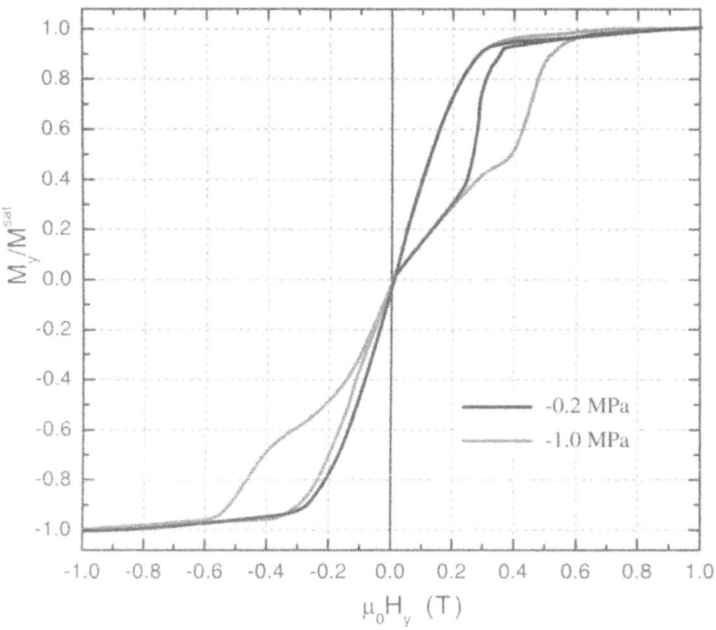

Fig. 7.13. Relative magnetization response in Ni-Mn-Ga by Heczko [69].

equally unfavorable with respect to the applied field (see Fig. 7.10), domain wall motion does not occur. Here it is assumed that the direction of the applied field is perfectly perpendicular to the easy axis of variant 1. In reality, this may not be the case and some domain wall motion may occur. More experimental evidence is needed to clarify this point.

Once the critical field for variant reorientation has been reached, the magnetic-field-favored variant 2 nucleates and a sharp change in the slope of the magnetization curves occurs. Fig. 7.13 clearly shows evidence of the stress dependence of the critical field to initiate the variant reorientation. In the reorientation region, the magnetization change is nonlinear and similar to the evolution of the magnetic-field-induced strain observed in the same field regime. During the reorientation process, magnetic domains and martensitic variants coexist in the arrangement qualitatively depicted in Fig. 7.2b. Such configurations have been observed experimentally in Ni-Mn-Ga [70–73]. Corresponding micrographs are shown in Fig. 7.14 on the next page[3]. Sullivan and Chopra also reported that more complex magnetic domain structures can exist in twinned Ni-Mn-Ga martensite under certain conditions [72, 73].

In this configuration, the third mechanism, the magnetic domain wall motion, is activated, because the magnetization vectors in some of the magnetic domains in variant 2 oppose the applied field. It is generally believed

[3] Reprinted from *Journal of Applied Physics*, Vol. 96, Issue 4, Ge, Y., Heczko, O., Söderberg, O. and Lindroos, V.K., pp. 2159–2163, Copyright 2004, with permission from AIP.

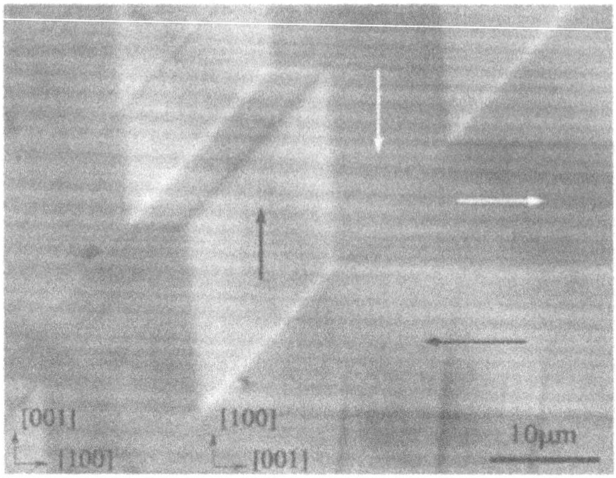

Fig. 7.14. Scanning electron microscopy (SEM) images by Ge et al. [71].

that these unfavorable magnetic domains are almost simultaneously eliminated with activation of the reorientation process [57, 65], such that the magnetization change in the reorientation region is mainly governed by the rearrangement of variants and the magnetization rotation.

At high fields, the stress-favored variant is completely eliminated and the material is magnetically saturated in the field direction. The resulting single variant, single domain configuration at magnetic saturation is illustrated in (c) of Fig. 7.2.

At higher stress levels, however, magnetic saturation can be reached through magnetization rotation before variant 1 is eliminated, and the variant reorientation process is not completed. This explains why at these stress levels only a fraction of the theoretically possible reorientation strain can be induced by the magnetic field.

At stress levels above the blocking stress, the critical field for the activation of the reorientation process is larger than the field needed to fully rotate the magnetization vector toward the direction of the applied field. The magnetization process then corresponds to the magnetization of the constrained crystal along the magnetic hard axis described earlier in this section. The value of the blocking stress depends on the magnetocrystalline anisotropy energy [58]. This explanation of the stress-level dependence of the achievable magnetic-field-induced strain will further be discussed in the context of analyzing model predictions in Sect. 7.4.

The magnetization curves of Fig. 7.13 exhibit large hystereses, whereas the magnetization curves of the constrained single crystal of Figs. 7.11 and 7.12 exhibit negligible hysteresis. Therefore, one can conclude that the only source of internal dissipation is the variant reorientation process, which then leads to the hysteretic nature of the magnetic-field-induced strain curves in Figs. 7.7 and 7.8, and the magnetization curves in Fig. 7.13.

The analysis of the experimental field-induced strain and magnetization response data, such as those presented in this section, is of great importance for the development of a constitutive model. To properly capture the complex magnetomechanical behavior of MSMAs, the model has to take all three of the described mechanisms into account. In the particular approach taken here, this is done through the evolution of internal state variables as will be described in Sect. 7.3.

7.3 Derivation of a Phenomenological Constitutive Model for Magnetic SMAs

Several constitutive models for MSMAs have been proposed in the literature [2, 26, 27, 67, 74–83]. Some of the most prominent models have been reviewed in [84–86]. The phenomenological MSMA model presented here relies on the thermodynamically-consistent derivation of constitutive equations from a free energy function in which dissipative effects related to the evolution of microstructure are incorporated through internal state variables. This can thus be viewed as a direct extension of the models presented for conventional SMAs in the preceding chapters. To address the constitutive response of MSMAs, the independent variable state space of the free energy is extended to include the magnetic field strength vector. The specific form of the Gibbs free energy function incorporates elastic, magnetic and magnetoelastic coupling terms.

Furthermore, additional internal state variables are introduced to describe changes in the free energy caused by the motion of magnetic domain walls and the rotation of local magnetization vectors away from magnetic easy axes, which are the mechanisms that were identified in Sect. 7.2 as the governing micro-scale mechanisms that cause the macroscopic magnetomechanical response characteristic for MSMAs. The evolution of these internal state variables will be shown to be governed by equations analogous to the transformation functions and criteria previously introduced for SMAs.

7.3.1 Extended Thermodynamic Framework

Following the above discussion, the Gibbs free energy for the thermodynamic description of a general continuous, deformable and magnetizable material is of the general form

$$G = G(T, \boldsymbol{\sigma}, \mathbf{H}, \boldsymbol{\zeta}) \,. \tag{7.3.1}$$

where $\mathbf{H}$ is the magnetic field strength and $\boldsymbol{\zeta}$ is the set of internal variables. The Gibbs free energy is related to the internal energy $u = u(s, \boldsymbol{\varepsilon}, \mathbf{M}, \boldsymbol{\zeta})$ through the Legendre transformation [87] (see Table 3.2)

$$G = u - sT - \frac{1}{\rho}\,\boldsymbol{\sigma} : \boldsymbol{\varepsilon} - \frac{\mu_0}{\rho}\,\mathbf{H} \cdot \mathbf{M} \,, \tag{7.3.2}$$

where s is the specific entropy. The constitutive relations for the entropy s, strain $\boldsymbol{\varepsilon}$, and magnetization $\mathbf{M}$ then follow directly (see Sect. 3.2.3) from taking partial derivatives of the specific Gibbs free energy function with respect to the independent state variables temperature T, mechanical stress $\boldsymbol{\sigma}$ and magnetic field strength $\mathbf{H}$, i. e.,

$$s = -\frac{\partial G}{\partial T} , \tag{7.3.3a}$$

$$\boldsymbol{\varepsilon} = -\rho \frac{\partial G}{\partial \boldsymbol{\sigma}} , \tag{7.3.3b}$$

$$\mu_0 \mathbf{M} = -\rho \frac{\partial G}{\partial \mathbf{H}} . \tag{7.3.3c}$$

Through the constitutive relations, the dependent state variables also depend on the internal state variables $\boldsymbol{\zeta}$, whose evolution accounts for the loading history dependence of the material behavior. The Clausius-Planck inequality (see Sect. 3.2.3) in reduced form is given by

$$-\rho \frac{\partial G}{\partial \boldsymbol{\zeta}} \cdot \dot{\boldsymbol{\zeta}} \geq 0 . \tag{7.3.4}$$

7.3.2 Choice of Internal State Variables

Motivated by the experimentally observed arrangement of martensitic twins and magnetic domains in MSMAs, as previously discussed in Sect. 7.2, specifically in the context of Fig. 7.14, an idealized microstructure is assumed. Fig. 7.15[1] shows its schematic representation. Other experiments have indicated that the actual microstructure of MSMA can be much more complex [72, 73], but this chosen representation has proven sufficient to explain all of the main characteristics of the MSMA response. Many researchers have proposed a similar representation [2, 67, 74, 77].

Fig. 7.15 depicts the coexistence of two martensitic variants and two magnetic domain types. The third martensitic variant, and thus the third magnetic domain type, are thought to have been eliminated by proper load application. This idealized microstructure was also indicated in Fig. 7.2 for a generic intermediate applied magnetic field level. In this sketch, however, the possibility of the rotation of the local magnetization away from the magnetic easy axes is included and defined by the rotation angles θ_i, with $i = 1, \ldots, 4$ and $0 \leq \theta_i \leq \frac{\pi}{2}$. Variant 1, as previously defined in Fig. 7.1, is magnetized along the x, [100]-direction, whereas variant 2 has its magnetic easy axis along the y, [010]-direction. The magnetic domains are defined by the preferred orientation of their magnetization vectors. In magnetic domain 1, the magnetization is oriented in the negative direction of the respective coordinate axis, and in domain 2, it points in the opposite direction. The two variants are separated

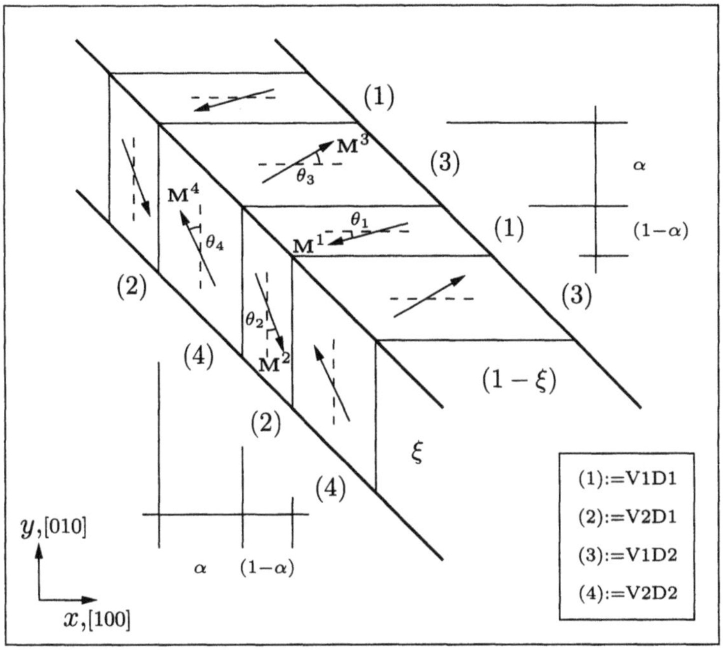

Fig. 7.15. Schematic representation of the microstructure showing the coexistence of martensitic variants and magnetic domains. The abbreviation V1D2, for example, stands for "variant 1, domain 2". The four distinct subdomains are numbered (1)–(4) to simplify the notation.

by a twin boundary shown here at a 45° angle. The twin boundary coincides with the 90° domain walls between adjacent magnetic domains. Both are assumed to have zero thickness on the considered scale. Within each variant, 180° domain walls separate the domains. The local magnetization vectors and their associated rotation angles are, by definition, considered constant within each domain. Note that the rotations shown in the sketch do not necessarily correspond to a realistic arrangement of magnetization vectors, but illustrate the nomenclature. Compatibility of domains along twin planes, which separates different variants, has also been assumed, such that the domain walls of neighboring twins meet on the twin boundary in a compatible manner.

Based on this discussion, the scalar volume fractions ξ and α for the martensitic variant 2 and the magnetic domain 2 (cf. Fig. 7.15), respectively, are introduced as internal state variables. The volume fraction of variant 1 is equal to $(1-\xi)$, and that of domain 1 is similarly given by $(1-\alpha)$, such that no additional variables need to be introduced. Additionally the inelastic reorientation strain tensor $\boldsymbol{\varepsilon}^r$ is formally identified as an internal state variable, although its evolution is later connected to the rate of ξ. It is implied that for a small strain approximation the additive decomposition

$$\varepsilon = \varepsilon^{th} + \varepsilon^{in} = \varepsilon^{th} + \varepsilon^{r} \tag{7.3.5}$$

is assumed to be sufficiently accurate (c.f. (3.3.39)). The set of internal state variables is completed by the rotation angles of the magnetization vectors θ_i depicted in Fig. 7.15.

7.3.3 Formulation of the Specific Gibbs Free Energy

Certainly the free energy expression must contain an elastic, a magnetic and a magnetoelastic part to capture the characteristic features of the constitutive response of MSMAs. Motivated by the discussed micro-scale arrangement of martensitic twins in MSMAs, the Gibbs free energy is constructed as a weighted average between the contributions of each variant. This assumes an appropriate *separation of scales*, which means that the characteristic internal features of the material, such as the martensitic variants and magnetic domains in MSMAs, are thought to be small enough to allow the definition of such average quantities at each material point. The deviation of the free energy from the weighted average of the individual contributions, as caused by the interaction of martensitic variants, is captured by a mixing term $G^{\xi-mix}$. In accordance with the general constitutive assumption of (7.3.1), the choice of internal state variables made in the previous section, and the above discussion about the general approach to formulating the Gibbs free energy, the following expression is proposed:

$$\begin{aligned} G &= G(T, \boldsymbol{\sigma}, \mathbf{H}, \boldsymbol{\varepsilon}^r, \xi, \alpha, \theta_j) \\ &= \xi\, G^{V2}(T, \boldsymbol{\sigma}, \mathbf{H}, \boldsymbol{\varepsilon}^r, \alpha, \theta_2, \theta_4) + (1-\xi)\, G^{V1}(T, \boldsymbol{\sigma}, \mathbf{H}, \boldsymbol{\varepsilon}^r, \alpha, \theta_1, \theta_3) \\ &\quad + G^{\xi-mix}(\xi)\,. \end{aligned} \tag{7.3.6}$$

Since MSMAs are typically used at constant temperatures, isothermal conditions are assumed and thermal as well as thermoelastic and thermomagnetic coupling terms are not explicitly included. If one is interested in temperature changes occurring in the martensitic phase, it is fairly straightforward to include the relevant thermal energy terms in the total Gibbs free energy expression of (7.3.11). This has been demonstrated by the authors in [85] where the temperature change per magnetic cycle due to the dissipation associated with the variant reorientation is computed.

The contribution of each martensitic variant $i = \{1, 2\}$ to the total Gibbs free energy, (7.3.6), is then proposed to be given by

$$\begin{aligned} G^{Vi} = &-\frac{1}{2\rho}\boldsymbol{\sigma}:\mathcal{S}^{Vi}\boldsymbol{\sigma} - \frac{1}{\rho}\boldsymbol{\sigma}:\boldsymbol{\varepsilon}^r + (1-\alpha)\left[-\frac{\mu_0}{\rho}\mathbf{M}^{ViD1}\cdot\mathbf{H} + G^{an,ViD1}\right] \\ &+ \alpha\left[-\frac{\mu_0}{\rho}\mathbf{M}^{ViD2}\cdot\mathbf{H} + G^{an,ViD2}\right] + G^{\alpha-mix}(\alpha) + G_0(T)\,. \end{aligned} \tag{7.3.7}$$

Here, $\mathcal{S}^{Vi}$ denotes the elastic compliance tensor of variant i. The local magnetization vectors $\mathbf{M}^{ViDj}$ have been introduced in Fig. 7.15. The first term in (7.3.7) describes the energy stored in the material due to elastic deformation. The second term is due to the interaction of the stress and inelastic strain. For the magnetic terms, the contributions of each magnetic domain are again taken into account in a weighted average sense, with the correcting mixing term $G^{\alpha-mix}$. It has been assumed that the density, the elastic compliance tensor, the reorientation strain tensor, and the reference value of the free energy are the same in both magnetic domain types.

The first magnetic energy term in (7.3.6) is the Zeeman energy for which the individual contributions are of the form $-\mathbf{M}^{ViDj}\cdot\mathbf{H}$, where the magnetization vectors of the different variant-domain combinations, cf. Fig. 7.15, are defined as

$$\mathbf{M}^{(1)}(\theta_1) = \mathbf{M}^{V1D1} = -M^{sat}\big(\cos(\theta_1)\,\mathbf{e}_x + \sin(\theta_1)\,\mathbf{e}_y\big)\,, \tag{7.3.8a}$$

$$\mathbf{M}^{(2)}(\theta_2) = \mathbf{M}^{V2D1} = M^{sat}\big(\sin(\theta_2)\,\mathbf{e}_x - \cos(\theta_2)\,\mathbf{e}_y\big)\,, \tag{7.3.8b}$$

$$\mathbf{M}^{(3)}(\theta_3) = \mathbf{M}^{V1D2} = M^{sat}\big(\cos(\theta_3)\,\mathbf{e}_x + \sin(\theta_3)\,\mathbf{e}_y\big)\,, \tag{7.3.8c}$$

$$\mathbf{M}^{(4)}(\theta_4) = \mathbf{M}^{V2D2} = M^{sat}\big(-\sin(\theta_4)\,\mathbf{e}_x + \cos(\theta_4)\,\mathbf{e}_y\big)\,. \tag{7.3.8d}$$

In the above relations, M^{sat} is the saturation magnetization, and $\mathbf{e}_x, \mathbf{e}_y$ are unit vectors in the respective coordinate directions. If the magnetization vectors are assumed to be fixed to the respective magnetic easy axes (i. e., $\theta_i = 0$), these expressions reduce to the formulation proposed by Hirsinger and Lexcellent [77]. The Zeeman energy accounts for the interaction of the local magnetization with the magnetic field [61]. The Zeeman energy difference across the twin boundary is the main magnetic driving force for variant reorientation. An important point is that $\mathbf{H}$ is the internal field at a generic material point, not the applied field.

The second magnetic term in (7.3.7) is the magnetocrystalline anisotropy energy, whose contributions are denoted $G^{an,ViDj}$. This is the energy stored in the material due to the work done by the magnetic field in rotating the magnetization vectors away from the magnetic easy axes. Magnetoelastic interactions in the form of ordinary magnetostriction have also been neglected since the associated strains are at least two orders of magnitude smaller in MSMAs than the strains caused by the variant rearrangement [56, 68].

An explicit form of the magnetocrystalline anisotropy energy for *uniaxial symmetry*, i. e., for crystals that exhibit only one preferred direction of magnetization, as for example the short magnetic easy axis c in tetragonal MSMA martensite [56, 74], is typically expressed in the form of a trigonometric power series [15, 61, 88]

$$G^{an,i}(\theta_i) = \sum_{n=1}^{N} K_n^i \sin^{2n}(\theta_i)\,, \tag{7.3.9}$$

where K_n are coefficients to be determined from magnetization measurements [25, 57] and θ_i are again the rotation angles of the local magnetization vectors. Only the first term (i.e. $N = 1$) and equal properties in each variant (i.e. $K_1^i = K_1$) will be considered in the current model. Note that since the rotation angles θ_i have been introduced as independent variables, the magnetocrystalline anisotropy energy contributions $G^{an,i}(\theta_i)$, generally take on distinct values.

The influence of the interaction of evolving martensitic variants or magnetic domains on the free energy, introduced in (7.3.6) by the mixing terms $G^{\xi-mix}$ and $G^{\alpha-mix}$, respectively, will be represented by a hardening function following the same concept first introduced in Chapter 3. In generic form, the mixing terms are then given by

$$G^{\xi-mix} = \frac{1}{\rho} f^{\xi}(\xi) \,, \tag{7.3.10a}$$

$$G^{\alpha-mix} = \frac{1}{\rho} f^{\alpha}(\alpha) \,. \tag{7.3.10b}$$

The scalar constant G_0 is the reference state value of the Gibbs free energy.

The combination of (7.3.6), (7.3.7) and (7.3.10) yields the following explicit form of the total Gibbs free energy:

$$\begin{aligned} G &= G(T, \boldsymbol{\sigma}, \mathbf{H}, \boldsymbol{\varepsilon}^r, \xi, \alpha, \theta_j) \qquad (7.3.11)\\ &= -\frac{1}{2\rho}\boldsymbol{\sigma}:\mathcal{S}(\xi)\boldsymbol{\sigma} - \frac{1}{\rho}\boldsymbol{\sigma}:\boldsymbol{\varepsilon}^r + \xi\Bigg[(1-\alpha)\left[-\frac{\mu_0}{\rho}\mathbf{M}^{(2)}(\theta_2)\cdot\mathbf{H} + G^{an,2}(\theta_2)\right] \\ &\quad + \alpha\left[-\frac{\mu_0}{\rho}\mathbf{M}^{(4)}(\theta_4)\cdot\mathbf{H} + G^{an,4}(\theta_4)\right]\Bigg] \\ &\quad + (1-\xi)\Bigg[(1-\alpha)\left[-\frac{\mu_0}{\rho}\mathbf{M}^{(1)}(\theta_1)\cdot\mathbf{H} + G^{an,1}(\theta_1)\right] \\ &\quad + \alpha\left[-\frac{\mu_0}{\rho}\mathbf{M}^{(3)}(\theta_3)\cdot\mathbf{H} + G^{an,3}(\theta_3)\right]\Bigg] + \frac{1}{\rho}f^{\xi}(\xi) + \frac{1}{\rho}f^{\alpha}(\alpha) + G_0(T)\,. \end{aligned}$$

The density has been assumed as identical in both martensitic variants. The effective compliance tensor is defined by $\mathcal{S} = \mathcal{S}^{V1} + \xi\Delta\mathcal{S} = \mathcal{S}^{V1} + \xi(\mathcal{S}^{V2} - \mathcal{S}^{V1})$.

According to (7.3.3), constitutive equations for the entropy, the strain and the magnetization follow from taking partial derivatives of (7.3.11).

$$s = -\frac{\partial G}{\partial T} = -\frac{\partial G_0}{\partial T} , \tag{7.3.12a}$$

$$\boldsymbol{\varepsilon} = -\rho \frac{\partial G}{\partial \boldsymbol{\sigma}} = \mathcal{S}(\xi)\boldsymbol{\sigma} + \boldsymbol{\varepsilon}^r , \tag{7.3.12b}$$

$$\begin{aligned} \mathbf{M} &= -\frac{\rho}{\mu_0}\frac{\partial G}{\partial \mathbf{H}} \\ &= \xi\Big[(1-\alpha)\mathbf{M}^{(2)}(\theta_2)+\alpha\mathbf{M}^{(4)}(\theta_4)\Big]+(1-\xi)\Big[(1-\alpha)\mathbf{M}^{(1)}(\theta_1)+\alpha\mathbf{M}^{(3)}(\theta_3)\Big], \end{aligned} \tag{7.3.12c}$$

where the local magnetization vectors $\mathbf{M}^{(i)}$ have been defined in (7.3.8). According to (7.3.12a), the entropy is only a function of temperature; therefore, it must be constant throughout the reorientation process if isothermal conditions are assumed. This means that the entropy changes due to dissipative effects must be balanced by suitable heat transfer to maintain a constant temperature. For adiabatic conditions, the entropy increases due to the dissipation associated with the reorientation of variants. It has been shown by the authors in [85] how the model must be extended slightly to capture adiabatic loading paths.

By taking derivatives of the total free energy, (7.3.11), with respect to the internal state variables, the following thermodynamic driving forces are derived

$$\boldsymbol{\pi}^r = -\rho\frac{\partial G}{\partial \boldsymbol{\varepsilon}^r} = \boldsymbol{\sigma} , \tag{7.3.13a}$$

$$\begin{aligned} \tilde{\pi}^\xi &= -\rho\frac{\partial G}{\partial \xi} \\ &= \frac{1}{2}\boldsymbol{\sigma}:\Delta\mathcal{S}\boldsymbol{\sigma} - \rho(1-\alpha)\Big[-\frac{\mu_0}{\rho}\big(\mathbf{M}^{(2)}-\mathbf{M}^{(1)}\big)\cdot\mathbf{H} + G^{an,2}(\theta_2) - G^{an,1}(\theta_1)\Big] \\ &\quad - \rho\alpha\Big[-\frac{\mu_0}{\rho}\big(\mathbf{M}^{(4)}-\mathbf{M}^{(3)}\big)\cdot\mathbf{H} + G^{an,4}(\theta_4) - G^{an,3}(\theta_3)\Big] - \frac{\partial f^\xi}{\partial \xi} , \end{aligned} \tag{7.3.13b}$$

$$\begin{aligned} \pi^\alpha &= -\rho\frac{\partial G}{\partial \alpha} \\ &= -\rho(1-\xi)\Big[-\frac{\mu_0}{\rho}\big(\mathbf{M}^{(3)}-\mathbf{M}^{(1)}\big)\cdot\mathbf{H} + G^{an,3}(\theta_3) - G^{an,1}(\theta_1)\Big] \\ &\quad - \rho\xi\Big[-\frac{\mu_0}{\rho}\big(\mathbf{M}^{(4)}-\mathbf{M}^{(2)}\big)\cdot\mathbf{H} + G^{an,4}(\theta_4) - G^{an,2}(\theta_2)\Big] - \frac{\partial f^\alpha}{\partial \alpha} . \end{aligned} \tag{7.3.13c}$$

Similarly, the driving forces for rotation of the magnetization vectors are given by

$$\pi^{\theta_1} = -\rho\frac{\partial G}{\partial \theta_1} \tag{7.3.14a}$$

$$= -\rho(1-\xi)(1-\alpha)\left[\frac{\mu_M^{sat}}{\rho}\left[-\sin(\theta_1)H_x + \cos(\theta_1)H_y\right] + \frac{\partial G^{an,1}}{\partial \theta_1}\right];$$

$$\pi^{\theta_2} = -\rho\frac{\partial G}{\partial \theta_2} \tag{7.3.14b}$$

$$= -\rho\xi(1-\alpha)\left[-\frac{\mu_0 M^{sat}}{\rho}\left[\cos(\theta_2)H_x + \sin(\theta_2)H_y\right] + \frac{\partial G^{an,2}}{\partial \theta_2}\right];$$

$$\pi^{\theta_3} = -\rho\frac{\partial G}{\partial \theta_3} \tag{7.3.14c}$$

$$= -\rho(1-\xi)\alpha\left[-\frac{\mu_0 M^{sat}}{\rho}\left[-\sin(\theta_3)H_x + \cos(\theta_3)H_y\right] + \frac{\partial G^{an,3}}{\partial \theta_3}\right];$$

$$\pi^{\theta_4} = -\rho\frac{\partial G}{\partial \theta_4} \tag{7.3.14d}$$

$$= -\rho\xi\alpha\left[-\frac{\mu_0 M^{sat}}{\rho}\left[-\cos(\theta_4)H_x - \sin(\theta_4)H_y\right] + \frac{\partial G^{an,4}}{\partial \theta_4}\right],$$

where the definitions of the magnetization vectors, (7.3.8), have been utilized. The defined quantities $\boldsymbol{\pi}^r$, $\tilde{\pi}^\xi$, π^α and π^{θ_i} are driving forces that are thermodynamically conjugate to the internal state variables such that, using the above definitions, the Clausius-Planck inequality can be rewritten as

$$\boldsymbol{\pi}^r : \dot{\boldsymbol{\varepsilon}}^r + \tilde{\pi}^\xi\dot{\xi} + \pi^\alpha\dot{\alpha} + \sum_{i=1}^{4}\pi^{\theta_i}\dot{\theta}_i \geq 0\,. \tag{7.3.15}$$

7.3.4 Evolution Equations and Activation Conditions

The effective number of internal state variables can be reduced by enforcing additional thermodynamic constraints, based on the following considerations. From physical observations, one can reasonably assume that the motion of magnetic domain walls and the rotation of the magnetization vectors are thermodynamically reversible processes [14, 61] and do not contribute to the entropy production. The single variant sample magnetization curves of Fig. 7.12, for example, in which only the mechanisms of domain wall motion (easy axis) and magnetization rotation (hard axis) are active, exhibit almost no hysteresis. However, the magnetization curves in Fig. 7.13 exhibit significant hysteresis due to the dissipation associated with the variant rearrangement. If the internal dissipation production is entirely attributed to the reorientation process, this implies that no dissipation is produced by the motion of domain walls or the rotation of the magnetization vectors, i. e., $\mathcal{D}^\alpha = \pi^\alpha \dot{\alpha} = 0$ and $\mathcal{D}^\theta = \sum_i^4 \pi^{\theta_i} \dot{\theta}_i = 0$. Since, in general, $\dot{\alpha} \neq 0$ and $\dot{\theta}_i \neq 0$, it follows from (7.3.15)

$$\pi^\alpha = -\rho \frac{\partial G}{\partial \alpha} = 0\,, \tag{7.3.16a}$$

$$\pi^{\theta_i} = -\rho \frac{\partial G}{\partial \theta_i} = 0\,. \tag{7.3.16b}$$

This does not mean that the domain walls do not move, nor that the magnetization vectors do not rotate, but rather that no dissipation is associated with these processes. Equations (7.3.16) represent a set of five relations, which can be solved to determine evolution equations, or more precisely direct functional dependencies, of the domain volume fraction α and the four rotation angles θ_i on the independent state variables, as well as the loading history through the remaining internal state variables. With (7.3.16) the inequality (7.3.15) takes the reduced form

$$\boldsymbol{\pi}^r : \dot{\boldsymbol{\varepsilon}}^r + \tilde{\pi}^\xi \dot{\xi} \geq 0\,. \tag{7.3.17}$$

In the modeling of conventional shape memory behavior, the transformation strain is usually related to the martensitic volume fraction [62, 89, 90]. Following this approach, the evolution of the reorientation strain associated with the magnetic shape memory effect is proposed as proportional to the rate of the martensitic variant volume fraction

$$\dot{\boldsymbol{\varepsilon}}^r = \boldsymbol{\Lambda}^r \dot{\xi}\,. \tag{7.3.18}$$

The reorientation strain is then no longer an independent internal state variable. In the equation above, $\boldsymbol{\Lambda}^r$ is the reorientation tensor defining the direction in which the reorientation strain develops. An explicit form of the tensor will be given in Sect. 7.4 for a specific example. For the special case of constant $\boldsymbol{\Lambda}^r$, (7.3.18) can be integrated to yield the reorientation strain, so that the total strain, using (7.3.12b), is then given by

$$\boldsymbol{\varepsilon} = \mathcal{S}\boldsymbol{\sigma} + \boldsymbol{\Lambda}^r \xi\,. \tag{7.3.19}$$

If $\boldsymbol{\Lambda}^r$ is not constant, for instance in non-proportional loading, (7.3.18) has to be evaluated incrementally [91].

One can further define the total thermodynamic driving force for the twin boundary motion associated with the variant reorientation process as

$$\pi^{\xi} = \boldsymbol{\pi}^{r} : \boldsymbol{\Lambda}^{r} + \tilde{\pi}^{\xi} = \boldsymbol{\sigma} : \boldsymbol{\Lambda}^{r} - \rho \frac{\partial G}{\partial \xi} , \tag{7.3.20}$$

where the definitions (7.3.13a) and (7.3.13b) have been utilized. Then, using (7.3.18) and (7.3.20), the Clausius-Planck inequality (7.3.17) can finally be written as

$$\pi^{\xi} \dot{\xi} \geq 0 . \tag{7.3.21}$$

The rate-independent nature of the formulation motivates the introduction of the following *reorientation function*:

$$\Phi^{\xi}(\boldsymbol{\sigma}, \mathbf{H}, \xi) = \begin{cases} \pi^{\xi} - Y^{\xi} , & \dot{\xi} > 0 \\ -\pi^{\xi} - Y^{\xi} , & \dot{\xi} < 0 \end{cases} , \qquad \Phi^{\xi} \leq 0 , \tag{7.3.22}$$

which defines the activation threshold for variant reorientation of conversely the elastic regime. The proposed reorientation function is similar to transformation functions used in the modeling of rate independent phenomenological modeling of conventional shape memory behavior (cf. (3.3.58)). The positive scalar quantity Y^{ξ} is physically related to internal dissipation associated with twin boundary motion. It is assumed that the reorientation process is subject to constraints derived from a principle of maximum dissipation, which can be expressed in terms of the Kuhn-Tucker type reorientation conditions [92]

$$\Phi^{\xi}(\boldsymbol{\sigma}, \mathbf{H}, \xi) \leq 0, \quad \Phi^{\xi} \dot{\xi} = 0 . \tag{7.3.23}$$

Note that in the elastic regime where $\Phi^{\xi} < 0$, the conditions (7.3.23) require $\dot{\xi} = 0$. The forward reorientation process is characterized by $\Phi^{\xi} = 0$ and $\dot{\xi} > 0$, whereas for the reverse process the conditions $\Phi^{\xi} = 0$ and $\dot{\xi} < 0$ exist. Since the thermodynamic driving force π^{ξ} and, consequently, the reorientation function Φ^{ξ} contain the hardening function f^{ξ}, the reorientation conditions (7.3.23) depend on the martensitic variant volume fraction and thereby on the loading history, not just the current values of stress and magnetic field.

To complete the formulation of the constitutive model, the generically introduced hardening functions (7.3.10) need to be specified. Several types of hardening functions were introduced and compared in Chapter 3, particularly in Sect. 3.4, in the context of conventional SMAs. These functions account for all micro-scale interactions that result in macroscopically observed hardening. Such hardening behavior is also clearly observed in MSMAs, for example, in the strain response curves of Fig. 7.7 on page 334, where a much higher magnetic field is needed to finish the reorientation process than is required to start it. Even though the martensitic twins are compatible across a twin boundary, the presence of one variant clearly influences the growth of the others. Effects such as the trapping of the progressing twin boundary at *pinning sites* as proposed by Faidley et al. [2], which influence the macroscopic evolution of the reorientation strain, can also be accounted for in this manner.

7.4 MSMA Response Under Specific Magnetomechanical Loading

In the previous section, the constitutive model was introduced in its most general form. In this section, a number of simplifying assumptions are made and specific loading conditions are considered, such that reduced versions of the constitutive equations be obtained.

7.4.1 Prediction of Magnetic-Field-Induced Variant Reorientation at Constant Stress (Fixed Domain Structure)

A loading case is considered for which σ_{xx} = const. ≤ 0 is the only non-zero component of the stress and H_y is the non-zero component of the magnetic field at a generic material point. For the chosen coordinate system, the non-zero components of the proposed reorientation tensor are given by[4]

$$\Lambda^r_{xx} = -\Lambda^r_{yy} = \varepsilon^{r,max} \,, \qquad \Lambda^r_{zz} = \Lambda^r_{xy} = \Lambda^r_{xz} = \Lambda^r_{yz} = 0 \,. \qquad (7.4.24)$$

The maximum strain value can be measured experimentally or is often approximated as $\varepsilon^{r,max} = (a-c)/a$, where a and c are the lattice parameters of the tetragonal martensite. More general reorientation strain tensors have been discussed in the literature [26, 28, 53].

Typically, the motion of magnetic domain walls is neglected in the modeling of MSMAs [2, 74, 77, 78] since it is assumed that unfavorable domains are eliminated at relatively low fields [57, 65] such that they do not significantly influence the magnetic-field-induced variant reorientation process and, therefore, do not justify the formulation of even more complex constitutive models.

If the motion of magnetic domain walls at low fields is neglected, the domain volume fraction takes the value of $\alpha = 1$, for $H_y > 0$, and $\alpha = 0$, for $H_y < 0$. Fig. 7.16[1,5] represents a modification of Fig. 7.2 for the assumption $\alpha = 1$.

Reduced Model Equations

For the described magnetomechanical loading and the assumed fixed domain configuration, a reduced set of constitutive equations is now derived.[6] The constraint (7.3.16b) on θ_3, utilizing (7.3.14c), takes the form

[4] Typically, specimens are cut such that the <100>$_m$-directions of the tetragonal martensite align with the direction of the applied loads.

[5] Reprinted from *Proceedings of SPIE Smart Structures and Materials Conference*, Vol. 5761, Kiefer, B. and Lagoudas, D.C., pp. 454–465, Copyright 2005, with permission from SPIE.

[6] All of the reduced equations presented in this section are derived for the case of $\alpha = 1$. It is straightforward to derive their counterparts for $\alpha = 0$ by using the general expressions provided in Sect. 7.3.

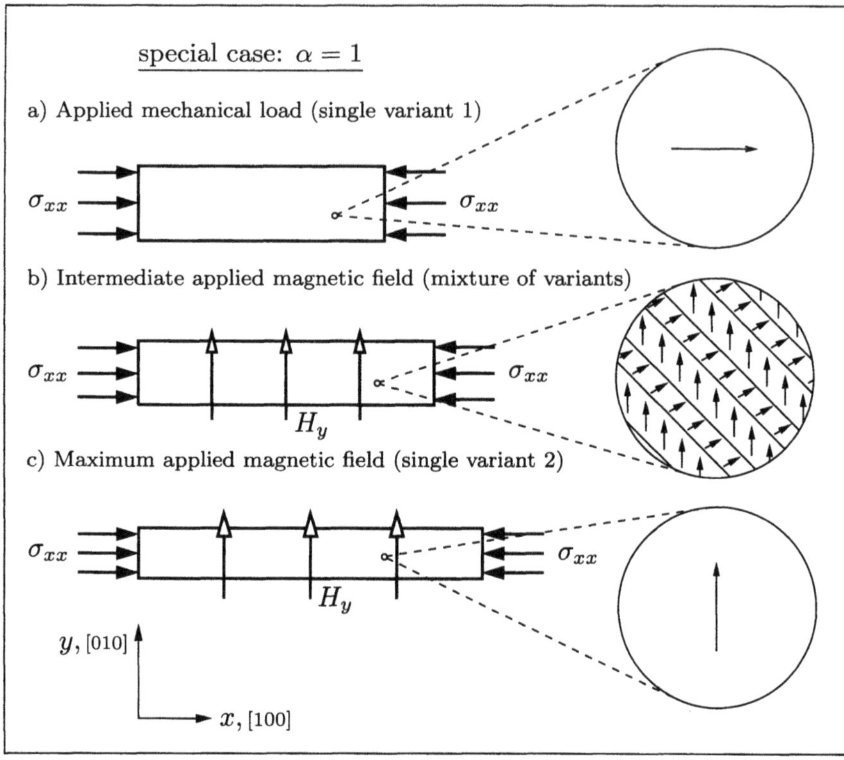

Fig. 7.16. The variant reorientation process with fixed domain structure $\alpha = 1$.

$$\pi^{\theta_3} = (1-\xi)\left[\mu_0 M^{sat} H_y - 2\rho K_1 \sin(\theta_3)\right]\cos(\theta_3) = 0\,. \tag{7.4.25}$$

From this, it is concluded that

$$\sin(\theta_3) = \frac{\mu_0 M^{sat}}{2\rho K_1} H_y\,, \tag{7.4.26}$$

for $0 \leq \theta_3 < \frac{\pi}{2}$ and $0 \leq \xi < 1$. Since the easy axis of variant 2 is aligned with the direction of the applied field, the magnetization vector does not rotate, such that the corresponding constraint $\pi^{\theta_4} = 0$ is identically satisfied.

For the considered loading case, using (7.3.8c), (7.3.8d), (7.3.9), (7.3.13b), and (7.3.20), the driving force for variant rearrangement reduces to

$$\pi^{\xi} = \sigma_{xx}\varepsilon^{r,max} + \mu_0 M^{sat}\left[1-\sin(\theta_3)\right]H_y + \rho K_1 \sin^2(\theta_3) - \frac{\partial f^{\xi}}{\partial \xi}\,, \tag{7.4.27}$$

where the difference in the elastic compliance of the variants ΔS has been considered small and thus neglected. If one further utilizes (7.4.26) to eliminate θ_3, (7.4.27) can be rewritten as

$$\pi^{\xi} = \sigma_{xx}\varepsilon^{r,max} + \mu_0 M^{sat} H_y - \frac{(\mu_0 M^{sat})^2}{4\rho K_1} H_y^2 - \frac{\partial f^{\xi}}{\partial \xi}\,, \tag{7.4.28}$$

In this example the trigonometric hardening function f^{ξ} given by (3.4.92) is employed. For the assumed fixed magnetic domain structure, the driving force π^{α} need not be considered.

Combining the reorientation function (7.3.22) with the driving force (7.4.28) and enforcing the Kuhn-Tucker loading conditions (7.3.23), the evolution equations for the martensitic variant volume fraction are derived as follows:

For the *forward reorientation process* (Variant 1 $\rightarrow$ Variant 2, $\dot{\xi}>0$):

$$\Phi^{\xi}\dot{\xi} = 0 \quad \Rightarrow \quad \Phi^{\xi} = 0 \quad \Rightarrow \quad \pi^{\xi} = Y^{\xi,c}\,. \tag{7.4.29}$$

$$\pi^{\xi} = \sigma_{xx}\varepsilon^{r,max} + \mu_0 M^{sat} H_y - \frac{(\mu_0 M^{sat})^2}{4\rho K_1} H_y^2 + a_c^{V2}\left[\pi - \cos^{-1}(2\xi - 1)\right] \tag{7.4.30}$$

$$- \mu_1^c - \mu_2^c = Y^{\xi,c}\,.$$

Since (7.4.30) contains only one scalar internal variable, the usual procedure of enforcing consistency conditions [62, 92, 93] is not necessary, and one can solve for ξ, which can be viewed as the equivalent of a plastic multiplier, directly in closed-form to find

$$\xi^{(1,2)} = \frac{1}{2}\cos\left(-\frac{1}{a_c^{V2}}\left[-\sigma_{xx}\varepsilon^{r,max} - \mu_0 M^{sat} H_y + \frac{(\mu_0 M^{sat})^2}{4\rho K_1} H_y^2 \right.\right. \tag{7.4.31}$$

$$\left.\left. + \mu_1^c + \mu_2^c + Y^{\xi,c}\right] + \pi\right) + \frac{1}{2}\,.$$

Similarly, it follows for the *reverse reorientation process* (Variant 2 $\rightarrow$ Variant 1, $\dot{\xi}<0$):

$$\Phi^{\xi}\dot{\xi} = 0 \quad \Rightarrow \quad \Phi^{\xi} = 0 \quad \Rightarrow \quad \pi^{\xi} = -Y^{\xi,c}\,. \tag{7.4.32}$$

$$\pi^{\xi} = \sigma_{xx}\varepsilon^{r,max} + \mu_0 M^{sat} H_y - \frac{(\mu_0 M^{sat})^2}{4\rho K_1} H_y^2 + a_c^{V1}\left[\pi - \cos^{-1}(2\xi - 1)\right] \tag{7.4.33}$$

$$- \mu_1^c + \mu_2^c = -Y^{\xi,c}\,,$$

so that

$$\xi^{(2,1)} = \frac{1}{2}\cos\left(-\frac{1}{a_c^{V1}}\left[-\sigma_{xx}\varepsilon^{r,max} - \mu_0 M^{sat} H_y + \frac{(\mu_0 M^{sat})^2}{4\rho K_1} H_y^2 + \mu_1^c - \mu_2^c - Y^{\xi,c} \right] + \pi \right) + \frac{1}{2} . \tag{7.4.34}$$

From the integration of (7.3.18) and the kinematic assumptions of (7.4.24) the components of the reorientation strain follow as

$$\varepsilon_{xx}^r = \varepsilon^{r,max}\xi , \quad \varepsilon_{yy}^r = -\varepsilon_{xx}^r , \quad \varepsilon_{zz}^r = \varepsilon_{xy}^r = \varepsilon_{xz}^r = \varepsilon_{yz}^r = 0 \tag{7.4.35}$$

The components of the magnetization vector, using (7.3.8) and (7.4.26) in (7.3.12c), are found to be

$$M_x = (1-\xi)M^{sat}\cos(\theta_3) = (1-\xi)M^{sat}\sqrt{1-\left(\frac{\mu_0 M^{sat}}{2\rho K_1}H_y\right)^2} , \tag{7.4.36a}$$

$$M_y = \xi M^{sat} + (1-\xi)M^{sat}\sin(\theta_3) = \xi M^{sat} + (1-\xi)\,\frac{\mu_0 (M^{sat})^2}{2\rho K_1} H_y , \tag{7.4.36b}$$

$$M_z = 0 . \tag{7.4.36c}$$

Again, the activation of the reorientation process is governed by the reorientation function (7.3.22) and the conditions (7.3.23). The evolution of ξ in (7.4.35) and (7.4.36) are then again described by evolution equations (7.4.31) and (7.4.34).

Calibration of the Model Parameters

The material parameters consist of the magnetocrystalline anisotropy constant ρK_1, the saturation magnetization M^{sat}, and the maximum reorientation strain $\varepsilon^{r,max}$, which follow from standard experiments described in the literature [50, 59]. Additionally, the critical magnetic field values $H_y^{s(1,2)}$, $H_y^{f(1,2)}$, $H_y^{s(2,1)}$ and $H_y^{f(2,1)}$ must be specified, which denote the start and finish of the forward and reverse magnetic-field-induced reorientation process, respectively. Fig. 7.17 shows how the critical magnetic fields and the maximum reorientation strain $\varepsilon^{max}\xi^{crit}$ are estimated from one positive magnetic-field-induced strain cycle at the calibration stress level, denoted σ_{xx}^*, which can be arbitrarily chosen in the range between zero and the blocking stress. The plotted experimental data was published by Heczko et al. [46] for the $Ni_{50.7}Mn_{28.4}Ga_{20.9}$ composition. In general, first estimates of the critical fields can be obtained by drawing tangent lines to the hysteresis curves, as indicated in Fig. 7.17. The calibration can then be refined by adjusting these parameters

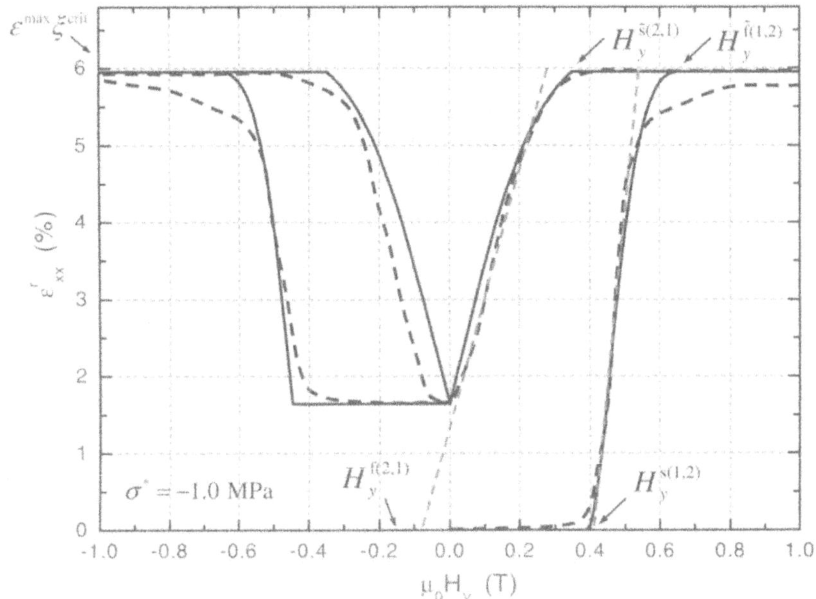

Fig. 7.17. Calibration of the model parameters using the experimental MFIS curve at -1.0 MPa. Solid line—model simulation; dashed line—experimental data [46].

such that the best simulation of the magnetic-field-induced strain response is obtained at the calibration stress level. The value of $H_y^{f(2,1)}$ was chosen to obtain the correct residual reorientation strain in the simulation.

The general relations between the material and the model parameters are listed in Table 7.1. They have been derived by evaluating the evolution equations for the martensitic variant volume fraction in the case of the forward

Table 7.1. Relations between material constants and model parameters

$$a_c^{V2} = \frac{\mu_0 M^{sat}}{\left(\pi - \cos^{-1}\left(2\xi^{crit} - 1\right)\right)}\left(H_y^{s(1,2)} - H_y^{\bar{f}(1,2)}\right) - \frac{(\mu_0 M^{sat})^2}{4\pi\rho K_1}\left[\left(H_y^{s(1,2)}\right)^2 - \left(H_y^{\bar{f}(1,2)}\right)^2\right]$$

$$\mu_1^c = \frac{1}{2}\mu_0 M^{sat}\left(H_y^{s(1,2)} + H_y^{f(2,1)}\right) - \frac{(\mu_0 M^{sat})^2}{8\rho K_1}\left[\left(H_y^{s(1,2)}\right)^2 + \left(H_y^{f(2,1)}\right)^2\right] + \sigma^* \varepsilon^{r,max}$$

$$\mu_2^c = \frac{\pi}{4}\left(a_c^{V2} - a_c^{V1}\right)$$

$$a_c^{V1} = \frac{\mu_0 M^{sat}}{\left(\pi - \cos^{-1}\left(2\xi^{crit} - 1\right)\right)}\left(H_y^{f(2,1)} - H_y^{\bar{s}(2,1)}\right) - \frac{(\mu_0 M^{sat})^2}{4\pi\rho K_1}\left[\left(H_y^{f(2,1)}\right)^2 - \left(H_y^{\bar{s}(2,1)}\right)^2\right]$$

$$Y^{\xi,c} = \frac{1}{2}\mu_0 M^{sat}\left(H_y^{s(1,2)} - H_y^{f(2,1)}\right) - \frac{(\mu_0 M^{sat})^2}{8\rho K_1}\left[\left(H_y^{s(1,2)}\right)^2 - \left(H_y^{f(2,1)}\right)^2\right] - \mu_2^c$$

(7.4.31) and reverse (7.4.34) reorientation process at $\xi = 0$ and $\xi = \xi^{crit}$, respectively, and enforcing continuity of the hardening function at $\xi = 1$. ξ^{crit} is the maximum volume fraction of variant 2 obtained at the stress level σ^*, which can be estimated by relating the maximum reorientation strain for this stress level to the maximum achievable reorientation strain $\varepsilon^{r,max}$. The magnetic field $H_y(\xi^{crit})$ at which the forward reorientation strain is terminated prior to its completion, because the magnetization vectors in both variants have aligned with the applied field, is denoted $H_y^{\tilde{f}(1,2)}$ (see (7.4.39) for the computation of its magnitude). The magnetic field at which the reverse reorientation process is then activated for $\xi = \xi^{crit} < 1$ is denoted $H_y^{\tilde{s}(2,1)}$. For the case of complete reorientation, ξ^{crit} is equal to 1, $H_y^{\tilde{f}(1,2)} = H_y^{f(1,2)}$ and $H_y^{\tilde{s}(2,1)} = H_y^{s(2,1)}$, such that the listed relations simplify. The parameter set listed in Table 7.2 was calibrated by employing the described methodology and utilizing the relations of Table 7.1[7] (refer also to Example 7.1).

In Fig. 7.17, the choice of parameters for the model simulation at the calibration stress level was observed to agree well with the experimental data. For model predictions at other stress levels, the parameters are, of course, not adjusted, as this would then result in a pure curve-fitting exercise. The details of the model predictions will be discussed in the following sections.

Although the outlined procedure itself is straightforward, calibrating material parameters can be a tedious effort due to the complexity of the magnetomechanical response of MSMA and the fact that real materials never behave as ideally as the model assumes. The model calibration is especially cumbersome for MSMA because the demagnetization effect makes experimental measurements difficult to interpret [94]. Measured data has to be corrected to account for the specimen shape dependence. The accurate interpretation of experimental results based on magnetostatic analysis for MSMA has been

Table 7.2. Material parameters for the considered $Ni_{50.7}Mn_{28.4}Ga_{20.9}$ composition [46], and the resulting hardening and hysteresis parameters calibrated at −1.0 MPa

Material Parameters						Model Parameters		
Quantity	Value	Unit	Quantity	Value	Unit	Quantity	Value	Unit
ρK_1	167.0	kJm^{-3}	$\mu_0 H_y^{s(1,2)}$	0.39	T	a_c^{V2}	−9.747	kPa
M^{sat}	514.0	kAm^{-1}	$\mu_0 H_y^{\tilde{f}(1,2)}$	0.65†	T	μ_1^c	−34.847	kPa
$\varepsilon^{r,max}$	6.2	%	$\mu_0 H_y^{\tilde{s}(2,1)}$	0.35	T	μ_2^c	54.700	kPa
σ^*	−1.0	MPa	$\mu_0 H_y^{f(2,1)}$	−0.15	T	a_c^{V1}	−79.394	kPa
ξ^{crit}	0.96					$Y^{\xi,c}$	58.451	kPa

† $= \mu_0 H^{crit}$

[7] Reprinted from *Proceedings of AIAA 2006*, paper 1766, Kiefer, B. and Lagoudas, D.C., pp. 1–15, Copyright 2006, with permission from American Institute of Aeronautics and Astronautics, Inc.

discussed in [85, 94, 95]. Advanced parameter optimization techniques, such as constrained optimization methods [2, 96], may also be employed to find more accurate sets of parameters.

The Reorientation Diagram

With a complete model at hand, a novel variant reorientation diagram is proposed as shown in Fig. 7.18, which is the graphical representation of the activation surfaces for variant reorientation. This diagram has been constructed based on the reduced constitutive equations presented in Sect. 7.4.1 and the set of model parameters listed in Table 7.2, specifically by evaluating the reorientation conditions (7.4.30) and (7.4.33) at $\xi = 0$ and $\xi = 1$, respectively.

Analogous visualizations of phase transformation surfaces in phase diagrams are common for conventional shape memory alloys [89, 97]. A reorientation diagram for MSMA was previously proposed by Kiefer and Lagoudas [84] for an earlier version of the constitutive model, in which the magnetization was assumed to be fixed to the respective magnetic easy axes of the martensitic variants and a normalized martensitic volume fraction was used. Other activation diagrams have been proposed by Tickle et al. [98] for variant

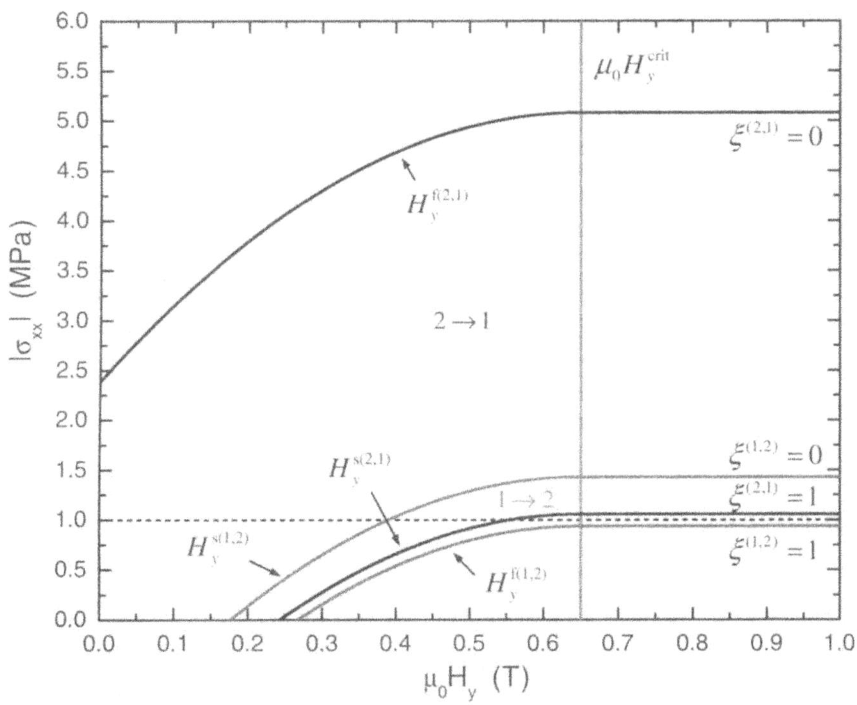

Fig. 7.18. Numerical $\mu_0 H_y - |\sigma_{xx}|$ variant reorientation diagram. The dashed line represents the magnetic loading path at the constant stress levels of $-1.0\,\text{MPa}$.

reorientation and by Yamamoto et al. [28] for martensitic phase transformation in MSMAs.

In accordance with the reorientation function (7.3.22) and the Kuhn-Tucker loading conditions (7.3.23), the forward reorientation process ($\dot{\xi} > 0$) in the variant reorientation diagram is activated at the line $\xi^{(1,2)} = 0$ and completed at the line $\xi^{(1,2)} = 1$. The reverse process ($\dot{\xi} < 0$) is activated at the line $\xi^{(2,1)} = 1$ and terminates at the line $\xi^{(2,1)} = 0$. The activation of each process takes place only if the activation lines are crossed in the proper loading direction, as determined by the Kuhn-Tucker loading conditions, and the value of the variant volume fraction is such that reorientation can occur.

The slopes of the activation lines in the reorientation diagram can be calculated from (7.4.30) or (7.4.33), respectively, and are given by the following *Clausius-Clapeyron*-type relation:

$$\frac{\mathrm{d}\sigma_{xx}}{\mathrm{d}H_y} = -\frac{1}{\varepsilon^{r,max}} \left[\mu_0 M^{sat} - \frac{(\mu_0 M^{sat})^2}{2\rho K_1} H_y \right] . \tag{7.4.37}$$

A similar expression was derived by Kiefer and Lagoudas [84] for the earlier model, which does not account for the rotation of the magnetization vectors. Tickle et al. [98], who also assumed that the magnetization vectors do not rotate, proposed the relation

$$\frac{\mathrm{d}\sigma_{xx}}{\mathrm{d}H_y} = -\frac{\mu_0 M^{sat}}{\varepsilon^{r,max}} , \tag{7.4.38}$$

which has been translated into the notation used here. The expressions (7.4.37) and (7.4.38) for the slope of the reorientation activation lines coincide at $H_y = 0$, i.e., the intersection with the stress axis, since the magnetization vectors at this field level are aligned with the easy axes in both models.

As depicted in the reorientation diagram of Fig. 7.18, the loading path at the calibration stress level of $\sigma^* = -1.0$ MPa intersects only the reorientation activation lines $H_y^{s(1,2)}(\sigma_{xx})$ and $H_y^{s(2,1)}(\sigma_{xx})$. The corresponding critical values were specified in Table 7.2. The line $H_y^{s(1,2)}(\sigma_{xx})$ is not crossed, such that only partial variant reorientation occurs at this stress level. In other words, the stress-independent critical value $\mu_0 H^{crit} = 0.65$ T, at which the magnetization in both variants has fully aligned with the applied field, is reached before the reorientation process is completed. This issue will be discussed in detail in the following section in the context of the interpretation of the model predictions. On the basis of the experimental data [46], it was assumed for the parameter calibration that only 96% of the second variant is produced during the forward reorientation process (cf. Table 7.2). Likewise, variant 1 is not completely recovered at this stress level as the magnetic field is removed, since the line $H_y^{f(2,1)}(\sigma_{xx})$ is not intersected. Is it important to note that for cases of incomplete reorientation ($0 < \xi < 1$), the appropriate activation lines lie within the regions bounded by the lines for $\xi = 0$ and $\xi = 1$ depicted in the reorientation diagram.

Several characteristic features can be concluded from the variant reorientation diagram for this particular set of parameters:

1. The blocking stress is predicted to be $-1.43\,\mathrm{MPa}$. For higher stresses, the forward reorientation activation line at $\xi = 0$ is not intersected;
2. Only in the range of 0 to $-0.94\,\mathrm{MPa}$ is the maximum strain obtained by complete variant reorientation. For higher stress levels, the reorientation is only partial, since the forward reorientation termination line is not intersected;
3. For this set of parameters, the model predicts that at least some of variant 1 is recovered at low stress levels because the activation line for the reverse process is intersected as the field reduces to zero;
4. Variant 1 can, in this case, not be fully recovered under any constant stress level, since the appropriate level is above the blocking stress.

Example 7.1. Determination of critical fields for variant reorientation

The critical fields for full forward reorientation $H_y^{s(1,2)}$ and $H_y^{f(1,2)}$ are determined by evaluating the reorientation conditions (7.4.30) at $\xi=0$ and $\xi=1$, respectively. Likewise, the critical fields for the reverse process, $H_y^{s(2,1)}$ and $H_y^{f(2,1)}$, follow from (7.4.33). This leads to the following set of equations:

$$\sigma_{xx}\varepsilon^{r,max} + \mu_0 M^{sat} H_y^{s(1,2)} - \frac{(\mu_0 M^{sat})^2}{4\rho K_1}\left(H_y^{s(1,2)}\right)^2 - \mu_1^c - \mu_2^c - Y^{\xi,c} = 0\,,$$

$$\sigma_{xx}\varepsilon^{r,max} + \mu_0 M^{sat} H_y^{f(1,2)} - \frac{(\mu_0 M^{sat})^2}{4\rho K_1}\left(H_y^{f(1,2)}\right)^2 + a_c^{V2}\pi - \mu_1^c - \mu_2^c - Y^{\xi,c} = 0\,,$$

$$\sigma_{xx}\varepsilon^{r,max} + \mu_0 M^{sat} H_y^{s(2,1)} - \frac{(\mu_0 M^{sat})^2}{4\rho K_1}\left(H_y^{s(2,1)}\right)^2 + a_c^{V1}\pi - \mu_1^c + \mu_2^c + Y^{\xi,c} = 0\,,$$

$$\sigma_{xx}\varepsilon^{r,max} + \mu_0 M^{sat} H_y^{f(2,1)} - \frac{(\mu_0 M^{sat})^2}{4\rho K_1}\left(H_y^{f(2,1)}\right)^2 - \mu_1^c + \mu_2^c + Y^{\xi,c} = 0\,.$$

This set of equations must be solved for the critical magnetic fields at each stress level. In this example, the critical values are determined for the calibration stress level of $-1\,\mathrm{MPa}$, which has the advantage that the critical fields are known *a priori* and the calculation can thus be used for verification. All necessary material and model parameters were specified in Table 7.2. Considering the physically meaningful roots, one obtains the following values:

$$\mu_0 H_y^{s(1,2)} = 0.390000\,\mathrm{T}$$

$$\mu_0 H_y^{f(1,2)} = (0.649805 + 0.099623i)\,\mathrm{T}$$

$$\mu_0 H_y^{s(2,1)} = 0.554718\,\mathrm{T}$$

$$\mu_0 H_y^{f(2,1)} = -0.150000\,\mathrm{T}$$

The computed values for $H_y^{s(1,2)}$ and $H_y^{f(2,1)}$ are consistent with the critical values specified in Table 7.2. The imaginary value of $H_y^{f(1,2)}$ indicates that the forward reorientation process is not completed at this stress level. This process actually terminates when the magnetization in both variants has fully aligned with the applied magnetic field. The corresponding field is given by $\mu_0 H^{crit} = \mu_0 H_y^{\tilde{f}(1,2)} = 0.65\,\mathrm{T}$ (cf. 7.4.39). Substituting $H_y = H^{crit}$ into (7.4.31) and solving for ξ reveals that 96% of variant 2 was produced during the forward reorientation. The reverse reorientation process then does not start at $\xi = 1$, and thus $H_y^{s(2,1)} = 0.55\,\mathrm{T}$, but rather at $\xi^{crit} = 0.96$. The corresponding field is computed by substituting ξ^{crit} into (7.4.33) and solving for H_y. This yields $H_y^{\tilde{s}(2,1)} = 0.35\,\mathrm{T}$, as was also specified in Table 7.2.

Numerical Results

Example simulations and predictions of the magnetic-field-induced strain and magnetization response of MSMAs produced by the proposed model are now considered.

The Magnetic-Field-Induced Strain Response

Based on the activation lines specified in the variant reorientation diagram of Fig. 7.18, and the evolution of the variant volume fraction and the reorientation strain given by (7.4.31), (7.4.34) and (7.4.35), respectively, magnetic-field-induced reorientation strain curves have been computed for three different stress levels. Fig. 7.19[8,9] displays the results.

All of the characteristic features, i.e. the nonlinear and hysteretic nature of the strain response, the reduction of the obtainable reorientation strain and the delay of the onset of reorientation with an increase in the stress level (the latter has already been discussed in the context of the proposed reorientation diagram of Fig. 7.18), which are typically observed in experiments [1, 13, 57, 58, 67], are captured by the model predictions. The curve at −1.0 MPa is the simulation of response from which the model parameters where calibrated (cf. Table 7.2). The two other curves at −1.1 MPa and −1.2 MPa are true model predictions, for which the same set of parameters was used.

[8] Reprinted from *Proceedings of ASME International Mechanical Engineering Congress and Exposition*, paper IMECE2006-15296, Lagoudas, D.C., Kiefer, B. and Broederdorf, A.J., pp. 1–11, Copyright 2006, with permission from ASME.

[9] Reprinted from *Proceedings of COMP07*, paper COMP2007-033, Lagoudas, D.C., Kiefer, B. and Broederdorf, A.J., pp. 1–8, Copyright 2007, with permission from COMP07.

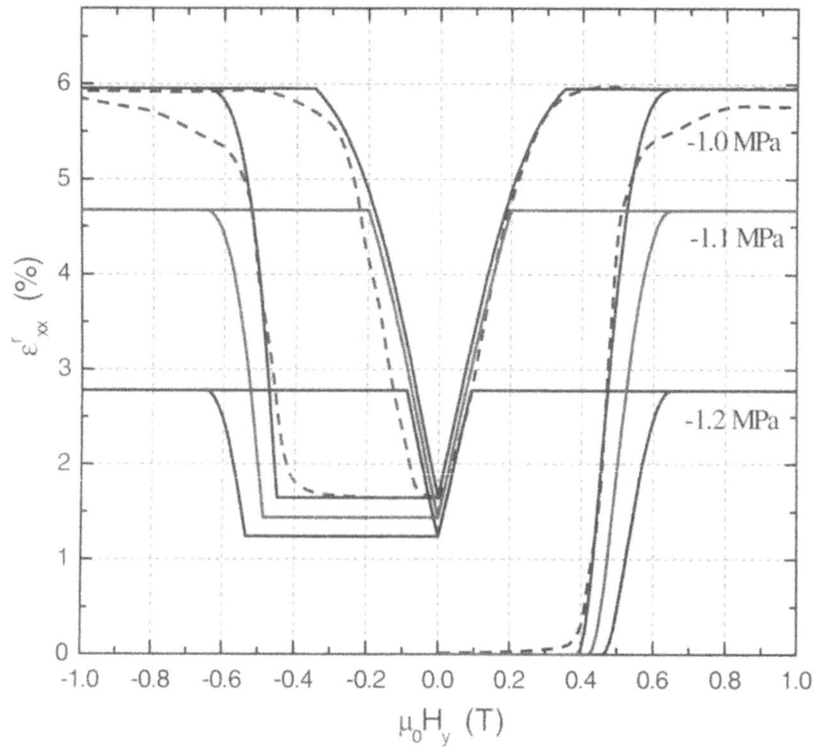

Fig. 7.19. Predicted MFIS hysteresis curves at different stress levels (solid lines) and comparison to experimental data (dashed line) [46].

Example 7.2. Evaluation of magnetic-field-induced strain curve

To illustrate the procedure, the magnetic-field-induced strain response curve at -1 MPa is plotted in this example. The field values that signify the end points of the reorientation regimes were calculated in Example 7.1. Using (7.4.35), the axial reorientation strain in the different regions is given by

$$\varepsilon^r_{xx} = \begin{cases} 0, & 0 < H_y < H_y^{s(1,2)}, \\ \xi^{(1,2)}\varepsilon^{max}, & H_y^{s(1,2)} < H_y < H^{crit} \quad (\text{since } H^{crit} < H_y^{f(1,2)}), \\ \xi^{crit}\varepsilon^{max}, & H^{crit} < H_y \leq H^{max}, \\ \xi^{crit}\varepsilon^{max}, & H^{max} \geq H_y > H_y^{\tilde{s}(2,1)}, \\ \xi^{(2,1)}\varepsilon^{max}, & H_y^{\tilde{s}(2,1)} > H_y > 0 \quad (\text{since } H_y^{f(2,1)} < 0). \end{cases}$$

As listed in Table 7.2 (see also Example 7.1), the forward reorientation only produces 96% of the second variant at -1 MPa, i.e. $\xi^{crit} = 0.96$, which corresponds to 5.95% magnetic-field-induced strain. The evolution of the volume

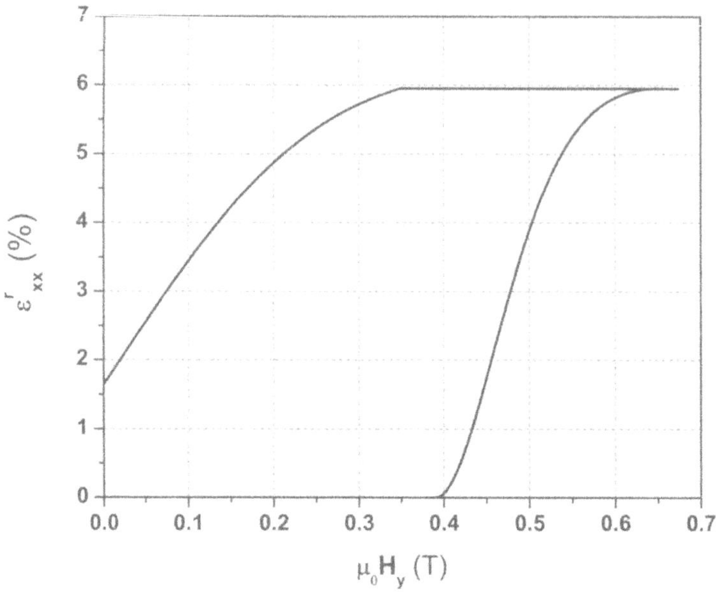

Fig. 7.20. Plot of the magnetic-field-induced reorientation strain as a function of applied magnetic field, as discussed in Example 7.2.

fraction in the reorientation regions is described by (7.4.31) and (7.4.34). The needed material and model parameters were again specified in Table 7.2. Fig. 7.20 shows the plot obtained for reorientation strain as a function of the magnetic field applied.

Another characteristic feature of the MSMA response that is captured by the presented model is the *first cycle effect*, i.e. an unsymmetrical response of the strain-magnetic field curves for the first cycle of positive and subsequent negative applied magnetic fields, as shown in Fig. 7.19. Subsequent cycles are predicted to be symmetric, which is in agreement with experimental observations (cf. Fig. 7.8).

To explain the predicted evolution of the magnetic-field-induced reorientation strain in more detail, Fig. 7.21 takes a closer look at the hysteresis loop under $-1.0\,\mathrm{MPa}$. Different characteristic configurations along the loading path have been numbered. For each of them, Table 7.3 shows a schematic representation of the variant volume fraction and the magnetization rotation as well as a list of the corresponding values of the applied magnetic field and the internal state variables, namely the reorientation strain, the variant volume fraction, and the magnetization rotation angle. It should be emphasized that the schematics shown in Table 7.3 are designed to illustrate only the connection of the macroscopic behavior to the evolution of the internal

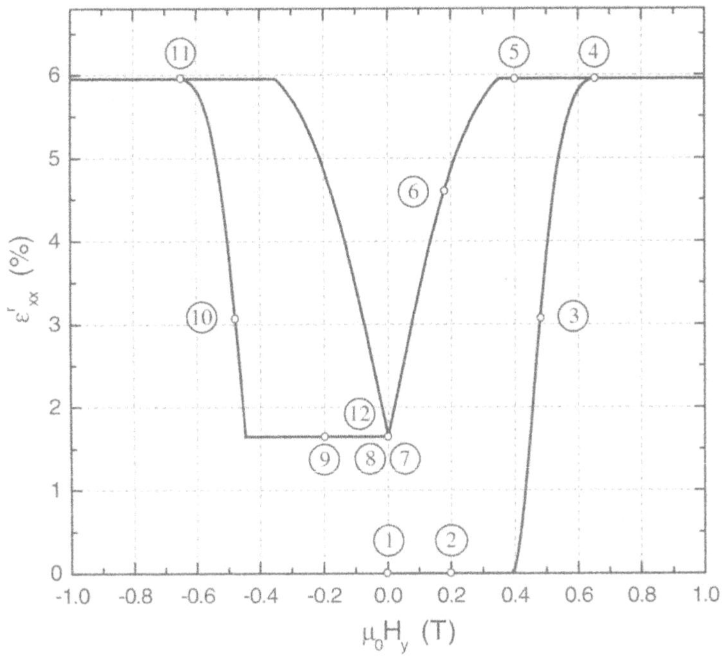

Fig. 7.21. Detail: MFIS hysteresis loop under −1.0 MPa.

state variables. They do not represent the actual distribution of the variants throughout a single crystal specimen.

Since it has been assumed that the error made by neglecting the magnetic domain wall motion at low magnetic fields is small, the single variant configuration 1 consists of only a single magnetic domain. The magnetic field of 0.2 T in configuration 2 is not sufficient to initiate the rearrangement of variants against the mechanical stress and the internal resistance to twin boundary motion. This field causes the magnetization vector in variant 1 to rotate by 17.9°, changing the magnetization of the specimen, but not producing any reorientation strain.

By increasing the magnetic field to the critical value, which is 0.39 T in this case (cf. Fig. 7.18 and Table 7.2), the reorientation process is initiated and variant 2 nucleates. In configuration 3, at 0.48 T, the variant rearrangement has produced 50% of variant 2, or 3.1% MFIS, while the magnetization has rotated by 47.6°. The stress of −1.0 MPa, which favors variant 1 and therefore counteracts the reorientation process, is higher than the resistance against 90° rotation of the magnetization in variant 1, as dictated by the magnetocrystalline anisotropy energy. Thus, the magnetization in variant 1 aligns with the external field before the reorientation process is completed, as indicated in schematic 4. Only 96% of variant 2 and therefore 5.95% reorientation strain can be magnetically induced at this stress level. The physical

Table 7.3. Configuration schematics and data for the strain hysteresis curve at −1.0 MPa

#	Schematic	$\mu_0 H_y$	ε^r_{xx}	ξ	$\theta_3(\theta_1)$
1		0.0 T	0.0 %	0.0	0.0°
2		0.2 T	0.0 %	0.0	17.9°
3		0.48 T	3.1 %	0.5	47.6°
4	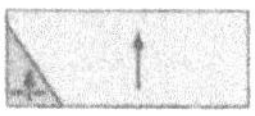	0.65 T	5.95 %	0.96	90.0°
5		0.4 T	5.95 %	0.96	38.0°
6	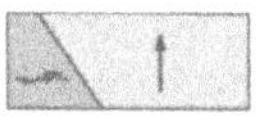	0.18 T	4.65 %	0.75	16.1°
7		0.0^+ T	1.65 %	0.27	0.0°
8		0.0^- T	1.65 %	0.27	0.0°
9		−0.2 T	1.65 %	0.27	17.9°
10	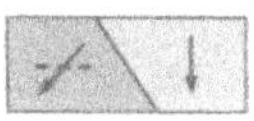	−0.5 T	3.1 %	0.5	47.6°
11	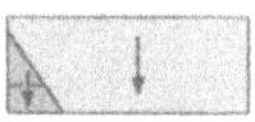	−0.65 T	5.95 %	0.96	90.0°
12	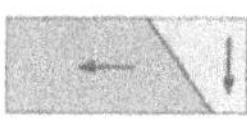	0.0^- T	1.65 %	0.27	0.0°

justification for this effect is given by the consideration that when the magnetization in variant 1 has completely aligned with the magnetic field, the Zeeman energy difference across the twin boundary vanishes and the driving force (7.4.28) no longer depends on the magnetic field. Thus, the reorientation process is terminated prematurely. This mechanism explains the reduction of the maximum magnetic-field-induced strain with increasing stress levels in the presented modeling approach.

According to (7.4.26), the critical magnetic field at which the magnetization in variant 1 has fully rotated is in the limit of $\theta_3 \to \frac{\pi}{2}$ given by

$$\mu_0 H_y^{crit} = \frac{2\rho K_1}{M^{sat}} = 0.65\ \mathrm{T}\,. \tag{7.4.39}$$

Note that the critical field is independent of the applied stress. The relative position of the critical field and the activation and termination fields for the forward reorientation process determine the amount of strain produced at each stress level. As previously observed from the variant reorientation diagram in Fig. 7.18, the magnetic-field-induced strain ranges from 0% at the blocking stress to its theoretical maximum of 6.2% for stresses below $|\sigma_{xx}|$=0.94 MPa.

When the magnetic field is subsequently decreased below H_y^{crit} to 0.4 T in configuration 5, for example, the magnetization in variant 1 rotates back towards the magnetic easy axis, but the MFIS stays constant. It must be emphasized that the activation field for the reverse reorientation process is not at 0.55 T, as the variant reorientation diagram in Fig. 7.18 suggests for $\xi = 1$, but rather at 0.35 T, which is the appropriate activation field for $\xi = 0.96$. By further lowering the field, variant 2 is reduced to 75% at 0.18 T in configuration 6, while the magnetization rotation angle in variant 1 has decreased to 16.1°. For this stress level, not all of variant 1 is recovered, even after complete removal of the magnetic field, and a residual MFIS of 1.65% remains in configuration 7. As evident from Fig. 7.19 as well as the variant reorientation diagram, higher compressive stresses help to recover a greater amount of variant 1. At zero field in configuration 7, the magnetization vectors in both variants are aligned with their respective easy axes. Since the effect of domain wall motion at low magnetic fields has been neglected, a remnant macroscopic magnetization is predicted by the model.

With the application of a negative magnetic field, the magnetization in both variants is assumed to instantaneously switch directions as indicated in schematics 7 and 8. In configuration 9, the ratio of variants remains unchanged, but the magnetization has rotated by 17.9°. Due to the mixture of variants, the activation of the reorientation process under a negative magnetic field is slightly delayed compared to the positive field hysteresis loop and occurs at -0.45 T, which is the appropriate activation value for $\xi = 0.27$. After the activation of the reorientation process through the magnetic field in the negative y-direction, the evolution of the magnetic-field-induced strain is symmetric to its positive counterpart described above in detail, and is consistent with the microstructure schematically shown in configurations 10 and 11. In terms of the residual strain, configuration 12 at the end of the negative loop is identical to configurations 7 and 8 at the end of the positive loop and the beginning of the negative loop, respectively. Unless a single variant configuration is purposely restored by temporarily raising the stress level, the reorientation strain that is obtainable in subsequent cycles is limited to 4.3%, which is the difference in strain between the configurations 7 (or 8, 9, 12) and 4 (or 5, 11). The reduction of the obtainable MFIS is thus limited to the first cycle, hence the term *first cycle effect.* It should also be clear that if the negative magnetic field had been applied first, the resulting hysteresis loop had been the mirror image of the presented one.

The Magnetization Response

The strong coupling between the deformation and changes in the magnetization that are characteristic of the MSMA constitutive response is made evident by considering the corresponding nonlinear magnetization hysteresis curves. The experimentally observed magnetization response was introduced in Fig. 7.13. Its unique characteristics will in the following paragraphs again, just as the macroscopic strain response before, be connected to the magnetic-field-induced activation of the main micro-scale mechanisms, namely the reorientation of variants and the local rotation of magnetization vectors, and in some cases, also magnetic domain wall motion.

The magnetic-field-induced magnetization curves plotted in Fig. 7.22[8,9] have been computed using (7.4.36), in addition to the evolution equations for the variant volume fraction, (7.4.31) and (7.4.34). All of the depicted curves represent model predictions, since the model parameters were entirely calibrated using information from experimental strain curves.

The schematics of Table 7.3 again prove helpful in understanding the connection between the evolution of the internal variables and the macroscopic material response, especially for the magnetization curve at −1.0 MPa. At low

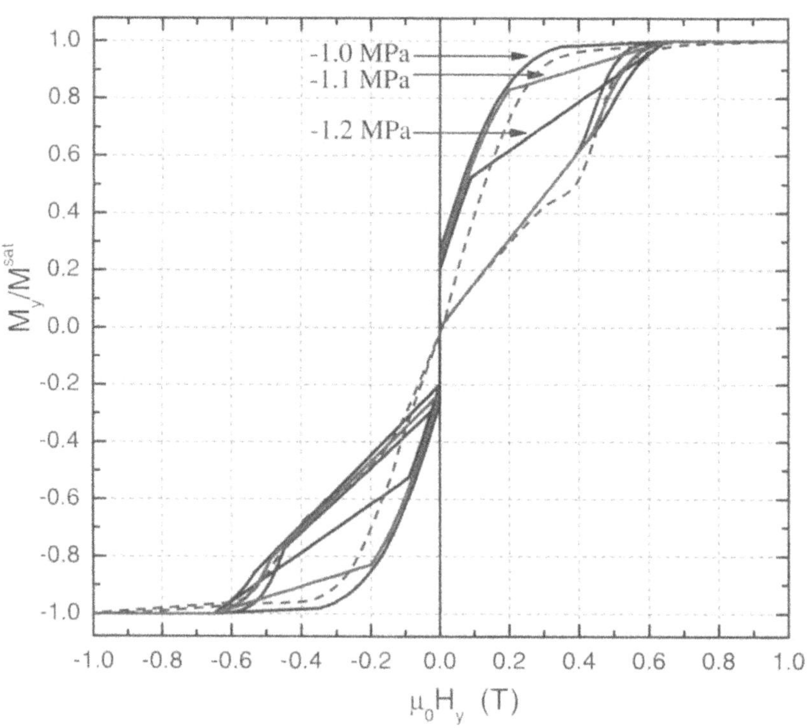

Fig. 7.22. Predicted magnetization hysteresis curves at different stress levels (solid lines) and comparison to experimental data [46] at −1.0 MPa (dashed line).

magnetic fields, when the material is in its initial single variant state, the M_y-curve represents the magnetization of variant 1 along its magnetic hard axis, which occurs via magnetization rotation. The magnetization M_y, according to (7.4.36), is predicted to have a linear dependence on H_y in this region, since only the first term in the expansion of the anisotropy energy (7.3.9) has been considered, which agrees well with the experimental observations. From (7.4.36), a slope of 1.53 T^{-1} has been calculated, a result in excellent agreement with the experimental data (dashed line) measured by Heczko et al. [46]. According to (7.4.26), the magnetization rotation is independent of the applied stress and all curves coincide in this initial region. The abrupt deviation from linearity of the magnetization curves occurs when the stress-dependent critical magnetic field is reached and the variant rearrangement is initiated. The magnetization in this region changes via the mechanism of variant rearrangement as well as magnetization rotation. The influence of the variant 2 magnetization becomes more prominent as the reorientation process progresses. When the critical magnetic field for full magnetization rotation in variant 1 has been reached, the reorientation process is terminated and the material is magnetized to saturation in the direction of the applied magnetic field.

Since the forward reorientation process is not completed for this stress level, the magnetization rotation in the residual variant 1 is reduced when the magnetic field is subsequently decreased below H^{crit}, resulting again in a linear variation of the magnetization. The slope, however, is different from the one initially observed for variant 1 at low magnetic fields, since 96% of the material still consists of variant 2, whose magnetization remains unaffected by the decrease of the magnetic field. Another abrupt nonlinear change in the magnetization occurs when the reverse reorientation process is activated.

Example 7.3. Evaluation of a magnetization curve

The magnetization response curve at -1 MPa is plotted in this example. The critical magnetic fields were determined in Example 7.1 and the corresponding magnetic field-induced strain curve was evaluated in Example 7.2. Using (7.4.36b), the y-component of the magnetization in the different regions is given by

$$M_y = \begin{cases} \dfrac{\mu_0 (M^{sat})^2}{2\rho K_1} H_y, & 0 < H_y < H_y^{s(1,2)} \,, \\[2ex] (1-\xi^{(1,2)}) \dfrac{\mu_0 (M^{sat})^2}{2\rho K_1} H_y + M^{sat}\xi^{(1,2)}, & H_y^{s(1,2)} < H_y < H^{crit} \,, \\[2ex] M^{sat}, & H^{crit} < H_y \leq H^{max} \,, \\[2ex] (1-\xi^{crit}) \dfrac{\mu_0 (M^{sat})^2}{2\rho K_1} H + M^{sat}\xi^{crit}, & H^{crit} > H_y > H_y^{\bar{s}(2,1)} \,, \\[2ex] (1-\xi^{(2,1)}) \dfrac{\mu_0 (M^{sat})^2}{2\rho K_1} H + M^{sat}\xi^{(2,1)}, & H_y^{\bar{s}(2,1)} > H_y > 0 \,. \end{cases}$$

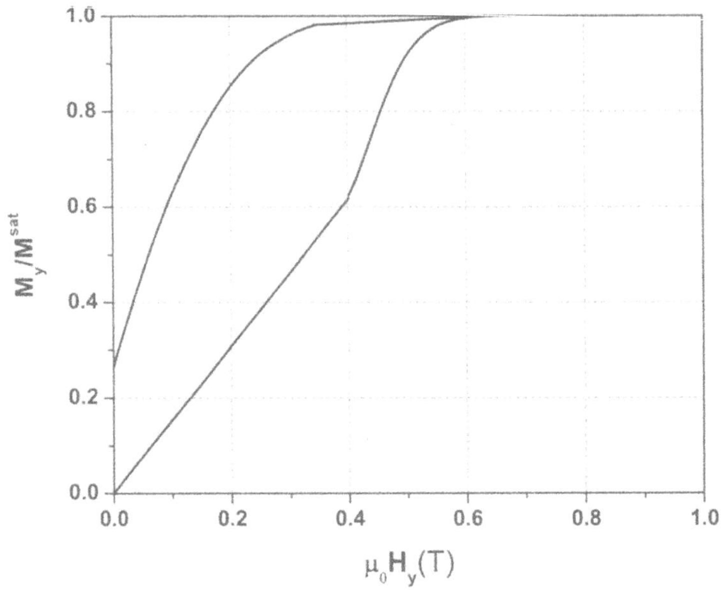

Fig. 7.23. Plot of normalized magnetization in the y−direction as a function of applied magnetic field as discussed in Example 7.3.

The evolution of the volume fraction in the reorientation regions is again defined by (7.4.31) and (7.4.34), respectively, and all necessary material and model parameters are specified in Table 7.2. The magnetization curve is shown in Fig. 7.23.

Due to the residual variant 2 volume fraction of 27%, a remnant magnetization is predicted at zero applied field, even though the magnetization in variant 1 has rotated back to its reference configuration. The model also predicts a jump of the magnetization curve as the applied field switches sign. As discussed, this discontinuity is a direct consequence of neglecting the mechanism of domain wall motion. The prediction of magnetization curves in this region thus deviates from the experimental measurements. However, the importance of this difference has not been deemed sufficient to justify raising the level of complexity of the model by accounting for the complicated evolution of the magnetic domains. This is only necessary if the main goal is to predict the MSMA magnetization response at low magnetic fields. It is concluded that for actuator applications the assumption of a fixed magnetic domain structure yields sufficiently accurate predictions of the magnetic-field-induced strain and magnetization response of MSMAs. Model predictions which account for the magnetization change due to magnetic domain wall motion at very low stress levels and low magnetic fields are presented next.

7.4.2 Prediction of Magnetic-Field-Induced Variant Reorientation at Constant Stress (Variable Domain Structure)

The previous section demonstrated that the assumption of a fixed magnetic domain structure leads to reasonable predictions of the magnetization behavior for stress levels at which most of the initial variant is recovered. It is observed in experiments [1, 46, 67], however, that low compressive stresses can be insufficient to even partially restore the stress-favored variant 1. If no reverse reorientation occurs, a residual strain of magnitude ε^{max} remains at zero applied field and this mechanism cannot contribute to changing the magnetization of the material. The remaining variant 2 has its easy axis aligned with the applied field such that no local rotation of its magnetization occurs. Therefore, the vanishing of the macroscopic magnetization as the applied field is removed, cf. Fig. 7.13, can be explained at low stresses only by magnetic domain wall motion. With the assumption of a fixed magnetic domain structure as used in the previous section, the model would, however, predict a remnant magnetization of M^{sat}. At such low stress levels, the evolution of magnetic domains, therefore, cannot be neglected. This mechanism is already incorporated in the general form of the constitutive model introduced in Sect. 7.3. Reduced model equations for the considered loading case with the possibility of magnetic domain wall motion shall now be derived.

Reduced Model Equations

Enforcing the constraint (7.3.16b) on θ_1 and utilizing (7.3.14a) leads to

$$\pi^{\theta_1} = -(1-\xi)(1-\alpha)\left[\mu_0 M^{sat} H_y + 2\rho K_1 \sin(\theta_1)\right]\cos(\theta_1) = 0\,. \tag{7.4.40}$$

It follows that

$$\sin(\theta_1) = -\frac{\mu_0 M^{sat}}{2\rho K_1} H_y\,, \tag{7.4.41}$$

for $0 \leq \theta_1 < \frac{\pi}{2}$, $0 \leq \xi < 1$ and $0 \leq \alpha < 1$. The equivalent relation for θ_3 was derived as (7.4.26) in the previous section. Note that

$$\sin(\theta_1) = -\sin(\theta_3)\,. \tag{7.4.42}$$

The corresponding constraints on the remaining rotation angles, with (7.3.14b) and (7.3.14d), are identically satisfied for $\theta_2 = \theta_4 = 0$.

For the considered loading conditions, the constraint (7.3.16a) on the driving force for magnetic domain wall motion, (7.3.13c), takes the form

$$\pi^{\alpha} = (1-\xi)\Big[\mu_0 M^{sat}\left[\sin(\theta_1) + \sin(\theta_3)\right] H_y - \rho K_1\left[\sin^2(\theta_1) - \sin^2(\theta_3)\right]\Big] + 2\xi\mu_0 M^{sat} H_y - \frac{\partial f^{\alpha}}{\partial \alpha} = 0\,. \tag{7.4.43}$$

Again assuming a hardening behavior of the type $\frac{\partial f^\alpha}{\partial \alpha} = a\alpha + b$, and utilizing (7.4.42), (7.4.43) can be written as

$$a\alpha + b = 2\xi\mu_0 M^{sat} H_y \,. \tag{7.4.44}$$

With the conditions $\alpha(H_y = 0) = 1/2$ and $\alpha(H_y = H^{crit,\alpha}) = 1$ the evolution equation for the magnetic domain volume fraction is derived from (7.4.44) as

$$\alpha = \frac{\xi H_y + H^{crit,\alpha}}{2H^{crit,\alpha}} \,. \tag{7.4.45}$$

Recall that $H^{crit,\alpha}$ has been defined to be the critical field at which magnetic saturation is achieved through domain wall motion.

The driving force for variant rearrangement, using (7.3.13a), (7.3.13b) and (7.3.20), takes the form

$$\pi^\xi = \sigma_{xx}\varepsilon^{r,max} - \mu_0 M^{sat}(1-2\alpha)H_y - \frac{(\mu_0 M^{sat})^2}{4\rho K_1} H_y^2 - \frac{\partial f^\xi}{\partial \xi} \,. \tag{7.4.46}$$

Note that for $\alpha = 1$, (7.4.46) appropriately reduces to (7.4.28). For the forward reorientation process, with trigonometric hardening (cf. (3.4.92)) and substituting (7.4.45), it follows that

$$\pi^\xi = \sigma_{xx}\varepsilon^{r,max} + \frac{\mu_0 M^{sat}}{H^{crit,\alpha}}\xi H_y - \frac{(\mu_0 M^{sat})^2}{4\rho K_1} H_y^2 + a_c^{V2}\left[\pi - \cos^{-1}(2\xi - 1)\right]$$
$$- \mu_1^c - \mu_2^c = Y^{\xi,c} \,. \tag{7.4.47}$$

This equation describes the evolution of the variant volume fraction under the simultaneous rotation of the magnetization vectors and the motion of magnetic domain walls in both variants. Due to its transcendental nature, it cannot be solved for ξ in closed-form. If one assumes a quadratic polynomial harding behavior of the form (3.3.63), rather than the trigonometric hardening assumed here, a closed-form solution for ξ can be found. However, the implementation of the model in which all three mechanisms for magnetization change are simultaneously active has proven to be very difficult. Preliminary analysis with a quadratic polynomial suggests that a simple linear variation of the term $\frac{\partial f^\alpha}{\partial \alpha}$ leads to an unsatisfactory evolution of the magnetic-field-induced strain in the reorientation region. The use of other hardening functions, which lead to an exponential evolution of the magnetic domain volume fraction, may yield the desired result.

Here, the motion of magnetic domain walls will be considered in magnetic field regimes in which no variant reorientation occurs. This assumption is very reasonable for those low stress levels at which no reverse reorientation is induced. Initially, the magnetic domain volume fraction takes the value of

$\alpha = 0.5$, which is consistent with (7.4.47) when evaluated at $\xi = 0$, and leads to a macroscopic magnetization of zero. For the forward reorientation process, which occurs at relatively high magnetic fields, it is assumed as in the previous section that $\alpha = 1$. Since the reverse reorientation process does not occur at the considered stress levels, no further assumptions on the evolution of the magnetic domain wall motion have to be made. The evolution of α is then properly described by (7.4.45).

The reorientation strain equations (7.4.35) are unaffected by the evolution of α, since cases are considered only if it is reasonable to assume that the variant reorientation and magnetic domain wall motion do not occur simultaneously. The y-component of the magnetization, using (7.3.8), (7.4.26) and (7.4.41) in (7.3.12c), is derived to be

$$M_y = \xi M^{sat}(2\alpha - 1) + (1 - \xi)\,\frac{\mu_0 (M^{sat})^2}{2\rho K_1} H_y \,. \tag{7.4.48}$$

The evolution of α for fixed ξ is given by (7.4.45). The evolution of ξ is coupled to the condition $\alpha = 1$ and is then described by (7.4.31). Alternatively, the same expression follows from solving (7.4.47) for ξ at $\alpha = 1$.

Calibration of the Model Parameters

In principle, the same set of parameters as that listed in Table 7.2 could be used here; however, as can be deduced from the variant reorientation diagram of Fig. 7.18, these parameters lead to the prediction of a partial recovery of variant 1 at the stress level of -0.2 MPa, which is inconsistent with the experimental strain data at this stress level [46]. To obtain a more accurate account of the magnetic-field-induced strain response at low stress levels, the alternative parameter set specified in Table 7.4 is used, which was obtained from a calibration at -0.2 MPa.

The Reorientation Diagram

The reorientation diagram in Fig. 7.24 was computed based on the set of parameters listed in Table 7.4 that where determined from a calibration at -0.2 MPa. This diagram is the equivalent of that shown in Fig. 7.18 for which the model parameters were found from the calibration at -1.0 MPa. In principle, one could have chosen to use just one calibration for all loading cases; however, the model calibration at -1.0 MPa has proven to be fairly accurate in a relatively wide range of stresses, while the calibration at -0.2 MPa leads to a more accurate prediction of the residual strain at very small stresses, which are considered in this section.

In the considered loading case, the magnetic field is applied at a constant stress of -0.2 MPa, as indicated by a dashed line in the reorientation diagram.

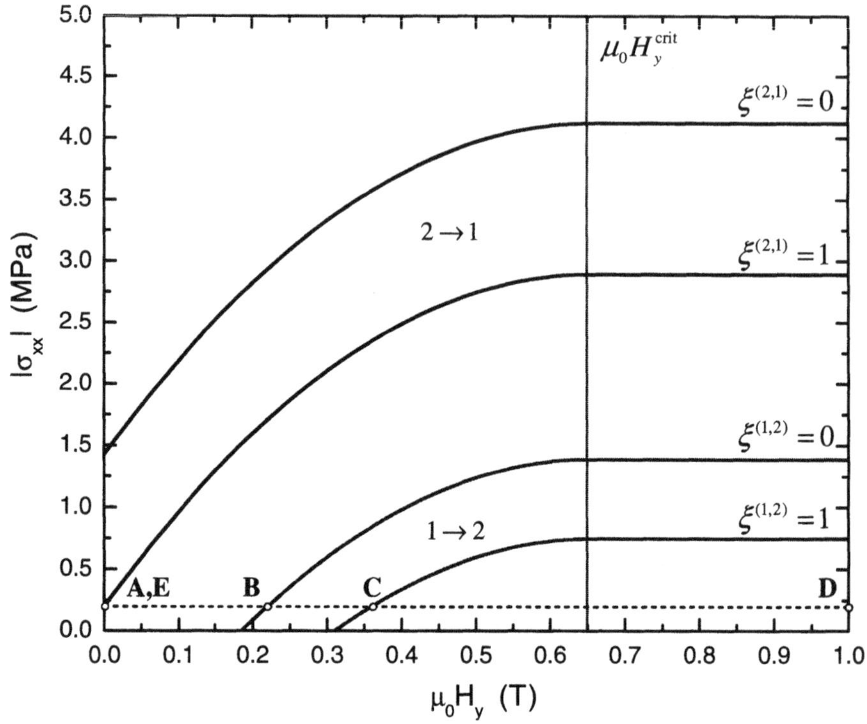

Fig. 7.24. Variant reorientation diagram for the parameters listed in Table 7.4. Loading path for model calibration A → E.

Several characteristic points in this loading sequence have been labeled A → E. Following the discussion from the beginning of this section, the magnetic domain wall motion is only considered at low stress levels and low magnetic fields, for which the reorientation process does not occur simultaneously.

Table 7.4. Material parameters for the $Ni_{50.7}Mn_{28.4}Ga_{20.9}$ composition [46], and the resulting hardening and hysteresis parameters when calibrated at −0.2 MPa

Material Parameters						Model Parameters		
Quantity	Value	Unit	Quantity	Value	Unit	Quantity	Value	Unit
ρK_1	167.0	$\mathrm{kJm^{-3}}$	$\mu_0 H_y^{s(1,2)}$	0.22	T	a_c^{V2}	−12.683	kPa
M^{sat}	514.0	$\mathrm{kAm^{-1}}$	$\mu_0 H_y^{f(1,2)}$	0.36	T	μ_1^c	−3.730	kPa
$\varepsilon^{r,max}$	6.2	%	$\mu_0 H_y^{s(2,1)}$	0.0	T	μ_2^c	9.188	kPa
σ^*	−0.2	MPa	$\mu_0 H_y^{f(2,1)}$	−0.135	T	a_c^{V1}	−24.382	kPa
ξ^{crit}	1.0		$\mu_0 H^{crit,\alpha}$	0.22	T	$Y^{\xi,c}$	76.080	kPa

Model Predictions

The magnetic-field-induced reorientation strain response is computed by evaluating (7.4.31) and (7.4.35) with the reorientation conditions given by (7.3.23) and utilizing the parameters set listed in Table 7.4. The magnetization in the direction of the applied field follows from additionally evaluating the relations (7.4.31), (7.4.45) and (7.4.48). The results have been plotted in Figs. 7.25 and 7.26.

The simulation of the magnetic-field-induced strain evolution, as well as the prediction of the magnetization response, are observed to agree rather well with the experimental data. The connection between the macroscopic response and the evolution of the internal variables, which capture the main characteristic features of the crystallographic and magnetic microstructure,

Table 7.5. Configuration schematics and data for the strain hysteresis curve at −0.2 MPa with partial magnetic domain wall motion

#	Schematic	$\mu_0 H_y$	ε^r_{xx}	ξ	θ_1, θ_3	α
1	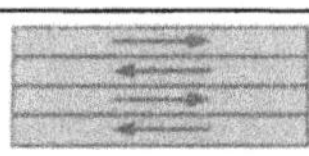	0.0 T	0.0 %	0.0	$0.0°, 0.0°$	0.5
2	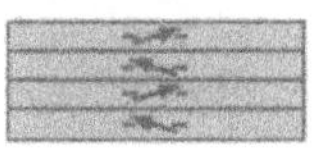	0.2 T	0.0 %	0.0	$-17.9°, 17.9°$	0.5
3	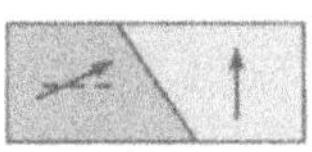	0.284 T	3.1 %	0.5	—, 25.9°	1.0
4	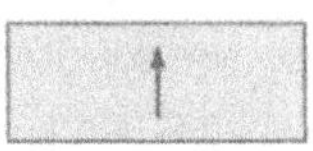	0.36 T	6.2 %	1.0	—	1.0
5	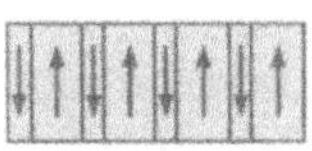	0.11 T	6.2 %	1.0	—	0.75
6	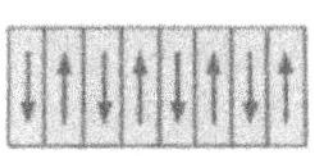	0.0 T	6.2 %	1.0	—	0.5
7	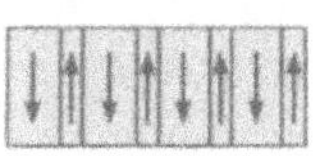	−0.11 T	6.2 %	1.0	—	0.25
8	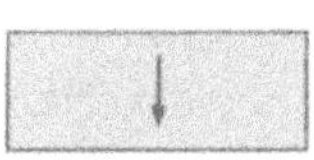	−0.22 T	6.2 %	1.0	—	0.0
9		0.0 T	6.2 %	1.0	—	0.5

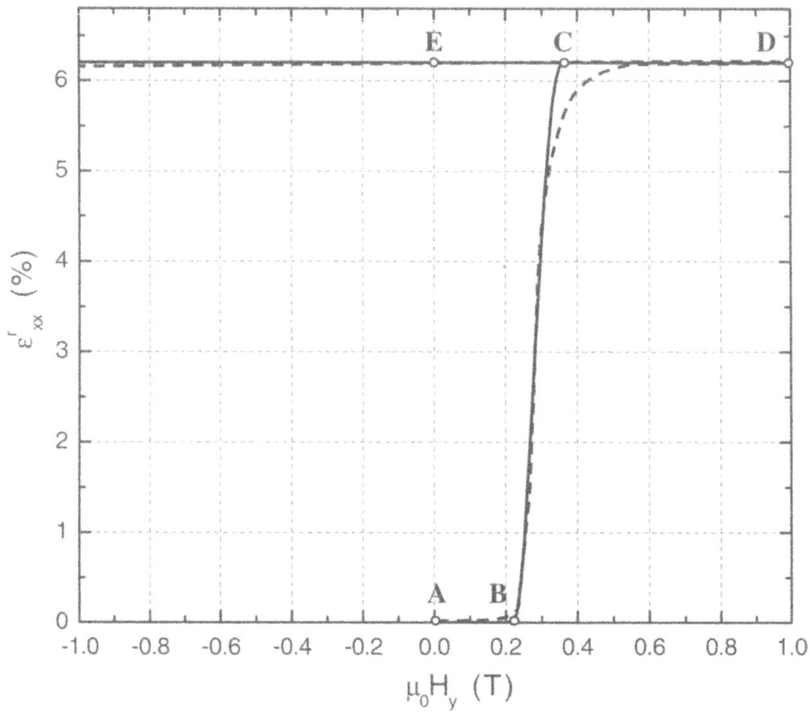

Fig. 7.25. Simulation of the magnetic-field-induced strain response at −0.2 MPa. Solid line—model, dashed line—experiment [46].

shall again be discussed in detail. Table 7.5 shows schematics for the sequence of numbered configurations defined in Fig. 7.26. These are analogous to the schematics presented in Table 7.3, with the difference that in this case the motion of magnetic domain walls has not been neglected.

In configuration 1, the magnetization vectors are arranged along the easy axis of variant 1, and both domain types are of equal volume fraction, such that the macroscopic magnetization is zero. As the magnetic field is applied perpendicularly to this easy axis in configuration 2, the equally unfavorably oriented magnetization vectors of both domains rotate toward the direction of the applied field. Upon reaching the critical field of 0.22 T, the forward reorientation process is initiated and unfavorable magnetic domains have again been assumed to be eliminated instantaneously (see discussion in Sect. 7.2.2). Configuration 3 is therefore predicted to consist of a mixed variant arrangement, but only of magnetic domain 2.

The critical field value of 0.36 T has been reached in configuration 4, and the forward reorientation process is completed, which results in the depicted single variant, single domain configuration. For example, when the magnetic field is subsequently lowered below the critical value of $\mu_0 H^{crit,\alpha} = 0.22$ T

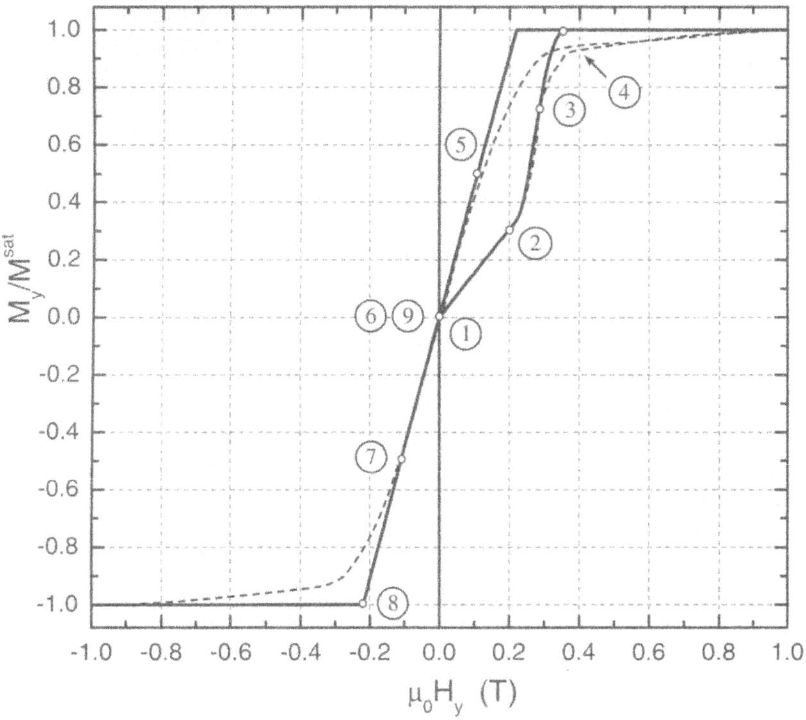

Fig. 7.26. Prediction of the magnetization response at −0.2 MPa when accounting for partial magnetic domain wall motion. Solid line—model, dashed line—experiment [46].

to 0.11 T in configuration 5, magnetic domain wall motion initiates and the macroscopic magnetization is reduced, even though variant reorientation does not occur. At zero field in configuration 6, the overall magnetization is again zero, but this time it transpires because the magnetization vectors in both domains are oriented along the easy axis of variant 2, with opposing orientations. As a negative magnetic field is applied, no variant reorientation occurs since the material already consists entirely of the magnetic-field-favored variant 2. The magnetic domain wall motion such as depicted in configuration 7, however, leads to magnetic saturation in configuration 8, at which point the critical value of $-H^{crit,\alpha}$ has been reached. When the magnetic field is removed, configuration 9 is obtained, which is identical to configuration 6. Thus, the macroscopic magnetization again vanishes, while the field-induced strain remains.

From the discussed example it is clearly evident that allowing for magnetic domain wall motion significantly improves the accuracy of the magnetization predictions at low stresses.

7.4.3 Prediction of Stress-Induced Variant Reorientation at Constant Magnetic Field

The second loading case considered is the magnetic-field-biased stress-induced reorientation of martensitic variants. Experimentally the magnetomechanical response of MSMAs under such loading conditions has been investigated by Müllner et al. [19], Straka and Heczko [99, 100] and Karaca et al. [59]. The investigated change in magnetization due to stress-induced variant reorientation, which in Müllner et al. [19] is referred to as the *reverse effect*, may lead to active sensor applications of MSMAs. This section demonstrates that the model introduced in Sect. 7.3 and 7.4, is capable of also predicting such response *without* adjustment of the model parameters.

Since the same components of the stress and magnetic field as in loading case 1 are applied here, with now the stress variable and the magnetic field constant, the same set of reduced model equations derived in Sect. 7.4.1 apply. Partial magnetic domain wall motion is again taken into account at magnetic fields below $H^{crit,\alpha}$, as was done in Sect. 7.4.2 so, the constitutive response is still described by (7.4.29)–(7.4.36).

Special attention must be paid to the response caused by mechanical loading at magnetic field levels below $H^{crit,\alpha}$, because magnetic domain wall motion does occur at these field levels. At $\mu_0 H_y = 0\,\mathrm{T}$, the magnetic domain volume fraction α takes the value of 0.5 according to (7.4.45). The driving force for reorientation π^ξ, (7.3.20), however, is independent of α at $\mu_0 H_y = 0\,\mathrm{T}$, such that

$$\pi^\xi = \sigma_{xx}\varepsilon^{r,max} - \frac{\partial f^\xi}{\partial \xi} \, . \tag{7.4.49}$$

The stress-induced evolution of ξ at 0 T can thus still be described by (7.4.34), which was derived for $\alpha = 1$.

It is again emphasized that the formulation of the model is such that the thermodynamic driving force for variant reorientation π^ξ incorporates contributions of the stress and the magnetic field (see for example (7.4.27)). The model parameters can thus be calibrated from a strain-magnetic field curve at constant applied stress or, alternatively, from a stress-strain curve at constant magnetic field. Ideally these calibrations should lead to the same model parameters, though in reality they are typically slightly different (this problem of course also arises for conventional shape memory alloys where one observes the model parameters calibrated from an isothermal tension test to be slightly different than those obtained from a thermal cycle at constant stress). The resistance to twin boundary motion, which on the micro-scale is governed by the *detwinning stress*, is in this phenomenological model captured by the reorientation function. This function depends not only on the macroscopic stress, but also the magnetic field and the internal state variables. The change in the detwinning stress during reorientation is captured phenomenologically by the hardening function.

All modeling results for the stress-induced response in this section are *exclusively* based on the parameters calibrated from the magnetic-field-induced reorientation strain curve at the constant stress of −0.2 MPa as presented in Sect. 7.4.2 (cf. Table 7.4).[10] No experimental data from the response under mechanical loading at a constant magnetic field was used in the calibration of the model. So, the chosen modeling approach is not merely a curve-fitting exercise.

Different loading paths for which the response caused by the magnetic-field-biased stress-induced reorientation of martensitic variants is predicted are indicated by dashed lines in the variant reorientation diagram shown in Fig. 7.27. This reorientation diagram is identical to that shown in Fig. 7.24, since, as discussed above, the same set of equations and parameters were used to determine it. As depicted in the diagram, in all loading cases the magnetic field is initially raised to 1 T under the constant stress of −0.2 MPa (A→D) and then lowered to the magnetic field H_y^{exp} at which the compressive stress cycle is to be conducted. This ensures a well-defined (single variant 2)

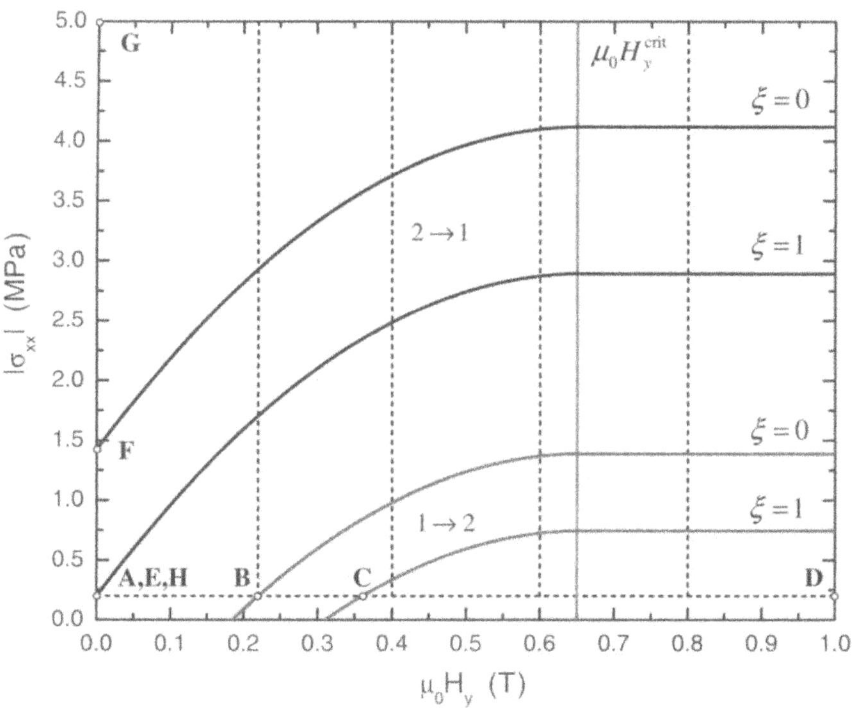

Fig. 7.27. Loading paths in the $\mu_0 H_y$–$|\sigma_{xx}|$ variant reorientation diagram. Example loading path A → H.

[10] This set of parameters was chosen here because it is again important to accurately predict the residual strain at low stress levels.

starting configuration for the mechanical cycle. The same is commonly done in experiments [99]. At the respective field levels of $H_y^{exp}=0$, 0.22, 0.4, 0.6 or 0.8 T the compressive stress is increased to -5.0 MPa, and then lowered back to -0.2 MPa. Finally the magnetic field is removed under constant stress. The case of stress loading at 0 T has been labeled with the letters E→H (A→E for the initial magnetic cycle) for illustration purposes.

Corresponding predicted stress-induced reorientation strain curves at different magnetic field levels are shown in Fig. 7.28 (solid lines). This three dimensional plot also contains the initial magnetic loading sequence. Characteristic points along the example loading path introduced in the reorientation diagram of Fig. 7.27 have again been labeled A→H. An interesting feature of this three-dimensional plot is that it also contains two of the earlier plots. Its projection onto the plane $\sigma_{xx}=0$ produces the same strain-magnetic field plot as the one depicted in Fig. 7.25. If one projects the start and finish points for the reorientation process onto the plane $\varepsilon^r_{xx}=0$ and connects them with lines, one obtains the variant reorientation diagram of Fig. 7.27. Additionally, the projection of the strain curves onto the plane $\mu_0 H_y=0$, produces a two-dimensional strain-stress plot, in which the response curves are parameterized by the value of the constant field $\mu_0 H_y$. The result of this last projection is shown in Fig. 7.29, where, for convenience, the induced reorientation strain has been plotted as a function of the absolute value of the stress.

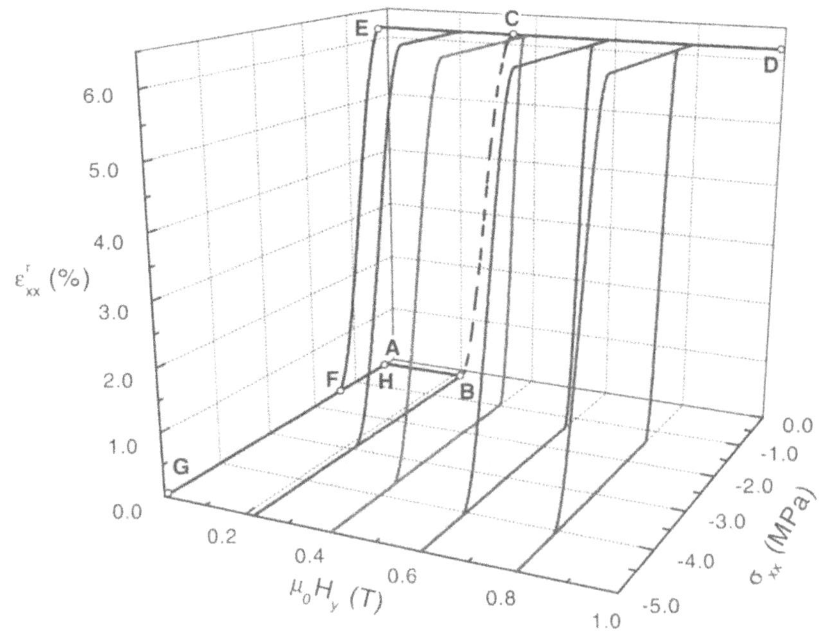

Fig. 7.28. Stress-induced reorientation strain vs. stress and magnetic field. Initial constant stress loop at -0.2 MPa.

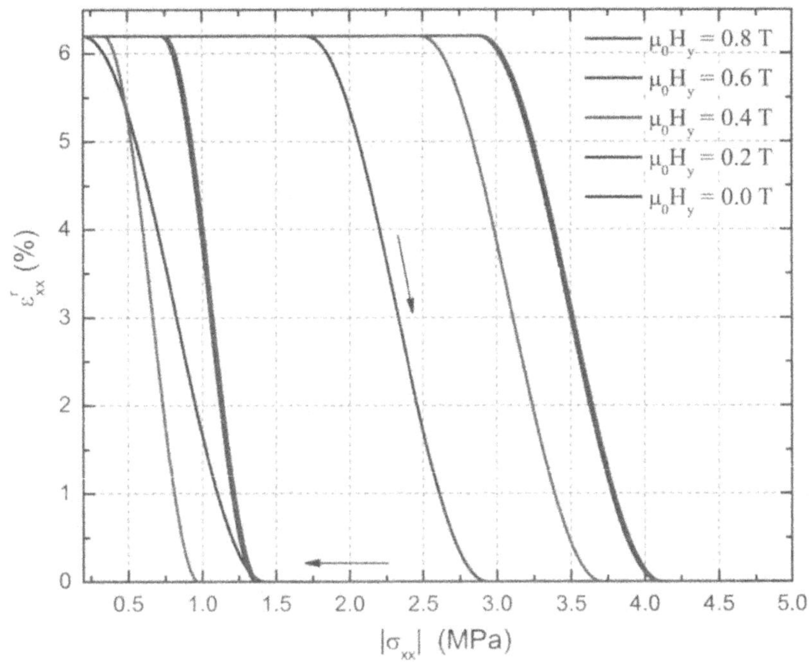

Fig. 7.29. Stress-induced reorientation strain vs. stress at different levels of $\mu_0 H_y$.

It can be stated that the predicted behavior is clearly consistent with the main features of the strain response observed by Straka and Heczko [99] in their experiments on the stress-induced reorientation of initially single variant Ni–Mn–Ga (five-layered modulated tetragonal) martensite:

1. The higher the magnetic field level, the higher the compressive stress needed to start the reorientation from the magnetic-field-favored to the stress-favored variant. This effect is of course directly evident from the variant reorientation diagram in Fig. 7.27.
2. At high magnetic fields (0.4, 0.6 and 0.8 T) the initial magnetic-field-favored is completely recovered as the stress is lowered back to the initial −0.2 MPa. This is the magnetic-field-biased *super-*, *magneto-* or *pseudoelastic effect* depending on preferred terminology [19, 59, 99].
3. Small magnetic fields (0 and 0.22 T) are not sufficient to return the stress-favored variant to the magnetic-field-favored variant configuration at the end of the stress loading cycle. This effect is often referred to as *superplasticity* [99, 100] or *magnetoplasticity* [19, 59, 101]. It is of course, unlike actual plastic deformation, a (crystallographically not thermodynamically) reversible effect and is the equivalent of the *one-way-shape-memory-effect* in SMAs.

4. At intermediate fields (in the region of 0.22–0.36 T in this case) only partial recovery of the magnetic-field-favored variant is reported in the referenced experiments. The constitutive model has been tested for these field levels and indeed predicts a mixture of both variants after mechanical loading, which corresponds to the observed partial recovery of the initial reorientation strain. In this case, however, the response to the removal of the magnetic field at the end of the loading sequence is quite complex. For the sake of clarity of the presented material the results of the modeling of partial stress-induced strain recovery followed by magnetic unloading have not been included here.

Further observations include that in the case of stress-driven reorientation the full induction of the stress-favored variant can be achieved at all magnetic field levels. Recall that in the magnetic field-driven case the induction of the field-favored variant was only partial for most stress levels and even completely suppressed at the blocking stress (cf. Fig. 7.19). There does not exists an equivalent *blocking magnetic field*. This statement can also be directly confirmed by consulting the variant reorientation diagram of Fig. 7.27. A magnetic loading path at constant stress only crosses the variant 1 $\rightarrow$ variant 2 reorientation band at stress levels below the blocking stress, whereas a stress loading path at any constant magnetic field crosses the variant 2 $\rightarrow$ variant 1 reorientation band.

This loading case also nicely illustrates the loading history dependence of the constitutive response, which the model is able to capture by accounting for the evolution of the microstructure through internal state variables. It is observed in the strain response of Fig. 7.28 that the sequence of loading significantly influences the response and determines the value of the residual strain. Note that, although all loading path start and end at the stress and magnetic field values of -0.2MPa and 0 T, the resulting strain is either zero or the maximum of 6.2% depending on the magnetic field level at which the stress cycle is applied (and in general takes any value in between at intermediate magnetic fields). The material response therefore clearly depends not only on the current values of the independent state variables stress and magnetic field, but also on the current configuration of the microstructure and thereby on the loading history, which must be taken into account when modeling these materials as was done in the proposed model.

Fig. 7.30 takes a closer look at the stress-induced reorientation strain response at the example magnetic field level of 0.4 T. The configuration schematics of Table 7.6 once more illustrate the connection between the evolution of the internal state variables and the observed macroscopic response. Recall that due to the prior magnetic loading, the material in configuration 1 consists entirely of the second magnetic-field-favored variant 2. Raising the compressive stress level initiates the reverse variant reorientation process at the stress of -2.49 MPa, corresponding to configuration 2. At -3.1 MPa, in

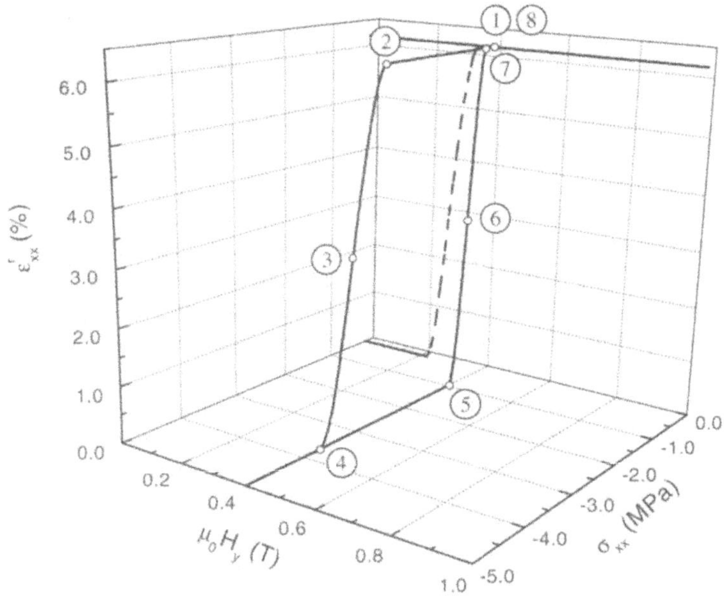

Fig. 7.30. Detail: stress-induced reorientation strain hysteresis loop at 0.4 T.

Table 7.6. Configuration schematics and data for the stress-induced strain hysteresis curve at 0.4 T

#	Schematic	σ_{xx}	ε^r_{xx}	ξ	θ_3
1	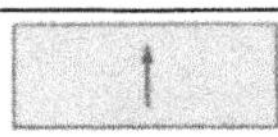	−0.20 MPa	6.2 %	1.0	—
2		−2.49 MPa	6.2 %	1.0	—
3	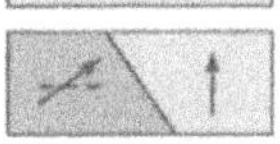	−3.10 MPa	3.1 %	0.5	38°
4	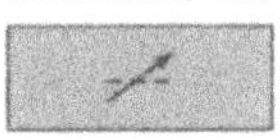	−3.71 MPa	0.0 %	0.0	38°
5		−0.98 MPa	0.0 %	0.0	38°
6		−0.66 MPa	3.1 %	0.5	38°
7	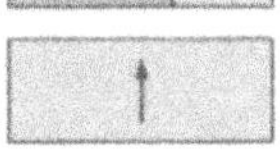	−0.34 MPa	6.2 %	1.0	—
8	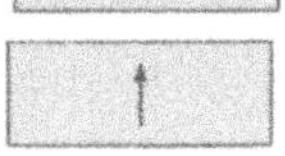	−0.20 MPa	6.2 %	1.0	—

configuration 3, half of variant 1 has been recovered. The process is completed in configuration 4. At this elevated magnetic field level of 0.4 T, however, the stress-induced variant is not sustained, and the forward reorientation process is again initiated and completed in configurations 5–7, as the stress is subsequently lowered. Configuration 8 at the end of the mechanical loading cycle is therefore identical to configuration 1 and entirely consists of variant 2. Similar explanations of the response apply to the other loading paths.

The associated stress-induced magnetization curves ("reverse effect") are shown in Fig. 7.31. The projection of these curves onto the plane $H_y = 0$ is plotted in Fig. 7.32. Just as for the strain response curves of Fig. 7.28, if the critical points of the magnetization curve are projected and connected in the plane $M_y = 0$ one obtains the reorientation diagram.

The response to the initial magnetic field cycle at -0.2 MPa, which has previously been shown in Fig. 7.26, is again depicted as a dashed line. A comparison of the results reveals that the nature of the stress-induced magnetization change is substantially different in nature than the magnetic-field-induced magnetization change of loading case 1 (cf. Fig. 7.22). In the magnetic-field-induced case, the magnetization always reaches saturation, whereas under

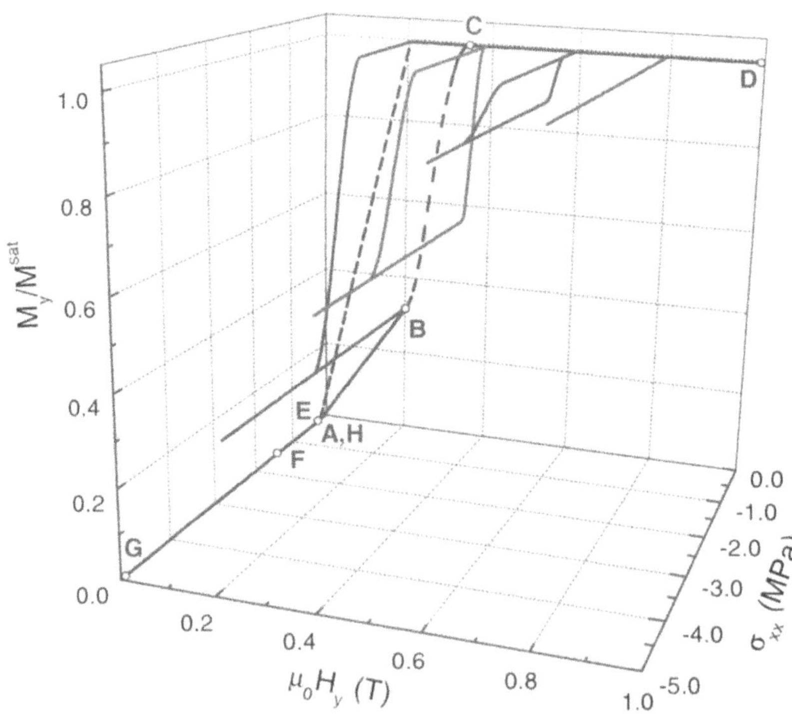

Fig. 7.31. Normalized magnetization vs. stress and magnetic field. Initial constant stress loop at -0.2 MPa.

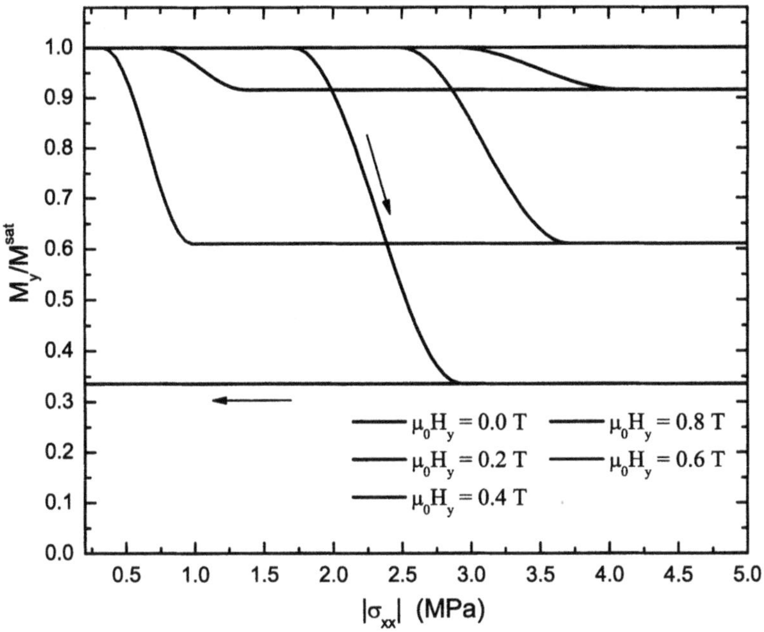

Fig. 7.32. Normalized magnetization vs. stress at different levels of $\mu_0 H_y$.

stress loading, the induced magnetization varies for different magnetic field levels. At low fields large changes in the relative magnetization occur; at large fields the changes are small. This prediction is consistent with the experimental observations of Straka and Heczko [100].

Referring back to the induced microstructural changes illustrated in Table 7.6, the explanation for this effect is that since the stress is applied at a constant magnetic field, the mechanism of magnetization rotation is not activated. The magnetization vector in variant 1 remains constant at the rotation angle associated with the magnetic field level, and the magnetization vector of variant 2 is fixed to its magnetic easy axis. At magnetic field levels above $\mu_0 H^{crit,\alpha} = 0.22\,\mathrm{T}$, where magnetic domain wall motion can be neglected, the only mechanism that leads to changes in the magnetization under mechanical loading at constant magnetic field, is thus the stress-induced reorientation of variants. In loading case 1, however, the mechanism of magnetization rotation was also active during variant reorientation, which explains the different nature of the induced magnetization response. At 0.8 T, which is above H^{crit}, the magnetization in both variants is already fully aligned with the applied field and no change in the magnetization is induced, even though it is clear from the strain response shown in Fig. 7.28 that variant reorientation occurs. The prediction is also consistent with the observation of Straka and Heczko [100]. Likewise, no change in the magnetization is observed at 0 T, but the reason in this case is that, in the absence of an external field, the arrangement of magnetic domains is such that the overall magnetization is zero.

From the comparison of the predicted magnetization response with the experimental findings of Straka and Heczko [100] and Müllner et al. [101], one observes, however, that the model apparently overpredicts the magnetization hysteresis. They report that, although the stress-strain hysteresis is significant, the stress magnetic field hysteresis is small. In fact Müllner et al. [101] even conclude that since in their observations the "reverse effect" is almost free of hysteresis, MSMAs show promise of usability in sensor applications. In the presented model, however, the hysteresis of both the strain and the magnetization response are directly tied to dissipative effects caused by twin boundary motion during the reorientation process. Thus if there exists significant hysteresis in the strain response the model must predict significant hysteresis in the magnetization response. This issue must be the subject of future investigations.

7.5 Summary

This chapter discussed the magneto-mechanical constitutive modeling of magnetic shape memory alloys, by connecting their macroscopic response to micro-scale mechanisms. The constitutive model followed the same approach used in previous chapters for conventional SMAs. Reorientation diagrams were constructed to visualize the variant reorientation conditions for various magneto-mechanical loading paths. Special consideration was given to magnetic loading at constant stress and mechanical loading at constant magnetic field. Predictions of the constitutive model were validated by comparison of results with experimental observations.

7.6 Problems

7.1. Consider the MSMA reorientation diagram shown in Fig. 7.33, as predicted by the model that accounts for the reorientation of martensitic variants as well as the local rotation of magnetization vectors. Of the full reorientation diagram (cf. Fig. 7.18) only the section relevant for magnetic field loading at constant stress is shown here.

(a) Define the physical significance of the field H^{crit}.
(b) Draw qualitative magnetic-field-induced strain and magnetization response curves for the three stress levels indicated in Fig. 7.33. Identify all critical values in the sketch.

7.2. Consider again the special loading case $\sigma_{xx} = \text{const.} \leq 0$ and $H_y \geq 0$ ($\alpha = 1$) discussed in Sect. 7.4.1.

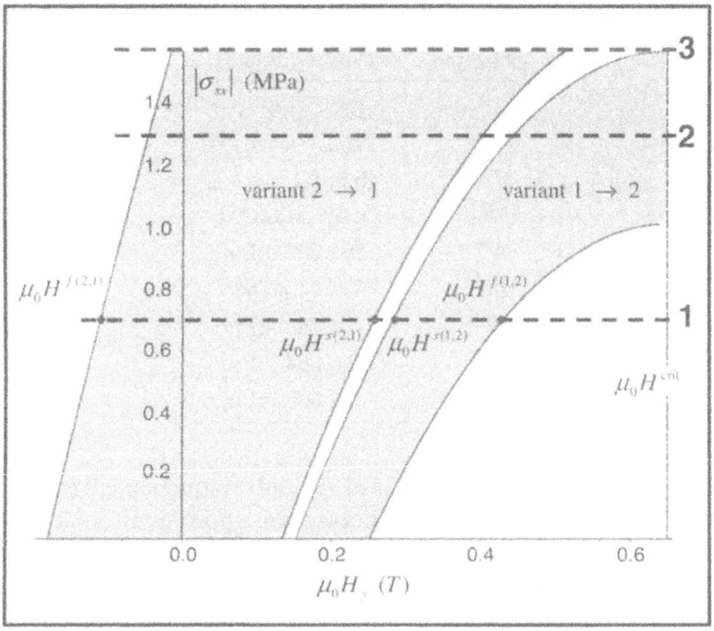

Fig. 7.33. Typical predicted MSMA reorientation diagram.

(a) Derive the driving force expression (7.4.27).
(b) Derive (7.4.25) by enforcing the constraint on π^{θ_3}. Using this relation deduce (7.4.28) from (7.4.27).
(c) Derive the expressions for $\xi^{(1,2)}$ and $\xi^{(2,1)}$ for power law hardening, which are the equivalent of (7.4.31) and (7.4.34). Use the following derivative of the *power law hardening function* [58]

$$\frac{\partial f^{\xi,pl}}{\partial \xi} = \begin{cases} A^{pl}\xi^{n-1} + B_1^{pl} + B_2^{pl} & , \quad \dot{\xi} > 0 \\ C^{pl}\Big[1 - (1-\xi)^{n-1}\Big] + B_1^{pl} - B_2^{pl}, & \dot{\xi} < 0 \end{cases} \tag{7.6.50}$$

(d) Derive the expression for $\mu_0 H^{crit}$, i.e. (7.4.39).
(e) Plot the magnetic-field-induced strain and magnetization curves for the stress levels -2MPa and -3MPa, respectively, using the parameters listed in the Table 7.7 (also see Examples 7.1, 7.2 and 7.3).

7.3. Based on the material parameters of Problem 7.2 above, calculate the actuation energy per unit volume of a MSMA as a function of applied stress. Compare with the actuation energy per unit volume of a typical SMA. Compare also the actuation energy per unit mass of a MSMA with a typical SMA, using $\rho_{MSMA} = 8020\,\mathrm{kg/m^3}$ and $\rho_{SMA} = 6500\,\mathrm{kg/m^3}$ as the corresponding mass densities.

Table 7.7. MSMA material parameters for Problem 7.2

Symbol	Value [unit]
μ_0	1.256 [μmTA^{-1}]
M^{sat}	742.4 [kAm^{-1}]
$\varepsilon^{m,max}$	5.6 [%]
ρK_1	0.58 [MJm^{-3}]
A^{pl}	204.807 [kPa]
B_1^{pl}	55.194 [kPa]
B_2^{pl}	174.763 [kPa]
C^{pl}	486.064 [kPa]
$Y^{\xi,c}$	22.508 [kPa]

7.4. MSMAs could potentially be used as high frequency actuators. Identify the physical mechanism that will dictate an upper limit in the frequency response of MSMAs. Consider effects both due to a fast changing magnetic field and also the effects of energy dissipation due to hysteresis.

7.5. Describe the feasibility of combining the shape memory effect of an SMA with the magnetic-field-induced reorientation strain of a MSMA to design a sensor-actuator combination responsive to thermo-mechano-magnetic fields.

7.6. Following the methods of Chapter 3, Sect. 3.5.1, derive a 1-D reduction of the model presented in this chapter. Using this result, merge this 1-D MSMA model with the 1-D reduction for the unified model given in Chapter 3 to create a consistent thermo-mechano-magnetic model for MSMAs.

References

[1] S. J. Murray, M. Marioni, S. M. Allen, R. C. O'Handley, 6% magnetic-field-induced strain by twin-boundary motion in ferromagnetic Ni-Mn-Ga, Applied Physics Letters 77 (6) (2000) 886–888.

[2] L. E. Faidley, M. J. Dapino, G. N. Washington, T. A. Lograsso, Reversible strain in Ni-MnGa with collinear field and stress, Proceedings of SPIE, Smart Structures and Materials: Active Materials: Behavior and Mechanics, San Diego, CA, 6–10 March 2005. Vol. 5761 (2005) 501–512.

[3] L. Hirsinger, N. Creton, C. Lexcellent, Stress-induced phase transformation in Ni-Mn-Ga alloys: Experiments and modelling, Material Science & Engineering A 378 (2004) 365–369.

[4] K. Ullakko, J. K. Huang, C. Kantner, R. C. O'Handley, V. V. Kokorin, Large magnetic-field-induced strains in Ni_2MnGa single crystals, Applied Physics Letters 69 (13) (1996) 1966–1968.

[5] R. A. Kellogg, A. B. Flatau, A. E. Clark, M. Wun-Fogle, T. A. Lograsso, Temperature and stress dependencies of the magnetic and magnetostrictive properties of $Fe_{0.81}Ga_{0.19}$, Journal of Applied Physics 91 (10) (2002) 7821–7823.

[6] I. Chopra, Review of state of art of smart structures and integrated systems, AIAA Journal 40 (11) (2002) 2145–2187.

[7] V. A. Chernenko, V. V. Kokorin, O. M. Babii, I. K. Zasimchuk, Phase diagrams in the Ni-Mn-Ga system under compression, Intermetallics 6 (1) (1998) 29–34.

[8] J. Pons, V. A. Chernenko, E. Cesari, V. A. L'vov, Stress-strain-temperature behaviour for martensitic transformation in Ni-Mn-Ga single crystals compressed along $\langle 001 \rangle$ and $\langle 110 \rangle$ axes, Journal de Physique IV France 112 (2003) 939–942.

[9] V. V. Martynov, V. V. Kokorin, The crystal structure of thermally- and stress-induced martensites in Ni_2MnGa single crystals, Journal de Physique III France 2 (1992) 739–749.

[10] H. E. Karaca, I. Karaman, D. C. Lagoudas, H. J. Maier, Y. I. Chumlyakov, Recoverable stress-induced martensitic transformation in a ferromagnetic CoNiAl alloy, Scripta Materialia 49 (2003) 831–836.

[11] J. Tellinen, I. Suorsa, I. Jääskeläinen, Aaltio, K. Ullakko, Basic properties of magnetic shape memory actuators, Proceedings of the 8th International Conference ACTUATOR 2002, Bremen, Germany, 10-12 June 2002 (2002) 566–569.

[12] A. Bhattacharyya, D. C. Lagoudas, Y. Wang, V. K. Kinra, On the role of thermoelectric heat transfer in the design of SMA actuators: Theoretical modeling and experiment, Smart Materials and Structures 4 (1995) 252–263.

[13] O. Heczko, A. Sozinov, K. Ullakko, Giant field-induced reversible strain in magnetic shape memory NiMnGa alloy, IEEE Transactions on Magnetics 36 (5) (2000) 3266–3268.

[14] B. D. Cullity, Introduction to Magnetic Materials, Addison-Wesley, Reading, MA, 1972.

[15] R. C. O'Handley, Modern Magnetic Materials, John Wiley & Sons, New York, 2000.

[16] C. Kittel, Introduction to Solid State Physics, 7th Edition, John Wiley & Sons, New York, 1996.

[17] M. Pasquale, Mechanical sensors and actuators, Sensors and Actuators A 106 (2003) 142–148.

[18] N. Sarawate, M. Dapino, Experimental characterization of the sensor effect in ferromagnetic shape memory Ni-Mn-Ga, Applied Physics Letters 88 (2006) 121923–1–3.

[19] P. Müllner, V. A. Chernenko, G. Kostorz, Stress-induced twin rearrangement resulting in change of magnetization in a Ni–Mn–Ga ferromagnetic martensite, Scripta Materialia 49 (2003) 129–133.

[20] I. Suorsa, J. Tellinen, K. Ullakko, E. Pagounis, Voltage generation induced by mechanical straining in magnetic shape memory materials, Journal of Applied Physics 95 (12) (2004) 8054–8058.
[21] P. Müllner, V. A. Chernenko, M. Wollgarten, Kostorz, Large cyclic deformations of a Ni-Mn-Ga shape memory alloy induced by magnetic fields, Journal of Applied Physics 92 (11) (2002) 6708–6713.
[22] P. J. Webster, K. R. A. Ziebeck, S. L. Town, M. S. Peak, Magnetic order and phase transformation in Ni_2MnGa, Philosophical Magazine B 49 (3) (1984) 295–310.
[23] I. K. Zasimchuk, V. V. Kokorin, V. V. Martynov, A. V. Tkachenko, V. A. Chernenko, Crystal structure of martensite in Heusler alloy Ni_2MnGa, Physics of Metals and Metallography 69 (6) (1990) 104–108.
[24] A. Sozinov, A. A. Likhachev, N. Lanska, K. Ullakko, Giant magnetic-field-induced strain in NiMnGa seven-layered martensitic phase, Applied Physics Letters 80 (10) (2002) 1746–1748.
[25] J. Cui, T. W. Shield, R. D. James, Phase transformation and magnetic anisotropy of an iron-palladium ferromagnetic shape-memory alloy, Acta Materialia 52 (2004) 35–47.
[26] R. D. James, M. Wuttig, Magnetostriction of martensite, Philosophical Magazine A 77 (5) (1998) 1273–1299.
[27] T. W. Shield, Magnetomechanical testing machine for ferromagnetic shape-memory alloys, Review of Scientific Instruments 74 (9) (2003) 4077–4088.
[28] T. Yamamoto, M. Taya, Y. Sutou, Y. Liang, T. Wada, L. Sorensen, Magnetic field-induced reversible variant rearrangement in Fe-Pd single crystals, Acta Materialia 52 (17) (2004) 5083–5091.
[29] S. J. Murray, R. Hayashi, M. Marioni, S. M. Allen, R. C. O'Handley, Magnetic and mechanical properties of FeNiCoTi and NiMnGa magnetic shape memory alloys, Proceedings of SPIE 3675 (1999) 204–211.
[30] A. Fujita, K. Fukamichi, F. Gejima, R. Kainuma, K. Ishida, Magnetic properties and large magnetic-field-induced strains in off-stoichiometric Ni-Mn-Al heusler alloys, Applied Physics Letters 77 (19) (2000) 3054–3056.
[31] T. Kakeshita, T. Takeuchi, T. Fukuda, M. Tsujiguchi, T. Saburi, R. Oshima, S. Muto, Giant magnetostriction in an ordered Fe_3Pt single crystal exhibiting a martensitic transformation, Applied Physics Letters 77 (10) (2000) 1502–1504.
[32] M. Wuttig, L. Liu, K. Tsuchiya, R. D. James, Occurrence of ferromagnetic shape memory alloys (invited), Journal of Applied Physics 87 (9) (2000) 4707–4711.
[33] M. Wuttig, J. Li, C. Craciunescu, A new ferromagnetic shape memory alloy system, Scripta Materialia 44 (2001) 2393–2397.
[34] A. Sozinov, A. A. Likhachev, N. Lanska, O. Söderberg, K. Ullakko, V. K. Lindroos, Effect of crystal structure on magnetic-field-induced strain in

Ni-Mn-Ga, Proceedings of SPIE, Symposium on Smart Structures and Materials 5053 (2003) 586–594.

[35] H. Morito, A. Fujita, R. Kainuma, K. Ishida, K. Oikawa, Magnetocrystalline anisotropy in single-crystal Co-Ni-Al ferromagnetic shape-memory alloy, Applied Physics Letters 81 (9) (2002) 1657–1659.

[36] T. Sakamoto, T. Fukuda, T. Kakeshita, T. Takeuchi, K. Kishio, Magnetic field-induced strain in iron-based ferromagnetic shape memory alloys, Journal of Applied Physics 93 (10) (2003) 8647–8649.

[37] A. A. Cherechukin, I. E. Dikshtein, D. I. Ermakov, A. V. Glebov, V. V. Koledov, D. A. Kosolapov, V. G. Shavrov, A. A. Tulaikova, E. P. Krasnoperov, T. Takagi, Shape memory effect due to magnetic field-induced thermoelastic martensitic transformation in polycrystalline Ni-Mn-Fe-Ga alloy, Physics Letters A 291 (2001) 175–183.

[38] S. Jeong, K. Inoue, S. Inoue, K. Koterazawa, M. Taya, K. Inoue, Effect of magnetic field on martensite transformation in a polycrystalline Ni_2MnGa, Material Science & Engineering A 359 (2003) 253–260.

[39] S. J. Murray, M. Farinelli, C. Kantner, J. K. Huang, A. M. Allen, R. C. O'Handley, Field-induced strain under load in Ni-Mn-Ga magnetic shape memory materials, Journal of Applied Physics 83 (11) (1998) 7297–7299.

[40] K. Ullakko, Y. Ezer, A. Sozinov, G. Kimmel, P. Yakovenko, V. K. Lindroos, Magnetic-field-induced strains in polycrystalline Ni-Mn-Ga at room temperature, Scripta Materialia 44 (2001) 475–480.

[41] T. Wada, Y. Liang, H. Kato, T. Tagawa, M. Taya, T. Mori, Structural change and straining in Fe-Pd polycrystals by magnetic field, Material Science & Engineering A 361 (2003) 75–82.

[42] M. A. Marioni, R. C. O'Handley, S. A. Allen, Analytical model for field-induced strain in ferromagnetic shape-memory alloy polycrystals, Journal of Applied Physics 91 (10) (2002) 7807–7809.

[43] T. Kakeshita, T. Takeuchi, T. Fukuda, T. Saburi, R. Oshima, S. Muto, K. Kishio, Magnetic field-induced martensitic transformation and giant magnetostriction in Fe-Ni-Co-Ti and ordered Fe_3Pt shape memory alloys, Materials Transactions, JIM 41 (8) (2000) 882–887.

[44] A. N. Vasil'ev, A. D. Bozhko, V. V. Khovailo, I. E. Dikshtein, V. G. Shavrov, V. D. Buchelnikov, M. Matsumoto, S. Suzuki, T. Takagi, J. Tani, Structural and magnetic phase transitions in shape-memory alloys $Ni_{2+x}Mn_{1-x}Ga$, Physical Review B 59 (2) (1999) 1113–1119.

[45] I. Karaman, H. E. Karaca, B. Basaran, D. C. Lagoudas, Y. I. Chumlyakov, H. J. Maier, Stress-assisted reversible magnetic field-induced phase transformation in Ni_2MnGa magnetic shape memory alloys, Scripta Materialia 55 (4) (2006) 403–406.

[46] O. Heczko, L. Straka, K. Ullakko, Relation between structure, magnetization process and magnetic shape memory effect of various martensites occuring in Ni-Mn-Ga alloys, Journal de Physique IV France 112 (2003) 959–962.

[47] A. A. Likhachev, A. Sozinov, K. Ullakko, Different modeling concepts of magnetic shape memory and their comparison with some experimental results obtained in Ni-Mn-Ga, Material Science & Engineering A 378 (2004) 513–518.

[48] F. Albertini, L. Pareti, A. Paoluzi, L. Morellon, P. A. Algarabel, M. R. Ibarra, L. Righi, Composition and temperature dependence of the magnetocrystalline anisotropy in $Ni_{2+x}Mn_{1+y}Ga_{1+z}$ $(x+y+z=0)$ Heusler alloys, Applied Physics Letters 81 (21) (2002) 4032–4034.

[49] F. Albertini, F. Canepa, S. Cirafici, E. A. Franceschi, M. Napoletano, A. Paoluzi, L. Pareti, M. Solzi, Composition dependence of magnetic and magnetothermal properties of Ni-Mn-Ga shape memory alloys, Journal of Magnetism and Magnetic Materials 272–276 (Part 3) (2004) 2111–2112.

[50] O. Heczko, L. Straka, Compositional dependence of structure, magnetization and magnetic anisotropy in Ni-Mn-Ga magnetic shape memory alloys, Journal of Magnetism and Magnetic Materials 272–276 (Part 3) (2004) 2045–2046.

[51] S. J. Murray, M. Marioni, A. M. Kukla, J. Robinson, R. C. O'Handley, S. M. Allen, Large field induced strain in single crystalline Ni-Mn-Ga ferromagnetic shape memory alloy, Journal of Applied Physics 87 (9) (2000) 5774–5776.

[52] S. J. Murray, Magneto-mechanical properties and applications of Ni-MnGa ferromagnetic shape memory alloy, Ph.D. thesis, Massachusetts Institute of Technology (February 2000).

[53] L. Hirsinger, N. Creton, C. Lexcellent, From crystallographic properties to macroscopic detwinning strain and magnetisation of Ni-Mn-Ga magnetic shape memory alloys, Journal de Physique IV France 115 (2004) 111–120.

[54] O. Heczko, N. Lanska, O. Soderberg, K. Ullakko, Temperature variation of structure and magnetic properties of Ni-Mn-Ga magnetic shape memory alloys, Journal of Magnetism and Magnetic Materials 242–245 (2002) 1446–1449.

[55] A. A. Likhachev, K. Ullakko, Quantitative model of large magnetostrain effect in ferromagnetic shape memory alloys, The European Physical Journal B 14 (2) (2000) 263–267.

[56] R. Tickle, R. D. James, Magnetic and magnetomechanical properties of Ni_2MnGa, Journal of Magnetism and Magnetic Materials 195 (3) (1999) 627–638.

[57] R. Tickle, Ferromagnetic shape memory materials, Ph.D. thesis, University of Minnesota (May 2000).

[58] B. Kiefer, H. E. Karaca, D. C. Lagoudas, I. Karaman, Characterization and modeling of the magnetic field-induced strain and work output in Ni_2MnGa shape memory alloys, Journal of Magnetism and Magnetic Materials 312 (1) (2007) 164–175.

[59] H. E. Karaca, I. Karaman, B. Basaran, Y. I. Chumlyakov, H. J. Maier, Magnetic field and stress induced martensite reorientation in NiMnGa ferromagnetic shape memory alloy single crystals, Acta Materialia 54 (1) (2006) 233–245.

[60] N. A. Spaldin, Magnetic Materials: Fundamentals and Device Applications, Cambridge University Press, 2003.

[61] C. Kittel, Physical theory of ferromagnetic domains, Reviews of Modern Physics 21 (4) (1949) 541–583.

[62] J. G. Boyd, D. C. Lagoudas, A thermodynamical constitutive model for shape memory materials. Part I. The monolithic shape memory alloy, International Journal of Plasticity 12 (6) (1996) 805–842.

[63] P. Popov, D. C. Lagoudas, A 3-D constitutive model for shape memory alloys incorporating pseudoelasticity and detwinning of self-accommodated martensite, International Journal of Plasticity 23 (10–11) (2007) 1679–1720.

[64] K. Otsuka, C. M. Wayman (Eds.), Shape Memory Materials, Cambridge University Press, Cambridge, 1999.

[65] R. C. O'Handley, S. M. Allen, D. I. Paul, C. P. Henry, M. Marioni, D. Bono, C. Jenkins, A. Banful, R. Wager, Keynote address: Magnetic field-induced strain in single crystal Ni-Mn-Ga, Proceedings of SPIE, Symposium on Smart Structures and Materials 5053 (2003) 200–206.

[66] A. Hubert, R. Schäfer, Magnetic Domains, Springer-Verlag, New York, 2001.

[67] A. A. Likhachev, K. Ullakko, Magnetic-field-controlled twin boundaries motion and giant magneto-mechanical effects in Ni-Mn-Ga shape memory alloy, Physics Letters A 275 (2000) 142–151.

[68] O. Heczko, Determination of ordinary magnetostriction in Ni-Mn-Ga magnetic shape memory alloy, Journal of Magnetism and Magnetic Materials 290–291 (2005) 846–849.

[69] O. Heczko, Magnetic shape memory effect and magnetization reversal, Journal of Magnetism and Magnetic Materials 290–291 (2) (2005) 787–794.

[70] A. A. Likhachev, A. Sozinov, K. Ullakko, Optimizing work output in Ni-Mn-Ga and other ferromagnetic shape memory alloys, Proceedings of SPIE, Symposium on Smart Structures and Materials 4699 (2002) 553–563.

[71] Y. Ge, O. Heczko, O. Söderberg, V. K. Lindroos, Various magnetic domain structures in Ni-Mn-Ga martensite exhibiting magnetic shape memory effect, Journal of Applied Physics 96 (4) (2004) 2159–2163.

[72] M. R. Sullivan, H. D. Chopra, Temperature- and field-dependent evolution of micromagnetic structure in ferromagnetic shape-memory-alloys, Physical Review B 70 (2004) 094427–(1–8).

[73] M. R. Sullivan, A. A. Shah, H. D. Chopra, Pathways of structural and magnetic transition in ferromagnetic shape-memory alloys, Physical Review B 70 (2004) 094428–(1–8).

[74] R. C. O'Handley, Model for strain and magnetization in magnetic shape-memory alloys, Journal of Applied Physics 83 (6) (1998) 3263–3270.

[75] S. J. Murray, R. C. O'Handley, S. M. Allen, Model for discontinuous actuation of ferromagnetic shape memory alloy under stress, Journal of Applied Physics 89 (2) (2001) 1295–1301.

[76] R. D. James, K. F. Hane, Martensitic transformations and shape-memory materials, Acta Materialia 48 (1) (2000) 197–222.

[77] L. Hirsinger, C. Lexcellent, Modelling detwinning of martensite platelets under magnetic and (or) stress actions on NiMnGa alloys, Journal of Magnetism and Magnetic Materials 254–255 (2003) 275–277.

[78] L. Hirsinger, C. Lexcellent, Internal variable model for magneto-mechanical behaviour of ferromagnetic shape memory alloys Ni-Mn-Ga, Journal de Physique IV France 112 (2003) 977–980.

[79] N. I. Glavatska, A. A. Rudenko, I. N. Glavatskiy, V. A. L'vov, Statistical model of magnetostrain effect in martensite, Journal of Magnetism and Magnetic Materials 265 (2) (2003) 142–151.

[80] V. D. Buchelnikov, S. I. Bosko, The kinetics of phase transformations in ferromagnetic shape memory alloys Ni-Mn-Ga, Journal of Magnetism and Magnetic Materials 258–259 (2003) 497–499.

[81] V. A. Chernenko, V. A. L'vov, P. Müllner, G. Kostorz, T. Takagi, Magnetic-field-induced superelasticity of ferromagnetic thermoelastic martensites: Experiments and modeling, Physical Review B 69 (2004) 134410–(1–8).

[82] P. Müllner, V. A. Chernenko, G. Kostorz, A microscopic approach to the magnetic-field-induced deformation of martensite (magnetoplasticity), Journal of Magnetism and Magnetic Materials 267 (2003) 325–334.

[83] K.-L. Lee, S. Seelecke, A model for ferromagnetic shape memory thin film actuators, Proceedings of SPIE, Smart Structures and Materials: Modeling, Signal Processing, and Control 5757 (2005) 302–313.

[84] B. Kiefer, D. C. Lagoudas, Magnetic field-induced martensitic variant reorientation in magnetic shape memory alloys, Philosophical Magazine Special Issue: Recent Advances in Theoretical Mechanics, in Honor of SES 2003 A.C. Eringen Medalist G.A. Maugin 85 (33-35) (2005) 4289–4329.

[85] B. Kiefer, A phenomenological constitutive model for magnetic shape memory alloys, Ph.D. thesis, Texas A&M University (December 2006).

[86] J. Kiang, L. Tong, Modelling of magneto-mechanical behavior of Ni-Mn-Ga single crystals, Journal of Magnetism and Magnetic Materials 292 (2005) 394–412.

[87] K. Hutter, A. A. F. van de Ven, Field Matter Interactions in Thermoelastic Solids, Vol. 88 of Lecture Notes in Physics, Springer-Verlag, New York, 1978.

[88] C. Kittel, J. K. Galt, Ferromagnetic domain theory, Solid State Physics 3 (1956) 437–564.

[89] D. C. Lagoudas, Z. Bo, M. A. Qidwai, A unified thermodynamic constitutive model for SMA and finite element analysis of active metal matrix composites, Mechanics of Composite Materials and Structures 3 (1996) 153–179.

[90] Z. Bo, D. C. Lagoudas, Thermomechanical modeling of polycrystalline SMAs under cyclic loading, Part I: Theoretical Derivations, International Journal of Engineering Science 37 (1999) 1089–1140.

[91] M. A. Qidwai, D. C. Lagoudas, Numerical implementation of a shape memory alloy thermomechanical constitutive model using return mapping algorithms, International Journal for Numerical Methods in Engineering 47 (2000) 1123–1168.

[92] J. C. Simo, T. J. R. Hughes, Computational Inelasticity, Vol. 7 of Interdisciplinary Applied Mathematics, Springer-Verlag, New York, 1998.

[93] A. S. Khan, S. Huang, Continuum Theory of Plasticity, John Wiley & Sons, New York, 1995.

[94] B. Kiefer, D. Lagoudas, Application of a magnetic sma constitutive model in the analysis of magnetomechanical boundary value problems, Proceedings of SPIE, Smart Structures and Materials: Active Materials: Behavior and Mechanics, San Diego, CA, 26 February–2 March 2006 6170 (2006) 330–341.

[95] D. C. Lagoudas, B. Kiefer, A. J. Broederdorf, Accurate interpretation of magnetic shape memory alloy experiments utilizing coupled magnetostatic analysis, Proceedings of ASME, International Mechanical Engineering Congress and Exposition, Chicago, IL, 5–10 November 2006 IMECE2006-15296 (2006) 1–11.

[96] C. Mueller, Thermodynamic modeling of polycrystalline shape memory alloys at finite strains, Ph.D. thesis, Ruhr-Universität Bochum (August 2003).

[97] A. Bekker, L. C. Brinson, Phase diagram based description of the hysteresis behavior of shape memory alloys, Acta Materialia 46 (10) (1998) 3649–3665.

[98] R. Tickle, R. D. James, T. Shield, P. Schumacher, M. Wuttig, V. V. Kokorin, Ferromagnetic shape memory in the NiMnGa system, IEEE Transactions on Magnetics 35 (5) (1999) 4301–4310.

[99] L. Straka, O. Heczko, Superelastic response of Ni–Mn–Ga martensite in magnetic fields and a simple model, IEEE Transactions on Magnetics 39 (5) (2003) 3402–3404.

[100] L. Straka, O. Heczko, Magnetization changes in Ni–Mn–Ga magnetic shape memory single crystal during compressive stress reorientation, Scripta Materialia 54 (2006) 1549–1552.

[101] P. Müllner, V. A. Chernenko, G. Kostorz, A microscopic approach to the magnetic-field-induced deformation of martensite (magnetoplasticity), Journal of Magnetism and Magnetic Materials 267 (2003) 325–334.

A

Generalized Framework for Modeling of SMAs

M. A. SIDDIQ QIDWAI AND D. C. LAGOUDAS

The purpose of this appendix is to extend the thermodynamic framework presented in Chapter 3 by formally including the principle of maximum transformation dissipation. A generalized phase transformation function, motivated by experimental results is also discussed, extending the transformation functions introduced in Chapters 3 and 4. The maximum dissipation principle imposes constraints on the form of internal state variable evolution equations and establishes the shape of the transformation function in generalized thermodynamic force space. A Green-Lagrange strain measure is employed in modeling to account for large deformations. The work in this appendix follows the work of Qidwai and Lagoudas [1].

A.1 Description of SMAs: Thermodynamic Potentials in the Lagrangian Formulation

For purposes of development, we prescribe a general Gibbs free energy potential, G, for a polycrystalline SMA given by

$$G = G(\mathbf{S}, T, \boldsymbol{\Upsilon}), \tag{A.1.1}$$

where $\boldsymbol{\Upsilon}$ is a set of Lagrangian *internal state variables* (defined in the reference configuration), $\mathbf{S}$ represents a stress measure defined in the reference configuration (e.g. the second Piola-Kirchoff stress tensor), and T is the thermodynamic temperature.

The phase transformation in SMAs is a dissipative process that involves an entropy increase. Dissipative processes require the application of the laws of thermodynamics to obtain constraints on the material response. The most popular methodology is the one proposed by Coleman and Noll [2] for inelastic materials as described in Chapter 3. In this methodology, the second law of thermodynamics is assumed to be represented by the strong form of the Clausius-Duhem inequality to derive constraints on the material behavior, which is given in the form of the Clausius-Plank inequality by the following expression:

$$-\rho_0 \, \dot{G} - \rho_0 s \dot{T} - \mathbf{E} : \dot{\mathbf{S}} \geq 0, \tag{A.1.2}$$

where ρ_0 is the mass density in the reference configuration, s is the entropy and $\mathbf{E}$ is the total Green-Lagrange strain tensor. For small strains, the corresponding equation in Chapter 3 is (3.2.21).

Taking into account (A.1.1), the rate of change of the Gibbs free energy is given by

$$\dot{G} = \frac{\partial G}{\partial \mathbf{S}} : \dot{\mathbf{S}} + \frac{\partial G}{\partial T}\dot{T} + \frac{\partial G}{\partial \boldsymbol{\Upsilon}} : \dot{\boldsymbol{\Upsilon}} . \tag{A.1.3}$$

Substituting (A.1.3) into (A.1.2), we obtain

$$-(\rho_0 \frac{\partial G}{\partial \mathbf{S}} + \mathbf{E}) : \dot{\mathbf{S}} - \rho_0 (\frac{\partial G}{\partial T} + s) : \dot{T} - \rho_0 \frac{\partial G}{\partial \boldsymbol{\Upsilon}} : \dot{\boldsymbol{\Upsilon}} \geq 0 . \tag{A.1.4}$$

In the thermoelastic region, $\dot{\boldsymbol{\Upsilon}}$ is zero and since the above relation should be valid for all $\dot{\mathbf{S}}$ and $\dot{T}$, we obtain the following constitutive relations:

$$\mathbf{E} = -\rho_0 \frac{\partial G}{\partial \mathbf{S}} , \tag{A.1.5}$$

$$s = -\frac{\partial G}{\partial T} . \tag{A.1.6}$$

These constitutive relations are also assumed to be valid everywhere at the boundary of the thermoelastic region. A detailed discussion on this assumption is presented in [3], where the authors employed the notion of a family of thermoelastic responses associated with the overall inelastic material behavior.

Now, using (A.1.4), (A.1.5) and (A.1.6), we define the transformation dissipation, $\mathcal{D}^t$, as

$$\mathcal{D}^t = -\rho_0 \frac{\partial G}{\partial \boldsymbol{\Upsilon}} : \dot{\boldsymbol{\Upsilon}} \geq 0 . \tag{A.1.7}$$

If we define

$$\boldsymbol{\Gamma} = -\rho_0 \frac{\partial G}{\partial \boldsymbol{\Upsilon}} \tag{A.1.8}$$

as the set of *generalized thermodynamic forces* conjugate to $\dot{\boldsymbol{\Upsilon}}$, then the transformation dissipation is simply given by

$$\mathcal{D}^t(\boldsymbol{\Gamma}; \dot{\boldsymbol{\Upsilon}}) = \boldsymbol{\Gamma} : \dot{\boldsymbol{\Upsilon}} \geq 0 . \tag{A.1.9}$$

This relation is the *rate of work done by phase transformation* due to changes in microstructure induced by $\dot{\boldsymbol{\Upsilon}}$ in the generalized thermodynamic force space.

A.1.1 Phase Transformation Function

To determine the thermoelastic region, one assumes the existence of a *transformation function*, $\Phi(\boldsymbol{\Gamma})$, which is used to define the boundary of the thermoelastic response, by the equation

$$\Phi(\boldsymbol{\Gamma}) = 0 . \tag{A.1.10}$$

Connectivity with the thermoelastic domain for at least one loading path requires

$$\Phi(\boldsymbol{\Gamma}(\mathbf{S}, T, \boldsymbol{\Upsilon})) \leq 0 . \tag{A.1.11}$$

A.1.2 Principle of Maximum Transformation Dissipation

The principle of maximum transformation dissipation is based on the premise that if the SMA material transforms at some point during its thermomechanical loading, then this transformation will be characterized by maximum dissipation. The maximum dissipation principle has been applied within the framework of plasticity [4, 5] to demonstrate the convexity of the yield criterion in stress-internal state variable space, the associativity of the internal state variables evolution equations, and the presence of Kuhn-Tucker conditions. The discussion here is more general in the sense that a broader generalized thermodynamic force internal state variable space is considered. It will be shown that the principle implies associative evolution equations and a convex thermoelastic region in the larger space of thermodynamic force internal state variables, at the same time allowing for non-associative evolution equations and non-convexity of the thermoelastic region in the stress-temperature space.

The principle of maximum transformation dissipation for SMAs is expressed by the following statement:

> *For a given set of internal state variables, $\boldsymbol{\Upsilon}$, among the set of all admissible generalized thermodynamic forces, $\boldsymbol{\Gamma}_{\mathrm{all}}$, satisfying the transformation criterion (A.1.11), the transformation dissipation, $\mathcal{D}^t$, defined by (A.1.9) is the global maximum for the* actual *generalized thermodynamic forces, $\boldsymbol{\Gamma}$.*

We can mathematically write the above maximum transformation dissipation principle as

$$\mathcal{D}^t(\boldsymbol{\Gamma};\dot{\boldsymbol{\Upsilon}}) = \max_{\boldsymbol{\Gamma}_{\mathrm{a}}\in\boldsymbol{\Gamma}_{\mathrm{all}}} \mathcal{D}^t(\boldsymbol{\Gamma}_{\mathrm{a}};\dot{\boldsymbol{\Upsilon}}) , \tag{A.1.12}$$

where

$$\boldsymbol{\Gamma}_{\mathrm{a}} = \left(-\rho_0 \frac{\partial G}{\partial \boldsymbol{\Upsilon}}\right)\bigg|_{\mathrm{admissible}} , \tag{A.1.13}$$

are the *admissible* generalized thermodynamic forces and

$$\boldsymbol{\Gamma}_{\mathrm{all}} = \{\boldsymbol{\Gamma}_{\mathrm{a}} \mid \Phi(\boldsymbol{\Gamma}_{\mathrm{a}}) \leq 0\} , \tag{A.1.14}$$

is the set of all admissible generalized thermodynamic forces.

A.1.3 Consequences of the Application of the Principle of Maximum Transformation Dissipation

To show the impact of the principle of maximum transformation dissipation, the Lagrange multiplier method is employed. Following the standard optimization method [6, 7], the maximum dissipation principle is transformed into a minimization principle by changing the sign of the objective function to $-\mathcal{D}^t$. Then using (A.1.9) and (A.1.12), the extended Lagrangian associated with the unconstrained optimization problem is given by

$$\mathcal{L}^t(\mathbf{\Gamma}_{\mathrm{a}}, \lambda; \dot{\mathbf{\Upsilon}}) = -\mathbf{\Gamma}_{\mathrm{a}} : \dot{\mathbf{\Upsilon}} + \lambda \Phi(\mathbf{\Gamma}_{\mathrm{a}}) \,, \tag{A.1.15}$$

where λ is the Lagrange multiplier. The solution to the above optimization problem is then given by $\{\mathbf{\Gamma}, \lambda\}$, satisfying the following relations:

$$\left.\frac{\partial \mathcal{L}^t(\mathbf{\Gamma}_{\mathrm{a}}, \lambda; \dot{\mathbf{\Upsilon}})}{\partial \mathbf{\Gamma}_{\mathrm{a}}}\right|_{\mathbf{\Gamma}_{\mathrm{a}}=\mathbf{\Gamma}} = -\dot{\mathbf{\Upsilon}} + \lambda \frac{\partial \Phi(\mathbf{\Gamma})}{\partial \mathbf{\Gamma}} = 0 \,, \tag{A.1.16}$$

along with Kuhn-Tucker optimality conditions given by

$$\lambda \geq 0, \qquad \Phi(\mathbf{\Gamma}) \leq 0, \qquad \lambda \Phi(\mathbf{\Gamma}) = 0 \,. \tag{A.1.17}$$

The relations (A.1.16)-(A.1.17) are necessary but not sufficient conditions for the existence of a global minimum for the dissipation potential, $-\mathcal{D}^t$. Sufficiency is guaranteed for a linear function if the constraint region is convex, its global minimum being always directly found at the boundary. Note that the dissipation potential, -$\mathcal{D}^t$, is a linear function and for its global minimum to exist the transformation function, $\Phi(\mathbf{\Gamma}(\mathbf{S}, T, \mathbf{\Upsilon}))$, has to be convex in the generalized thermodynamic force space ($\mathbf{\Gamma}$-space). Non-convexity in the stress-temperature $\{\mathbf{S}, T\}$ space is still allowed.

The requirement of convexity of Φ for the stationary point to be a global minimum is now derived. Based on a result found in [4] (Lemma 2.6.1), the function $\Phi(\mathbf{\Gamma})$ is convex if and only if

$$\Phi(\mathbf{\Gamma}_{\mathrm{a}}) - \Phi(\mathbf{\Gamma}) \geq (\mathbf{\Gamma}_{\mathrm{a}} - \mathbf{\Gamma}) : \frac{\partial \Phi(\mathbf{\Gamma})}{\partial \mathbf{\Gamma}} . \tag{A.1.18}$$

The satisfaction of the above inequality is shown as follows. Based on the definition (A.1.12), we have

$$\begin{aligned} \mathcal{D}^t(\mathbf{\Gamma}; \dot{\mathbf{\Upsilon}}) &\geq \mathcal{D}^t(\mathbf{\Gamma}_{\mathrm{a}}; \dot{\mathbf{\Upsilon}}) \\ \mathbf{\Gamma} : \dot{\mathbf{\Upsilon}} &\geq \mathbf{\Gamma}_{\mathrm{a}} : \dot{\mathbf{\Upsilon}} \,. \end{aligned} \tag{A.1.19}$$

Substituting the relation from (A.1.16) and bringing all the terms on the right-hand side, we obtain

$$\lambda \, (\mathbf{\Gamma}_{\mathrm{a}} - \mathbf{\Gamma}) : \frac{\partial \Phi(\mathbf{\Gamma})}{\partial \mathbf{\Gamma}} \leq 0 \,. \tag{A.1.20}$$

Then, using the Kuhn-Tucker condition, i.e., $\lambda \Phi(\mathbf{\Gamma}_{\mathrm{a}}) = 0$, we can write

$$\lambda \, (\mathbf{\Gamma}_{\mathrm{a}} - \mathbf{\Gamma}) : \frac{\partial \Phi(\mathbf{\Gamma})}{\partial \mathbf{\Gamma}} \leq \lambda \Phi(\mathbf{\Gamma}_{\mathrm{a}}) \,. \tag{A.1.21}$$

For the non-trivial case, when $\lambda > 0$ and noting that $\Phi(\mathbf{\Gamma}) = 0$, we obtain

$$(\mathbf{\Gamma}_{\mathrm{a}} - \mathbf{\Gamma}) : \frac{\partial \Phi(\mathbf{\Gamma})}{\partial \mathbf{\Gamma}} \leq \Phi(\mathbf{\Gamma}_{\mathrm{a}}) - \Phi(\mathbf{\Gamma}) . \tag{A.1.22}$$

Which coincides with (A.1.18), and hence convexity in the generalized thermodynamic force space follows.

A.2 Constitutive Modeling of Polycrystalline SMAs: Lagrangian Formulation

We now present a constitutive model based on the thermomechanical framework developed in Sect. A.1 and the earlier work of Lagoudas and coworkers [8, 9]. The model is capable of modeling the pseudoelastic and the shape memory effects. However, the modeling of reorientation requires an extended model, for example a model similar to the one presented in Chapter 6. The Gibbs free energy potential, G, for such an SMA material is proposed as follows:

$$\begin{aligned} G(\mathbf{S}, T, \mathbf{E}^t, \xi) = &- \frac{1}{2\rho_0}\mathbf{S} : \mathcal{S} : \mathbf{S} - \frac{1}{\rho_0}\mathbf{S} : \left[\boldsymbol{\alpha}(T - T_0) + \mathbf{E}^t\right] + \\ & c\left[T - T_0 - T\ln\left(\frac{T}{T_0}\right)\right] - s_0 T + u_0 + \frac{1}{\rho_0} f(\xi) \end{aligned} \tag{A.2.23}$$

where $\mathcal{S}$, $\boldsymbol{\alpha}$, c, s_0 and u_0 are the effective compliance tensor, effective thermal expansion coefficient tensor, effective specific heat, effective specific entropy at the reference state and effective specific internal energy at the reference state, respectively similar to material properties introduced in Chapter 3. Also, the changes in the microstructure of the material are represented by the set of internal state variables, $\boldsymbol{\Upsilon} = \{\mathbf{E}^t, \xi\}$, where $\mathbf{E}^t$ is the Lagrangian transformation strain tensor and ξ is the martensitic volume fraction. The function, $f(\xi)$, in (A.2.23) accounts for transformation hardening due to interaction between martensite and austenite and the martensitic variants themselves as described in Chapter 3.

The constitutive relations are obtained using (A.1.5) and (A.1.6) with the specific form of G given by (A.2.23), i.e.,

$$\mathbf{E} = \mathcal{S} : \mathbf{S} + \boldsymbol{\alpha}(T - T_0) + \mathbf{E}^t, \tag{A.2.24}$$

$$s = \frac{1}{\rho_0}\mathbf{S} : \boldsymbol{\alpha} + c\ln\left(\frac{T}{T_0}\right) + s_0. \tag{A.2.25}$$

Additive strain decomposition appears in (A.2.24) and the difference $\mathbf{E}^e = \mathbf{E} - \boldsymbol{\alpha}(T - T_0) - \mathbf{E}^t$ can be identified as the elastic strain. Note that the elastic strain depends on the degree of phase transformation through the dependence of the elastic compliance on ξ. The nonlinearity in the constitutive behavior is due to the use of nonlinear kinematics and the evolution of $\mathbf{E}^t$.

Using (A.1.9), the transformation dissipation, $\mathcal{D}^t$, generalized thermodynamic forces, $\boldsymbol{\Gamma}$, and their generalized conjugate internal variable rates, $\dot{\boldsymbol{\Upsilon}}$, respectively, are given by

$$\mathcal{D}^t\left(\mathbf{S}, p; \dot{\mathbf{E}}^t, \dot{\xi}\right) = \mathbf{S} : \dot{\mathbf{E}}^t + p\dot{\xi} \geq 0\,, \tag{A.2.26}$$

$$\boldsymbol{\Gamma} = \{\mathbf{S}, p\}\,, \tag{A.2.27}$$

$$\dot{\boldsymbol{\Upsilon}} = \left\{ \dot{\mathbf{E}}^t, \dot{\xi} \right\}. \tag{A.2.28}$$

In the above equations, the generalized thermodynamic force $\mathbf{S}$ is found by applying (A.1.8), i.e.,

$$-\rho_0 \frac{\partial G}{\partial \mathbf{E}^t} = \mathbf{S}, \tag{A.2.29}$$

and the generalized thermodynamic force p is also derived from (A.1.8) with the following evaluation:

$$\begin{aligned} p &= -\rho_0 \frac{\partial G}{\partial \xi} \\ &= \frac{1}{2}\mathbf{S} : \Delta\mathcal{S} : \mathbf{S} + \Delta\boldsymbol{\alpha} : \mathbf{S}(T - T_0) - \rho_0\,\Delta c\left[T - T_0 - T\ln\left(\frac{T}{T_0}\right)\right] \\ &\quad + \rho_0\,\Delta s_0\,T - \frac{\partial f(\xi)}{\partial \xi} - \rho_0\,\Delta u_0\,. \end{aligned} \tag{A.2.30}$$

The terms $\Delta\mathcal{S}, \Delta\boldsymbol{\alpha}, \Delta c, \Delta s_0, \Delta u_0$ in (A.2.30) indicate the difference of compliance tensor, thermal expansion coefficient tensor, specific heat, reference entropy and reference internal energy between the martensitic and austenitic phases, respectively, as defined in Chapter 3.

At this point, we utilize the assumption of the existence of a thermoelastic region defined by a transformation function $\Phi(\mathbf{S}, p)$ with its boundary being determined by the relation

$$\Phi(\mathbf{S}, p) = \Phi(\boldsymbol{\Gamma}) = 0\,. \tag{A.2.31}$$

It is further assumed that the construction of Φ will be such that all admissible material states will satisfy the constraint

$$\Phi(\mathbf{S}, p) \leq 0\,. \tag{A.2.32}$$

Then, in light of (A.1.16) and (A.1.17), the application of the principle of maximum transformation dissipation results in the following transformation evolution equations and Kuhn-Tucker conditions:

$$\dot{\mathbf{E}}^t = \lambda \frac{\partial \Phi(\mathbf{S}, p)}{\partial \mathbf{S}}, \tag{A.2.33}$$

$$\dot{\xi} = \lambda \frac{\partial \Phi(\mathbf{S}, p)}{\partial p}, \tag{A.2.34}$$

$$\lambda \geq 0, \qquad \Phi(\mathbf{S}, p) \leq 0, \qquad \lambda\Phi(\mathbf{S}, p) = 0\,. \tag{A.2.35}$$

Since (A.2.34) is a scalar equation, it is obvious that the Lagrange multiplier, λ, is proportional to the rate of martensitic volume fraction, $\dot{\xi}$. An important observation is obtained by substituting the expression for λ from (A.2.34) into (A.2.33), with the following result:

$$\dot{\mathbf{E}}^t = \dot{\xi}\,\frac{1}{\frac{\partial \Phi(\mathbf{S},p)}{\partial p}}\frac{\partial \Phi(\mathbf{S},p)}{\partial \mathbf{S}}\,. \tag{A.2.36}$$

In words, the above equation states the following: *The application of the principle of maximum transformation dissipation implies that for an SMA constitutive model, which admits the martensitic volume fraction as an internal state variable, the evolution of the remaining internal state variables is directly proportional to the evolution of the martensitic volume fraction.*

Recall now the expression for transformation dissipation given by (A.2.26) and substitute the transformation strain evolution equation given by (A.2.36) into it to obtain

$$\mathcal{D}^t = \left(\frac{1}{\frac{\partial \Phi(\mathbf{S},p)}{\partial p}}\mathbf{S} : \frac{\partial \Phi(\mathbf{S},p)}{\partial \mathbf{S}} + p \right) \dot{\xi} = \pi \dot{\xi} \geq 0\,. \tag{A.2.37}$$

The above expression states that all the dissipation is caused by a change in the martensitic volume fraction. From this development, an effective thermodynamic force, π, is defined, which is conjugate to ξ and a combination of the generalized thermodynamic forces $\{\mathbf{S}, p\}$.

We now propose the following general form of the transformation function

$$\begin{aligned}\Phi(\mathbf{S},p) &= [\hat{\Phi}(\mathbf{S}) + p]^2 - Y^2 \\ &= [\hat{\Phi}(\mathbf{S}) + p - Y][\hat{\Phi}(\mathbf{S}) + p + Y]\,,\end{aligned} \tag{A.2.38}$$

where Y is the measure of internal dissipation due to microstructural changes during phase transformation.

Substituting (A.2.38) into (A.2.33) and (A.2.34), we obtain the following set of equations:

$$\dot{\mathbf{E}}^t = 2\lambda[\hat{\Phi}(\mathbf{S}) + p]\frac{\partial \hat{\Phi}(\mathbf{S})}{\partial \mathbf{S}}\,, \tag{A.2.39}$$

$$\dot{\xi} = 2\lambda[\hat{\Phi}(\mathbf{S}) + p]\,. \tag{A.2.40}$$

Inverting (A.2.40) for λ and substituting it in (A.2.39), the transformation strain evolution is shown to depend on the evolution of the martensitic volume fraction as

$$\dot{\mathbf{E}}^t = \dot{\xi}\frac{\partial \hat{\Phi}(\mathbf{S})}{\partial \mathbf{S}}\,. \tag{A.2.41}$$

Different functional forms of the phase transformation function $\hat{\Phi}(\mathbf{S})$ are presented in the next sections.

A.2.1 J_2 Transformation Function

Let $\hat{\Phi}(\mathbf{S})$ be defined by

$$\hat{\Phi}(\mathbf{S}) = \alpha\sqrt{3J_2}\,, \tag{A.2.42}$$

where the material parameter α corresponds to the maximum transformation strain obtained during forward transformation [8–10], $J_2 = \frac{1}{2}\mathbf{S}' : \mathbf{S}' = \frac{1}{2}\|\mathbf{S}'\|^2$ is the second deviatoric stress invariant and $\mathbf{S}' = \mathbf{S} - \frac{1}{3}\,\mathrm{tr}\,(\mathbf{S})$ is the deviatoric stress. The transformation strain evolution equation is then obtained using (A.2.41) as follows

$$\dot{\mathbf{E}}^t = \sqrt{\frac{3}{2}}\alpha\frac{\mathbf{S}'}{\|\mathbf{S}'\|}\,\dot{\xi} = \mathbf{\Lambda}\,\dot{\xi}\,, \tag{A.2.43}$$

where $\mathbf{\Lambda} = \sqrt{\frac{3}{2}}\alpha\frac{\mathbf{S}'}{\|\mathbf{S}'\|}$. Using the transformation tensor $\mathbf{\Lambda}$, $\hat{\Phi}(\mathbf{S})$ can also be described as

$$\hat{\Phi}(\mathbf{S}) = \mathbf{S} : \mathbf{\Lambda} = \mathbf{S} : \frac{\partial\hat{\Phi}(\mathbf{S})}{\partial\mathbf{S}}\,. \tag{A.2.44}$$

A.2.2 $J_2 - I_1$ Transformation Function

For this type of phase transformation function, we define $\hat{\Phi}(\mathbf{S})$ to be

$$\hat{\Phi}(\mathbf{S}) = \beta\sqrt{3J_2} + \gamma\, I_1\,, \tag{A.2.45}$$

where $I_1 = \mathrm{tr}\,(\mathbf{S})$ is the first stress invariant [11]. In (A.2.45), β and γ are material parameters, which are used to demonstrate the tension-compression asymmetry. The transformation strain evolution equation is then given by

$$\dot{\mathbf{E}}^t = \left(\sqrt{\frac{3}{2}}\beta\frac{\mathbf{S}'}{\|\mathbf{S}'\|} + \gamma\mathbf{1}\right)\dot{\xi} = \mathbf{\Lambda}\,\dot{\xi}\,, \tag{A.2.46}$$

where $\mathbf{1}$ is the second order identity tensor.

A.2.3 $J_2 - J_3 - I_1$ Transformation Function

Another choice of $\hat{\Phi}(\mathbf{S})$ that captures the tension-compression asymmetry in addition to the volumetric transformation strain is proposed as a general transformation function and is given by:

$$\hat{\Phi}(\mathbf{S}) = \kappa\sqrt{3J_2}\left[1 + \nu\frac{J_3}{(3J_2)^{\frac{3}{2}}}\right]^{\zeta} + \omega\, I_1\,, \tag{A.2.47}$$

where $J_3 = \mathrm{Det}\,\mathbf{S}'$ is the third deviatoric stress invariant, κ and ν are related to tension compression asymmetry, ω is introduced to capture the volumetric effect and ζ is another material constant. For $\nu = 0$ (or $\zeta = 0$) and $\omega = 0$, the form of J_2 function is recovered. For $\nu = 0$ (or $\zeta = 0$), the form of $J_2 - I_1$ transformation function is recovered. For $\zeta = \frac{1}{2}$ and $\omega = 0$, the transformation function proposed in [12], based on $J_2 - J_3$ is obtained. The transformation strain evolution equation compatible with (A.2.47) and for $\zeta = 1$ is given by the following expression:

$$\dot{\mathbf{E}}^t = \left\{ \kappa \left[\frac{3\mathbf{S}'}{2\sqrt{3J_2}} + \nu \frac{\sqrt{3J_2}\left(\mathbf{S}'\mathbf{S}' - \frac{2}{3}J_2\mathbf{1}\right) - 3J_3\mathbf{S}'}{(3J_2)^2} \right] + \omega\mathbf{1} \right\} \dot{\xi}, \tag{A.2.48}$$

References

[1] M. A. Qidwai, D. C. Lagoudas, On the thermodynamics and transformation surfaces of polycrystalline NiTi shape memory alloy material, International Journal of Plasticity 16 (2000) 1309–1343.

[2] B. D. Coleman, W. Noll, The thermodynamics of elastic materials with heat conduction and viscosity, Archive for Rational Mechanics and Analysis 13 (1963) 167–178.

[3] K. R. Rajagopal, A. R. Srinivasa, Mechanics of the inelastic behavior of materials. Part I: Theoretical underpinnings, International Journal of Plasticity 14 (10–11) (1998) 945–967.

[4] J. C. Simo, T. J. R. Hughes, Computational Inelasticity, Vol. 7 of Interdisciplinary Applied Mathematics, Springer-Verlag, New York, 1998.

[5] K. R. Rajagopal, A. R. Srinivasa, Mechanics of the inelastic behavior of materials. Part II: Inelastic response, International Journal of Plasticity 14 (10–11) (1998) 969–995.

[6] G. S. Beveridge, R. S. Schechter, Optimization: Theory and Practice, McGraw-Hill, Inc., New York, 1970.

[7] C. McMillan, Mathematical programming: an introduction to the design and application of optimal decision machines, John Wiley & Sons, Inc., New York, 1970.

[8] J. G. Boyd, D. C. Lagoudas, A thermodynamical constitutive model for shape memory materials. Part I. The monolithic shape memory alloy, International Journal of Plasticity 12 (6) (1996) 805–842.

[9] D. C. Lagoudas, Z. Bo, M. A. Qidwai, A unified thermodynamic constitutive model for SMA and finite element analysis of active metal matrix composites, Mechanics of Composite Materials and Structures 3 (1996) 153–179.

[10] Z. Bo, D. C. Lagoudas, Thermomechanical modeling of polycrystalline SMAs under cyclic loading, Part I: Theoretical Derivations, International Journal of Engineering Science 37 (1999) 1089–1140.

[11] F. Auricchio, R. L. Taylor, J. Lubliner, Shape-memory alloys: Macromodelling and numerical simulations of the superelastic behavior, Computer Methods in Applied Mechanics and Engineering 146 (1997) 281–312.

[12] Y. Gillet, E. Patoor, M. Berveiller, Calculation of pseudoelastic elements using a non-symmetrical thermomechanical transformation criterion and associated rule, Journal of Intelligent Material Systems and Structures 9 (5) (1998) 366–378.

B

Numerical Solutions to Boundary Value Problems

P. Popov

This appendix is intended as a brief summary of a standard, displacement-based Finite Element Method (FEM) for problems with material nonlinearity and an associated strain-driven return mapping algorithm for resolving the material response. Here we focus on displacement-based FEM methods,which are best suited for static boundary value problems of structures whose materials do not exhibit large jumps in material coefficients. A strain-driven return mapping algorithm is then a natural choice for incorporating the material response into the FEM. This combination generalizes easily to semi-discrete, time-dependent FEM discretizations. The FEM discretization is considered first in Sect. B.1. Based on this, a suitable return mapping algorithm is then presented in Sect. B.2.

B.1 Displacement-Based Finite Element Methods for Nonlinear Problems

In general, when solving boundary problems for nonlinear hysteretic materials, such as SMAs, one is presented with two challenges:

- Select an appropriate discretization of the underlying partial differential equations.
- Integrate the constitutive equation in the selected discretization scheme.

While these tasks are quite interrelated, one usually selects the discretization method first based on the real-life problem that needs simulation. Then, a suitable numerical algorithm is devised, which incorporates the nonlinear material behavior into the discretization.

There are many factors to consider when selecting a discretization scheme. For example, if one wishes to solve static problems, by far the optimal family of methods is the Petrov-Galerkin discretizations using finite elements [1, 2]. If, on the other hand, one wishes to solve time-dependent problems, there is no method that works best, and depending on the problem, the simulator has to select an appropriate method. To name a few choices, it can be a particular conservative finite volume method [3]; a specific explicit finite difference scheme [4]; or one of the multitude of standard, mixed, hybrid or discontinuous Galerkin finite element methods [2, 5–7]. The selection is guided by the type of time-dependent problem - linear or nonlinear, monotone stress dependence or non-monotone (e.g. material that allows softening), and structures with several materials that have large differences in material properties, etc.

It is assumed that the reader is familiar with discretization of partial differential equations. Thus, we outline only the most basic parts of a discretization of a boundary value problem in inelasticity with internal variable to account for changes in microstructure due to the phase transformation. The notation is the following: $\mathbf{x}$ denotes spatial and $\mathbf{p}$ denotes material coordinates, $\mathbf{u}$ is the displacement vector, a material field, and the Piola-Kirchoff stress tensor $\mathbf{S}$, also a material field. It is convenient to introduce the symmetric part of the gradient operator:

$$\mathbf{e}(\mathbf{w}) = \frac{1}{2}\left(\nabla \mathbf{w} + (\nabla \mathbf{w})^T\right),$$

for some field $\mathbf{w}$. Observe that the infinitesimal strain tensor $\boldsymbol{\varepsilon}(\mathbf{p})$ is given by:

$$\boldsymbol{\varepsilon}(\mathbf{p}) = \mathbf{e}(\mathbf{u}).$$

Constitutive model. Consider a nonlinear, hysteretic body occupying a (reference) domain Ω. For this material body we are given a constitutive model. Let this constitutive model be formulated in terms of small strains. Next, assume for the moment, that this constitutive model provides an explicit (closed form), path-dependent relationship between the Piola-Kirchoff stress tensor $\mathbf{S}(\mathbf{p})$ and the strain $\boldsymbol{\varepsilon}(\mathbf{p})$ and internal variables $\boldsymbol{\xi}(\mathbf{p})$, that is at some instance of time:

$$\mathbf{S}(\mathbf{p}) = \mathbf{S}(\mathbf{e}(\mathbf{u}(\mathbf{p})); \boldsymbol{\xi}(\mathbf{p})). \tag{B.1.1}$$

The semicolon denotes that $\mathbf{S}$ depends, possibly, on the *entire* history of the internal variables (cf. Chapters 3-6).

Loading process. We now consider a loading process of this body. Assume its material state (displacements, strains, stresses, internal variables) is well known at some initial time t_0 and that at this initial (reference) state, the balance of linear momentum is satisfied in Ω. Consider further a quasistatic process (all transient effects are negligible) in which this body is loaded at discrete instances of time $t_1, t_2, ..., t_{n+1}$. The material body is subject to a distributed force $\mathbf{b}(\mathbf{p})$ and the loading includes prescribed displacements $\hat{\mathbf{u}}_i$, $i = 1, 2, ..., n+1$ on part of the boundary of $\Gamma_D \subseteq \partial\Omega$ (Dirichlet data) and/or prescribed tractions $\hat{\mathbf{s}}_i$, $i = 1, 2, ..., n+1$ on another part of the boundary $\Gamma_N \subseteq \partial\Omega$ (Neumann data). The mixed boundary value problem for such quasistatic loading is: *Find* $\mathbf{u}_i(\mathbf{p})$ *and* $\boldsymbol{\xi}_i(\mathbf{p})$ *such that balance of linear momentum is satisfied:*

$$\nabla \cdot \mathbf{S}_i + \mathbf{b}_0 = \mathbf{0} \textit{ in } \Omega, \tag{B.1.2}$$

with Dirichlet

$$\mathbf{u}_i = \hat{\mathbf{u}}_i \text{ on } \Gamma_D \tag{B.1.3}$$

and/or Neumann

$$\mathbf{S}_i\mathbf{n} = \hat{\mathbf{s}}_i \text{ on } \Gamma_N \tag{B.1.4}$$

boundary data at each instance of time $t_1, ..., t_n, t_{n+1}$. *Here* $\mathbf{S}_i$ *is a shorthand notation for* $\mathbf{S}(\mathbf{e}(\mathbf{u}_i(\mathbf{p})); \boldsymbol{\xi}(\mathbf{p}))$ *and* $\mathbf{n}$ *is the outward normal to the boundary* $\partial\Omega$. Note that since we are loading at discrete intervals of time, we have

$$\mathbf{S}_i(\mathbf{p}) = \mathbf{S}\left(\mathbf{e}(\mathbf{u}_i(\mathbf{p})), \boldsymbol{\xi}_0(\mathbf{p}), ..., \boldsymbol{\xi}_{i-1}(\mathbf{p})\right).$$

For simplicity, through the rest of this section we will omit the path dependence in $\mathbf{S}$ and will simply write

$$\mathbf{S}(\mathbf{e}(\mathbf{u}_i(\mathbf{p}))).$$

Discretization. We begin with a triangulation $\mathcal{T}^h$ with a mesh parameter, h, which partitions the domain Ω. Given two material vector fields $\mathbf{v}$ and $\mathbf{w}$ let us define the bilinear form:

$$(\mathbf{v}, \mathbf{w})_\Omega = \int_\Omega \mathbf{v}(\mathbf{p}) \cdot \mathbf{w}(\mathbf{p})\, d\mathbf{p},$$

and given a material tensor field $\mathbf{A}$ and a vector field $\mathbf{w}$

$$(\mathbf{A}, \mathbf{w})_\Omega = \int_\Omega \mathbf{A}(\mathbf{p}) : \mathbf{e}(\mathbf{w}(\mathbf{p}))d\mathbf{p}.$$

This shorthand notation is quite common in the mathematical literature and the reader is referred, for example, to the classical text [2] for more details. Also, let $H^1(\Omega)$ be the usual Hilbert space and $H^1_D(\Omega) \subseteq H^1(\Omega)$ be the subset of functions vanishing on Γ_D. Finally, let d be the spatial dimension of the problem ($d = 2, 3$).

The starting point of any Petrov-Galerkin numerical method is a weak form of the underlying differential equation. Taking the conservation of linear momentum (B.1.2) at some time step t_i and integrating by parts once, we obtain the weak form (the subscript, indicating the loading step is omitted for clarity):

$$(\mathbf{S}(\mathbf{e}(\mathbf{u})), \mathbf{e}(\mathbf{w}))_\Omega = (\mathbf{b}, \mathbf{w})_\Omega + (\hat{\mathbf{s}}, \mathbf{w})_{\Gamma^N}, \quad \forall \mathbf{w} \in \left[H^1_D(\Omega)\right]^d, \tag{B.1.5}$$

$$\mathbf{u} = \hat{\mathbf{u}}, \quad \text{on } \Gamma^D. \tag{B.1.6}$$

Next, in a *displacement-based* method, a discrete space $\mathcal{V}$ spanning some (vector) basis function $\boldsymbol{\psi}^{(i)}(\mathbf{p})$, $i = 1, .., N$, is constructed:

$$\mathcal{V} = span\left\{\boldsymbol{\psi}^{(1)}, ..., \boldsymbol{\psi}^{(N)}\right\}.$$

The basis functions, for example, in conforming linear methods, can be identified with each node in the triangulation $\mathcal{T}^h$ in a standard way (cf. e.g. [2]). The displacements are then discretized as follows:

$$\mathbf{u}^h = \mathbf{u}^h(\mathbf{U}, \mathbf{p}) = \sum_{i=1}^{N} U_i \psi^{(i)}(\mathbf{p}) = \mathbf{U} \cdot \boldsymbol{\Psi}(\mathbf{p}),$$

where $\mathbf{u}^h \in \mathcal{V}$ is the discretized displacement field, $\mathbf{U} = (U_1, ..., U_N)^T$ is the column vector of the nodal values of the displacements and $\boldsymbol{\Psi} = \left(\psi^{(1)}, ..., \psi^{(N)}\right)$ is the column vector of all the basis functions. The discrete strain $\boldsymbol{\varepsilon}^h$ then becomes:

$$\boldsymbol{\varepsilon}^h = \boldsymbol{\varepsilon}^h(\mathbf{U}, \mathbf{p}) = \mathbf{e}(\mathbf{u}^h) = \sum_{i=1}^{N} U_i \mathbf{e}\left(\psi^{(i)}(\mathbf{p})\right). \tag{B.1.7}$$

Observe that $\boldsymbol{\varepsilon}^h$ is material field (it is a function of $\mathbf{p}$) and it is also a function of the unknowns nodal values $\mathbf{U}$.

Given this discretization, one now writes (again omitting the subscript indicating the loading step) the discrete equivalent of the weak form (B.1.5), (B.1.6):

$$(\mathbf{S}(\mathbf{e}(\mathbf{u}^h)), \mathbf{e}(\mathbf{w}^h))_\Omega = (\mathbf{b}, \mathbf{w}^h)_\Omega + (\hat{\mathbf{s}}, \mathbf{w}^h)_{\Gamma^N}, \qquad \forall \mathbf{w}^h \in \mathcal{V}, \tag{B.1.8}$$

$$\mathbf{u}^h = \hat{\mathbf{u}}, \qquad \text{on } \Gamma^D. \tag{B.1.9}$$

Recall that we are dealing with a quasistatic loading process consisting of the loading steps $t_1, ..., t_n, t_{n+1}$. At the initial time t_0 the entire material is its initial state, which is known. Assume next that the material state (displacements, strains, stresses, temperature, internal variables) is known for all instances $t_1, ..., t_n$ and at these instances the corresponding weak forms (B.1.8) and Dirichlet boundary conditions (B.1.9) are satisfied. Our goal is to design a numerical procedure that will produce the material state at t_{n+1} and this state will satisfy equations (B.1.8) and (B.1.9). We will refer to the instance of time t_{n+1} as the current loading step and to any other instance of time $t_1, ..., t_n$ as a previous loading step.

For the current (t_{n+1}) loading step, $\hat{\mathbf{u}}_{n+1}$ and $\hat{\mathbf{s}}_{n+1}$ are given and the (discrete) fields $\mathbf{u}^h_{n+1}$, $\boldsymbol{\varepsilon}^h_{n+1}$, $\mathbf{S}^h_{n+1}$ and $\boldsymbol{\xi}^h_{n+1}$ have to be computed. For simplicity, the superscript $n+1$ will be dropped in the following discussion.

The discrete weak formulation (B.1.8), (B.1.9) consists of the nonlinear system of equations (B.1.8) and the constraint on the displacements (B.1.9). To solve the nonlinear system of equations (B.1.8), let $\mathbf{w}^{(1)}, ..., \mathbf{w}^{(N)}$ be N linearly independent test functions (usually chosen to be the basis function themselves, that is, $\mathbf{w}^{(i)} \equiv \psi^{(i)}$, $i = 1, ..., N$) and define the residual function $\mathbf{F}^h(\mathbf{U}) = \left(F^h_1(\mathbf{U})), ..., F^h_N(\mathbf{U})\right)^T$ as follows:

$$F_i^h(\mathbf{U}) = \left(\mathbf{S}\left(\mathbf{e}\left(\mathbf{u}^h\right)\right), \mathbf{w}^{(i)}\right)_\Omega - \left(\mathbf{b}, \mathbf{w}^{(i)}\right)_\Omega - \left(\hat{\mathbf{s}}, \mathbf{w}^{(i)}\right)_{\Gamma^N}. \tag{B.1.10}$$

Observe that all quantities in the integrals on the right-hand side are material fields (they all depend on $\mathbf{p}$). On the other hand, $\mathbf{S}$ also depends implicitly on $\mathbf{U}$, through the strain (cf. equation (B.1.7)). Thus, after integration (recall the definition of the bilinear forms), the right-hand side will depend on $\mathbf{U}$ only. This residual function $\mathbf{F}^h(\mathbf{U})$ provides a convenient way of rewriting equation (B.1.8). Clearly, equation (B.1.8) is satisfied if and only if:

$$\mathbf{F}^h(\mathbf{U}) = \mathbf{0}. \tag{B.1.11}$$

The last system is a nonlinear system of N equations with N unknowns, so one has to use some iterative method to solve it. Let us use, for example, Newton-Raphson's method applied to the last equation (B.1.11). Newton-Raphson's method attempts to find the solution $\mathbf{U}$ as the limit of a simple iterative procedure:

$$\mathbf{U} = \lim_{k\to\infty} \mathbf{U}^{(k)},$$

where the $k+1$ iterate is defined by the recursive formula:

$$\mathbf{F}^h(\mathbf{U}^{(k)}) + \frac{\partial \mathbf{F}^h(\mathbf{U}^{(k)})}{\partial \mathbf{U}}(\mathbf{U}^{(k+1)} - \mathbf{U}^{(k)}) = \mathbf{0}. \tag{B.1.12}$$

Denoting by $\mathbf{L}$ the Jacobian in the above equation:

$$\mathbf{L}(\mathbf{U}) = \frac{\partial \mathbf{F}^h(\mathbf{U})}{\partial \mathbf{U}}, \tag{B.1.13}$$

the recursive relation (B.1.12) can be written as

$$\mathbf{U}^{(k+1)} = \mathbf{U}^{(k)} - \mathbf{L}^{-1}(\mathbf{U}^{(k)})\mathbf{F}^h(\mathbf{U}^{(k)}). \tag{B.1.14}$$

In view of the last equation, it is more than clear that the execution of a Newton step requires one to evaluate $\mathbf{F}^h(\mathbf{U}^{(k)})$ and $\mathbf{L}(\mathbf{U}^{(k)})$, then solve a linear system with the Jacobian $\mathbf{L}$ and add the result to $\mathbf{U}^{(k)}$. Unfortunately, not everything that is clear is also simple.

Computing the residual. Let us first discuss how to evaluate $\mathbf{F}^h(\mathbf{U}^{(k)})$. In order to evaluate it, we need to compute the three integrals in equation (B.1.10). The second and third integral are trivial (all involved quantities do not depend on $\mathbf{U}^{(k)}$). To evaluate the first integral, recall that all quantities involved are at the current loading step, that is t_{n+1}. Assume we have already computed successfully k Newton steps. This means that we know the *guess* nodal displacements $\mathbf{U}^{(k)}$, and thus the displacements field $\mathbf{u}^h\left(\mathbf{U}^{(k)}, \mathbf{p}\right)$ (which does not satisfy the residual equation (B.1.11) with sufficient accuracy).

Now, since one knows $\mathbf{U}^{(k)}$, then he (or she) can directly compute the discrete strains ${\boldsymbol\varepsilon^h}^{(k)}(\mathbf{p})$ at any material point $\mathbf{p}$ using equation (B.1.7):

$$\boldsymbol{\varepsilon}^{h(k)}(\mathbf{p}) = \mathbf{e}\left(\mathbf{u}^h\left(\mathbf{U}^{(k)}, \mathbf{p}\right)\right).$$

The process is assumed (for simplicity) isothermal, therefore the temperature is also known. Now, the constitutive behavior of the material (SMA in our case) is known, therefore, given the strain $\boldsymbol{\varepsilon}^{h(k)}$ and temperature (some fixed value T) at a material point $\mathbf{p}$, one can compute, in principle, the remaining field variables (stress and internal variables). How this is done numerically will be presented in Sect. B.2.

For the time being, assume we have a numerical algorithm, (usually called *return mapping*) which, given a strain $\boldsymbol{\varepsilon}^h(\mathbf{p})$, temperature T, and the previous material states at $t_1, ..., t_n$ at a material point $\mathbf{p}$ computes the stress $\mathbf{S}(\mathbf{p})$ and the internal variables $\boldsymbol{\xi}(\mathbf{p})$ at this same material point. It is now clear how to proceed with first integral on the right-hand side of equation (B.1.10):

$$\left(\mathbf{S}\left(\boldsymbol{\varepsilon}^{h(k)}\right), \mathbf{w}\right)_\Omega = \sum_{\tau \in T^h} \int_\tau \mathbf{S}\left(\boldsymbol{\varepsilon}^{h(k)}\right) : \mathbf{e}\left(\mathbf{w}^h\right) d\mathbf{p}.$$

The integrals over each element τ are evaluated using some numerical quadrature employing a finite amount of spatial points:

$$\int_\tau \mathbf{S}\left(\boldsymbol{\varepsilon}^{h(k)}\right) : \mathbf{e}\left(\mathbf{w}^h\right) d\mathbf{p} = \sum_{i=1}^{M} w_i \mathbf{S}\left(\boldsymbol{\varepsilon}^{h(k)}\right)|_{(\mathbf{p}_i)} : \mathbf{e}\left(\mathbf{w}^h\right)|_{(\mathbf{p}_i)},$$

where M is the number of quadrature points $\mathbf{p}_i$ with weights w_i. Therefore, for a given displacement $\mathbf{u}^h\left(\mathbf{U}^{(k)}, \mathbf{p}\right)$ and thus strain $\boldsymbol{\varepsilon}^{h(k)}$, the *return mapping* algorithm is called for each quadrature point $\mathbf{p}_i$, the corresponding value $\mathbf{S}(\mathbf{p}_i)$ of the stress at that integration point is computed and the integral over τ evaluated. For further practical details on computing element integrals the reader is referred to [2, 8].

Computing the Jacobian. Evaluating the Jacobian matrix $\mathbf{L}$ in equation (B.1.14) requires some further manipulations. Observe that:

$$\begin{aligned}
\mathbf{L} = \frac{\partial \mathbf{F}^h(\mathbf{U})}{\partial \mathbf{U}} &= \int_\Omega \frac{\partial \mathbf{S}\left(\boldsymbol{\varepsilon}^h\right)}{\partial \mathbf{U}} : \mathbf{e}\left(\mathbf{w}^h\right) d\mathbf{p} \\
&= \int_\Omega \left(\frac{\partial \mathbf{S}\left(\boldsymbol{\varepsilon}^h\right)}{\partial \boldsymbol{\varepsilon}^h} : \frac{\partial \boldsymbol{\varepsilon}^h}{\partial \mathbf{U}}\right) : \mathbf{e}\left(\mathbf{w}^h\right) d\mathbf{p} \\
&= \int_\Omega \left(\mathcal{L}(\mathbf{u}^h) : \frac{\partial \boldsymbol{\varepsilon}^h}{\partial \mathbf{U}}\right) : \mathbf{e}\left(\mathbf{w}^h\right) d\mathbf{p}.
\end{aligned} \tag{B.1.15}$$

The last row in the above equation was obtained by defining $\mathcal{L}$ to be the derivative:

$$\mathcal{L} = \frac{\partial \mathbf{S}\left(\boldsymbol{\varepsilon}; \boldsymbol{\xi}_0, ..., \boldsymbol{\xi}_n\right)}{\partial \boldsymbol{\varepsilon}}. \tag{B.1.16}$$

Here we have explicitly written the previously suppressed path dependence (B.1.1) of the Piola-Kirchhoff stress $\mathbf{S}$. $\mathcal{L}$ depends on the entire loading path and, since the stress $\mathbf{S}$ is a function of the small strains $\boldsymbol{\varepsilon}$, it can be viewed as a function of either displacements or strains. Thus, $\mathcal{L}\left(\mathbf{u}^h\right)$ in the last row of equation (B.1.15) stands for:

$$\mathcal{L}\left(\mathbf{u}^h;\boldsymbol{\xi}_0^h,...,\boldsymbol{\xi}_n^h\right)=\mathcal{L}\left(\boldsymbol{\varepsilon}^h;\boldsymbol{\xi}_0^h,...,\boldsymbol{\xi}_n^h\right)=\frac{\partial\mathbf{S}}{\partial\boldsymbol{\varepsilon}}\left(\boldsymbol{\varepsilon}^h;\boldsymbol{\xi}_0^h,...,\boldsymbol{\xi}_n^h\right).$$

This derivative is usually named *algorithmic tangent stiffness*. Given a material point $\mathbf{p}$, a strain $\boldsymbol{\varepsilon}^h$, and a *fixed* loading history $\boldsymbol{\xi}_0^h,...,\boldsymbol{\xi}_n^h$ it gives the change in $\mathbf{S}$ due to an infinitesimal change in $\boldsymbol{\varepsilon}^h$:

$$\mathcal{L}\left(\boldsymbol{\varepsilon}^h;\boldsymbol{\xi}_0^h,...,\boldsymbol{\xi}_n^h\right):\left(\delta_k\otimes\delta_l\right)$$
$$=\lim_{\Delta\to0}\frac{\mathbf{S}\left(\boldsymbol{\varepsilon}^h+\Delta\left(\delta_k\otimes\delta_l\right);\boldsymbol{\xi}_0^h,...,\boldsymbol{\xi}_n^h\right)-\mathbf{S}\left(\boldsymbol{\varepsilon};\boldsymbol{\xi}_0^h,...,\boldsymbol{\xi}_n^h\right)}{\Delta}. \tag{B.1.17}$$

This derivative has to be provided by the return mapping algorithm, the details of which are given in Sect. B.2.

The term algorithmic tangent deserves some further clarification. In general, $\boldsymbol{\varepsilon}^h$ and $\boldsymbol{\varepsilon}_n^h$ are different. So, one can, for example, compute the increment of $\mathbf{S}$ around the point $\boldsymbol{\varepsilon}^h$ and with respect to that same material state, e.g. $\left(\boldsymbol{\varepsilon}^h,\mathbf{S}^h,\boldsymbol{\xi}^h,...\right)$. That is, one loads the material from the state $\left(\boldsymbol{\varepsilon}_n^h,\mathbf{S}_n^h,\boldsymbol{\xi}_n^h,...\right)$ to the state $\left(\boldsymbol{\varepsilon}^h,\mathbf{S}^h,\boldsymbol{\xi}^h,...\right)$, after which computes the derivative $\tilde{\mathcal{L}}$:

$$\tilde{\mathcal{L}}=\frac{\partial\mathbf{S}\left(\boldsymbol{\varepsilon}^h;\boldsymbol{\xi}_0,...,\boldsymbol{\xi}_n,\boldsymbol{\xi}\right)}{\partial\boldsymbol{\varepsilon}^h}. \tag{B.1.18}$$

In other words:

$$\tilde{\mathcal{L}}:\left(\delta_k\otimes\delta_l\right)=\lim_{\Delta\to0}\frac{\mathbf{S}\left(\boldsymbol{\varepsilon}^h+\Delta\left(\delta_k\otimes\delta_l\right);\boldsymbol{\xi}_0^h,...,\boldsymbol{\xi}_n^h,\boldsymbol{\xi}\right)-\mathbf{S}\left(\boldsymbol{\varepsilon}^h;\boldsymbol{\xi}_0^h,...,\boldsymbol{\xi}_n^h,\boldsymbol{\xi}\right)}{\Delta}. \tag{B.1.19}$$

This second derivative is called continuum tangent stiffness and it, in general, is different from (B.1.16). The two would coincide only in the limit $\boldsymbol{\varepsilon}^h\to\boldsymbol{\varepsilon}_n$.

In fact, $\mathcal{L}$ also depends on the chosen evolution scheme which evolves the internal state variables from the last loading step t_n to the current one. Indeed, since we are dealing with a discrete set of loading steps at $t_1,t_2,...$, one has to assume some sort of continuous path (in material state space) that connects the material states (at a given material point $\mathbf{p}$). This itself is another implicit discretization (of the evolution of the material state at that material point) that leads to the fact that $\mathcal{L}$ depends both on the material response (constitutive model) and the selected numerical scheme which evolves the material state. The details are given in the next Sect. B.2.

Coming back to the Jacobian matrix, a further simple calculation shows that

$$\frac{\partial \varepsilon^h}{\partial \mathbf{U}} = \frac{\partial}{\partial \mathbf{U}} \left(\mathbf{e} \left(\mathbf{U} \cdot \boldsymbol{\Psi}(\mathbf{p}) \right) \right) = \frac{\partial}{\partial \mathbf{U}} \left(\mathbf{U} \cdot \mathbf{e} \left(\boldsymbol{\Psi}(\mathbf{p}) \right) \right) = \mathbf{e} \left(\boldsymbol{\Psi}(\mathbf{p}) \right), \tag{B.1.20}$$

where $\mathbf{e}(\boldsymbol{\Psi}(\mathbf{p}))$ is the vector-column $\left(\mathbf{e}\left(\boldsymbol{\Psi}^{(1)}(\mathbf{p})\right), ...\mathbf{e}\left(\boldsymbol{\Psi}^{(N)}(\mathbf{p})\right)\right)^T$ consisting of the strain fields generated by basis functions. Combining equations (B.1.15) and (B.1.20), the components of the Jacobian matrix (B.1.13) are found to be:

$$L_{ij} = \int_{\Omega} \mathbf{e}\left(\boldsymbol{\Psi}^{(i)}(\mathbf{p})\right) : \mathcal{L}(\mathbf{U} \cdot \boldsymbol{\Psi}(\mathbf{p})) : \mathbf{e}\left(\boldsymbol{\Psi}^{(j)}(\mathbf{p})\right) d\mathbf{p}.$$

Now, the Jacobian (B.1.13) required in (B.1.12) is completely defined in terms of the selected FEM basis functions $\boldsymbol{\Psi}$ and the algorithmic tangent stiffness $\mathcal{L}$. Again, as in the case of $\mathbf{F}^h(\mathbf{U}^{(k)})$, evaluating the integrals in B.1 is done by quadrature formulae and thus the return mapping algorithm needs to provide $\mathcal{L}(\mathbf{u}^h)$ at only the necessary quadrature points.

The reader may perform the above computations componentwise for the shape functions using symbol notation for the stress, strains, etc. Working out these details will give further clarity in understanding the topic. It is also a useful exercise to write a flow-chart of the entire process of loading the material, solving the global balance of linear momentum (B.1.9) at the discrete instances and applying the Newton iteration (B.1.12) at each load step. While the whole thing looks complex, actually most of the difficulties are associated with book-keeping the history dependence. In fact the procedure described in this section is nothing more than a simple exercise in taking derivatives and carefully applying the chain rule (cf. equation (B.1.15)). Note however, that the numerical analysis of such boundary value problems and discretizations (existence and uniqueness of continuous solutions, stability of the discrete weak forms, convergence of the Newton iteration, etc.) are far from trivial topics.

B.2 Numerical Implementation of an SMA Constitutive Model

As we have seen in the previous section, a displacement-based finite element method requires a numerical algorithm that performs the following task: At a given material point, the history as well as the current material state (strain, temperature, internal variables) is known and new values are given for the strain and temperature. One has to find numerically the corresponding stress and internal variables at this particular material point. Thus, the selection of the finite element discretization (displacement-based) determines the type of numerical implementation needed for a given SMA model.

In this section, a numerical implementation that does exactly that is outlined for the three-phase model of Chapter 6. The implementation is an extension of the closest point projection algorithm presented in Chapter 4 (see also

[9]). The complex name of this algorithm is motivated by a certain geometric interpretation and should in no way be a cause for alarm to the reader. The algorithm is in fact a straightforward implicit backward Euler method along with a Newton iteration (the time stepping is implicit). The onset and termination of phase transformation introduces a few additional twists which, after proper care, by no means complicate the algorithm.

The implementation in Chapter 4 is for the class of SMA models of Boyd and Lagoudas [10] which have only one active transformation surface (either $A \to M^d$ or $M^d \to A$). The numerical scheme described in this section is adapted to the multiple transformation surfaces present in the current model. The scheme belongs to the general family of return mapping algorithms [11–13], which couple in a natural with displacement-based finite element methods. In this section, the major steps of the numerical implementation of the SMA constitutive model are presented. Note that unlike Chapters 3-6, in order to be consistent with the previous section, we use the Piola-Kirchoff stress $\mathbf{S}$ instead of the Cauchy stress $\boldsymbol{\sigma}$. Since the two are asymptotically the same for small strains, this does not change the following presentation in any way.

Consider a single material point. First, rewrite equation (6.4.24) as:

$$\boldsymbol{\varepsilon} = \boldsymbol{\mathcal{S}}(\boldsymbol{\xi}) : \mathbf{S} + \boldsymbol{\alpha}(\boldsymbol{\xi})(T - T_0) + \boldsymbol{\varepsilon}^{in}. \tag{B.2.21}$$

The evolution equations (6.4.8) and (6.4.9) and the decomposition (6.4.7) imply that the total inelastic strain $\boldsymbol{\varepsilon}^{in}$ can be written as

$$\dot{\boldsymbol{\varepsilon}}^{in} = \boldsymbol{\Lambda}^t(\mathbf{S})\dot{\xi}_2 + \boldsymbol{\Lambda}^d(\mathbf{S})\dot{\xi}_3, \tag{B.2.22}$$

where $\boldsymbol{\Lambda}^t(\mathbf{S})$ and $\boldsymbol{\Lambda}^d(\mathbf{S})$ are defined by equations (6.4.45) and (6.4.44) respectively. It is also convenient to write the consistency conditions imposed by the transformation surfaces (6.4.38)–(6.4.43) for the evolution of the internal variables $\boldsymbol{\xi}$ in the following compact form:

$$\dot{\xi}_1 \geq 0, \quad \Phi_1^+ \leq 0, \quad \Phi_1^+\dot{\xi} = 0, \tag{B.2.23a}$$

$$\dot{\xi}_1 \leq 0, \quad \Phi_1^- \leq 0, \quad \Phi_1^-\dot{\xi} = 0, \tag{B.2.23b}$$

$$\dot{\xi}_2 \geq 0, \quad \Phi_2^+ \leq 0, \quad \Phi_2^+\dot{\xi} = 0, \tag{B.2.23c}$$

$$\dot{\xi}_2 \leq 0, \quad \Phi_2^- \leq 0, \quad \Phi_2^-\dot{\xi} = 0, \tag{B.2.23d}$$

$$\dot{\xi}_3 \geq 0, \quad \Phi_3 \leq 0, \quad \Phi_3\dot{\xi} = 0. \tag{B.2.23e}$$

Thus, at each material point, the state variables satisfy the nonlinear system of differential-algebraic equations (B.2.21), (B.2.22) along with the constraints (B.2.23).

B.2.1 The Loading Step

Assume now, that the history of all field and internal variables at the material point is known. In particular, the values of $\boldsymbol{\varepsilon}_n$, T_n, $\mathbf{S}_n$, $\boldsymbol{\varepsilon}_n^{in}$, and $\boldsymbol{\xi}_n$ are known.

The subscript n denotes the loading step (see the previous section). The new values of ε_{n+1} and T_{n+1} for the strain and temperature respectively are also given.[1] Since the steps are discontinuous events, it is assumed that the continuous loading path that the material follows between step n and $n+1$ is characterized by

$$\varepsilon_{n+\lambda} = (1-\lambda)\varepsilon_n + \lambda\varepsilon_{n+1}, \quad \text{for } \forall\lambda \in [0,1], \tag{B.2.24}$$

and

$$T_{n+\lambda} = (1-\lambda)T_n + \lambda T_{n+1}, \quad \text{for } \forall\lambda \in [0,1]. \tag{B.2.25}$$

Here, $\lambda \in [0,1]$ is a loading parameter.

Closest Point Projection. The Closest Point Projection Return Mapping Algorithm is a numerical method that computes the values for $\mathbf{S}_{n+1}$, ε^{in}_{n+1} and $\boldsymbol{\xi}_{n+1}$ by solving equations (B.2.21)–(B.2.22) along with the constraints (B.2.23). This is done by first discretizing the evolution equation (B.2.22):

$$\varepsilon^{in}_{n+1} = \varepsilon^{in}_n + (\xi_{2n+1} - \xi_{2n})\mathbf{\Lambda}^t(\mathbf{S}_{n+1}) + (\xi_{3n+1} - \xi_{3n})\mathbf{\Lambda}^d(\mathbf{S}_{n+1}). \tag{B.2.26}$$

For a geometric interpretation of this backward Euler discretization, see Chapter 4. The stress-strain relation (B.2.21) is then written as:

$$\mathbf{S}_{n+1} = \mathcal{S}(\boldsymbol{\xi}_{n+1})^{-1} : \left(\varepsilon_{n+1} - \varepsilon^{in}_{n+1} - \boldsymbol{\alpha}(\boldsymbol{\xi}_{n+1})(T_{n+1} - T_0)\right). \tag{B.2.27}$$

In order to solve the discrete system (B.2.26), (B.2.27) subject to the constraints (B.2.23), first substitute ε^{in}_{n+1} from equation (B.2.26) into (B.2.27) and then rearrange the terms to obtain:

$$\begin{aligned} &\mathcal{S}(\boldsymbol{\xi}_{n+1})\mathbf{S}_{n+1} - \varepsilon_{n+1} + \alpha(\boldsymbol{\xi}_{n+1})\left(T_{n+1} - T_0\right) \\ &+\varepsilon^{in}_n + (\xi_{2n+1} - \xi_{2n})\mathbf{\Lambda}^t(\mathbf{S}_{n+1}) + (\xi_{3n+1} - \xi_{3n})\mathbf{\Lambda}^d(\mathbf{S}_{n+1}) = \mathbf{0}. \end{aligned} \tag{B.2.28}$$

Note that in the above equation, all members with subscript n as well as ε_{n+1} and T_{n+1} have known values. It is convenient to introduce the residual $\mathbf{F}$:

$$\begin{aligned} \mathbf{F}(\mathbf{S},\boldsymbol{\xi}) =& \mathcal{S}(\boldsymbol{\xi})\mathbf{S} - \varepsilon_{n+1} + \boldsymbol{\alpha}(\boldsymbol{\xi})\left(T_{n+1} - T_0\right) \\ &+ \varepsilon^{in}_n + (\xi_2 - \xi_{2n})\mathbf{\Lambda}^t(\mathbf{S}) + (\xi_3 - \xi_{3n})\mathbf{\Lambda}^d(\mathbf{S}). \end{aligned} \tag{B.2.29}$$

Observe that the system (B.2.26), (B.2.27) is now equivalent to

$$\mathbf{F}(\mathbf{S}_{n+1}, \boldsymbol{\xi}_{n+1}) = \mathbf{0}. \tag{B.2.30}$$

The Closest Point Projection method (see Chapter 4 for the geometric interpretation which lead to this name), like most return mapping algorithms, first performs a linear thermoelastic loading using equation (B.2.27), called

[1] Alternatively, the increments $\Delta\varepsilon_{n+1} = \varepsilon_{n+1} - \varepsilon_n$ and $\Delta T_{n+1} = T_{n+1} - T_n$ may be given, which, of course, is equivalent to knowing ε_{n+1} and T_{n+1}.

thermoelastic prediction. It then determines, using (B.2.23), if phase transformation occurs or not. If it does not, then the solution is accepted. If it does, it determines which one and performs a *transformation correction.* Without loss of generality, suppose that during the loading step the forward stress-induced phase transformation occurs and the rest of the phase transitions do not happen. This implies $\xi_{2n+1} - \xi_{2n} > 0$ and (B.2.23) reduces to

$$\Phi_2^+(\mathbf{S}_{n+1}, T_{n+1}, \boldsymbol{\xi}_{n+1}) = 0. \tag{B.2.31}$$

The Closest Point Projection method then does nothing else but solve (B.2.30) and (B.2.31) by Newton's method in order to obtain a consistent material state. These two steps are explained in details below. Both the predictor and corrector steps can be viewed as part of an iterative process which solves the nonlinear algebraic system of equations (B.2.30), subject to the constraints (B.2.23), by constructing a converging sequence

$$\mathbf{S}_{n+1}^{(k)} \underset{k\to\infty}{\longrightarrow} \mathbf{S}_{n+1}, \quad \boldsymbol{\varepsilon}_{n+1}^{in(k)} \underset{k\to\infty}{\longrightarrow} \boldsymbol{\varepsilon}_{n+1}^{in}, \quad \boldsymbol{\xi}_{n+1}^{(k)} \underset{k\to\infty}{\longrightarrow} \boldsymbol{\xi}_{n+1}. \tag{B.2.32}$$

B.2.2 Thermoelastic Prediction

As the first step, a *thermoelastic prediction* is performed during which the internal variables do not change:

$$\boldsymbol{\varepsilon}_{n+1}^{in(0)} = \boldsymbol{\varepsilon}_n^{in}, \tag{B.2.33}$$

$$\boldsymbol{\xi}_{n+1}^{(0)} = \boldsymbol{\xi}_n, \tag{B.2.34}$$

$$\mathbf{S}_{n+1}^{(0)} = \boldsymbol{\mathcal{S}}(\boldsymbol{\xi}_n)^{-1} : \left[\boldsymbol{\varepsilon}_{n+1} - \boldsymbol{\varepsilon}_n^{in} - \boldsymbol{\alpha}(\boldsymbol{\xi}_n)(T_{n+1} - T_0)\right]. \tag{B.2.35}$$

It should be noted, that this first step corresponds to purely thermoelastic loading without any transformation ($\dot{\boldsymbol{\xi}} = 0$), hence its name thermoelastic prediction. The corresponding values of the five transformation functions are then evaluated:

$$\Phi_\alpha^{(0)} = \Phi_\alpha(\mathbf{S}_{n+1}^{(0)}, T_{n+1}, \boldsymbol{\xi}_{n+1}^{(0)}). \tag{B.2.36}$$

The subscript α is understood in the sense $\Phi_\alpha \in \{\Phi_1^+, \Phi_1^-, \Phi_2^+, \Phi_2^-, \Phi_3\}$. If the value of all transformation functions satisfy $\Phi_\alpha^{(0)} \leq 0$ then all equations and constraints are satisfied and the iteration is terminated for $k = 0$.

B.2.3 Transformation Correction

The predictor step assumed that $\boldsymbol{\xi}_n = \boldsymbol{\xi}_{n+1}$, hence (B.2.23) are satisfied if and only if all $\Phi_\alpha \leq 0$. Therefore, if at least one of the transformation functions $\Phi_\alpha > 0$ then the corresponding consistency condition is violated. Such surfaces will be referred to as *inconsistent.* The existence of inconsistent surfaces implies that during the loading step, phase transformation takes place and a *transformation correction* is needed. During this step, the stress and the internal variables are modified in accordance with the transformation evolution equations so that the consistency conditions are satisfied.

The consistency condition(s) that correspond to the phase transformation(s) taking place will be called *active*. The same term will be used for the respective transformation surfaces. The consistency conditions that the elastic predictor violates are not necessarily the active ones, nor are they necessarily the only active ones. For examples of such cases, the reader is referred to [14].

For the time being, assume that it is known which transformation(s) are active during the load step. We will consider the case of a single or two simultaneous active surfaces.

Single active transformation. Suppose first that only one transformation is active, say $\varPhi_\alpha$. This implies that the corresponding volume fraction, denoted also by ξ_α, has nonzero rate.[2] That is, $\dot{\xi}_\alpha \neq 0$, and the corresponding consistency conditions (B.2.23) is satisfied, if and only if:

$$\varPhi_\alpha(\mathbf{S}_{n+1}, T_{n+1}, \boldsymbol{\xi}_{n+1}) = 0. \tag{B.2.37}$$

Therefore, during the transformation correction, one has to solve (B.2.30) along with the last equation. This is done by Newton's method: *For the given k-th iterate of $\mathbf{S}_{n+1}^{(k)}$, $\boldsymbol{\varepsilon}_{n+1}^{in(k)}$ and $\boldsymbol{\xi}_{n+1}^{(k)}$, find the $k+1$ iterates by linearizing $\mathbf{F}$ and $\varPhi_\alpha$ around $\left(\mathbf{S}_{n+1}^{(k)}, \boldsymbol{\xi}_{n+1}^{(k)}\right)$ and requiring that:*

$$\mathbf{F}^{(k)} + \frac{\partial \mathbf{F}^{(k)}}{\partial \mathbf{S}} : \Delta \mathbf{S}^{(k)} + \frac{\partial \mathbf{F}^{(k)}}{\partial \xi_\alpha} \Delta \xi_\alpha^{(k)} = 0, \tag{B.2.38}$$

$$\varPhi_\alpha^{(k)} + \frac{\partial \varPhi_\alpha^{(k)}}{\partial \mathbf{S}} : \Delta \mathbf{S}^{(k)} + \frac{\partial \varPhi_\alpha^{(k)}}{\partial \xi_\alpha} \cdot \Delta \xi_\alpha^{(k)} = 0. \tag{B.2.39}$$

When the increments $\Delta\mathbf{S}^{(k)}$ and $\Delta\xi_\alpha^{(k)}$ are determined from the above system of linear equations, the stress and the internal variable are updated according to

$$\mathbf{S}_{n+1}^{(k+1)} = \mathbf{S}_{n+1}^{(k)} + \Delta \mathbf{S}^{(k)}, \quad \xi_{\alpha n+1}^{(k+1)} = \xi_{\alpha n+1}^{(k)} + \Delta \xi_\alpha^{(k)},$$

and $\boldsymbol{\varepsilon}_{n+1}^{in(k)}$ is updated according to equation (B.2.26).

Observe that in equations (B.2.38) and (B.2.39) the following shorthand notation is used for $\mathbf{F}^{(k)}$, $\varPhi_\alpha^{(k)}$ and all their derivatives:

$$\mathbf{F}^{(k)} = \mathbf{F}\left(\mathbf{S}_{n+1}^{(k)}, \boldsymbol{\xi}_{n+1}^{(k)}\right), \qquad \varPhi_\alpha^{(k)} = \varPhi_\alpha\left(\mathbf{S}_{n+1}^{(k)}, T_{n+1}, \xi_{n+1}^{(k)}\right),$$

$$\frac{\partial \mathbf{F}^{(k)}}{\partial \mathbf{S}} = \frac{\partial \mathbf{F}}{\partial \mathbf{S}}\left(\mathbf{S}_{n+1}^{(k)}, \boldsymbol{\xi}_{n+1}^{(k)}\right), \qquad \frac{\partial \varPhi_\alpha^{(k)}}{\partial \mathbf{S}} = \frac{\partial \varPhi_\alpha}{\partial \mathbf{S}}\left(\mathbf{S}_{n+1}^{(k)}, T_{n+1}, \xi_{n+1}^{(k)}\right),$$

$$\frac{\partial \mathbf{F}^{(k)}}{\partial \xi_\alpha} = \frac{\partial \mathbf{F}}{\partial \xi_\alpha}\left(\mathbf{S}_{n+1}^{(k)}, \boldsymbol{\xi}_{n+1}^{(k)}\right), \qquad \frac{\partial \varPhi_\alpha^{(k)}}{\partial \xi_\alpha} = \frac{\partial \varPhi_\alpha}{\partial \xi_\alpha}\left(\mathbf{S}_{n+1}^{(k)}, T_{n+1}, \xi_{n+1}^{(k)}\right).$$

[2] If $\varPhi_\alpha \in \{\varPhi_1^+, \varPhi_1^-\}$ then the internal variable responsible is $\xi_\alpha = \xi_1$, if $\varPhi_\alpha \in \{\varPhi_2^+, \varPhi_2^-\}$ then $\xi_\alpha = \xi_2$ and if $\varPhi_\alpha = \varPhi_3$ then $\xi_\alpha = \xi_3$.

Two simultaneous active transformations. If two of the transformations are active, say Φ_α and Φ_β, then during the correction, equation (B.2.30) along with

$$\Phi_\alpha(\mathbf{S}_{n+1}, T_{n+1}, \boldsymbol{\xi}_{n+1}) = 0, \tag{B.2.40}$$

$$\Phi_\beta(\mathbf{S}_{n+1}, T_{n+1}, \boldsymbol{\xi}_{n+1}) = 0, \tag{B.2.41}$$

is being solved, again by Newton's method: *For the given k-th iterate of* $\mathbf{S}_{n+1}^{(k)}$, $\boldsymbol{\varepsilon}_{n+1}^{in(k)}$ *and* $\boldsymbol{\xi}_{n+1}^{(k)}$, *find the* $k+1$ *iterates by linearizing* $\mathbf{F}$, Φ_α *and* Φ_β *around* $\left(\mathbf{S}_{n+1}^{(k)}, \boldsymbol{\xi}_{n+1}^{(k)}\right)$ *and requiring that:*

$$\mathbf{F}^{(k)} + \frac{\partial \mathbf{F}^{(k)}}{\partial \mathbf{S}} : \Delta \mathbf{S}^{(k)} + \frac{\partial \mathbf{F}^{(k)}}{\partial \xi_\alpha} \Delta \xi_\alpha^{(k)} + \frac{\partial \mathbf{F}^{(k)}}{\partial \xi_\beta} \Delta \xi_\beta^{(k)} = 0, \tag{B.2.42}$$

$$\Phi_\alpha^{(k)} + \frac{\partial \Phi_\alpha^{(k)}}{\partial \mathbf{S}} : \Delta \mathbf{S}^{(k)} + \frac{\partial \Phi_\alpha^{(k)}}{\partial \xi_\alpha} \cdot \Delta \xi_\alpha^{(k)} + \frac{\partial \Phi_\alpha^{(k)}}{\partial \xi_\beta} \cdot \Delta \xi_\beta^{(k)} = 0, \tag{B.2.43}$$

$$\Phi_\beta^{(k)} + \frac{\partial \Phi_\beta^{(k)}}{\partial \mathbf{S}} : \Delta \mathbf{S}^{(k)} + \frac{\partial \Phi_\beta^{(k)}}{\partial \xi_\alpha} \cdot \Delta \xi_\alpha^{(k)} + \frac{\partial \Phi_\beta^{(k)}}{\partial \xi_\beta} \cdot \Delta \xi_\beta^{(k)} = 0. \tag{B.2.44}$$

When the increments $\Delta \mathbf{S}^{(k)}$, $\Delta \xi_\alpha^{(k)}$ *and* $\Delta \xi_\beta^{(k)}$ *are determined from the above system of linear equations, the stress and the internal variable are updated according to*

$$\mathbf{S}_{n+1}^{(k+1)} = \mathbf{S}_{n+1}^{(k)} + \Delta \mathbf{S}^{(k)}, \quad \xi_{\alpha n+1}^{(k+1)} = \xi_{\alpha n+1}^{(k)} + \Delta \xi_\alpha^{(k)}, \quad \xi_{\beta n+1}^{(k+1)} = \xi_{\beta n+1}^{(k)} + \Delta \xi_\beta^{(k)},$$

and $\boldsymbol{\varepsilon}_{n+1}^{in(k)}$ *is updated according to equation (B.2.26).*

This completes the outline for the numerical implementation of the model. The details of solving the linear system (B.2.38)-(B.2.39) or (B.2.42)-(B.2.44), including the functional form of the derivatives involved are omitted.

It is important to note that when $\Phi_2^\pm$ is the only active surface, the iteration (B.2.38),(B.2.39) reduces to the Closest Point Projection method of Qidwai and Lagoudas [9], see also Chapter 4. In the latter work, the algorithm is formulated by defining a residual for the evolution equation (B.2.26), instead of (B.2.28). It is easy to show that the two lead to the same algorithm. The current approach has the advantage that by taking the residual of Hooke's law the algorithm generalizes for the twinning transformation $A \leftrightarrow M^t$ in which no transformation strain is generated.

B.2.4 Active Surfaces and Other Implementation Details

Returning to the question of which transformation surfaces are active during the correction, observe that at any time during the loading step, they can be separated into two groups, $\mathcal{F} = \{\Phi_1^+, \Phi_2^+, \Phi_3\}$ and $\mathcal{B} = \{\Phi_1^-, \Phi_2^-, \Phi_3\}$. If two phase transformations surfaces are active, say Φ_α and Φ_β, then either

$\Phi_\alpha, \Phi_\beta \in \mathcal{F}$ or $\Phi_\alpha, \Phi_\beta \in \mathcal{B}$. The inconsistent surfaces from the predictor state need not follow the same rule. However, observe that the elastic prediction is a continuous mapping of $\mathbf{S}_{n+1}$ with respect to T_{n+1} and $\boldsymbol{\varepsilon}_{n+1}$ and the loading step $n+1$ is part of a continuous loading path (B.2.24)-(B.2.25). Then, for some sufficiently small loading parameter, λ, the thermoelastic predictor for the state $n+\lambda$ will satisfy the above criterion. Since the thermoelastic prediction is a computationally inexpensive process, if the current prediction cannot determine whether the loading step belongs to $\mathcal{F}$ or $\mathcal{B}$, the simplest practical approach is to take $\lambda = \frac{1}{2}, \frac{1}{4}, \ldots$ until this can be determined.

Once this is done, in order to find which transformation(s) are active, the natural way is to attempt a correction of the ones that are violated first. It may happen that after the correction, some other transformation surface becomes inconsistent, or the increment of the corresponding volume fraction is inconsistent (the consistency conditions also specify the sign of $\dot{\boldsymbol{\xi}}$, that is, if active surface is Φ_1^-, then $\xi_{1n} > \xi_{1n+1}$, etc.). In such a case the brute force approach of attempting all possible single and double transformations from the active set is used.

The last important detail is how to terminate the transformations, that is how to impose the constrains (6.4.2). Again, without loss of generality, suppose the correction step was restoring consistency of Φ_1^-, and either

$$c_{3n+1} < 0,$$

or

$$c_{1n+1} > 1.$$

Suppose $c_{1n+1} > 1$. The volume fraction can be treated as a continuous, monotonous function of the loading parameter λ and, which is more, $c_{1n} < 1$, therefore the equation

$$c_{1n+\lambda} = 1$$

has a root λ in the interval $\lambda \in [0, 1]$. Given such precise information about the location of the root and that the explicit form of $c_{1n+\lambda}$ as a function of λ does not seem easy to determine, one can just use a modified secant's method to determine the root of the above equation. Each evaluation of $c_{1n+\lambda}$ consists of performing the transformation correction for the intermediate loading step $t_{n+\lambda}$.

B.2.5 Algorithmic Tangent Stiffness (Jacobian)

So far, the return-mapping algorithm described in this section calculates the state variables at the current time step $n+1$ at a material point. The algorithmic tangent stiffness and thermal moduli will now be defined. Recall in Sect. B.2.1, that the loading step is defined by specifying the strain $\boldsymbol{\varepsilon}_{n+1}$ and temperature, T_{n+1}. All the remaining state variables $\mathbf{S}_{n+1}$, $\boldsymbol{\xi}_{n+1}$, $\dot{\boldsymbol{\xi}}_{n+1}$ and $\boldsymbol{\varepsilon}^{in}_{n+1}$ are determined using the system of equations and constraints

(B.2.26), (B.2.27) and (B.2.23). Thus, they can be treated as implicit functions of $\boldsymbol{\varepsilon}_{n+1}$ and T_{n+1}, and in particular:

$$\mathbf{S}_{n+1} = \tilde{\mathbf{S}}_{n+1}\left(\boldsymbol{\varepsilon}_{n+1}, T_{n+1}\right). \tag{B.2.45}$$

The tangent stiffness and thermal tangent moduli are defined by

$$\mathcal{L} = \frac{\partial \tilde{\mathbf{S}}_{n+1}\left(\boldsymbol{\varepsilon}_{n+1}, T_{n+1}\right)}{\partial \boldsymbol{\varepsilon}_{n+1}} \tag{B.2.46}$$

and

$$\mathcal{M} = \frac{\partial \tilde{\mathbf{S}}_{n+1}\left(\boldsymbol{\varepsilon}_{n+1}, T_{n+1}\right)}{\partial T_{n+1}}, \tag{B.2.47}$$

respectively. The tangent stiffness is a fourth-order tensor while the thermal tangent is a second-order tensor.

Our goal is to arrive at an analytical expression for $\mathcal{L}$, given our inelastic SMA constitutive theory and the selected return mapping algorithm. In general it depends on which transformation surfaces are active. The subscript $n+1$ will be omitted for the rest of the derivation, and all state variable without subscript will, by default, be considered at time step $n+1$. The thermal tangent moduli are listed for completeness only, since they are required when a coupled, thermo-mechanical problem is solved and such problems are not considered in this work. The derivation of $\mathcal{M}$ follows the same procedure that is used for $\mathcal{L}$ given below, so it is left to the reader.

To derive $\mathcal{L}$, first, it is useful to notice that if a volume fraction ξ_α is active during the loading step, then its derivative with respect to strain can be evaluated. Indeed, suppose first the only one transformation surface is active during the loading step and ξ_α is the active variable. This implies that

$$\Phi_\alpha(\mathbf{S}, \xi_\alpha) = 0,$$

so ξ_α is an implicit function of $\mathbf{S}$. The remaining variables which Φ_α is a function of do not change during the loading step, so they are suppressed. In view of equation (B.2.45) the last equation can be differentiated by $\boldsymbol{\varepsilon}$ to obtain:

$$\frac{\partial \Phi_\alpha}{\partial \mathbf{S}} : \frac{\partial \mathbf{S}}{\partial \boldsymbol{\varepsilon}} + \frac{\partial \Phi_\alpha}{\partial \xi_\alpha}\frac{\partial \xi_\alpha}{\partial \boldsymbol{\varepsilon}} = \mathbf{0}.$$

Therefore, if the second-nd order tensor $\boldsymbol{\Xi}_\alpha$ is defined as

$$\boldsymbol{\Xi}_\alpha = -\frac{\partial \Phi_\alpha}{\partial \mathbf{S}} \Big/ \frac{\partial \Phi_\alpha}{\partial \xi_\alpha}, \tag{B.2.48}$$

then

$$\frac{\partial \xi_\alpha}{\partial \boldsymbol{\varepsilon}} = \boldsymbol{\Xi}_\alpha : \frac{\partial \mathbf{S}}{\partial \boldsymbol{\varepsilon}}$$

is the derivative of ξ_α with respect to $\boldsymbol{\varepsilon}$. Note that transformation surface and its derivatives have a well-defined functional form, so this quantity can be evaluated directly for any values of $\mathbf{S}$, T and $\boldsymbol{\xi}$.

If, on the other hand, two surfaces are active, say Φ_α and Φ_{α}, then the appropriate derivatives are calculated in the same way, but this time a 2×2 linear system has to be solved:

$$\begin{bmatrix} \boldsymbol{\Xi}_\alpha \\ \boldsymbol{\Xi}_\beta \end{bmatrix} = - \begin{bmatrix} \dfrac{\partial \Phi_\alpha}{\partial \xi_\alpha} & \dfrac{\partial \Phi_\alpha}{\partial \xi_\beta} \\ \dfrac{\partial \Phi_\beta}{\partial \xi_\alpha} & \dfrac{\partial \Phi_\beta}{\partial \xi_\beta} \end{bmatrix}^{-1} \begin{bmatrix} \dfrac{\partial \Phi_\alpha}{\partial \mathbf{S}} \\ \dfrac{\partial \Phi_\beta}{\partial \mathbf{S}} \end{bmatrix}, \tag{B.2.49}$$

and

$$\frac{\partial \xi_\alpha}{\partial \boldsymbol{\varepsilon}} = \boldsymbol{\Xi}_\alpha : \frac{\partial \mathbf{S}}{\partial \boldsymbol{\varepsilon}}, \tag{B.2.50}$$

$$\frac{\partial \xi_\beta}{\partial \boldsymbol{\varepsilon}} = \boldsymbol{\Xi}_\beta : \frac{\partial \mathbf{S}}{\partial \boldsymbol{\varepsilon}}. \tag{B.2.51}$$

Finally, let

$$\boldsymbol{\Xi}_\alpha = \mathbf{0}, \tag{B.2.52}$$

whenever Φ_α is not an active surface.

To summarize, $\boldsymbol{\Xi}_\alpha$ is given by equation (B.2.48) if only one transformation is active, by (B.2.49) if two are active and by (B.2.52) if Φ_α is not an active transformation surface.

With these definitions, the tangent stiffness $\mathcal{L}$ is given by the formula:

$$\mathcal{L} = \Big[\big(\Delta\boldsymbol{\mathcal{S}} : \mathbf{S} + \Delta\boldsymbol{\alpha}(T_{n+1} - T_0) \big) \otimes (\boldsymbol{\Xi}_1 + \boldsymbol{\Xi}_2) + \boldsymbol{\mathcal{S}}(\boldsymbol{\xi}) + \boldsymbol{\Lambda}^t \otimes \boldsymbol{\Xi}_2 \\ + \boldsymbol{\Lambda}^d \otimes \boldsymbol{\Xi}_3 + (\xi_2 - \xi_{2n}) \frac{\partial \boldsymbol{\Lambda}^t}{\partial \mathbf{S}} + (\xi_3 - \xi_{3n}) \frac{\partial \boldsymbol{\Lambda}^d}{\partial \mathbf{S}} \Big]^{-1} . \tag{B.2.53}$$

The formula will be proven only for the case when Φ_1 and Φ_2 are active. Checking all other possibilities of active transformation surfaces is left to the reader. Note that the above formula includes derivatives of the transformation flow tensors. Since both $\boldsymbol{\Lambda}^t$ and $\boldsymbol{\Lambda}^d$ are J_2 based, the reader is referred to [9] for the functional form of these derivatives, as well as the derivation itself.

So, assume, that $\xi_1 - \xi_{1n} \neq 0$, $\xi_2 - \xi_{2n} \neq 0$ and $\xi_3 = \xi_{3n}$ and differentiate equation (B.2.28) with respect to $\boldsymbol{\varepsilon}$. Note that for fixed T, all the quantities in this equation are either implicit functions of $\boldsymbol{\varepsilon}$ or enter as constants, e.g. independent of $\boldsymbol{\varepsilon}$. By applying the product rule and, very carefully, the chain rule, the following result is obtained:

$$
\begin{aligned}
\mathbf{0} =& \frac{\partial}{\partial \boldsymbol{\varepsilon}_{n+1}} \left(\boldsymbol{\mathcal{S}}(\boldsymbol{\xi})\mathbf{S}\right) - \mathbf{I} + \frac{\partial}{\partial \boldsymbol{\varepsilon}_{n+1}} \left(\boldsymbol{\alpha}(\boldsymbol{\xi}) \left(T_{n+1} - T_0\right)\right) + \frac{\partial}{\partial \boldsymbol{\varepsilon}_{n+1}} \left((\xi_2 - \xi_{2n})\boldsymbol{\Lambda}^t(\mathbf{S})\right) \\
=& \frac{\partial \boldsymbol{\mathcal{S}}(\boldsymbol{\xi})}{\partial \boldsymbol{\varepsilon}_{n+1}} : \mathbf{S} + : \boldsymbol{\mathcal{S}}(\boldsymbol{\xi}) : \frac{\partial \mathbf{S}}{\partial \boldsymbol{\varepsilon}_{n+1}} - \mathbf{I} + \frac{\partial \boldsymbol{\alpha}(\boldsymbol{\xi})}{\partial \boldsymbol{\varepsilon}_{n+1}} \left(T_{n+1} - T_0\right) \\
&+ \left(\frac{\partial \xi_2}{\partial \boldsymbol{\varepsilon}_{n+1}}\right) \otimes \boldsymbol{\Lambda}^t(\mathbf{S}) + \left((\xi_2 - \xi_{2n}) \frac{\partial \boldsymbol{\Lambda}^t(\mathbf{S})}{\partial \boldsymbol{\varepsilon}_{n+1}}\right) \\
=& \left(\frac{\partial \boldsymbol{\mathcal{S}}(\boldsymbol{\xi})}{\partial \boldsymbol{\xi}} \cdot \frac{\partial \boldsymbol{\xi}}{\partial \boldsymbol{\varepsilon}_{n+1}}\right) : \mathbf{S} + \left(T_{n+1} - T_0\right) \frac{\partial \boldsymbol{\alpha}(\boldsymbol{\xi})}{\partial \boldsymbol{\xi}} \cdot \frac{\partial \boldsymbol{\xi}}{\partial \boldsymbol{\varepsilon}_{n+1}} + \boldsymbol{\mathcal{S}}(\boldsymbol{\xi}) : \frac{\partial \mathbf{S}}{\partial \boldsymbol{\varepsilon}_{n+1}} - \mathbf{I} \\
&+ \left(\boldsymbol{\Xi}_2 : \frac{\partial \mathbf{S}}{\partial \boldsymbol{\varepsilon}_{n+1}}\right) \otimes \boldsymbol{\Lambda}^t(\mathbf{S}) + \left((\xi_2 - \xi_{2n}) \frac{\partial \boldsymbol{\Lambda}^t(\mathbf{S})}{\partial \mathbf{S}} : \frac{\partial \mathbf{S}}{\partial \boldsymbol{\varepsilon}_{n+1}}\right), \qquad \text{(B.2.54)}
\end{aligned}
$$

where $\mathbf{I}$ is the fourth-order identity tensor defined by 4.3.29. Now, using equation (6.4.15), (B.2.50) and (B.2.51), observe that:

$$
\begin{aligned}
&\left(\frac{\partial \boldsymbol{\mathcal{S}}(\boldsymbol{\xi})}{\partial \boldsymbol{\xi}} \cdot \frac{\partial \boldsymbol{\xi}}{\partial \boldsymbol{\varepsilon}_{n+1}}\right) : \mathbf{S} + \left(T_{n+1} - T_0\right) \frac{\partial \boldsymbol{\alpha}(\boldsymbol{\xi})}{\partial \boldsymbol{\xi}} \cdot \frac{\partial \boldsymbol{\xi}}{\partial \boldsymbol{\varepsilon}_{n+1}} = \\
&\left(\Delta \boldsymbol{\mathcal{S}} : \mathbf{S} + \left(T_{n+1} - T_0\right) \Delta \boldsymbol{\alpha}\right) \otimes \left(\frac{\partial \xi_1}{\partial \boldsymbol{\varepsilon}_{n+1}} + \frac{\partial \xi_2}{\partial \boldsymbol{\varepsilon}_{n+1}}\right) = \\
&\left[\left(\Delta \boldsymbol{\mathcal{S}} : \mathbf{S} + \left(T_{n+1} - T_0\right) \Delta \boldsymbol{\alpha}\right) \otimes \left(\boldsymbol{\Xi}_1 + \boldsymbol{\Xi}_2\right)\right] : \frac{\partial \mathbf{S}}{\partial \boldsymbol{\varepsilon}}. \qquad \text{(B.2.55)}
\end{aligned}
$$

Therefore, by combining with (B.2.54), rearranging the terms, and factoring $\partial \mathbf{S}/\partial \boldsymbol{\varepsilon}$ the following, hopefully correct, identity is obtained:

$$
\begin{aligned}
\mathbf{I} = \Big[&\left(\Delta \boldsymbol{\mathcal{S}} : \mathbf{S} + \left(T_{n+1} - T_0\right) \Delta \boldsymbol{\alpha}\right) \otimes \left(\boldsymbol{\Xi}_1 + \boldsymbol{\Xi}_2\right) + \boldsymbol{\mathcal{S}}(\boldsymbol{\xi}) \\
&+ \boldsymbol{\Lambda}^t \otimes \boldsymbol{\Xi}_2 + (\xi_2 - \xi_{2n}) \frac{\partial \boldsymbol{\Lambda}^t(\mathbf{S})}{\partial \mathbf{S}}\Big] : \frac{\partial \mathbf{S}}{\partial \boldsymbol{\varepsilon}}. \qquad \text{(B.2.56)}
\end{aligned}
$$

After inverting the fourth-order tensor, the final result is obtained:

$$
\begin{aligned}
\frac{\partial \mathbf{S}}{\partial \boldsymbol{\varepsilon}} = \Big[&\left(\Delta \boldsymbol{\mathcal{S}} : \mathbf{S} + \left(T_{n+1} - T_0\right) \Delta \boldsymbol{\alpha}\right) \otimes \left(\boldsymbol{\Xi}_1 + \boldsymbol{\Xi}_2\right) \\
&+ \boldsymbol{\mathcal{S}}(\boldsymbol{\xi}) + \boldsymbol{\Lambda}^t \otimes \boldsymbol{\Xi}_2 + (\xi_2 - \xi_{2n}) \frac{\partial \boldsymbol{\Lambda}^t(\mathbf{S})}{\partial \mathbf{S}}\Big]^{-1}. \qquad \text{(B.2.57)}
\end{aligned}
$$

Remember that $\xi_3 = \xi_{3n}$, that is Φ_3 is not an active surface and hence by the definition (B.2.52), $\boldsymbol{\Xi}_3$, it is identically zero. Therefore, (B.2.53) reduces to (B.2.57) when Φ_1 and Φ_2 are the active surfaces.

As a final remark, observe that the discretized evolution equation (B.2.26) enters in the expression for $\mathcal{L}$ and as a result the latter includes a dependency $\xi_{in+1} - \xi_{in}$, $i = 2, 3$, as well as the derivative of the transformation flow tensors. A different discretization of (B.2.22) would have resulted in a different evolution equation, so $\mathcal{L}$ is algorithmic-specific, hence the name *algorithmic tangent*.

References

[1] O. Axelsson, V. Barker, Finite Element Solution of Boundary Value Problems- Theory and Computations, Academic Press, London, 1984.

[2] P. Ciarlet, The Finite Element Method for Elliptic Problems, no. 40 in Classics in Applied Mathematics, SIAM, Philadelphia, 2002.

[3] R. LeVeque, Finite Volume Methods for Hyperbolic Problems, Cambridge University Press, 2002.

[4] A. Samarskii, The Theory of Difference Schemes, Pure and Applied Mathematics, CRC, 2001.

[5] V. Girault, P. Raviart, Finite Element Methods for Navier-Stokes Equations, Springer-Verlag, Berlin, 1986.

[6] F. Brezzi, F. M., Mixed and Hybrid Finite Element Methods, Springer Series in Computational Mathematics, Springer Verlag, 1991.

[7] B. Cockburn, K. George, C. Shu, Discontinuous Galerkin Methods: Theory, Compuration and Applications, Springer, 2000.

[8] J. N. Reddy, An Introduction to the Finite Element Method, McGraw-Hill, 1993.

[9] M. A. Qidwai, D. C. Lagoudas, Numerical implementation of a shape memory alloy thermomechanical constitutive model using return mapping algorithms, International Journal for Numerical Methods in Engineering 47 (2000) 1123–1168.

[10] J. G. Boyd, D. C. Lagoudas, A thermodynamic constitutive model for the shape memory materials. Part I. The monolithic shape memory alloys, International Journal of Plasticity 12 (6) (1996) 805–842.

[11] M. Ortiz, E. P. Popov, Accuracy and stability of integration algorithms for elastoplastic constitutive relations, International Journal for Numerical Methods in Engineering 21 (1985) 1561–1576.

[12] J. C. Simo, T. J. R. Hughes, General return mapping algorithms for rate-independent plasticity, in: C. S. Desai (Ed.), Constitutive Laws for Engineering Materials: Theory and Applications, Elsevier Science Publishing Co., Inc., 1987, pp. 221–231.

[13] J. C. Simo, T. J. R. Hughes, Computational Inelasticity, Vol. 7, Springer-Verlag New York Inc, New York, 1998.

[14] P. Popov, Constitutive modelling of shape memory alloys andupscaling of deformable porous media, Ph.D. thesis, Texas A&M University (2005).

C

Numerical Implementation of Transformation Induced Plasticity in SMAs

P. B. Entchev and D. C. Lagoudas

The numerical implementation of the thermomechanical constitutive model for transformation induced plasticity in SMAs is presented here, supplementing the material in Chapter 5. The structure of the governing equations of the SMA constitutive model is very similar to those characterizing rate-independent plasticity in metals. The approach taken in this work is to implement a return-mapping algorithm for the SMA constitutive model, presented in Chapter 5. The development of the algorithm is based on the work of Qidwai and Lagoudas [1] *with special care taken to account for the development of transformation induced plastic strains. The algorithm presented in this Appendix is an implementation of the 'closest point projection return mapping algorithm' and follows closely the work of Entchev and Lagoudas* [2] *and Entchev* [3].

C.1 Summary of the SMA Constitutive Model Equations

For convenience, the SMA constitutive equations presented in Chapter 5 are summarized here. The strain ε is given by

$$\boldsymbol{\varepsilon} = \mathcal{S} : \boldsymbol{\sigma} + \boldsymbol{\alpha}(T - T_0) + \boldsymbol{\varepsilon}^t + \boldsymbol{\varepsilon}^p. \tag{C.1.1}$$

The evolution equations for the transformation strain $\boldsymbol{\varepsilon}^t$ and the plastic strain $\boldsymbol{\varepsilon}^p$ are given by

$$\dot{\boldsymbol{\varepsilon}}^t = \boldsymbol{\Lambda}\dot{\xi}, \tag{C.1.2}$$

$$\dot{\boldsymbol{\varepsilon}}^p = \widetilde{\boldsymbol{\Lambda}}^p\dot{\zeta}^d, \tag{C.1.3}$$

where $\boldsymbol{\Lambda}$ and $\widetilde{\boldsymbol{\Lambda}}^p$ are defined as

$$\boldsymbol{\Lambda} = \begin{cases} \frac{3}{2}H^{\mathrm{cur}}\frac{\boldsymbol{\sigma}^{\mathrm{eff}\prime}}{\bar{\sigma}^{\mathrm{eff}}} \; ; & \dot{\xi} > 0, \\ H^{cur-r}\dfrac{\varepsilon^{t-r}}{\bar{\varepsilon}^{t-r}} \; ; & \dot{\xi} < 0, \end{cases} \tag{C.1.4}$$

$$\widetilde{\boldsymbol{\Lambda}}^p = \begin{cases} \frac{3}{2}C_1^p \exp\left[-\frac{\zeta^d}{C_2^p}\right]\frac{\boldsymbol{\sigma}^{\mathrm{eff}\prime}}{\bar{\sigma}^{\mathrm{eff}}}, & \dot{\xi} > 0, \\ C_1^p \exp\left[-\frac{\zeta^d}{C_2^p}\right]\frac{\boldsymbol{\varepsilon}^t_{\mathrm{max}}}{\bar{\varepsilon}^t_{\mathrm{max}}}, & \dot{\xi} < 0, \end{cases} \tag{C.1.5}$$

and ζ^d is given by

$$\zeta^d = \int_0^t |\dot{\xi}^d(\tau)| \mathrm{d}\tau. \tag{C.1.6}$$

The effective stress $\boldsymbol{\sigma}^{\text{eff}}$ is defined in terms of the applied stress $\boldsymbol{\sigma}$ and the back stress $\boldsymbol{\beta}$ as

$$\boldsymbol{\sigma}^{\text{eff}} = \boldsymbol{\sigma} + \boldsymbol{\beta}. \tag{C.1.7}$$

The transformation function Φ is defined as

$$\Phi = \begin{cases} \pi - Y, & \dot{\xi} > 0, \\ -\pi - Y, & \dot{\xi} < 0. \end{cases} \tag{C.1.8}$$

In the above equation, π is the thermodynamic force conjugate to ξ and is given by

$$\begin{aligned} \pi = {} & \frac{1}{2}\boldsymbol{\sigma} : \Delta\, \mathcal{S} : \boldsymbol{\sigma} + \boldsymbol{\sigma} : \Delta\boldsymbol{\alpha}(T - T_0) + \boldsymbol{\sigma} : \frac{\partial \varepsilon^t}{\partial \xi} + \boldsymbol{\beta} : \frac{\partial \varepsilon^t}{\partial \xi} + \eta \\ & - \rho\Delta c \left[T - T_0 - T \ln\left(\frac{T}{T_0}\right) \right] + \rho\Delta s_0 (T - M^{0s}) + Y. \end{aligned} \tag{C.1.9}$$

Expressions for the back stress $\boldsymbol{\beta}$ and the drag stress η are given in Sect. 5.2.1. The evolution of the martensitic volume fraction ξ is constrained by the Kuhn–Tucker conditions

$$\begin{aligned} \dot{\xi} \geq 0, \quad \Phi \leq 0, \quad \Phi\dot{\xi} = 0, \\ \dot{\xi} \leq 0, \quad \Phi \leq 0, \quad \Phi\dot{\xi} = 0. \end{aligned} \tag{C.1.10}$$

Thus, the final system of algebraic and differential equations consists of equations (C.1.1)–(C.1.3), which is constrained by the Kuhn-Tucker condition (C.1.10).

C.2 Closest Point Projection Return Mapping Algorithm

To proceed with the solution of the above problem, it is first assumed that the strain and temperature histories $\boldsymbol{\varepsilon}(t)$ and $T(t)$ are given. In addition, the initial conditions for ξ, $\boldsymbol{\varepsilon}^t$ and $\boldsymbol{\varepsilon}^p$ are also known. The main idea of the return mapping algorithm is to divide the problem into two parts. The first step is to obtain a thermoelastic prediction, assuming that the transformation and plastic strains do not change. During the second step, if the predicted thermoelastic state violates the transformation condition, a transformation correction problem is solved to restore consistency. During the transformation correction, the evolution equations for the transformation and plastic strains are discretized as

$$\boldsymbol{\varepsilon}^t_{n+1} = \boldsymbol{\varepsilon}^t_n + (\xi_{n+1} - \xi_n)\left[\lambda \boldsymbol{\Lambda}_{n+1} + (1-\lambda)\boldsymbol{\Lambda}_n\right], \tag{C.2.11}$$

$$\varepsilon^p_{n+1} = \varepsilon^p_n + (\zeta^d_{n+1} - \zeta^d_n)\left[\lambda\widetilde{\mathbf{\Lambda}}^p_{n+1} + (1-\lambda)\widetilde{\mathbf{\Lambda}}^p_n\right], \tag{C.2.12}$$

where $\lambda \in [0,1]$ and subscript n indicates functional evaluation at time t_n. Different values of λ will result in different integration algorithms. In this work, λ is chosen to be equal to 1, which results in the backward Euler integration rule. The resulting algorithm is referred to as *the closest point projection* return mapping algorithm. The stress-strain relation (C.1.1) for the case of $\lambda = 1$ is also written as

$$\boldsymbol{\sigma}_{n+1} = \mathcal{S}^{-1}_{n+1} : \left(\varepsilon_{n+1} - \varepsilon^t_{n+1} - \varepsilon^p_{n+1} - \boldsymbol{\alpha}_{n+1}(T_{n+1} - T_0)\right). \tag{C.2.13}$$

The main steps of the algorithm are described in detail in the following sections.

C.2.1 Thermoelastic Prediction

During the thermoelastic prediction step, the following evaluations are performed:

$$\varepsilon_{n+1} = \varepsilon_n + \Delta\varepsilon_{n+1}, \tag{C.2.14a}$$
$$T_{n+1} = T_n + \Delta T_{n+1}, \tag{C.2.14b}$$
$$\varepsilon^{t(0)}_{n+1} = \varepsilon^t_n, \tag{C.2.14c}$$
$$\varepsilon^{p(0)}_{n+1} = \varepsilon^p_n, \tag{C.2.14d}$$
$$\xi^{(0)}_{n+1} = \xi_n, \tag{C.2.14e}$$

where $\Delta\varepsilon_{n+1}$ and ΔT_{n+1} are the (known) total strain and temperature increments for the $(n+1)^{th}$ time step. The superscript is introduced to indicate the current iteration number. Parenthesis around zero means evaluation of the quantities from the previous time step. The value of the stress is obtained by

$$\boldsymbol{\sigma}^{(0)}_{n+1} = \mathcal{S}^{-1}_n : \left[\varepsilon_{n+1} - \varepsilon^{t(0)}_{n+1} - \varepsilon^{p(0)}_{n+1} - \boldsymbol{\alpha}_n(T_{n+1} - T_0)\right]. \tag{C.2.15}$$

The corresponding value of the transformation function Φ is evaluated as

$$\Phi^{(0)}_{n+1} = \Phi(\boldsymbol{\sigma}^{(0)}_{n+1}, T_{n+1}, \xi^{(0)}_{n+1}). \tag{C.2.16}$$

At this point, if the value of the transformation function satisfies the transformation criterion $\Phi^{(0)}_{n+1} \leq 0$ then no phase transformation occurs at this time increment. However, if the transformation criterion is violated, i. e., $\Phi^{(0)}_{n+1} > 0$, then a transformation correction is performed. The transformation correction is performed by solving equations (C.2.11)–(C.2.13) with initial guess given by equations (C.2.14), (C.2.15) and constrained by the Kuhn–Tucker conditions (C.1.10). During the transformation correction the values of total strain ε_{n+1} and temperature T_{n+1} do not change. Only the values of the internal state variables and the stress are updated.

C.2.2 Transformation Correction

To proceed with the transformation correction step, the following transformation and plastic residuals are defined for the kth iteration as

$$\mathbf{R}^{t(k)}_{n+1} = -\boldsymbol{\varepsilon}^{t(k)}_{n+1} + \boldsymbol{\varepsilon}^{t}_{n} + \boldsymbol{\Lambda}^{(k)}_{n+1}(\xi^{(k)}_{n+1} - \xi_n) = 0, \tag{C.2.17}$$

$$\mathbf{R}^{p(k)}_{n+1} = -\boldsymbol{\varepsilon}^{p(k)}_{n+1} + \boldsymbol{\varepsilon}^{p}_{n} + \widetilde{\boldsymbol{\Lambda}}^{p(k)}_{n+1}\left|\xi^{d(k)}_{n+1} - \xi^{d}_{n}\right| = 0. \tag{C.2.18}$$

The transformation (consistency) condition is given by

$$\Phi^{(k)}_{n+1} = \Phi(\boldsymbol{\sigma}^{(k)}_{n+1}, T_{n+1}, \xi^{(k)}_{n+1}) = 0. \tag{C.2.19}$$

Equations (C.2.17)–(C.2.19) are solved using the Newton–Raphson method. Keeping in mind that during the transformation correction the total strain and temperature are constant, the linearized forms of equations (C.2.17)–(C.2.19) using Taylor series expansion become

$$\mathbf{R}^{t(k)}_{n+1} - \Delta\boldsymbol{\varepsilon}^{t(k)}_{n+1} + \Delta\boldsymbol{\Lambda}^{(k)}_{n+1}(\xi^{(k)}_{n+1} - \xi_n) + \boldsymbol{\Lambda}_{n+1}\Delta\xi^{(k)}_{n+1} = 0, \tag{C.2.20}$$

$$\mathbf{R}^{p(k)}_{n+1} - \Delta\boldsymbol{\varepsilon}^{p(k)}_{n+1} + \Delta\widetilde{\boldsymbol{\Lambda}}^{p(k)}_{n+1}\left|\xi^{d(k)}_{n+1} - \xi^{d}_{n}\right| + \widetilde{\boldsymbol{\Lambda}}^{p}_{n+1}\Delta\left|\xi^{d(k)}_{n+1} - \xi^{d}_{n}\right| = 0, \tag{C.2.21}$$

$$\Phi^{(k)}_{n+1} + \partial_{\boldsymbol{\sigma}}\Phi^{(k)}_{n+1} : \Delta\boldsymbol{\sigma}^{(k)}_{n+1} + \partial_{\xi}\Phi^{(k)}_{n+1}\Delta\xi^{(k)}_{n+1} = 0. \tag{C.2.22}$$

Next, the increment of the stress-strain constitutive relation (C.2.13) is evaluated, which yields

$$\Delta\,\mathcal{S}_{n+1} : \boldsymbol{\sigma}^{(k)}_{n+1} + \mathcal{S}_{n+1} : \Delta\boldsymbol{\sigma}^{(k)}_{n+1} = -\Delta\boldsymbol{\alpha}_{n+1}(T_{n+1} - T_0) - \Delta\boldsymbol{\varepsilon}^{t(k)}_{n+1} - \Delta\boldsymbol{\varepsilon}^{p(k)}_{n+1}. \tag{C.2.23}$$

The increments of the elastic compliance tensor $\Delta\,\mathcal{S}_{n+1}$ and the thermal expansion coefficient tensor $\Delta\boldsymbol{\alpha}_{n+1}$ are expressed using their evaluations [cf. equation (3.3.46)] in terms of the increment of the martensitic volume fraction ξ as

$$\Delta\,\mathcal{S}_{n+1} = \Delta\,\mathcal{S}\Delta\xi^{(k)}_{n+1}, \quad \Delta\boldsymbol{\alpha}_{n+1} = \Delta\boldsymbol{\alpha}\Delta\xi^{(k)}_{n+1}. \tag{C.2.24}$$

Using equations (C.2.24) and (C.2.23) the following expression for the increment of the transformation strain $\Delta\boldsymbol{\varepsilon}^{t(k)}_{n+1}$ is obtained:

$$\Delta\boldsymbol{\varepsilon}^{t(k)}_{n+1} = -\,\mathcal{S}_{n+1} : \Delta\boldsymbol{\sigma}^{(k)}_{n+1} - \Delta\boldsymbol{\varepsilon}^{p(k)}_{n+1} - \left[\Delta\boldsymbol{\alpha}(T_{n+1} - T_0) + \Delta\,\mathcal{S} : \boldsymbol{\sigma}^{(k)}_{n+1}\right]\Delta\xi^{(k)}_{n+1}. \tag{C.2.25}$$

The increment of the transformation direction tensor $\Delta\boldsymbol{\Lambda}^{(k)}_{n+1}$ is given by

$$\Delta\boldsymbol{\Lambda}^{(k)}_{n+1} = \partial_{\boldsymbol{\sigma}}\boldsymbol{\Lambda}^{(k)}_{n+1} : \Delta\boldsymbol{\sigma}^{(k)}_{n+1}. \tag{C.2.26}$$

Equations (C.2.25) and (C.2.26) are substituted into equation (C.2.20) to obtain

$$\left[\ \mathcal{S}_{n+1}^{(k)} + \partial_{\boldsymbol{\sigma}} \boldsymbol{\Lambda}_{n+1}^{(k)} (\xi_{n+1}^{(k)} - \xi_n)\right] : \Delta\boldsymbol{\sigma}_{n+1}^{(k)} =$$
$$- \left(\mathbf{R}_{n+1}^{t(k)} + \Delta\boldsymbol{\varepsilon}_{n+1}^{p(k)} + \left[\Delta\boldsymbol{\alpha}(T_{n+1} - T_0) + \Delta\,\mathcal{S} : \boldsymbol{\sigma}_{n+1}^{(k)} + \boldsymbol{\Lambda}_{n+1}^{(k)}\right] \Delta\xi_{n+1}^{(k)}\right). \tag{C.2.27}$$

The increment of the plastic direction tensor $\Delta\widetilde{\boldsymbol{\Lambda}}_{n+1}^{p(k)}$, which enters equation (C.2.21) is given by

$$\Delta\widetilde{\boldsymbol{\Lambda}}_{n+1}^{p(k)} = \partial_{\boldsymbol{\sigma}} \widetilde{\boldsymbol{\Lambda}}_{n+1}^{p(k)} : \Delta\boldsymbol{\sigma}_{n+1}^{(k)} + \partial_{\xi} \widetilde{\boldsymbol{\Lambda}}_{n+1}^{p(k)} : \Delta\xi_{n+1}^{(k)}, \tag{C.2.28}$$

while the quantity $\Delta\left|\xi_{n+1}^{d(k)} - \xi_n^d\right|$ is given by

$$\begin{aligned} \Delta\left|\xi_{n+1}^{d(k)} - \xi_n^d\right| &= \Delta\left|\left(\frac{H^{\text{cur}}\xi}{H}\right)_{n+1}^{(k)} - \left(\frac{H^{\text{cur}}\xi}{H}\right)_n\right| \\ &= \text{sign}\left[\left(\frac{H^{\text{cur}}\xi}{H}\right)_{n+1}^{(k)} - \left(\frac{H^{\text{cur}}\xi}{H}\right)_n\right] \frac{H^{\text{cur}}}{H} \Delta\xi_{n+1}^{(k)}. \end{aligned} \tag{C.2.29}$$

Thus, using equations (C.2.28) and (C.2.29), expression (C.2.21) becomes

$$\mathbf{R}_{n+1}^{p(k)} - \Delta\boldsymbol{\varepsilon}_{n+1}^{p(k)} + \partial_{\boldsymbol{\sigma}} \mathbf{R}_{n+1}^{p(k)} : \Delta\boldsymbol{\sigma}_{n+1}^{(k)} + \partial_{\xi} \mathbf{R}_{n+1}^{p(k)} : \Delta\xi_{n+1}^{(k)} = 0, \tag{C.2.30}$$

where the quantities $\partial_{\boldsymbol{\sigma}} \mathbf{R}_{n+1}^{p(k)}$ and $\partial_{\xi} \mathbf{R}_{n+1}^{p(k)}$ are defined as

$$\partial_{\boldsymbol{\sigma}} \mathbf{R}_{n+1}^{p(k)} = \left|\xi_{n+1}^{d(k)} - \xi_n^d\right| \partial_{\boldsymbol{\sigma}} \widetilde{\boldsymbol{\Lambda}}_{n+1}^{p(k)}, \tag{C.2.31}$$

$$\partial_{\xi} \mathbf{R}_{n+1}^{p(k)} = v\left|\xi_{n+1}^{d(k)} - \xi_n^d\right| \partial_{\xi} \widetilde{\boldsymbol{\Lambda}}_{n+1}^{p(k)} + \frac{H^{\text{cur}}}{H} \text{sign}\left(\xi_{n+1}^{d(k)} - \xi_n^d\right) \widetilde{\boldsymbol{\Lambda}}_{n+1}^{p(k)}. \tag{C.2.32}$$

The increment of the plastic strain $\Delta\boldsymbol{\varepsilon}_{n+1}^{p(k)}$ is expressed using equation (C.2.30) and substituted into equation (C.2.27) to obtain

$$\begin{aligned} \left[\ \mathcal{S}_{n+1}^{(k)} + \partial_{\boldsymbol{\sigma}} \boldsymbol{\Lambda}_{n+1}^{(k)} (\xi_{n+1}^{(k)} - \xi_n) + \partial_{\boldsymbol{\sigma}} \mathbf{R}_{n+1}^{p(k)}\right] : \Delta\boldsymbol{\sigma}_{n+1}^{(k)} &= \\ - \left(\mathbf{R}_{n+1}^{t(k)} + \mathbf{R}_{n+1}^{p(k)}\right)& \\ - \left[\Delta\boldsymbol{\alpha}(T_{n+1} - T_0) + \Delta\,\mathcal{S} : \boldsymbol{\sigma}_{n+1}^{(k)} + \boldsymbol{\Lambda}_{n+1}^{(k)} + \partial_{\xi} \mathbf{R}_{n+1}^{p(k)}\right] \Delta\xi_{n+1}^{(k)}& \end{aligned} \tag{C.2.33}$$

To simplify further the derivations, the following tensorial quantities are defined:

$$\boldsymbol{\Xi}_{n+1}^{(k)} \equiv \left[\ \mathcal{S}_{n+1}^{(k)} + \partial_{\boldsymbol{\sigma}} \boldsymbol{\Lambda}_{n+1}^{(k)} (\xi_{n+1}^{(k)} - \xi_n) + \partial_{\boldsymbol{\sigma}} \mathbf{R}_{n+1}^{p(k)}\right]^{-1}, \tag{C.2.34}$$

$$\chi_{n+1}^{(k)} \equiv \Delta\boldsymbol{\alpha}(T_{n+1} - T_0) + \Delta\,\mathcal{S} : \boldsymbol{\sigma}_{n+1}^{(k)} + \boldsymbol{\Lambda}_{n+1}^{(k)} + \partial_{\xi} \mathbf{R}_{n+1}^{p(k)}, \tag{C.2.35}$$

which leads to the following expression for the increment of stress $\Delta\boldsymbol{\sigma}^{(k)}_{n+1}$:

$$\Delta\boldsymbol{\sigma}^{(k)}_{n+1} = -\boldsymbol{\Xi}^{(k)}_{n+1} : \left[\mathbf{R}^{t(k)}_{n+1} + \mathbf{R}^{p(k)}_{n+1} + \chi^{(k)}_{n+1}\Delta\xi^{(k)}_{n+1}\right]. \tag{C.2.36}$$

The only remaining unknown is the increment of the martensitic volume fraction $\Delta\xi^{(k)}_{n+1}$. The consistency condition (C.2.22) together with equation (C.2.36) is used to find

$$\Delta\xi^{(k)}_{n+1} = \frac{\partial_{\boldsymbol{\sigma}}\Phi^{(k)}_{n+1} : \boldsymbol{\Xi}^{(k)}_{n+1} : \left[\mathbf{R}^{t(k)}_{n+1} + \mathbf{R}^{p(k)}_{n+1}\right] - \Phi^{(k)}_{n+1}}{\partial_{\xi}\Phi^{(k)}_{n+1} - \partial_{\boldsymbol{\sigma}}\Phi^{(k)}_{n+1} : \boldsymbol{\Xi}^{(k)}_{n+1} : \chi^{(k)}_{n+1}}. \tag{C.2.37}$$

Once $\Delta\xi^{(k)}_{n+1}$ is found from equation (C.2.37) it is used to calculate the increment of stress using equation (C.2.36). Then the increments of plastic and transformation strain are found using equations (C.2.30) and (C.2.25).

Finally, the expressions for $\partial_{\boldsymbol{\sigma}}\Phi^{(k)}_{n+1}$, $\partial_{\xi}\Phi^{(k)}_{n+1}$, $\partial_{\boldsymbol{\sigma}}\boldsymbol{\Lambda}^{(k)}_{n+1}$, $\partial_{\boldsymbol{\sigma}}\widetilde{\boldsymbol{\Lambda}}^{p(k)}_{n+1}$, and $\partial_{\xi}\widetilde{\boldsymbol{\Lambda}}^{p(k)}_{n+1}$, which are used in the algorithm, are derived. The derivative of the transformation function with respect to stress can be obtained by differentiating equation (C.1.8). It follows that

$$\partial_{\boldsymbol{\sigma}}\Phi^{(k)}_{n+1} = \begin{cases} \partial_{\boldsymbol{\sigma}}\pi^{(k)}_{n+1}, & \dot{\xi} > 0, \\ -\partial_{\boldsymbol{\sigma}}\pi^{(k)}_{n+1}, & \dot{\xi} < 0. \end{cases} \tag{C.2.38}$$

Using equation (C.1.9), the derivative of the thermodynamic driving force π with respect to stress is given by

$$\partial_{\boldsymbol{\sigma}}\pi^{(k)}_{n+1} = \Delta\boldsymbol{\alpha}(T_{n+1} - T_0) + \Delta\,\mathcal{S} : \boldsymbol{\sigma}^{(k)}_{n+1} + \boldsymbol{\Lambda}^{(k)}_{n+1}. \tag{C.2.39}$$

Similarly the derivative of the transformation function with respect to the martensitic volume fraction is

$$\partial_{\xi}\Phi^{(k)}_{n+1} = \begin{cases} \partial_{\xi}\boldsymbol{\beta}^{(k)}_{n+1} : \boldsymbol{\Lambda}^{(k)}_{n+1} + \partial_{\xi}\eta^{(k)}_{n+1}, & \dot{\xi} > 0, \\ -\partial_{\xi}\boldsymbol{\beta}^{(k)}_{n+1} : \boldsymbol{\Lambda}^{(k)}_{n+1} - \partial_{\xi}\eta^{(k)}_{n+1}, & \dot{\xi} < 0. \end{cases} \tag{C.2.40}$$

The quantities $\partial_{\xi}\boldsymbol{\beta}^{(k)}_{n+1}$ and $\partial_{\xi}\eta^{(k)}_{n+1}$ are calculated using the expressions for $\boldsymbol{\beta}$ and η in Sect. 5.2.1.

Using equation (C.1.4) the derivative of the transformation direction tensor with respect to stress $\partial_{\boldsymbol{\sigma}}\boldsymbol{\Lambda}^{(k)}_{n+1}$ is evaluated for the case of forward phase transformation ($\dot{\xi} > 0$) as

$$\partial_{\boldsymbol{\sigma}}\boldsymbol{\Lambda}^{(k)}_{n+1} = \frac{3}{2}\partial_{\boldsymbol{\sigma}}H^{\mathrm{cur}} \otimes \frac{\boldsymbol{\sigma}^{\mathrm{eff}\prime}}{\bar{\sigma}^{\mathrm{eff}}} + \frac{3}{2}H^{\mathrm{cur}}\partial_{\boldsymbol{\sigma}}\left[\frac{\boldsymbol{\sigma}^{\mathrm{eff}\prime}}{\bar{\sigma}^{\mathrm{eff}}}\right], \tag{C.2.41}$$

where the iteration counter k and the time step number $(n+1)$ have been omitted for simplicity. The corresponding derivative for the case of reverse phase transformation is equal to zero. To calculate the derivative of the current maximum transformation strain with respect to stress $\partial_{\boldsymbol{\sigma}} H^{\text{cur}}$, recall equation (5.2.13). Differentiating both sides of equation (5.2.13) results in

$$\partial_{\boldsymbol{\sigma}} \bar{\sigma}' = \partial_{\boldsymbol{\sigma}} \left.\bar{\beta}\right|_{\xi=1} = \partial_{H^{\text{cur}}} \left.\bar{\beta}\right|_{\xi=1} \partial_{\boldsymbol{\sigma}} H^{\text{cur}}. \tag{C.2.42}$$

where $\bar{\sigma}'$ is defined in 3.3.54. Therefore, $\partial_{\boldsymbol{\sigma}} H^{\text{cur}}$ is given by

$$\partial_{\boldsymbol{\sigma}} H^{\text{cur}} = \frac{\partial_{\boldsymbol{\sigma}} \bar{\sigma}'}{\partial_{H^{\text{cur}}} \left.\bar{\beta}\right|_{\xi=1}}. \tag{C.2.43}$$

After performing standard calculations, the expression for $\partial_{\boldsymbol{\sigma}} \bar{\sigma}'$ is found to be

$$\partial_{\boldsymbol{\sigma}} \bar{\sigma}' = \frac{3}{2} \frac{\boldsymbol{\sigma}'}{\bar{\sigma}'}. \tag{C.2.44}$$

To calculate the second term of equation (C.2.41) the quantity $\partial_{\boldsymbol{\sigma}} \left[\frac{\boldsymbol{\sigma}^{\text{eff}\prime}}{\bar{\sigma}^{\text{eff}}} \right]$ must be found. Again, performing standard operations, the following expression is derived:

$$\partial_{\boldsymbol{\sigma}} \left(\frac{\boldsymbol{\sigma}^{\text{eff}\prime}}{\bar{\sigma}^{\text{eff}}} \right) = \frac{1}{\bar{\sigma}^{\text{eff}}} \left(\mathbf{I} - \frac{1}{3} \mathbf{1} \otimes \mathbf{1} - \frac{3}{2} \frac{\boldsymbol{\sigma}^{\text{eff}\prime}}{\bar{\sigma}^{\text{eff}}} \otimes \frac{\boldsymbol{\sigma}^{\text{eff}\prime}}{\bar{\sigma}^{\text{eff}}} \right), \tag{C.2.45}$$

where $\mathbf{I}$ and $\mathbf{1}$ are defined in 4.3.29 and 4.3.30 respectively.

Similarly, the derivative of the plastic direction tensor with respect to stress $\partial_{\boldsymbol{\sigma}} \widetilde{\boldsymbol{\Lambda}}^{p(k)}_{n+1}$ for the case of forward phase transformation is calculated as

$$\partial_{\boldsymbol{\sigma}} \widetilde{\boldsymbol{\Lambda}}^{p(k)}_{n+1} = \frac{3}{2} C^p_1 \exp\left(-\frac{\zeta^d}{C^p_2} \right) \partial_{\boldsymbol{\sigma}} \left(\frac{\boldsymbol{\sigma}^{\text{eff}\prime}}{\bar{\sigma}^{\text{eff}}} \right), \tag{C.2.46}$$

while the corresponding derivative for the case of reverse phase transformation is again equal to zero. The derivative of $\widetilde{\boldsymbol{\Lambda}}^{p(k)}_{n+1}$ with respect to ξ is given by

$$\partial_{\xi} \widetilde{\boldsymbol{\Lambda}}^{p(k)}_{n+1} = \begin{cases} -\frac{3}{2} \frac{C^p_1}{C^p_2} \exp\left[-\frac{\zeta^d}{C^p_2} \right] \frac{H^{\text{cur}}}{H} \frac{\boldsymbol{\sigma}^{\text{eff}\prime}}{\bar{\sigma}^{\text{eff}}}, & \dot{\xi} > 0, \\ \frac{C^p_1}{C^p_2} \exp\left[-\frac{\zeta^d}{C^p_2} \right] \frac{H^{\text{cur}}}{H} \frac{\boldsymbol{\varepsilon}^t_{\max}}{\bar{\varepsilon}^t_{\max}}, & \dot{\xi} < 0. \end{cases} \tag{C.2.47}$$

C.2.3 Consistent Tangent Stiffness and Thermal Moduli Tensors

The solution of the constitutive equations for the SMA involves calculating the stress increment for a given strain and temperature increment at each time step. This procedure makes use of the tangent stiffness tensor and the thermal tangent moduli tensor, appearing in the linearized problem. An example of such a situation is a displacement-based finite element method formulation that uses Newton's method to solve the resulting non-linear equations. As mentioned by Qidwai and Lagoudas [1] the tangent tensors used by the global

Newton's method must be obtained in a manner consistent with the integration algorithm. The use of the continuum tangent tensors would lead to a loss of the quadratic convergence of Newton's iterative method [4]. Therefore, in this section the consistent tangent moduli tensors which are calculated at the end of the transformation correction iterative procedure and are passed to the global Newton's method, are derived.

The total differential of the stress-strain relation (C.2.13) results in

$$\Delta\, \mathcal{S} : \boldsymbol{\sigma}_{n+1} \mathrm{d}\xi_{n+1} + \mathcal{S}_{n+1} : \mathrm{d}\boldsymbol{\sigma}_{n+1} = \mathrm{d}\boldsymbol{\varepsilon}_{n+1} - \mathrm{d}\boldsymbol{\varepsilon}^t_{n+1} - \mathrm{d}\boldsymbol{\varepsilon}^p_{n+1} - \boldsymbol{\alpha}_{n+1} \mathrm{d}T_{n+1} - \Delta\boldsymbol{\alpha}(T_{n+1} - T_0)\mathrm{d}\xi_{n+1}. \tag{C.2.48}$$

The differential of the evolution equations for the transformation and plastic strains [cf. equations (C.2.11), (C.2.12)] result in

$$\mathrm{d}\boldsymbol{\varepsilon}^t_{n+1} = \boldsymbol{\Lambda}_{n+1} \mathrm{d}\xi_{n+1} + (\xi_{n+1} - \xi_n)\partial_{\boldsymbol{\sigma}} \boldsymbol{\Lambda}_{n+1} : \mathrm{d}\boldsymbol{\sigma}_{n+1}, \tag{C.2.49}$$

$$\mathrm{d}\boldsymbol{\varepsilon}^p_{n+1} = \partial_{\boldsymbol{\sigma}} \mathbf{R}^p_{n+1} : \mathrm{d}\boldsymbol{\sigma}_{n+1} + \partial_{\xi} \mathbf{R}^p_{n+1} \mathrm{d}\xi_{n+1}, \tag{C.2.50}$$

where the quantities $\partial_{\boldsymbol{\sigma}} \mathbf{R}^p_{n+1}$ and $\partial_{\xi} \mathbf{R}^p_{n+1}$ are defined by (C.2.31) and (C.2.32). The substitution of (C.2.49) and (C.2.50) into (C.2.48) eventually leads to

$$\mathrm{d}\boldsymbol{\sigma} = \boldsymbol{\Xi}_{n+1} : \left[\mathrm{d}\boldsymbol{\varepsilon} - \boldsymbol{\alpha}_{n+1} \mathrm{d}T_{n+1} - \boldsymbol{\chi}_{n+1} \mathrm{d}\xi_{n+1}\right], \tag{C.2.51}$$

where $\boldsymbol{\Xi}_{n+1}$ and $\boldsymbol{\chi}_{n+1}$ are defined by (C.2.34) and (C.2.35). Next, the consistency condition is written in the following form:

$$\mathrm{d}\Phi_{n+1} = \partial_{\boldsymbol{\sigma}} \Phi_{n+1} : \mathrm{d}\boldsymbol{\sigma}_{n+1} + \partial_{\xi} \Phi_{n+1} \mathrm{d}\xi_{n+1} + \partial_T \Phi_{n+1} \mathrm{d}T_{n+1} = 0. \tag{C.2.52}$$

Equation (C.2.51) together with (C.2.52) give the following expression for $\mathrm{d}\xi_{n+1}$:

$$\mathrm{d}\xi_{n+1} = \frac{\partial_{\boldsymbol{\sigma}} \Phi_{n+1} : \boldsymbol{\Xi}_{n+1} : \left[\mathrm{d}\boldsymbol{\varepsilon} - \boldsymbol{\alpha}_{n+1} \mathrm{d}T_{n+1}\right] + \partial_T \Phi_{n+1} \mathrm{d}T_{n+1}}{\partial_{\boldsymbol{\sigma}} \Phi : \boldsymbol{\Xi}_{n+1} : \boldsymbol{\chi}_{n+1} - \partial_{\xi} \Phi}. \tag{C.2.53}$$

Upon substituting (C.2.53) into (C.2.51) the following expression for $\mathrm{d}\boldsymbol{\sigma}_{n+1}$ is obtained:

$$\mathrm{d}\boldsymbol{\sigma}_{n+1} = \left[\boldsymbol{\Xi}_{n+1} - \frac{\boldsymbol{\Xi}_{n+1} : \boldsymbol{\chi}_{n+1} \otimes \partial_{\boldsymbol{\sigma}} \Phi : \boldsymbol{\Xi}_{n+1}}{\partial_{\boldsymbol{\sigma}} \Phi : \boldsymbol{\Xi}_{n+1} : \boldsymbol{\chi}_{n+1} - \partial_{\xi} \Phi}\right] : \mathrm{d}\boldsymbol{\varepsilon}_{n+1} + \boldsymbol{\Xi}_{n+1} : \left[\boldsymbol{\chi}_{n+1} \frac{\partial_{\boldsymbol{\sigma}} \Phi : \boldsymbol{\Xi}_{n+1} : \boldsymbol{\alpha}_{n+1} - \partial_T \Phi_{n+1}}{\partial_{\boldsymbol{\sigma}} \Phi : \boldsymbol{\Xi}_{n+1} : \boldsymbol{\chi}_{n+1} - \partial_{\xi} \Phi} - \boldsymbol{\alpha}_{n+1}\right] \mathrm{d}T_{n+1}. \tag{C.2.54}$$

The consistent tangent stiffness tensor $\mathcal{L}_{n+1}$ and the consistent tangent thermal moduli tensor $\boldsymbol{\Theta}_{n+1}$ are defined as

Table C.1. Summary of the closest point projection numerical algorithm for the SMA constitutive model with transformation induced plasticity.

---

1. Let $k = 0$, $\boldsymbol{\varepsilon}_{n+1} = \boldsymbol{\varepsilon}_n + \Delta\boldsymbol{\varepsilon}_{n+1}$, $T_{n+1} = T_n + \Delta T_{n+1}$, $\boldsymbol{\varepsilon}^{t(0)}_{n+1} = \boldsymbol{\varepsilon}^t_n$, $\boldsymbol{\varepsilon}^{p(0)}_{n+1} = \boldsymbol{\varepsilon}^p_n$, $\xi^{(0)}_{n+1} = \xi_n$, $\mathcal{S}^{(0)}_{n+1} = \mathcal{S}_n$, $\boldsymbol{\alpha}^{(0)}_{n+1} = \boldsymbol{\alpha}_n$.
2. Calculate $\boldsymbol{\sigma}^{(k)}_{n+1}$, $\Phi^{(k)}_{n+1}$, $\mathbf{R}^{t(k)}_{n+1}$, $\mathbf{R}^{p(k)}_{n+1}$:

$$\begin{aligned}\boldsymbol{\sigma}^{(k)}_{n+1} &= \mathcal{S}^{(k)-1}_{n+1} : \left[\boldsymbol{\varepsilon}_{n+1} - \boldsymbol{\varepsilon}^{t(k)}_{n+1} - \boldsymbol{\varepsilon}^{p(k)}_{n+1} - \boldsymbol{\alpha}^{(k)}_{n+1}(T_{n+1} - T_0)\right]\\ \Phi^{(k)}_{n+1} &= \Phi(\boldsymbol{\sigma}^{(k)}_{n+1}, T_{n+1}, \xi^{(k)}_{n+1})\\ \mathbf{R}^{t(k)}_{n+1} &= -\boldsymbol{\varepsilon}^{t(k)}_{n+1} + \boldsymbol{\varepsilon}^t_n + \boldsymbol{\Lambda}^{(k)}_{n+1}(\xi^{(k)}_{n+1} - \xi_n)\\ \mathbf{R}^{p(k)}_{n+1} &= -\boldsymbol{\varepsilon}^{p(k)}_{n+1} + \boldsymbol{\varepsilon}^p_n + \widetilde{\boldsymbol{\Lambda}}^{p(k)}_{n+1}\left|\xi^{d(k)}_{n+1} - \xi^d_n\right|\end{aligned}$$

   If $\Phi^{(k)}_{n+1} < 0$ or $\left|\Delta\xi^{(k)}_{n+1}\right| <$ *tolerance* then go to step 8.
3. Calculate $\boldsymbol{\Xi}^{(k)}_{n+1}$ and $\chi^{(k)}_{n+1}$:

$$\begin{aligned}\boldsymbol{\Xi}^{(k)}_{n+1} &= \left[\mathcal{S}^{(k)}_{n+1} + \partial_{\boldsymbol{\sigma}}\boldsymbol{\Lambda}^{(k)}_{n+1}(\xi^{(k)}_{n+1} - \xi_n) + \partial_{\boldsymbol{\sigma}}\mathbf{R}^{p(k)}_{n+1}\right]^{-1}\\ \chi^{(k)}_{n+1} &= \Delta\boldsymbol{\alpha}(T_{n+1} - T_0) + \Delta\,\mathcal{S} : \boldsymbol{\sigma}^{(k)}_{n+1} + \boldsymbol{\Lambda}^{(k)}_{n+1} + \partial_\xi \mathbf{R}^{p(k)}_{n+1}\end{aligned}$$

4. Calculate the increments of the martensitic volume fraction $\Delta\xi^{(k)}_{n+1}$, stress $\Delta\boldsymbol{\sigma}^{(k)}_{n+1}$, transformation strain $\Delta\boldsymbol{\varepsilon}^{t(k)}_{n+1}$ and plastic strain $\Delta\boldsymbol{\varepsilon}^{p(k)}_{n+1}$:

$$\Delta\xi^{(k)}_{n+1} = \frac{\partial_{\boldsymbol{\sigma}}\Phi^{(k)}_{n+1} : \boldsymbol{\Xi}^{(k)}_{n+1} : \left[\mathbf{R}^{t(k)}_{n+1} + \mathbf{R}^{p(k)}_{n+1}\right] - \Phi^{(k)}_{n+1}}{\partial_\xi\Phi^{(k)}_{n+1} - \partial_{\boldsymbol{\sigma}}\Phi^{(k)}_{n+1} : \boldsymbol{\Xi}^{(k)}_{n+1} : \chi^{(k)}_{n+1}}$$

$$\begin{aligned}\Delta\boldsymbol{\sigma}^{(k)}_{n+1} &= -\boldsymbol{\Xi}^{(k)}_{n+1} : \left[\mathbf{R}^{t(k)}_{n+1} + \mathbf{R}^{p(k)}_{n+1} + \chi^{(k)}_{n+1}\Delta\xi^{(k)}_{n+1}\right]\\ \Delta\boldsymbol{\varepsilon}^{p(k)}_{n+1} &= -\mathbf{R}^{p(k)}_{n+1} - \partial_{\boldsymbol{\sigma}}\mathbf{R}^{p(k)}_{n+1} : \Delta\boldsymbol{\sigma}^{(k)}_{n+1} - \partial_\xi\mathbf{R}^{p(k)}_{n+1} : \Delta\xi^{(k)}_{n+1}\\ \Delta\boldsymbol{\varepsilon}^{t(k)}_{n+1} &= -\,\mathcal{S}_{n+1} : \Delta\boldsymbol{\sigma}^{(k)}_{n+1} - \Delta\boldsymbol{\varepsilon}^{p(k)}_{n+1} - \left[\Delta\boldsymbol{\alpha}(T_{n+1} - T_0) + \Delta\,\mathcal{S} : \boldsymbol{\sigma}^{(k)}_{n+1}\right]\Delta\xi^{(k)}_{n+1}\end{aligned}$$

5. Update martensitic volume fraction, transformation and plastic strains, elastic compliance tensor and thermal expansion coefficient tensor:

$$\begin{aligned}\xi^{(k+1)}_{n+1} &= \xi^{(k)}_{n+1} + \Delta\xi^{(k)}_{n+1}\\ \boldsymbol{\varepsilon}^{t(k+1)}_{n+1} &= \boldsymbol{\varepsilon}^{t(k)}_{n+1} + \Delta\boldsymbol{\varepsilon}^{t(k)}_{n+1}\\ \boldsymbol{\varepsilon}^{p(k+1)}_{n+1} &= \boldsymbol{\varepsilon}^{p(k)}_{n+1} + \Delta\boldsymbol{\varepsilon}^{p(k)}_{n+1}\\ \mathcal{S}^{(k+1)}_{n+1} &= \mathcal{S}^A + \xi^{(k+1)}_{n+1}\Delta\,\mathcal{S}\\ \boldsymbol{\alpha}^{(k+1)}_{n+1} &= \boldsymbol{\alpha}^A + \xi^{(k+1)}_{n+1}\Delta\boldsymbol{\alpha}\end{aligned}$$

6. Set iteration counter $k = k + 1$. Return to step 2.
7. Calculate consistent tangent stiffness tensor $\mathcal{L}_{n+1}$ and thermal moduli tensor $\boldsymbol{\Theta}_{n+1}$:

$$\mathcal{L}_{n+1} = \boldsymbol{\Xi}_{n+1} - \frac{\boldsymbol{\Xi}_{n+1} : \chi_{n+1} \otimes \partial_{\boldsymbol{\sigma}}\Phi : \boldsymbol{\Xi}_{n+1}}{\partial_{\boldsymbol{\sigma}}\Phi : \boldsymbol{\Xi}_{n+1} : \chi_{n+1} - \partial_\xi\Phi}$$

$$\boldsymbol{\Theta}_{n+1} = \boldsymbol{\Xi}_{n+1} : \left[\chi_{n+1}\frac{\partial_{\boldsymbol{\sigma}}\Phi : \boldsymbol{\Xi}_{n+1} : \boldsymbol{\alpha}_{n+1} - \partial_T\Phi_{n+1}}{\partial_{\boldsymbol{\sigma}}\Phi : \boldsymbol{\Xi}_{n+1} : \chi_{n+1} - \partial_\xi\Phi} - \boldsymbol{\alpha}_{n+1}\right]$$

8. Update ζ_{n+1} and ζ^d_{n+1}:

$$\begin{aligned}\zeta_{n+1} &= \zeta_n + |\Delta\xi_{n+1}|\\ \zeta^d_{n+1} &= \zeta^d_n + \tfrac{H^{\text{cur}}_{n+1}}{H}\,|\Delta\xi_{n+1}|\end{aligned}$$

   Exit and return to global iteration

$$\mathcal{L}_{n+1} \equiv \frac{\mathrm{d}\boldsymbol{\sigma}_{n+1}}{\mathrm{d}\boldsymbol{\varepsilon}_{n+1}} = \boldsymbol{\Xi}_{n+1} - \frac{\boldsymbol{\Xi}_{n+1} : \chi_{n+1} \otimes \partial_{\boldsymbol{\sigma}}\Phi : \boldsymbol{\Xi}_{n+1}}{\partial_{\boldsymbol{\sigma}}\Phi : \boldsymbol{\Xi}_{n+1} : \chi_{n+1} - \partial_{\xi}\Phi}, \tag{C.2.55}$$

$$\boldsymbol{\Theta}_{n+1} \equiv \frac{\mathrm{d}\boldsymbol{\sigma}_{n+1}}{\mathrm{d}T_{n+1}} = \boldsymbol{\Xi}_{n+1} : \left[\chi_{n+1} \frac{\partial_{\boldsymbol{\sigma}}\Phi : \boldsymbol{\Xi}_{n+1} : \boldsymbol{\alpha}_{n+1} - \partial_T \Phi_{n+1}}{\partial_{\boldsymbol{\sigma}}\Phi : \boldsymbol{\Xi}_{n+1} : \chi_{n+1} - \partial_{\xi}\Phi} - \boldsymbol{\alpha}_{n+1}\right]. \tag{C.2.56}$$

The material parameters are updated only at the points of reversal of the phase transformation. To update them, the evolution equations (5.2.24)–(5.2.27) are utilized. These equations require knowledge of two parameters, which are the accumulated martensitic volume fraction ζ and accumulated detwinned martensitic volume fraction ζ^d. Therefore, these variables must be made available and must be updated at each increment. To accomplish this, after the convergence of the return mapping iterative procedure, the values of ζ and ζ^d are calculated using their values from the previous increment and the newly found values of the increment of the martensitic volume fraction $\Delta\xi_{n+1}$ and the current maximum transformation strain H^{cur}_{n+1}.

C.2.4 Summary of the Numerical Algorithm for SMA Constitutive Model with Transformation Induced Plasticity

A summary of all steps of the numerical algorithm for an SMA undergoing transformation induced plasticity are presented in Table C.1.

References

[1] M. A. Qidwai, D. C. Lagoudas, On the thermodynamics and transformation surfaces of polycrystalline NiTi shape memory alloy material, International Journal of Plasticity 16 (2000) 1309–1343.

[2] D. C. Lagoudas, P. B. Entchev, Modeling of transformation-induced plasticity and its effect on the behavior of porous shape memory alloys. Part I: Constitutive model for fully dense SMAs, Mechanics of Materials 36 (9) (2004) 865–892.

[3] P. B. Entchev, Micromechanical modeling of porous shape memory alloys, Ph.D. thesis, Texas A&M University, Department of Aerospace Engineering (May 2002).

[4] J. Simo, R. Taylor, Consistent tangent operators for rate-independent elastoplacticity, Computer Methods in Applied Mechanics and Engineering 48 (1985) 101–118.

Index

Lightning Source UK Ltd.
Milton Keynes UK
UKHW020612061022
409990UK00003B/141

9 780387 476841